Advanced Mathematics

A unified course in pure mathematics,
mechanics and statistics

By the same authors (with F. J. Budden)
Advanced Mathematics 2

A list of the contents of Book 2 may be found on p. 584.

Advanced Mathematics 1

A unified course in pure mathematics, mechanics and statistics

L. K. Turner   D. Knighton

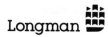
Longman

LONGMAN GROUP LIMITED
*Longman House, Burnt Mill, Harlow, Essex CM20 2JE, England
and Associated Companies throughout the world*

*Distributed in the United States by Longman Inc,
The Longman Building, 95 Church Street, White Plains, NY 10601*

First published in this format 1976
Second edition in this format 1986
Third impression 1992
ISBN 0 582 35513 3

Set in 10/12 pt Times New Roman

The publisher's policy is to use paper manufactured from sustainable forests.

*Produced by Longman Singapore Publishers Pte Ltd
Printed in Singapore*

Acknowledgements

We are grateful to the following Examinations Boards for permission
to reproduce questions from past examinations papers in mathematics:

The Associated Examining Board: University of Cambridge Local
Examinations Syndicate; Joint Matriculation Board; University of
London Schools Examinations Board; Oxford Colleges Admissions
Office; University of Oxford Delegacy of Local Examinations; Oxford
and Cambridge Schools Examinations Board; Welsh Joint Education
Committee.

Unfortunately, we have been unable to trace the copyright holders of
five questions in Miscellaneous problems 6 and of question 5 in the
revision exercises for the appendix, and we would appreciate any
information which would enable us to do so.

We would also like to take this opportunity to state that the above-
mentioned Boards accept no responsibility whatsoever for the accuracy
or method of working in the answers given.

Contents

Sections marked * may be delayed until a second reading.

Preface

Notation

1 Introduction   **1**
1.1   Logical symbols   1
1.2   Sets and Boolean algebra   2
1.3   Coordinates   5
1.4   Straight lines   10
1.5   Rational and irrational numbers   15
1.6   Surds   18
* 1.7   Linear transformations and matrices   19
* 1.8   Multiplication of matrices   24

2 Functions   **28**
2.1   Functions and their graphs   28
2.2   Quadratic functions   36
* 2.3   Polynomial functions and the remainder theorem   43
2.4   Indices and power functions   52
2.5   Logarithms and logarithmic functions   59
* 2.6   Continuity and limits   68
* 2.7   Rational functions   74

3 Rates of change: differentiation   **79**
3.1   Average velocity and instantaneous velocity   79
3.2   Differentiation   83
3.3   Standard methods of differentiation   88
3.4   Higher derivatives   90
3.5   Alternative notation: tangents and normals   93
3.6   Kinematics   98

3.7   Reverse processes   100
3.8   Stationary values: maxima and minima   104
* 3.9   Further differentiation   114
* 3.10 Implicit functions and parameters   119
* 3.11 Rates of change; approximations and errors   124
* 3.12 Further kinematics   129

4   **Areas: integration**   **136**
4.1   The area beneath a curve: the definite integral   136
4.2   The generation of area: indefinite integrals   140
4.3   Calculation of definite integrals   146
* 4.4   Numerical integration   152
* 4.5   Other summations: volumes of revolution   157
* 4.6   Mean values   162
* 4.7   Integration by substitution   164

5   **Trigonometric functions**   **171**
5.1   Sine, cosine and tangent   171
5.2   Simple equations   174
5.3   Cotangent, secant, cosecant and the use of Pythagoras' theorem   177
* 5.4   Sine and cosine rules   180
5.5   Compound angles: $a \cos \theta + b \sin \theta$ 183
5.6   Multiple angles   191
5.7   Factor formulae   195
5.8   Radians, general solutions and small angles   198
5.9   Differentiation of trigonometric functions   204
5.10 Integration of trigonometric functions   209
* 5.11 Inverse trigonometric functions   213

6   **Sequences and series**   **219**
6.1   Sequences and series   219
6.2   Arithmetic progressions   221
6.3   Geometric progressions   224
* 6.4   Finite series: the method of differences   227
* 6.5   Mathematical induction   231
* 6.6   Iteration: recurrence relations   234
* 6.7   Solution of equations: iterative methods   239
* 6.8   The Newton–Raphson method   247
* 6.9   Interlude: Fibonacci numbers and prime numbers   251
6.10 Permutations and combinations   254
6.11 Pascal's triangle: the binomial theorem   259

7 Probability and statistics   **264**

  7.1   Trials, events and probabilities   264

  7.2   Compound events   268

* 7.3   Conditional probability   274

* 7.4   Probability distributions: three standard types   279

  7.5   Statistics: location and spread   284

  7.6   Frequency distributions   293

  7.7   The Normal distribution   303

8 Introduction to vectors   **318**

  8.1   Vectors and vector addition   318

  8.2   Position vectors and components   324

  8.3   Scalar products   331

* 8.4   Lines, planes and angles   337

  8.5   Differentiation of vectors   348

  8.6   Motion under gravity: projectiles   351

  8.7   Motion in a circle   356

* 8.8   Relative motion   360

9 Introduction to mechanics   **368**

  9.1   Historical introduction: Newton's first law   368

  9.2   Force, mass and weight: Newton's second law   371

  9.3   Reactions: Newton's third law   382

  9.4   Friction   389

  9.5   Impulse and momentum   395

  9.6   Work, energy and power   405

10 Further calculus   **420**

  10.1   The logarithmic function, $\ln x$ 420

  10.2   The exponential function, e^x 431

  10.3   Interlude: infinite series   437

  10.4   Maclaurin series   440

* 10.5   Integration by parts: reduction formulae   450

* 10.6   Partial fractions   456

* 10.7   General integration   465

Appendix: introduction to coordinate geometry   **477**

  A1   Straight lines   477

  A2   Circles   481

* A3   Parabolas   485

* A4   Ellipses and hyperbolas   492
* A5   Polar coordinates   504

Revision exercises   **509**

Answers to exercises   **539**

Contents of Book 2   **584**

Index   **586**

Preface to the second edition

The aim of this course remains that of its first edition: to provide within two books an Advanced level programme in pure mathematics, mechanics and statistics which combines directness and economy with the maximum possible clarity.

At the same time we are aware of the growing number of students who wish to continue their work beyond Ordinary level, but not as far as Advanced level. To serve this purpose too, and also to ensure that the major part of A-level work is covered in the first year of such a course, the balance between the two books has been adjusted. As a result it has been possible to include within this volume the whole of the agreed 'core syllabus' in mathematics, particularly bearing in mind the new AS level examinations.

We have taken this opportunity to reflect the greater emphasis now given to numerical work by electronic calculators, and also to extend many of the exercises and the miscellaneous problems at the end of each chapter. By this means we hope to serve the needs not only of those who require more practice on simple exercises, but also of those who look for a demanding challenge.

Lastly, we thank all those who have assisted with this edition, and especially those who have written with helpful suggestions. We are deeply grateful, and trust that they and others will continue to do so.

L.K.T.
D.K.

Notation

Logical symbols

$p, q, r \ldots$	statements, or propositions
$\sim p$	negation of p
$\Rightarrow$	implies
$\Leftarrow$	is implied by
$\Leftrightarrow$	implies and is implied by
$\not\Rightarrow$ (etc.)	does not imply (etc.)

Equalities and inequalities

$=$	is equal to
$\neq$	is not equal to
$\approx$	is approximately equal to
$\equiv$	is always equal (or identical) to
$<$	is less than
$>$	is greater than

Sets

$x, y, \ldots$	elements of a set
$A, B, \ldots$	sets of elements
$n(A)$	number of elements in A
$\in$	belongs to
$\notin$	does not belong to
$\{x : x \in A\}$	set of elements x such that $x \in A$
$\varnothing$	empty set
$\mathscr{E}$	universal set
A'	complement of A within $\mathscr{E}$
	$= \{x : x \in \mathscr{E} \text{ and } x \notin A\}$
$A \cup B$	union of A and B
	$= \{x : x \in A \text{ and/or } x \in B\}$
$A \cap B$	intersection of A and B
	$= \{x : x \in A \text{ and } x \in B\}$
$\mathbb{N}$	{natural numbers: 0, 1, 2, 3, 4 . . .}
$\mathbb{Z}$	{all integers}
$\mathbb{Q}$	{all rational numbers}

$\mathbb{R}$ {all real numbers}

$\mathbb{Z}^+$ (etc.)   {all positive integers} (etc.)

Matrices and determinants

$\mathbf{A} = \begin{pmatrix} a & b \\ c & d \end{pmatrix}$ matrix $\mathbf{A}$

$\det \mathbf{A}$ or Δ determinant of $\mathbf{A}(= ad - bc)$

$\mathbf{A}^{-1}$ inverse matrix of $\mathbf{A}$ (when $\det \mathbf{A} \neq 0$)

Functions

f: $x \mapsto y$ function f maps x on to y

f(x) image of x under mapping f (or
                               output of f, when input is x)

f^{-1} inverse function of f

gf   function f followed by function g

$|x|$ magnitude of x

$[x]$ integral part of x (greatest integer not
                               greater than x)

$\log_a x$ logarithm to base a of x

$\lg x$ $\log_{10} x$

$\ln x$ $\log_e x$ ($e \approx 2.718$)

Limits, derivatives and integrals

$\rightarrow$ tends to

$\rightarrow \infty$ becomes unlimitedly large (tends to
                               infinity)

$lim_{x \to a} f(x)$ limit of f(x) as $x \to a$

$lim_{x \to a+} f(x)$ limit of f(x) as $x \to a$ from above

$lim_{x \to a-} f(x)$ limit of f(x) as $x \to a$ from below

$\delta x, \delta y \ldots$ small increments in $x, y \ldots$

f$'(x)$ or $\dfrac{dy}{dx}$ derivative

f$''(x)$ or $\dfrac{d^2 y}{dx^2}$ second derivative

$\displaystyle\int_a^b f(x)\,dx$ definite integral, or area beneath f(x)
                               from $x = a$ to $x = b$

$\displaystyle\int f(x)\,dx$ indefinite integral

Trigonometric functions

rad

radian $\left(1 \text{ rad} = \dfrac{180°}{\pi}\right)$

$\sin^{-1}$

inverse sine $(-\frac{1}{2}\pi \leqslant \sin^{-1} x \leqslant \frac{1}{2}\pi)$

$\cos^{-1}$

inverse cosine $(0 \leqslant \cos^{-1} x \leqslant \pi)$

$\tan^{-1}$

inverse tangent
$(-\frac{1}{2}\pi < \tan^{-1} x < \frac{1}{2}\pi)$

Series

$\displaystyle\sum_{r=1}^{n} u_r$ or $\displaystyle\sum_{1}^{n} u_r$

$u_1 + u_2 + \ldots + u_n$

$a + (a + d) + (a + 2d) + \ldots$

arithmetic progression (a.p.)

$a + ar + ar^2 + \ldots$

geometric progression (g.p.)

Permutations and combinations

$n!$

n factorial $(= n \times (n - 1) \times \ldots \times 3 \times 2 \times 1)$

^nP_r

number of permutations of n different objects, taken r at a time

^nC_r or $\dbinom{n}{r}$

number of combinations of n different objects, taken r at a time

Probability

$\mathscr{E}$

universal set of equally likely outcomes in a given trial

E

a particular event and the corresponding set of outcomes

$P(E)$

probability of $E = \dfrac{n(E)}{n(\mathscr{E})}$

$P(A \cup B)$

probability of event A and/or event B

$P(A \cap B)$

probability of event A and event B

$P(A|B)$

probability of A conditional upon B

A, B exclusive

$A \cap B = \varnothing$

A, B exhaustive

$A \cup B = \mathscr{E}$

A, B independent

$P(A \cap B) = P(A)P(B)$

p_r

probability of event E_r

Statistics

$x_1, x_2, \ldots$	values
m	mean $= \dfrac{1}{n}\sum x_r$
s^2	variance $= \dfrac{1}{n}\sum (x_r - m)^2$
	$= \dfrac{1}{n}\sum x_r^2 - m^2$
s	standard deviation
$f_1, f_2, \ldots$	frequencies of $x_1, x_2 \ldots$
	$\Rightarrow m = \dfrac{1}{n}\sum f_r x_r$
	and $s^2 = \dfrac{1}{n}\sum f_r(x_r - m)^2$
	$= \dfrac{1}{n}\sum f_r x_r^2 - m^2$

Vectors

AB, CD $\ldots$	displacement vectors
$\mathbf{a}, \mathbf{b} \ldots$	free vectors, or position vectors, with moduli $a, b, c \ldots$
$\mathbf{a} + \mathbf{b}, \mathbf{a} - \mathbf{b}$	vector sum and vector difference
$\mathbf{i}, \mathbf{j}, \mathbf{k}$	unit vectors, with right-hand set of mutually perpendicular axes
$\mathbf{r} = x\mathbf{i} + y\mathbf{j} + z\mathbf{k}$	vector $\mathbf{r}$ with components $x\mathbf{i}, y\mathbf{j}, z\mathbf{k}$
$\mathbf{a}.\mathbf{b}$	scalar product $ab \cos\theta$
$\mathbf{v} = \dfrac{d\mathbf{r}}{dt} = \dot{\mathbf{r}} = \dot{x}\mathbf{i} + \dot{y}\mathbf{j} + \dot{z}\mathbf{k}$	velocity $\uparrow\mathbf{v}$
$\mathbf{a} = \dfrac{d\mathbf{v}}{dt} = \ddot{\mathbf{r}} = \ddot{x}\mathbf{i} + \ddot{y}\mathbf{i} + \ddot{z}\mathbf{k}$	acceleration $\uparrow\mathbf{a}$
$\mathbf{v}_P(Q) = \dfrac{d}{dt}(\mathbf{PQ})$	velocity of Q relative to P
$\mathbf{a}_P(Q) = \dfrac{d}{dt}\mathbf{v}_P(Q)$	acceleration of Q relative to P
$\dot{\theta} = \omega$	angular velocity
$\ddot{\theta} = \dot{\omega}$	angular acceleration

Mechanics

g

acceleration due to gravity ($g \approx 9.8\,\text{m s}^{-2}$ on Earth, but frequently taken as $10\,\text{m s}^{-2}$)

F

force **F**

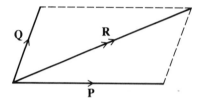

resultant **R**, with components **P**, **Q**

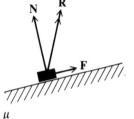

resultant reaction **R**, with normal component **N** and frictional component **F**

μ

coefficient of friction

λ

angle of friction

SI units

time	second	s	
length	metre	m	
frequency	hertz	Hz	($1\,\text{Hz} = 1\,\text{s}^{-1}$)
length	metre	m	
mass	kilogram	kg	
force	newton	N	($1\,\text{N} = 1\,\text{kg m s}^{-2}$)
work, energy	joule	J	($1\,\text{J} = 1\,\text{Nm}$)
power	watt	W	($1\,\text{W} = 1\,\text{J s}^{-1}$)

1 Introduction

1.1 Logical symbols

Every piece of mathematics is about the consequences of a particular set of rules or axioms. As with the basic laws of a country or the rules of a game, some sets of axioms prove to be more useful or more interesting than others; but their only essential quality is to be clear and consistent, without contradictions.

From these axioms we proceed to other mathematical statements. This usually begins with a process of speculation, frequently needing considerable imagination and insight. If our speculations can be proved correct, we call them *deductions* from our axioms, and one of our first needs is for a set of symbols to denote this process of *implication*.

Suppose that p and q are two such statements. We shall use the symbols $\Rightarrow$, $\Leftarrow$, and $\Leftrightarrow$ to denote implications as follows:

$$p \Rightarrow q \quad \text{means} \quad \text{'}p \text{ implies } q\text{'}$$
$$p \Leftarrow q \quad \text{means} \quad \text{'}p \text{ is implied by } q\text{' or '}q \text{ implies } p\text{'}$$
$$p \Leftrightarrow q \quad \text{means} \quad \text{'}p \text{ implies } q\text{' and '}q \text{ implies } p\text{'}$$

For example, if p is the statement 'ABCD is a square'
and $\qquad\quad$ q is the statement 'ABCD is a rectangle'
then $\qquad\qquad$ $p \Rightarrow q$

Similarly $\quad x^2 = 1 \quad \Leftarrow \quad x = 1$

and $\quad$ '$\triangle ABC$ is equilateral' $\Leftrightarrow$ '$\triangle ABC$ is equiangular'

In this last case, the two statements are said to be *equivalent*.

A line striking through any of these symbols inserts the word 'not'. So $p \not\Rightarrow q$ means 'p does not imply q'. This is not to say that q cannot be true, but simply that it is not a consequence of p.

For example $\quad x^2 = 1 \quad \not\Rightarrow \quad x = 1$

(for though x might be 1, it might equally well be -1. Such a case which shows the implication to be invalid is usually called a *counter-example*.)

If p is a given statement, we shall use $\sim p$ to indicate its opposite, or *negation*.†

---

† p' is sometimes used instead of $\sim p$.

So if p denotes 'John is English'
then $\sim p$ denotes 'John is not English'

If, further, q denotes 'John is a Yorkshireman'
then $\sim q$ denotes 'John is not a Yorkshireman'
and we see that

$$p \not\Rightarrow q, \quad p \Leftarrow q, \quad \sim p \Rightarrow \sim q, \quad \text{etc.}$$

Exercise 1.1

1   Use the symbols $\Rightarrow$, $\Leftarrow$, and $\Leftrightarrow$ to connect the statements p and q, where
   a) p: Mary lives in Scotland
       q: Mary lives in Glasgow
   b) p: ABCD is a square
       q: ABCD is a rhombus
   c) p: ABCD is a parallelogram
       q: ABCD is a rectangle
   d) p: In $\triangle ABC$, $\hat{A} = 90°$
       q: In $\triangle ABC$, $AB^2 + AC^2 = BC^2$
   e) p: Hexagon ABCDEF is equilateral
       q: Hexagon ABCDEF is equiangular

2   Use the symbols $\not\Rightarrow$, $\not\Leftarrow$, $\not\Leftrightarrow$ to connect the above statements, giving counter-examples to show why the implication is not valid.

3   Repeat no. **1** for the statements $\sim p$ and $\sim q$.

4   What do you know about $\sim p$ and $\sim q$ if
     a) $p \Rightarrow q$ **b)** $p \Leftarrow q$ **c)** $p \Leftrightarrow q$?

1.2 Sets and Boolean algebra

We now look briefly at the mathematics of collections, or sets. This may already be familiar to the reader and is of interest to us for two reasons: firstly because it leads to an algebra, named after its inventor George Boole (1815–64), which provides both striking similarities and also sharp contrasts with the algebra of ordinary numbers; and secondly because it gives us a language and a notation which will be particularly useful when we investigate (as in chapter 7) questions of probability.

   Sets may be of any kind: the pupils of a school, the species of animals in a zoo, or the numbers which are multiples of 5. The only requirement is that we should always be able to decide whether or not an object belongs to the set: those which do are called its *elements*. 'Belongs to' is abbreviated as $\in$, whilst the set itself is usually written in braces, { }, or denoted by a capital letter.

So if A is the set of even integers from 2 to 10, we write

$$A = \{\text{even integers from 2 to 10}\} = \{2, 4, 6, 8, 10\}$$

and $2 \in A, 4 \in A$, but $3 \notin A$

When our attention is entirely confined within a particular set of objects, we usually call it the *universal set*, which we shall represent by $\mathscr{E}$ (ensemble). Other sets of objects under consideration can then be regarded as *sub-sets* of $\mathscr{E}$.

If, for instance, we are interested only in the positive integers from 1 to 10, we can take

$$\mathscr{E} = \{1, 2, 3, 4, 5, 6, 7, 8, 9, 10\}$$

so $A = \{2, 4, 6, 8, 10\}$ is a sub-set of $\mathscr{E}$, and we write $A \subset \mathscr{E}$, or $\mathscr{E} \supset A$.

Using the symbol $\in$, we can also write

$$A = \{x : x \in \mathscr{E}, x \text{ is even}\}$$

Here the colon means 'such that'; so the whole sentence should be read:

'A is the set of every x belonging to $\mathscr{E}$ such that x is even.'

The *complement of A within* $\mathscr{E}$ is the set of all elements of $\mathscr{E}$ which do not belong to A, and is written A'.

So if $\mathscr{E} = \{1, 2, 3, 4, 5, 6, 7, 8, 9, 10\}$
and $A = \{2, 4, 6, 8, 10\}$
we see that $A' = \{1, 3, 5, 7, 9\}$

Sometimes a particular set has no members,

e.g. $\{x : x \in \mathscr{E}, x \text{ is a multiple of 11}\}$

Such a set is called *empty*, or a *null-set*, and written $\varnothing$.

When two sets A and B contain exactly the same elements they are said to be *equal*, and we write $A = B$.

It is also useful to have special names for four important sets of numbers:

$$\mathbb{N} = \{\text{natural numbers}\} = \{0, 1, 2, 3, 4, 5 \ldots\}$$

$$\mathbb{Z} = \{\text{all integers}\} \qquad = \{0, \pm 1, \pm 2, \pm 3 \ldots\}$$

$$\mathbb{Q} = \{\text{all rational numbers†}\}$$

$$\mathbb{R} = \{\text{all real numbers†}\}$$

The sets of all positive integers, positive rational numbers and positive real numbers are denoted by $\mathbb{Z}^+$, $\mathbb{Q}^+$, $\mathbb{R}^+$ respectively.

So $\mathbb{Z}^+ = \{x : x \in \mathbb{Z}, x > 0\}$,   etc.

† See page 16.

Exercise 1.2a

1 Simplify:
   a) $\{x : x \in \mathbb{Z}, -1 \leqslant x \leqslant 3\}$
   b) $\{x : x \in \mathbb{Z}^+, -1 \leqslant x \leqslant 3\}$
   c) $\{x : x \in \mathbb{Z}, x > 4\}$
   d) $\{x : x \in \mathbb{Z}^+, x \leqslant 2\}$
   e) $\{x : x \in \mathbb{Z}^+, x \leqslant 0\}$

2 Simplify: **a)** $\varnothing'$ **b)** $\mathscr{E}'$ **c)** $(A')'$

Union and intersection

If A and B are two sets, we now define two new sets:

a) the set of all elements belonging to A *or* B *or both* is called the *union* of A and B, and written $A \cup B$;

b) the set of all elements belonging to A *and* B is called the *intersection* of A and B, and written $A \cap B$.

So if $A = \{a, b, c, d, e\}$
and $B = \{a, e, i, o, u\}$
      $A \cup B = \{a, b, c, d, e, i, o, u\}$
and $A \cap B = \{a, e\}$

It will have been noticed that the operations of union and intersection are independent of the order of A and B, so that

$$A \cup B = B \cup A \quad \text{and} \quad A \cap B = B \cap A$$

Such operations are said to be *commutative*.
   They can also be illustrated very conveniently in *Venn diagrams*†:

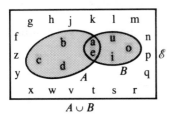
$A \cup B$

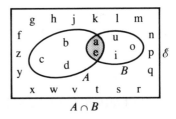
$A \cap B$

$\mathscr{E} = \{\text{the 26 letters of the alphabet}\}$
$A = \{\text{the first 5 letters of the alphabet}\}$
$B = \{\text{the vowels}\}$

† John Venn (1834–1925), English philosopher and logician.

Such diagrams can also be used to depict the complement of A:

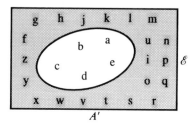

A'

Exercise 1.2b

1   Simplify:
   a) $A \cup \varnothing$ **b)** $A \cap \varnothing$ **c)** $A \cup \mathscr{E}$ **d)** $A \cap \mathscr{E}$ **e)** $A \cup A$
   f) $A \cap A$ **g)** $A \cup (A \cap B)$ **h)** $A \cap (A \cup B)$ **i)** $A \cup A'$ **j)** $A \cap A'$

2   Use Venn diagrams to show that union and intersection are *associative*, that is
   a) $A \cup (B \cup C) = (A \cup B) \cup C$ **b)** $A \cap (B \cap C) = (A \cap B) \cap C$
   (so that they can be written without ambiguity as $A \cup B \cup C$ and $A \cap B \cap C$ respectively).

3   Use Venn diagrams to show that *intersection is distributive over union* and that *union is distributive over intersection*, that is
   a) $A \cap (B \cup C) = (A \cap B) \cup (A \cap C)$
   b) $A \cup (B \cap C) = (A \cup B) \cap (A \cup C)$

4   Use Venn diagrams to verify *de Morgan's laws*†:
   a) $(A \cup B)' = A' \cap B'$ **b)** $(A \cap B)' = A' \cup B'$

5   Simplify and illustrate in Venn diagrams:
   a) $A \cap (A' \cup B)$ **b)** $A \cap B \cap (A' \cup B')$
   c) $(A \cup B)' \cup B'$ **d)** $(A \cup B) \cap (A \cup B')$

1.3 Coordinates

Suppose that a point O is marked on a plane, together with a pair of perpendicular lines, Ox and Oy, each with a uniform scale. The reader will be used to defining other points in the plane by means of *ordered pairs* of numbers (x, y), known as their *Cartesian* coordinates (after the French mathematician

---

† Augustus de Morgan (1806–71), first Professor of Mathematics, University of London.

and philosopher, René Descartes (1596–1650)). O is called the *origin* and Ox, Oy are the *coordinate axes*.

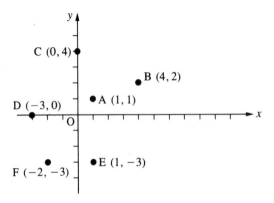

Mid-point, distance and gradient

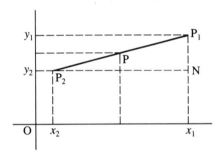

Suppose that P_1 is (x_1, y_1) and P_2 is (x_2, y_2). Hence the coordinates of the mid-point are given by

$$x = \tfrac{1}{2}(x_1 + x_2)$$
$$y = \tfrac{1}{2}(y_1 + y_2)$$

It is also seen that

$$P_1P_2^2 = P_2N^2 + NP_1^2$$
$$= (x_1 - x_2)^2 + (y_1 - y_2)^2$$
so $$P_1P_2 = \sqrt{\{(x_1 - x_2)^2 + (y_1 - y_2)^2\}}$$

Further the gradient of P_1P_2 is defined as

$$\frac{y_1 - y_2}{x_1 - x_2}$$

So, in our figure, the mid-points, lengths and gradients of AB, DE, EF, and AE are as follows:

midpoint	length	gradient
AB: $\left(\dfrac{1+4}{2}, \dfrac{1+2}{2}\right) = \left(\dfrac{5}{2}, \dfrac{3}{2}\right)$	$\sqrt{[(4-1)^2 + (2-1)^2]} = \sqrt{10}$	$\dfrac{2-1}{4-1} = \dfrac{1}{3}$
DE: $\left(\dfrac{-3+1}{2}, \dfrac{0-3}{2}\right) = \left(-1, -\dfrac{3}{2}\right)$	$\sqrt{[(1+3)^2 + (-3)^2]} = 5$	$\dfrac{-3-0}{1-(-3)} = -\dfrac{3}{4}$
EF: $\left(\dfrac{-2+1}{2}, \dfrac{-3-3}{2}\right) = \left(-\dfrac{1}{2}, -3\right)$	$\sqrt{[(1+2)^2 + 0^2]} = 3$	$\dfrac{-3+3}{1+2} = 0$
AE: $\left(\dfrac{1+1}{2}, \dfrac{1-3}{2}\right) = (1, -1)$	$\sqrt{[(1-1)^2 + (1+3)^2]} = 4$	$\dfrac{1-(-3)}{1-1} = \dfrac{4}{0}$

which is undefined.

Perpendicular lines

Suppose that P is a point, not on a coordinate axis, with coordinates (h, k).

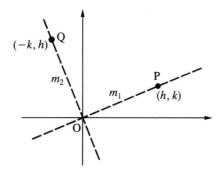

If OP is rotated anti-clockwise through a right angle, then P moves to Q, $(-k, h)$, and the gradients of OP and OQ are

$$m_1 = \frac{k}{h} \quad \text{and} \quad m_2 = -\frac{h}{k}$$

So OP, OQ perpendicular

$$\Rightarrow \quad m_1 m_2 = \frac{k}{h} \times -\frac{h}{k}$$

$$\Rightarrow \quad m_1 m_2 = -1$$

Conversely, if $m_1 m_2 = -1$ it is readily seen that the two lines must be perpendicular.

Example 1

In our original figure (p. 6), gradient of AB $= \dfrac{2-1}{4-1} = \dfrac{1}{3}$

$$\text{gradient of OE} = \dfrac{-3-0}{1-0} = -3$$

So AB and OE are perpendicular.

Exercise 1.3a

For nos. **1–3**, draw axes Ox, Oy and label them from $x = -5$ to $+5$ and $y = -5$ to $+5$.

1　Mark the points O $(0, 0)$,　A $(4, 5)$,　B $(3, 0)$,　C $(0, -2)$,　D $(-2, 1)$.

2　Find the mid-points, lengths and gradients of OA, OB, OC, OD. Name any two lines which are perpendicular.

3　Find the mid-points, lengths, and gradients of the six lines joining A, B, C, and D. Name any pairs which are parallel or perpendicular.

4　M is the mid-point of side BC of $\triangle$ABC. What do you know about AB² + AC² and 2(AM² + BM²)? [Let M, B, C, A be $(0, 0)$, $(a, 0)$, $(-a, 0)$, (x, y) respectively.]

Equations and inequalities: curves and regions

Sometimes we do not know the exact position of a point P but merely that its coordinates x, y are restricted by some equation or inequality. There is then a corresponding restriction on the position of P and the set of all possible positions is a sub-set of all the points of the plane.

Example 2

If $y = x^2$, possible pairs (x, y) are $(0, 0)$, $(1, 1)$, $(2, 4)$, $(\frac{1}{2}, \frac{1}{4})$, $(-0.3, +0.09)$, etc, [but *not* $(1, 2)$, $(2, 3)$, $(-1, -1)$, etc.], and it is readily seen that the set of all points is indicated by:

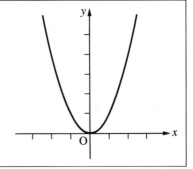

Example 3

Similarly, if $x + y > 2$, possible positions of (x, y) are $(2\frac{1}{2}, 0)$, $(1\frac{1}{4}, 1)$, $(-1, 4)$, $(5, -2.9)$, etc. [but *not* $(0, 1.9)$, $(-1, -2)$, $(1, 1)$, etc.]. So the set of all such points can be indicated by:

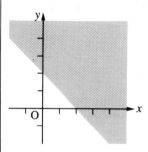

If we were told that $x + y \geqslant 2$, the region would also include the set of points for which $x + y = 2$ (i.e. the line $x + y = 2$):

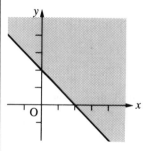

Example 4

Suppose (x, y) is restricted by $x + 3y = 9$ *and* $2x - y = 4$.

Then $\left.\begin{array}{l} 2x + 6y = 18 \\ 2x - y\ \ = 4 \end{array}\right\} \Rightarrow\ \ 7y = 14\ \ \Rightarrow\ \ y = 2\ \ \Rightarrow\ \ x = 3$
and

So $(3, 2)$ is the only possible position of (x, y) and this is the point of intersection of $x + 3y = 9$ and $2x - y = 4$.

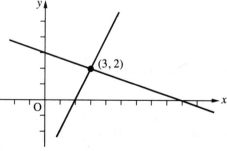

Exercise 1.3b

In each of nos. **1–10** draw axes Ox, Oy and label them from $x = -5$ to $+5$ and from $y = -5$ to $+5$. Then for each of the given equations or inequalities
a) name four points which satisfy the restriction;
b) name two points which do not satisfy the restriction;
c) indicate in your figure the set of points defined by the restriction, i.e. the corresponding curve or region.

1   **a)** $x + y < 4$ **b)** $x + y = 4$ **c)** $x + y > 4$

2   **a)** $y \leqslant 2x + 3$ **b)** $y \geqslant 2x + 3$

3   **a)** $x^2 + y^2 \leqslant 25$ **b)** $x^2 + y^2 > 25$

4   **a)** $xy = 12$ **b)** $xy \neq 12$

5   **a)** $xy > 0$ **b)** $xy = 0$

6   **a)** $x^2 + y^2 < 0$ **b)** $x^2 + y^2 = 0$ **c)** $x^2 + y^2 \geqslant 0$

7   **a)** $xy + 20 = 0$ **b)** $xy + 20 > 0$

8   **a)** $y = 2x$ **b)** $y = x$ **c)** $y = \frac{1}{2}x$ **d)** $y = 0x$ **e)** $y = -x$

9   **a)** $x^2 - y^2 > 0$ **b)** $x^2 - y^2 = 0$ **c)** $x^2 - y^2 < 0$

10   **a)** $y = x^3$ **b)** $4x^2 + 9y^2 = 36$

11   Illustrate the following, and in each case find and show in your illustration the points of intersection:
   a) $x + y = 4$ and $y = 2x + 1$
   b) $x + y = 4$ and $7x + y + 8 = 0$
   c) $y = 2x + 1$ and $7x + y + 8 = 0$
   d) $y = x + 2$ and $y = x^2$
   e) $y = x^3$ and $y = 4x$

12   Illustrate the region of the plane where
   a) $0 \leqslant y < x$
   b) $x^2 + y^2 \leqslant 1$ and $xy > 0$
   c) $x + 1 < y < 2x - 3$

1.4 Straight lines

We saw in the last section that when the coordinates (x, y) of a point P were restricted by an equation, the point P was usually confined to a particular curve. So we refer to the original restriction as the *equation of the curve*. There is therefore a two-way problem: to investigate the curve which arises from a given equation, and to find the equation of a given curve. We shall begin with the case of a 'curve' which is a straight line.

Example 1

Find the equation of the straight line with gradient $\frac{1}{3}$ which passes through the point $(1, 2)$.

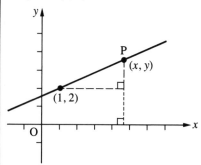

$P(x, y)$ is on this line

$$\Rightarrow \quad \frac{y - 2}{x - 1} = \frac{1}{3}$$

$$\Rightarrow \quad 3y - 6 = x - 1$$

$$\Rightarrow \quad x - 3y + 5 = 0$$

which is therefore the equation of the line.

More generally, the line with gradient m which passes through the point (x_1, y_1) has the equation

$$\frac{y - y_1}{x - x_1} = m$$

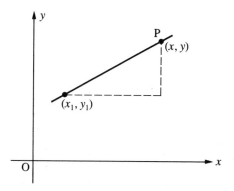

Example 2

Find the equation of the line which passes through (1, 3) and (3, 6).

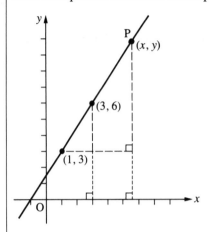

$P(x, y)$ is on this line

$$\Rightarrow \quad \frac{y - 6}{x - 3} \qquad = \frac{6 - 3}{3 - 1} = \frac{3}{2}$$

$$\Rightarrow \quad 2(y - 6) \qquad = 3(x - 3)$$

$$\Rightarrow \quad 3x - 2y + 3 = 0$$

More generally, the line which passes through (x_1, y_1) and (x_2, y_2) has the equation

$$\frac{y - y_1}{x - x_1} = \frac{y_2 - y_1}{x_2 - x_1}$$

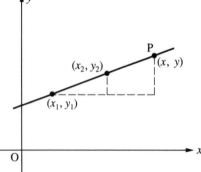

Example 3

A is (3, 4),   B is (7,5),   C is (5,1). Find the equation of the line through A which is perpendicular to BC.

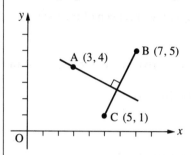

Gradient of BC $= \dfrac{5-1}{7-5} = \dfrac{4}{2} = 2$

$\Rightarrow$ required gradient $= -\dfrac{1}{2}$

So required line is $\dfrac{y-4}{x-3} = -\dfrac{1}{2}$ $\Rightarrow$ $x + 2y - 11 = 0$

Exercise 1.4a

In nos. **1–6**, O is (0, 0),   A is (3, 1),   B is (5, 5),   C is (9, 1). Find the equations of

1   a) OA   b) OB   c) OC

2   a) BC   b) CA   c) AB

3   the three *medians* (which join the vertices of △ABC to the mid-points of the opposite sides). Show that they all meet in a point G (the *centroid* of △ABC), and find the coordinates of G.

4   the three *altitudes* of △ABC (which pass through the vertices and are perpendicular to the opposite sides). Show that they all meet in a point H (the *orthocentre* of △ABC) and find the coordinates of H.

5   the three perpendicular bisectors of BC, CA, AB. Show that they all meet in a point X (the *circumcentre* of △ABC), and find the coordinates of X. Verify by coordinate geometry that XA = XB = XC.

6   Show that the centroid, the orthocentre and the circumcentre all lie on a straight line, and find its equation.

7 Show that the equation of the straight line with gradient m which passes through the point $(0, c)$ is $y = mx + c$.

The general equation of a straight line

In all these examples the equation of a straight line has been of the form

$$ax + by + c = 0$$

where a, b, c are constants and a, b are not both zero (since otherwise $c = 0$ and the equation would entirely collapse).

To examine this equation, we shall distinguish two cases:

Case 1 $b = 0$

Then $ax + c = 0 \iff x = -\dfrac{c}{a}$ (since $a \neq 0$)

and this is the equation of a straight line parallel to Oy.

Case 2 $b \neq 0$

Then $ax + by + c = 0 \iff y = -\dfrac{a}{b}x - \dfrac{c}{b}$ (since $b \neq 0$)

and (by no. **7** of the last exercise) this is the equation of a straight line of gradient $-a/b$ which goes through $(0, -c/b)$. So again the given equation represents a straight line.

Hence in every case $ax + by + c = 0$ represents a straight line.

The plotting of straight lines is frequently helped by concentrating on particular features:

Example 4

$3x + 4y - 12 = 0$

$x = 0 \implies 4y - 12 = 0 \implies y = 3$

$y = 0 \implies 3x - 12 = 0 \implies x = 4$

So the line passes through $(0, 3)$ and $(4, 0)$ and is therefore

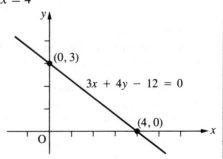

Example 5

$x + 2y = 0$

This line passes through $(0, 0)$ and has gradient $-\frac{1}{2}$ (since $y = -\frac{1}{2}x$). So the line is

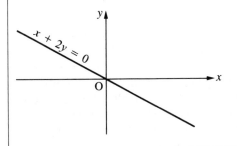

Exercise 1.4b

Sketch the lines:

1 a) $y = x$ **b)** $y = x + 1$ **c)** $y = x + 2$ **d)** $y = x - 1$

2 a) $y = -2x$ **b)** $y = -2x + 1$ **c)** $y = -2x + 2$ **d)** $y = -2x - 2$

Sketch the triangles defined by the following sets of lines and find the coordinates of their vertices:

3 a) $x - 4 = 0$ **b)** $x + y - 4 = 0$ **c)** $x - y - 2 = 0$

4 a) $x - 2y + 3 = 0$ **b)** $x + y = 0$ **c)** $y + 2 = 0$

5 a) $2x + y - 4 = 0$ **b)** $3x - y - 6 = 0$ **c)** $x + y = 0$

1.5 Rational and irrational numbers

Soon after human beings learned to count and to use the *natural numbers* 0, 1, 2, 3, 4, 5, ... they must have realised that they could add them and multiply them and obtain answers which always remained safely within the same set, usually written $\mathbb{N}$.

When, however, they asked inverse questions, such as

'What do I need to add to 4 in order to obtain 3?'

they found that they could not provide an answer from within the same set and were compelled to generate a new set of numbers, including negative whole numbers and zero. This we call the set of *integers*, usually written $\mathbb{Z}$. Within this set all questions about addition and subtraction could once again be answered.

When, however, they proceeded further to such questions as

'What number when multiplied by 4 provides 3?'

and 'How many times does -5 go into 7?'

once again it was necessary for them to expand their field of numbers, this time by adding 3/4, $-7/5$ and all other numbers which are the quotients of two integers. This enlarged set is called the field of *rational numbers*, usually written $\mathbb{Q}$, and within this field the four processes of addition and subtraction, multiplication and division (except division by 0) can all be carried out with the answer again lying safely within the field. Here at last, together with geometry, Pythagoras and his followers felt they had the apparatus for explaining the universe; and this they set to do, ranging from the movement of planets to the harmonies of music.

But the function $f: x \mapsto x^2$ soon leads to the question,

'Which number has 2 as its square?'

or 'What is the side of a square which has area 2?'

Now $(1.4)^2 = 1.96$, so it is clear that the answer is roughly 1.4.

For a better approximation, we note that

$$2 \div 1.4 \approx 1.428$$

$$\Rightarrow \quad 2 \approx 1.4 \times 1.428$$

Taking the average of 1.4 and 1.428 leads us to look at the number 1.414.

Now $\quad 2 \div 1.414 \approx 1.41443$

$$\Rightarrow \quad 2 \approx 1.414 \times 1.41443$$

and we next look at their average, 1.414215.

By this means we can approach as close as we like—1.4, 1.414, 1.414215 etc. —to the number whose square is 2, but the number itself remains strangely elusive.

Let us therefore try another approach and suppose that the required number is the fraction p/q, which is expressed in its lowest terms (i.e. with p and q as integers having no common factor).

Then $\quad (p/q)^2 = 2$

So $\quad\quad p^2 = 2q^2$, which is an even integer.

But the square of any odd integer is odd. So p cannot be odd, and must be even.

Let $\qquad p = 2m$ (where m is an integer)

Then $(2m)^2 = 2q^2$

$\Rightarrow \qquad q^2 = 2m^2$

So q^2 is even, and therefore q must be even.

Hence p and q are both even, and so are both divisible by 2. But this is a blatant contradiction of our original statement that p and q have no common factor. There must, therefore, be a slip in our working, a flaw in our argument or an inconsistency in our original assumptions.

Close inspection fails to reveal either a slip or a flaw, so we are inevitably led to the conclusion that the contradiction which we have reached must spring from an internal inconsistency within our assumptions.

But there were only two such assumptions: that x is the quotient of two integers and that $x^2 = 2$.

We therefore see that there *cannot* be a number x which is the quotient of two integers and whose square is 2.

So $\sqrt{2}$ is not a rational number.

This proof, by the exceedingly powerful method of *reductio ad absurdum*, was known to Euclid, and its implications were immediately clear: that the apparent completeness of the rational numbers is an illusion and that others exist (in fact, as we now know, completely 'outnumbering' the rationals).

Worse than this, the stage in the argument which we called a 'contradiction' or an 'absurdity' must have seemed to the Greeks more like a catastrophe. For here were two integers p and q which both *did* and *did not* have 2 as a common factor. The whole rationality of mathematics seemed imperilled and numbers like $\sqrt{2}$ were said by some to be 'unspeakable'. Rather less dramatically, we simply call numbers *irrational* if they cannot be expressed as the ratio of two integers. Together with the rational numbers, they form the field of *real numbers*.

Exercise 1.5

1 Decide, with proofs, whether the following numbers are rational or irrational:
   a) $\sqrt{3}$ **b)** $\sqrt{10}$ **c)** $\sqrt[3]{2}$ **d)** $\sqrt{\frac{4}{9}}$ **e)** $\sqrt[3]{\frac{1}{8}}$

2 Express as fractions:
   a) 0.3 **b)** 0.004 **c)** 0.426 **d)** 2.43
   Is every finite decimal a rational number?

3 Express as decimals:
   a) $\frac{1}{5}$ **b)** $\frac{1}{10}$ **c)** $\frac{1}{3}$ **d)** $\frac{1}{11}$ **e)** $\frac{2}{7}$
   Can every rational number be expressed as a finite or as an infinitely recurring decimal?

4 Show that
  a) $\frac{1}{9} = 0.\dot{1}$,   and so express $0.\dot{2}$ as a fraction;
  b) $\frac{1}{99} = 0.\dot{0}\dot{1}$,   and so express $0.\dot{1}\dot{3}$ as a fraction;
  c) $\frac{1}{999} = 0.\dot{0}0\dot{1}$,   and so express $0.\dot{4}6\dot{8}$ as a fraction.
  Are all infinitely recurring decimals bound to be rational numbers? Can an irrational number ever be expressed as an infinitely recurring decimal?

1.6 Surds

In the last section we encountered irrational numbers like $\sqrt{2}$ and $\sqrt[3]{5}$. These are usually known as *surds* and expressions containing them are so common that we must spend a little time learning how they can be simplified.

Example

Simplify

a) $\sqrt{200}$ **b)** $\dfrac{9}{\sqrt{3}}$

c) $(2\sqrt{5} - 3)(2\sqrt{5} + 3)$ **d)** $\dfrac{2\sqrt{5} + 1}{2\sqrt{5} + 3}$

a) $\sqrt{200} = \sqrt{(100 \times 2)} = \sqrt{100} \times \sqrt{2} = 10\sqrt{2}$

b) $\dfrac{9}{\sqrt{3}} = \dfrac{3 \times \sqrt{3} \times \sqrt{3}}{\sqrt{3}} = 3\sqrt{3}$

c) $(2\sqrt{5} - 3)(2\sqrt{5} + 3) = (2\sqrt{5})^2 - 3^2$
                                   $= 20 - 9 = 11$

d) To simplify $\dfrac{2\sqrt{5} + 1}{2\sqrt{5} + 3}$, we first observe from the last example that the denominator would be simplified if we were to multiply it by $2\sqrt{5} - 3$, usually known as its *conjugate*.

So we write

$$\frac{2\sqrt{5} + 1}{2\sqrt{5} + 3} = \frac{(2\sqrt{5} + 1)(2\sqrt{5} - 3)}{(2\sqrt{5} + 3)(2\sqrt{5} - 3)} = \frac{(2\sqrt{5})^2 - 6\sqrt{5} + 2\sqrt{5} - 3}{(2\sqrt{5})^2 - 3^2}$$

$$= \frac{17 - 4\sqrt{5}}{20 - 9} = \frac{17 - 4\sqrt{5}}{11}$$

This is known as *rationalising the denominator*.

Exercise 1.6

1 Express as simple surds:
  a) $\sqrt{8}$ **b)** $\sqrt{12}$ **c)** $\sqrt{18}$ **d)** $\sqrt{50}$ **e)** $\sqrt{72}$
  f) $\sqrt{98}$ **g)** $\sqrt{288}$ **h)** $\sqrt{300}$ **i)** $\sqrt{1000}$ **j)** $\sqrt{2400}$

2 Simplify:
  a) $\dfrac{2}{\sqrt{2}}$ **b)** $\dfrac{24}{\sqrt{8}}$ **c)** $\dfrac{9}{\sqrt{27}}$ **d)** $\dfrac{10}{\sqrt{5}}$

  e) $\sqrt{8} \times \sqrt{32}$ **f)** $\sqrt{10} \times \sqrt{1000}$ **g)** $\dfrac{\sqrt{20}}{\sqrt{10}}$ **h)** $\dfrac{\sqrt{200}}{\sqrt{10}}$

3 Simplify:
  a) $\sqrt{8} - \sqrt{2}$ **b)** $\sqrt{125} + \sqrt{5}$ **c)** $\sqrt{75} - \sqrt{48}$
  d) $\sqrt{3}(2 - \sqrt{3})$ **e)** $\sqrt{6}(\sqrt{24} - \sqrt{6})$ **f)** $(2 - \sqrt{2})^2$
  g) $(\sqrt{5} + \sqrt{3})(\sqrt{5} - \sqrt{3})$ **h)** $(2\sqrt{3} + 1)(2\sqrt{3} - 1)$

4 Rationalise the denominators and simplify:
  a) $\dfrac{1}{\sqrt{5}}$ **b)** $\dfrac{4}{\sqrt{8}}$ **c)** $\dfrac{1}{\sqrt{8} - \sqrt{2}}$ **d)** $\dfrac{1}{\sqrt{3} - 1}$

  e) $\dfrac{2}{2 + \sqrt{3}}$ **f)** $\dfrac{\sqrt{5} - \sqrt{2}}{\sqrt{5} + 1}$ **g)** $\dfrac{\sqrt{3} - \sqrt{2}}{\sqrt{3} + \sqrt{2}}$

  h) $\dfrac{1}{\sqrt{2} - 1} - \dfrac{1}{\sqrt{2} + 1}$

1.7 Linear transformations and matrices

Consider the equations

$$x' = 2x + y \qquad y' = 3x - 2y$$

We can use these equations to transform any point P with coordinates (x, y) into another point P' with coordinates (x', y'). For instance,

  (2, 1) is transformed into (5, 4)
  (0, 2) is transformed into (2, −4)

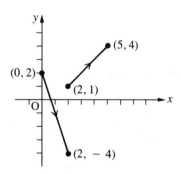

In this way we obtain a transformation of the entire plane, and this transformation is called 'linear' on account of the original equations.
More generally,

$$x' = ax + by$$
$$y' = cx + dy$$

where a, b, c, d are constants, is a *linear transformation*.

We shall now investigate the effects of a number of simple linear transformations.

Example 1

$$x' = x$$
$$y' = -y$$

transforms $(1, 1)$ into $(1, -1)$
$(1, 2)$ into $(1, -2)$, etc.

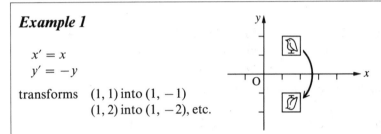

So we see that this linear transformation represents a *reflection* in the x-axis.

Example 2

$$x' = 2x$$
$$y' = 3y$$

transforms $(1, 1)$ into $(2, 3)$
$(2, 2)$ into $(4, 6)$, etc.,

so it represents   a stretch $\times$ 2 parallel Ox
and   a stretch $\times$ 3 parallel Oy

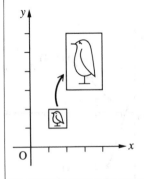

Example 3

$$x' = -x - y$$
$$y' = x - y$$

The square ABCD, where

A is $(1, 1)$, B $(2, 1)$, C $(2, 2)$, D $(1, 2)$

is transformed into A′B′C′D′ where

A′ is $(-2, 0)$, B′ $(-3, 1)$, C′ $(-4, 0)$, D′ $(-3, -1)$.

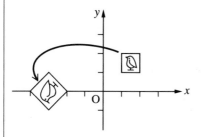

So the transformation represents an anticlockwise rotation through $135°$ combined with a 2-way stretch $\times \sqrt{2}$.

Example 4

$$x' = x + y$$
$$y' = y$$

transforms A, B, C, D into

A′ $(2, 1)$, B′ $(3, 1)$, C′ $(4, 2)$, D′ $(3, 2)$.

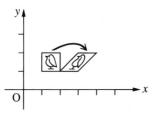

Such a transformation is called a *shear*.

It is clear from these examples that the linear transformation

$$x' = ax + by$$
$$y' = cx + dy$$

is defined by the values of the four elements in the square arrangement:

$$\begin{pmatrix} a & b \\ c & d \end{pmatrix}$$

This is called the *matrix* of the transformation, which can be written as

$$\begin{pmatrix} x' \\ y' \end{pmatrix} = \begin{pmatrix} a & b \\ c & d \end{pmatrix}\begin{pmatrix} x \\ y \end{pmatrix}$$

where $\begin{pmatrix} a & b \\ c & d \end{pmatrix}\begin{pmatrix} x \\ y \end{pmatrix} = \begin{pmatrix} ax + by \\ cx + dy \end{pmatrix}$

Example 5

What are the transformations defined by

a) $\begin{pmatrix} 3 & 0 \\ 0 & 1 \end{pmatrix}$ **b)** $\begin{pmatrix} 0 & 1 \\ -1 & 0 \end{pmatrix}$ **c)** $\begin{pmatrix} 2 & -2 \\ -1 & 1 \end{pmatrix}$?

a) $\begin{pmatrix} 3 & 0 \\ 0 & 1 \end{pmatrix}$ is the matrix of the transformation

$$\begin{pmatrix} x' \\ y' \end{pmatrix} = \begin{pmatrix} 3 & 0 \\ 0 & 1 \end{pmatrix}\begin{pmatrix} x \\ y \end{pmatrix} = \begin{pmatrix} 3x + 0y \\ 0x + 1y \end{pmatrix} = \begin{pmatrix} 3x \\ y \end{pmatrix}$$

and so represents a stretch × 3 parallel to Ox.

b) $\begin{pmatrix} 0 & 1 \\ -1 & 0 \end{pmatrix}$ is the matrix of

$$\begin{pmatrix} x' \\ y' \end{pmatrix} = \begin{pmatrix} 0 & 1 \\ -1 & 0 \end{pmatrix}\begin{pmatrix} x \\ y \end{pmatrix} = \begin{pmatrix} 0x + 1y \\ -1x + 0y \end{pmatrix}$$

$$= \begin{pmatrix} y \\ -x \end{pmatrix}$$

and so represents a clockwise rotation through 90°.

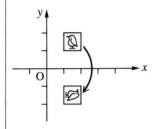

c) $\begin{pmatrix} 2 & -2 \\ -1 & 1 \end{pmatrix}$ is the matrix of

$$\begin{pmatrix} x' \\ y' \end{pmatrix} = \begin{pmatrix} 2 & -2 \\ -1 & 1 \end{pmatrix} \begin{pmatrix} x \\ y \end{pmatrix} = \begin{pmatrix} 2x - 2y \\ -x + y \end{pmatrix}$$

from which we see that for every point P (x, y), the point P' (x', y') is such that $x' = -2y'$.

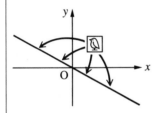

Hence every point P is transformed into a point of the line $x' = -2y'$, and our figure becomes squashed and unrecognisable.

Such a transformation and its corresponding matrix are called *singular*.

Exercise 1.7

Transform the points

O $(0, 0)$, U $(1, 0)$, V $(0, 1)$, W $(1, 1)$

by means of the following matrices, and in each case illustrate the effect of the transformation:

1 $\begin{pmatrix} 1 & 0 \\ 0 & 1 \end{pmatrix}$, usually called **I** (the *identity* or *unit* matrix).

2 $\begin{pmatrix} -1 & 0 \\ 0 & 1 \end{pmatrix}$ **3** $\begin{pmatrix} -1 & 0 \\ 0 & -1 \end{pmatrix}$ **4** $\begin{pmatrix} 0 & 1 \\ 1 & 0 \end{pmatrix}$ **5** $\begin{pmatrix} 2 & 0 \\ 0 & 2 \end{pmatrix}$

6 $\begin{pmatrix} 0 & 0 \\ 0 & 0 \end{pmatrix}$ **7** $\begin{pmatrix} 0 & -1 \\ 1 & 0 \end{pmatrix}$ **8** $\begin{pmatrix} 0 & 2 \\ -2 & 0 \end{pmatrix}$ **9** $\begin{pmatrix} 1 & 1 \\ 1 & 1 \end{pmatrix}$

10 $\begin{pmatrix} 1 & 0 \\ 1 & 1 \end{pmatrix}$

* 1.8 Multiplication of matrices

Suppose that a transformation with matrix $\mathbf{T} = \begin{pmatrix} a & b \\ c & d \end{pmatrix}$ is followed by a

transformation with matrix $\mathbf{U} = \begin{pmatrix} \alpha & \beta \\ \gamma & \delta \end{pmatrix}$.

If $\mathbf{T}$ transforms $\mathbf{P}\,(x, y)$ into $\mathbf{P}'\,(x', y')$
and $\mathbf{U}$ transforms $\mathbf{P}'\,(x', y')$ into $\mathbf{P}''\,(x'', y'')$

then $\begin{pmatrix} x' \\ y' \end{pmatrix} = \mathbf{T}\begin{pmatrix} x \\ y \end{pmatrix} = \begin{pmatrix} a & b \\ c & d \end{pmatrix}\begin{pmatrix} x \\ y \end{pmatrix}$ $\Leftrightarrow$ $\begin{aligned} x' &= ax + by \\ y' &= cx + dy \end{aligned}$

and $\begin{pmatrix} x'' \\ y'' \end{pmatrix} = \mathbf{U}\begin{pmatrix} x' \\ y' \end{pmatrix} = \begin{pmatrix} \alpha & \beta \\ \gamma & \delta \end{pmatrix}\begin{pmatrix} x' \\ y' \end{pmatrix}$ $\Leftrightarrow$ $\begin{aligned} x'' &= \alpha x' + \beta y' \\ y'' &= \gamma x' + \delta y' \end{aligned}$

So the combined transformation is

$$x'' = \alpha(ax + by) + \beta(cx + dy) = (\alpha a + \beta c)x + (\alpha b + \beta d)y$$
$$y'' = \gamma(ax + by) + \delta(cx + dy) = (\gamma a + \delta c)x + (\gamma b + \delta d)y$$

which has matrix $\begin{pmatrix} \alpha a + \beta c & \alpha b + \beta d \\ \gamma a + \delta c & \gamma b + \delta d \end{pmatrix}$

But this combination can also be written

$$\begin{pmatrix} x'' \\ y'' \end{pmatrix} = \mathbf{U}\begin{pmatrix} x' \\ y' \end{pmatrix} = \mathbf{UT}\begin{pmatrix} x \\ y \end{pmatrix}$$

The matrix of the combined transformation is therefore called the *product* of
$\mathbf{U}$ and $\mathbf{T}$ and written $\mathbf{UT}$.

So $\mathbf{UT} = \begin{pmatrix} \alpha & \beta \\ \gamma & \delta \end{pmatrix}\begin{pmatrix} a & b \\ c & d \end{pmatrix} = \begin{pmatrix} \alpha a + \beta c & \alpha b + \beta d \\ \gamma a + \delta c & \gamma b + \delta d \end{pmatrix}$

The reader should note very clearly that transformation $\mathbf{T}$ followed by $\mathbf{U}$ has
matrix $\mathbf{UT}$, and $\mathbf{U}$ followed by $\mathbf{T}$ has matrix $\mathbf{TU}$, and can easily show that these
two matrices are normally different.

Example 1

If $\quad \mathbf{A} = \begin{pmatrix} 1 & 2 \\ 3 & 4 \end{pmatrix}$ and $\mathbf{B} = \begin{pmatrix} 5 & 6 \\ 7 & 8 \end{pmatrix}$, find $\mathbf{AB}$ and $\mathbf{BA}$.

$$\mathbf{AB} = \begin{pmatrix} 1 & 2 \\ 3 & 4 \end{pmatrix}\begin{pmatrix} 5 & 6 \\ 7 & 8 \end{pmatrix} = \begin{pmatrix} 1 \times 5 + 2 \times 7 & 1 \times 6 + 2 \times 8 \\ 3 \times 5 + 4 \times 7 & 3 \times 6 + 4 \times 8 \end{pmatrix}$$

$$= \begin{pmatrix} 19 & 22 \\ 43 & 50 \end{pmatrix}$$

and $\quad \mathbf{BA} = \begin{pmatrix} 5 & 6 \\ 7 & 8 \end{pmatrix}\begin{pmatrix} 1 & 2 \\ 3 & 4 \end{pmatrix} = \begin{pmatrix} 5 \times 1 + 6 \times 3 & 5 \times 2 + 6 \times 4 \\ 7 \times 1 + 8 \times 3 & 7 \times 2 + 8 \times 4 \end{pmatrix}$

$$= \begin{pmatrix} 23 & 34 \\ 31 & 46 \end{pmatrix}$$

So $\quad \mathbf{AB} \neq \mathbf{BA}$

and we see that in general the multiplication of matrices is *not* commutative.

Example 2

If $\quad \mathbf{I} = \begin{pmatrix} 1 & 0 \\ 0 & 1 \end{pmatrix}$ and $\mathbf{A} = \begin{pmatrix} a & b \\ c & d \end{pmatrix}$, calculate $\mathbf{IA}$ and $\mathbf{AI}$.

$$\mathbf{IA} = \begin{pmatrix} 1 & 0 \\ 0 & 1 \end{pmatrix}\begin{pmatrix} a & b \\ c & d \end{pmatrix} = \begin{pmatrix} 1a + 0c & 1b + 0d \\ 0a + 1c & 0b + 1d \end{pmatrix} = \begin{pmatrix} a & b \\ c & d \end{pmatrix}$$

$$\mathbf{AI} = \begin{pmatrix} a & b \\ c & d \end{pmatrix}\begin{pmatrix} 1 & 0 \\ 0 & 1 \end{pmatrix} = \begin{pmatrix} a1 + b0 & a0 + b1 \\ c1 + d0 & c0 + d1 \end{pmatrix} = \begin{pmatrix} a & b \\ c & d \end{pmatrix}$$

So $\quad \mathbf{IA} = \mathbf{AI} = \mathbf{A}$

and we call $\mathbf{I}$ the *unit* (or *identity*) matrix.

Example 3

If $\mathbf{A} = \begin{pmatrix} 1 & 0 \\ 0 & -1 \end{pmatrix}$, $\mathbf{B} = \begin{pmatrix} -1 & 0 \\ 0 & 1 \end{pmatrix}$ and $\mathbf{C} = \begin{pmatrix} -1 & 0 \\ 0 & -1 \end{pmatrix}$, calculate $\mathbf{A}^2$ ($= \mathbf{AA}$), $\mathbf{B}^2$, and $\mathbf{C}^2$ and explain the meaning of your answers.

$$\mathbf{A}^2 = \begin{pmatrix} 1 & 0 \\ 0 & -1 \end{pmatrix}\begin{pmatrix} 1 & 0 \\ 0 & -1 \end{pmatrix} = \begin{pmatrix} 1 & 0 \\ 0 & 1 \end{pmatrix} = \mathbf{I}$$

$$\mathbf{B}^2 = \begin{pmatrix} -1 & 0 \\ 0 & 1 \end{pmatrix}\begin{pmatrix} -1 & 0 \\ 0 & 1 \end{pmatrix} = \begin{pmatrix} 1 & 0 \\ 0 & 1 \end{pmatrix} = \mathbf{I}$$

$$\mathbf{C}^2 = \begin{pmatrix} -1 & 0 \\ 0 & -1 \end{pmatrix}\begin{pmatrix} -1 & 0 \\ 0 & -1 \end{pmatrix} = \begin{pmatrix} 1 & 0 \\ 0 & 1 \end{pmatrix} = \mathbf{I}$$

Now **A**, **B**, **C** are the matrices which represent reflection in Ox, reflection in Oy and point-reflection in O, respectively; so it is clear that each of them, when repeated, leads to the *identity* transformation, which leaves every point unchanged and has matrix **I**.

Exercise 1.8

1   $\mathbf{A} = \begin{pmatrix} 1 & 0 \\ 0 & -1 \end{pmatrix}$, $\mathbf{B} = \begin{pmatrix} -1 & 0 \\ 0 & 1 \end{pmatrix}$, $\mathbf{C} = \begin{pmatrix} -1 & 0 \\ 0 & -1 \end{pmatrix}$. Calculate:

   a) **AB** and **BA**
   b) **AC** and **CA**
   c) **BC** and **CB**
   and hence construct a multiplication table for the matrices **I**, **A**, **B**, **C**. Explain the meaning of your answers.

2   If $\mathbf{P} = \begin{pmatrix} 0 & 1 \\ -1 & 0 \end{pmatrix}$ and $\mathbf{Q} = \begin{pmatrix} 0 & -1 \\ 1 & 0 \end{pmatrix}$, calculate:

   a) **PQ** and **QP**
   b) **P**^{2} and **Q**^{2}
   Explain the meaning of your answers.

3   If $\mathbf{U} = \begin{pmatrix} 0 & 1 \\ 1 & 0 \end{pmatrix}$ and $\mathbf{V} = \begin{pmatrix} -1 & 0 \\ 0 & 1 \end{pmatrix}$, calculate:

   a) **UV** and **VU**
   b) **U**^{2} and **V**^{2}
   Explain the meaning of your answers.

4   If $\mathbf{L} = \begin{pmatrix} a & b \\ c & d \end{pmatrix}$, $\mathbf{M} = \begin{pmatrix} e & f \\ g & h \end{pmatrix}$ and $\mathbf{N} = \begin{pmatrix} p & q \\ r & s \end{pmatrix}$, calculate **L(MN)** and (**LM)N**, and so show that matrix multiplication is *associative*.

5   If $\mathbf{A} \neq \begin{pmatrix} 0 & 0 \\ 0 & 0 \end{pmatrix}$, is it true that $\mathbf{AB} = \mathbf{AC} \;\Rightarrow\; \mathbf{B} = \mathbf{C}$?

Miscellaneous problems

1 Of three men, only two always tell the truth. One day the first said to the second about the third, 'He's always truthful'. Shortly afterwards the second said the same to the third about the first. Which wouldn't you trust?

2 The two straight lines

$$x + 2y + 6 = 0 \quad \text{and} \quad 3x - 4y - 5 = 0$$

meet at the point P.

Show that, whatever the values of λ and μ, the equation

$$\lambda(x + 2y + 6) + \mu(3x - 4y - 5) = 0$$

represents a straight line through P.

Hence, without finding the coordinates of P, find the equations of

a) the line OP

b) the line through P which is parallel to $3x + 2y - 7 = 0$

c) the line through P which is perpendicular to $x - 2y + 4 = 0$

3 Can $\sqrt[3]{7}$ be represented by a recurring decimal? Prove your answer.

4 If O, P and Q are the points $(0, 0)$, (a, b) and (c, d), find a formula for the area of $\triangle OPQ$.

5 A linear transformation has the matrix

$$\mathbf{A} = \begin{pmatrix} a & b \\ c & d \end{pmatrix}$$

Find:

a) the points into which the four corners of the square $(0, 0)$, $(1, 0)$, $(0, 1)$, $(1, 1)$ are transformed;

b) the factor by which the area of this square is magnified;

c) what happens to this factor when $ad - bc = 0$. Why?

2 Functions

2.1 Functions and their graphs

Mathematics is largely concerned with relationships: how things depend on one another. In particular, when we are dealing with quantities or numbers, *that which depends* is called a *function of that (or those) on which it depends*. Because we usually have some freedom, however restricted, to change these latter quantities, they are usually called *independent variables*; the former is the *dependent variable*.

For example, if we are told the radius r of a circle, its area A is automatically determined. So *A is a function of r*. This function, which we call f, is simply the name of a piece of machinery for converting an input r into an output A.

We write $f: r \mapsto A$ and $A = f(r)$

So f(1) is shorthand for *the area of a circle of radius* 1 and f(4.3) is shorthand for *the area of a circle of radius* 4.3.

We know, of course, that the particular machinery in this case is the formula πr^2, so we write

$$f(r) = \pi r^2$$

and we can calculate f(r) for any value of r.

Hence $f(1) = \pi 1^2 = \pi \approx 3.142$
$\qquad f(4.3) = \pi(4.3)^2 \approx 58.1$

The set of permitted values of the independent variable is called the *domain* of the function, and the corresponding set for the dependent variable is the *range* of the function. So, in this example, both domain and range of the function are the set of all positive numbers.

Exercise 2.1a

1   If $f(r) = \pi r^2$, where $\pi \approx 3.14$,
    a) calculate f(2), f($\frac{1}{2}$), f(10) and f(100).
    b) what is the value of f(0), and what is its meaning?

c) write in shorthand form the area of a circle of radius three metres. What is its value?

d) Repeat c) for a circle of radius one-third of a metre.

2   When a stone is thrown vertically upwards with a velocity $v\,\mathrm{m\,s^{-1}}$, its maximum height $h\,\mathrm{m}$ is given approximately by the function

$$h = f(v) = \frac{1}{20}v^2$$

a) Calculate its maximum height when thrown with velocity $10\,\mathrm{m\,s^{-1}}$, $20\,\mathrm{m\,s^{-1}}$, $30\,\mathrm{m\,s^{-1}}$.

b) At what speed must it be projected in order to reach the height of a roof $125\,\mathrm{m}$ above its point of projection?

3   When a heavy mass swings at the end of a light string of length $l\,\mathrm{m}$, its period of oscillation $T\,\mathrm{s}$ is given approximately by the function

$$T = f(l) = 2\sqrt{l}$$

a) What is the period of a pendulum of length $1\,\mathrm{m}$, $4\,\mathrm{m}$?

b) What is the length of a pendulum whose period is $1\,\mathrm{s}$, $3\,\mathrm{s}$?

4   When you look out to sea on a clear day from a height h metre, the distance d kilometre of the horizon is given approximately by the function

$$d = f(h) = 3.6\sqrt{h}$$

a) What is the distance of the horizon from a height of $1\,\mathrm{m}$, $4\,\mathrm{m}$, $100\,\mathrm{m}$ respectively?

b) How high would you have to go in order to see $18\,\mathrm{km}$, $36\,\mathrm{km}$?

In these examples, the functions are expressible by an algebraic formula. But the only essential for a function is the existence of machinery for converting an *input* (the independent variable) into an *output* (the dependent variable), and this machinery could equally take the form of a set of rules, e.g. the Income Tax tables, whereby a person's gross income I (and a welter of other information) is fed into the function-machine, called the Department of Inland Revenue, which all too certainly calculates the Income Tax T.

So $f: I \mapsto T$ and $T = f(I)$

The way this machine works will vary from person to person, depending on other information about size of family, pension contributions and a host of other items, but for a particular man $f(1000)$, $f(2000)$, etc. represent his tax payments in the event of his earning £1000, £2000, etc.

For relatively low incomes no tax is payable; so for relatively small values of I, the value of T is zero. For these values of I, what we called the dependent *variable* (T) is in fact *constant*, and we usually say that the function itself is constant over this part of the domain.

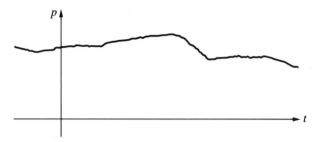

In another case, a function might exist in the form of a graph, e.g. the continuous trace of a barograph recording atmospheric pressure. Here the pressure p is a function of time t (measured, say, from midnight on a particular day). If we wish to know the pressure at a particular time, reference to the graph converts the input (t) into an output (p), and we write

$$f: t \mapsto p \quad \text{and} \quad p = f(t)$$

In the above examples there is no particular sanctity in the use of the letter f to describe the function machine, and if there is any danger of confusion between two functions different letters can be used.

So, for the sister of the man paying income tax, her function might be written

$$T = g(I)$$

and for atmospheric pressures at a different weather-station we might use the Greek letter ϕ (phi), and write

$$p = \phi(t)$$

Inverse and composite functions

If f is a function which converts an input x into an output $f(x)$, it can be represented by means of a *mapping* of the domain of f into its *codomain* (so that the range of a particular function is a subset of its codomain).

The only essential is that every element of the domain should be mapped on to a unique element (or *image*) in the codomain. Sometimes, as we have already seen, the reverse is not true, and a particular element of the codomain can be the image of more than one element of the domain. In our illustration, for instance, $f(c) = f(d)$; and if $g(x) = x^2$, then $g(a) = g(-a)$ for all values of a.

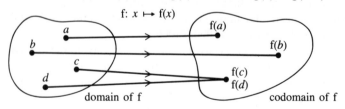

But if f is a function whose outputs all arise from unique inputs, we can regard the backwards mapping from output to input as also defining a function, which we call the *inverse function* f^{-1}.

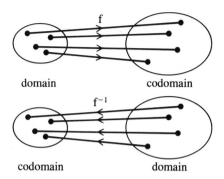

domain   codomain

codomain   domain

For example, if $f: x \mapsto 2x,$ then $f^{-1}: x \mapsto \frac{1}{2}x$
  $g: x \mapsto x + 1,$ then $g^{-1}: x \mapsto x - 1$
  $h: x \mapsto 3x + 2,$ then $h^{-1}: x \mapsto \frac{1}{3}(x - 2)$

Sometimes we can define such inverse functions where they would not otherwise exist, by restricting the domain of the original function.

If, for instance, $f: x \mapsto x^2$ then $+4$ is the image of both $+2$ and -2, so there is no inverse function.

But if we restrict the domain of f by stating that $x \geqslant 0$, then $f: x \mapsto x^2$ has the inverse function $f^{-1}: x \mapsto \sqrt{x}$.

Lastly, if f and g are two functions, we sometimes need to use the output of f as the input of g. We call this combination the *composite function* gf. For example, if $f: x \mapsto x^2$ and $g: x \mapsto x + 1$, then gf means 'square x, and then add one', and fg means 'add one to x, and then square'.

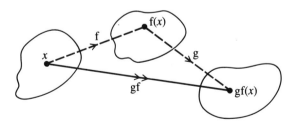

So $gf: x \mapsto x^2 + 1$ and $fg: x \mapsto (x + 1)^2$

Hence gf and fg are, in general, different functions; and (to make matters worse!) neither of them is related to the algebraic product

$$f(x)\, g(x) = x^2(x + 1)$$

Exercise 2.1b

1 $f: x \mapsto x + 1$ and $g: x \mapsto x^3$. Find the functions:
   a) fg   **b)** gf   **c)** f^{-1}
   d) g^{-1} **e)** $(fg)^{-1}$ **f)** $g^{-1}f^{-1}$

2 Repeat this question with $f: x \mapsto 3x$ and $g: x \mapsto x^2$. Is it necessary to restrict the domain of f or g?

3 What do you notice about f^{-1}, g^{-1} and $(fg)^{-1}$?

Graphs of functions

We have seen, as in the case of the barograph, that a function could be defined by means of a graph. Certainly it is always possible to illustrate a function in this way, and this usually displays most fully its particular features. Using two perpendicular axes, we adopt the convention that the *independent* variable (or *input*) is plotted *horizontally*, and the *dependent* variable (or *output*) is plotted *vertically*.

Example 1

$$f(h) = 3.6\sqrt{h}$$

This function can be calculated for different values of h:

h	0	1	2	3	4
$f(h)$	0	3.6	5.1	6.2	7.2

$\left(\begin{array}{l}\text{Domain of f: all non-negative numbers}\\ \text{Range of f: all non-negative numbers}\end{array}\right)$

and these values can be plotted as a graph:

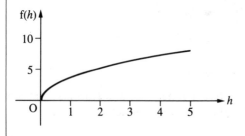

Example 2

f(x) = |x|

where |x| means the *numerical value of x*   (|7| = 7, |−2.5| = 2.5, etc.)

$$\begin{pmatrix} \text{Domain of f: all numbers} \\ \text{Range of f: all non-negative numbers} \end{pmatrix}$$

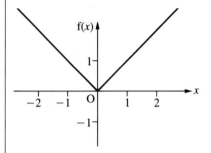

Example 3

$$g(x) = \begin{cases} x - 1 & \text{if} \quad x < 0 \\ 0 & \text{if} \quad x = 0 \\ x + 1 & \text{if} \quad x > 0 \end{cases}$$

(In our illustrations we shall use blobs to make clear the values of the function.)

$$\begin{pmatrix} \text{Domain of g: all numbers} \\ \text{Range of g: all numbers greater than 1, less then } -1, \text{ and zero} \end{pmatrix}$$

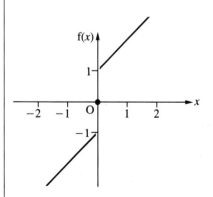

Example 4

$h(x) = [x]$

where [ ] means *the integral part of x* or *the greatest integer not greater than x.*

(So that $[2.3] = 2$, $[3] = 3$, $[-2.7] = -3$, etc.)

$\begin{pmatrix} \text{Domain of h: all numbers} \\ \text{Range of h: all integers} \end{pmatrix}$

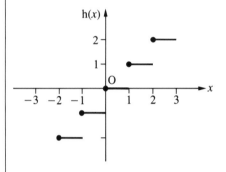

When the independent variable is an integer, we usually represent it by the letter n rather than x:

Example 5

$\phi(n) = (-1)^n$

$\begin{pmatrix} \text{Domain of } \phi \text{: all integers} \\ \text{Range of } \phi \text{: } \pm 1 \end{pmatrix}$

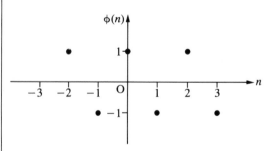

It will be noticed that in examples 2 and 5 the graph is symmetrical about the vertical axis. This can also be seen since

$$f(-x) \equiv (-x)^2 \equiv x^2 \equiv f(x)$$

and $\phi(-n) \equiv (-1)^{-n} \equiv (-1)^n \equiv \phi(n)$

Any such function, for which $f(-x) \equiv f(x)$, is called an *even* function.

Similarly, in example 3, the graph has point-symmetry about the origin, i.e. it is unaltered by rotation through two right angles or by successive reflections in the two axes. Here $g(-x) \equiv -g(x)$ and any such function is called an *odd* function.

It is, of course, rather special for a graph to have either kind of symmetry, and functions (unlike integers) are generally neither even nor odd.

Finally it will be recalled that if a and b are constants, the graph of $y = ax + b$ is a straight line. The corresponding function $f(x) = ax + b$ is therefore called a *linear function*.

So, $f(x) = 2x + 1, \quad 7 - x, \quad -3$, etc. are all linear functions.

Exercise 2.1c

In the case of each of the following functions (nos **1–13**),
a) calculate, if possible, f(0), f(3), f(2.5) and f(−1);
b) state whether it is an even function, an odd function, or neither;
c) state its domain and range, and sketch its graph.

1 $f(x) = 3x - 2$

2 $f(x) = 7$

3 $f(x) = \dfrac{1}{x^2}$

4 $f(x) = x^3$

5 $f(x) = \sqrt{x}$

6 $f(x) = \dfrac{1}{x} \text{ if } x \neq 0$
$0 \text{ if } x = 0$

7 $f(x) = x - [x]$

8 $f(x) = x^2 \quad \text{if } |x| < 1$
$\dfrac{1}{x^2} \quad \text{if } |x| \geq 1$

9 $f(x) = 1 \quad \text{if } x \text{ is rational}$
$0 \quad \text{if } x \text{ is irrational}$

10 $f(x) = x \quad \text{if } x \text{ is rational}$
$0 \quad \text{if } x \text{ is irrational}$

11 $f(n) = \dfrac{(-1)^n}{n}$

12   f(n) is the number of days in the *n*th month of a leap year.

13   f(n) is the digit in the *n*th decimal place of π.

14   Can we say anything about the value of f(0) if f(x) is
   a) an even function;
   b) an odd function?

15   For what values of k is f(x) = x^k
   a) an even function;
   b) an odd function?

16   What can be said about the constants a and b if the linear function
   f(x) = ax + b is
   a) an even function;
   b) an odd function?

17   If f(x) = x³, what are the following functions? Show in separate diagrams
   how their graphs are related to that of f(x) and in each case state in words
   how the graph of f(x) has been transformed.
   a) f(x) + 2   **b)** f(x) − 3   **c)** f(x − 1)   **d)** f(x + 4)   **e)** 2f(x)
   f) − 3f(x)   **g)** f(2x)   **h)** f(−x)

18   If f(x) is defined for all values of x, is either f(x) + f(−x) or f(x) − f(−x)
   an even or an odd function?

19   Show that if the function f(x) has an inverse function f⁻¹(x), their two
   graphs are reflections of each other in the line y = x.

2.2 Quadratic functions

These are especially important as the simplest type of non-linear function.

Example 1

$$f(x) = x^2 - 2x - 3$$

We can start by finding f(x) for various values of x and plotting the graph of
f(x):

x	−3	−2	−1	0	1	2	3	4	5
x^2	9	4	1	0	1	4	9	16	25
$-2x$	6	4	2	0	−2	−4	−6	−8	−10
-3	−3	−3	−3	−3	−3	−3	−3	−3	−3
$f(x) = x^2 - 2x - 3$	12	5	0	−3	−4	−3	0	5	12

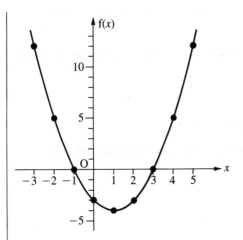

This curve clearly illustrates the important features:
a) $f(0) = -3$
b) As $x \rightarrow \pm \infty$, $f(x) \rightarrow +\infty$
c) $f(x)$ has its *minimum value* -4 when $x = 1$
d) $f(x) = 0$ when $x = -1$ or $+3$.

The first two of these could have been seen without the labour of plotting the graph. What about c) and d)?
We can write

$$f(x) = x^2 - 2x - 3 = (x - 1)^2 - 4 \quad \text{(known as } \textit{completing the square}\text{)}$$

Now $(x - 1)^2 \geqslant 0$

So $f(x) \geqslant -4$

and we see that $f(x)$ has its minimum value -4 when $x - 1 = 0 \Rightarrow x = 1$.

Also $f(x) = 0 \iff (x - 1)^2 = 4 \iff x - 1 = \pm 2 \iff x = -1 \text{ or } +3$

Alternatively, we can write

$$f(x) = x^2 - 2x - 3 = (x + 1)(x - 3)$$

So $f(x) = 0 \iff (x + 1)(x - 3) = 0 \iff x = -1 \text{ or } x = 3$
Hence $f(x) = 0$ when $x = -1$ or 3.

Example 2

Describe the important features of $f(x) = x^2 - 3x + 5$, and then sketch its graph.

a) $f(0) = 5$

b) As $x \to \pm \infty$, $f(x) \to + \infty$

c) $f(x) = x^2 - 3x + 5 = (x - \frac{3}{2})^2 + \frac{11}{4} \geqslant \frac{11}{4}$

So $f(x)$ has a minimum value $\frac{11}{4}$ when $x = \frac{3}{2}$, and is therefore never zero.

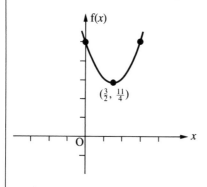

$(\frac{3}{2}, \frac{11}{4})$

Example 3

Describe the important features of $f(x) = 1 + x - x^2$, and then sketch its graph.

a) $f(0) = 1$

b) As $x \to \pm \infty$, $f(x) \to - \infty$

c) $f(x) = 1 + x - x^2$

$= \frac{5}{4} - (x - \frac{1}{2})^2 \leqslant \frac{5}{4}$

So $f(x)$ has a maximum value $\frac{5}{4}$ when $x = \frac{1}{2}$.

d) $f(x) = \frac{5}{4} - (x - \frac{1}{2})^2$

So $f(x) = 0 \Rightarrow \frac{5}{4} - (x - \frac{1}{2})^2 = 0$

$\Rightarrow (x - \frac{1}{2})^2 = \frac{5}{4}$

$\Rightarrow x - \frac{1}{2} = \pm \dfrac{\sqrt{5}}{2}$

$\Rightarrow x = \dfrac{1 \pm \sqrt{5}}{2}$

$\Rightarrow x = 1.62 \text{ or } -0.62$

So the graph of f(x) is

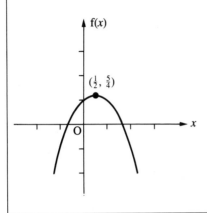

$(\frac{1}{2}, \frac{5}{4})$

Example 4

f(x) = $2x^2 + 3x - 5$

Here we can say   f(x) = $2(x^2 + \frac{3}{2}x - \frac{5}{2})$ and regard f(x) as a magnification of the function $x^2 + \frac{3}{2}x - \frac{5}{2}$, which we can investigate in similar fashion.

The general quadratic function

f(x) = $ax^2 + bx + c$ ($a \neq 0$)

a) f(0) = c

b) As $x \to \pm \infty$; if $a > 0$,   f(x) $\to + \infty$

$a < 0$,   f(x) $\to - \infty$

c) f(x) = $ax^2 + bx + c$

$$= a\left\{ x^2 + \frac{b}{a}x + \frac{c}{a} \right\}$$

$$= a\left\{ \left(x + \frac{b}{2a} \right)^2 + \frac{4ac - b^2}{4a^2} \right\}$$

$$= a\left(x + \frac{b}{2a} \right)^2 + \frac{4ac - b^2}{4a}$$

If $a > 0$, f(x) $\geqslant \dfrac{4ac - b^2}{4a}$, which is its minimum value $\left(\text{when } x = -\dfrac{b}{2a} \right)$

If $a < 0$, $f(x) \leqslant \dfrac{4ac - b^2}{4a}$, which is its maximum value $\left(\text{when } x = -\dfrac{b}{2a}\right)$

d) $f(x) = 0 \quad \Rightarrow \quad a\left\{\left(x + \dfrac{b}{2a}\right)^2 + \dfrac{4ac - b^2}{4a^2}\right\} = 0$

$$\Rightarrow \quad \left(x + \dfrac{b}{2a}\right)^2 = \dfrac{b^2 - 4ac}{4a^2}$$

$$\Rightarrow \quad x + \dfrac{b}{2a} = \pm \dfrac{\sqrt{(b^2 - 4ac)}}{2a}$$

$$\Rightarrow \quad \boxed{x = \dfrac{-b \pm \sqrt{(b^2 - 4ac)}}{2a}}$$

So $f(x) = 0$ has $\begin{array}{l}\text{two distinct} \\ \text{two equal} \\ \text{no}\end{array}$ real roots $\Leftrightarrow b^2 - 4ac \begin{array}{l}> \\ = 0 \\ <\end{array}$

Example 5

Solve the equation $\quad 3x^2 + 4x - 2 = 0$

Here $a = 3, b = 4, c = -2$

so $\quad x = \dfrac{-4 \pm \sqrt{(4^2 - 4 \times 3 \times -2)}}{2 \times 3} = \dfrac{-4 \pm \sqrt{40}}{6} \approx \dfrac{-4 \pm 6.324}{6}$

$$\approx \dfrac{+2.324}{6} \quad \text{or} \quad \dfrac{-10.324}{6}$$

$\Rightarrow \quad x \approx 0.39 \quad \text{or} \quad -1.72$

So $\quad 3x^2 + 4x - 2 = 0 \quad \Rightarrow \quad x \approx 0.39 \quad \text{or} \quad -1.72$

Example 6

$\quad 4x^2 - 20x + 25 = 0$

Here $\quad a = 4, b = -20, c = 25$

$$x = \dfrac{20 \pm \sqrt{(400 - 400)}}{8} = \dfrac{20 \pm 0}{8} = 2\tfrac{1}{2} \quad \text{or} \quad 2\tfrac{1}{2}$$

So $\quad 4x^2 - 20x + 25 = 0 \quad \Rightarrow \quad x = 2\tfrac{1}{2}$

Example 7

$f(x) = x^2 - x + 1 = 0$

Here $a = 1, b = -1, c = 1$

$\Rightarrow \quad x = \dfrac{1 \pm \sqrt{(1 - 4)}}{2} = \dfrac{1 \pm \sqrt{-3}}{2}$

So $x^2 - x + 1 = 0$ is true for no real value of x (and this can also be seen since $f(x) = (x - \frac{1}{2})^2 + \frac{3}{4} \geqslant \frac{3}{4})$.

Exercise 2.2

1   Solve the following equations and illustrate your answers by sketching graphs of the corresponding functions:
   a) $x^2 - 4 = 0$
   b) $x^2 - 4x = 0$
   c) $x^2 - 4x + 3 = 0$
   d) $x^2 - 4x + 4 = 0$
   e) $x^2 - 4x + 2 = 0$
   In each case check one of your answers and also calculate the sum of the roots. Do you notice anything special? What about the product of the roots?

2   Solve the following equations and again calculate the sum and product of the roots:
   a) $2x^2 + x = 1$
   b) $2x^2 + x = 2$
   c) $3x^2 = 2x + 5$
   d) $x^2 = px + q$

3   What is the range of the following functions, and for what values of x are the functions greatest or least?
   a) $x^2 - 6x$
   b) $4 - 3x - x^2$
   c) $(x - 2)(x - 4)$
   Illustrate by sketching their graphs.

4   For what values of x is
   a) $x^2 > x$
   b) $x^2 \leqslant x - 2$
   c) $x^2 \leqslant 1 - x$
   d) $x^2 > x - 1$?

5 Find the points of intersection of
a) $x + y = 3$ and $4xy = 5$
b) $y = x + 2$ and $y = 2x^2$
Illustrate by sketching graphs.

6 For what values of m does the line $y = mx$ meet the curve $y = x^2 - 4x + 3$ in   a) two points,   b) one point,   c) no points? Illustrate by a sketch.

7 The equation $ax^2 + bx + c = 0$ has roots α and β. Use the formula

$$x = \frac{-b \pm \sqrt{(b^2 - 4ac)}}{2a}$$

to verify that $\alpha + \beta = -b/a$ and $\alpha\beta = +c/a$. Then use these results to check your solutions of nos. **1** and **2**.

8 A stone thrown vertically upwards with velocity $20\,\mathrm{m\,s^{-1}}$ is at height y m after t s, where

$$y = 20t - 5t^2$$

For how long is it over 10 m above the ground?

9 A batsman hits a cricket ball so that its trajectory has the equation

$$y = x - \frac{1}{80}x^2$$

where x and y are horizontal and vertical distances measured in metres. How far away does it land, and what is its greatest height?

10 The sum of n terms of

$$1 + 4 + 7 + 10 + \cdots \text{ is }\quad \tfrac{1}{2}n\,(3n - 1)$$

How many terms have to be taken for this sum to exceed 1000?

11 A car factory produces n cars per day and makes a profit of £p per day, where

$$p = 2n^2 - 300n$$

How many cars need to be produced per day
a) for production to be profitable;
b) for a daily profit of £10 800;
c) for a daily loss of £11 200;
d) for worst possible loss?
Would you expect the given function to be valid for all values of n?

* The number i

In example 7 we asked whether there were any values of x for which $x^2 - x + 1 = 0$. Even more simply, we might ask if there are values for which $x^2 + 1 = 0$. When does $x^2 = -1$? What is $\sqrt{-1}$? Does $\sqrt{-1}$ actually exist? For that matter, does -1 actually exist? Or any other number?

Using the symbol i, we can readily express

$$\sqrt{-4} \quad \text{as} \quad \sqrt{4} \times \sqrt{-1} = 2i$$
$$\text{and} \quad \sqrt{-3} \quad \text{as} \quad \sqrt{3} \times \sqrt{-1} = \sqrt{3}i$$

Moreover, the solutions of example 7 can be written as $\dfrac{1 \pm \sqrt{3}i}{2}$ and we can easily verify that

$$f\left(\frac{1 + \sqrt{3}i}{2}\right) = \left(\frac{1 + \sqrt{3}i}{2}\right)^2 - \left(\frac{1 + \sqrt{3}i}{2}\right) + 1$$

$$= \frac{-2 + 2\sqrt{3}i}{4} - \frac{1 + \sqrt{3}i}{2} + 1 = 0$$

and similarly that $f\left(\dfrac{1 - \sqrt{3}i}{2}\right) = 0$

Having introduced the symbol i for $\sqrt{-1}$, it might be feared that further symbols would be required in order to answer such questions as $\sqrt{i}$ and $\sqrt[3]{-1}$. But is this so? Answer these questions by first calculating the squares of $\pm\dfrac{1 + i}{\sqrt{2}}$ and the cubes of -1 and $\dfrac{1 \pm \sqrt{3}i}{2}$.

It can be seen, therefore, that the introduction of the single symbol i has a great unifying power, enabling all quadratic equations and many others to be solved. We shall return to this in Book 2.

*2.3 Polynomial functions and the remainder theorem

We have seen that the general linear function is $ax + b$ and the general quadratic function is

$$ax^2 + bx + c$$

So we now naturally proceed to

$$ax^3 + bx^2 + cx + d \quad \text{(called a \textit{cubic})}$$
$$\text{and} \quad ax^4 + bx^3 + cx^2 + dx + e \quad \text{(called a \textit{quartic})}$$

More generally, any function

$$P(x) = a_0 x^n + a_1 x^{n-1} + a_2 x^{n-2} + \ldots + a_{n-1}x + a_n$$

(where n is a positive integer and a_0, a_1, ..., a_n are constants) is called a *polynomial of degree n*, and a_0, ... a_n are its *coefficients*.

So linear, quadratic, cubic and quartic polynomials have degree 1, 2, 3, 4 respectively.

With more complicated polynomials, just as with linear and quadratic functions, it is useful to be able to recognise the main features and to be able to sketch their graphs. But first we must look at the structure of such polynomials and the way in which they can be multiplied and divided.

Multiplication and division of polynomials

Two examples will be sufficient to show similarities with long multiplication and division in ordinary arithmetic:

Example 1

Multiply $3x^2 - 2x - 1$ by $2x + 3$

which can be written as:

	x^3	x^2	x	1
$3x^2 - 2x - 1$		3	-2	-1
$2x + 3$			2	3
$6x^3 - 4x^2 - 2x$	6	-4	-2	
$+9x^2 - 6x - 3$		9	-6	-3
$6x^3 + 5x^2 - 8x - 3$	6	5	-8	-3

Example 2

Divide $x^3 - 5x^2 + 11x - 6$ by $x - 2$

$$
\begin{array}{r}
x^2 - 3x + 5 \\
x - 2\overline{)x^3 - 5x^2 + 11x - 6} \\
x^3 - 2x^2 \\
\hline
-3x^2 + 11x \\
-3x^2 + 6x \\
\hline
5x - 6 \\
5x - 10 \\
\hline
+4
\end{array}
$$

or

	1	-3	5	
$1 - 2)1$	-5	11	-6	
1	-2			
	-3	11		
	-3	6		
		5	-6	
		5	-10	
			$+4$	

So when $x^3 - 5x^2 + 11x - 6$ is divided by $x - 2$, the quotient is $x^2 - 3x + 5$ and the remainder $+4$.

Now when, for example, 25 is divided by 7, quotient $= 3$ and remainder $= 4$

and $25 = 3 \times 7 + 4$

More generally,

number $=$ divisor $\times$ quotient $+$ remainder

So in our example, as can very easily be checked by multiplication,

$$x^3 - 5x^2 + 11x - 6 \equiv (x - 2)(x^2 - 3x + 5) + 4$$

Exercise 2.3a

1   Find the product of
   a) $x^2 + 3x - 2$ and $2x - 1$
   b) $x^2 - x + 1$ and $x + 1$
   c) $x^3 + x^2 + x + 1$ and $x - 1$
   d) $2x^3 + x^2 - 5x + 4$ and $2x - 3$

2   Find the quotient and remainder when
   a) $2x^3 + 3x^2 - 4x + 5$ is divided by $x + 2$
   b) $4x^3 - 6x^2 + 5$ is divided by $2x - 1$
   c) x^5 is divided by $x + 1$
   d) $x^4 - 1$ is divided by $x - 1$
   e) $x^4 + 2x^3 + 3x^2 + 4x + 5$ is divided by $x^2 + x + 1$
   Check your answers by multiplication.

The remainder theorem

It frequently happens that we wish to discover factors of a given polynomial, i.e. to find expressions which divide into the polynomial and leave zero remainder. Can this be done without going through the process of division? Is there any way, for instance, in example 2 of finding the remainder when $x^3 - 5x^2 + 11x - 6$ is divided by $x - 2$, *without performing the division*?

Let us suppose that the quotient is $Q(x)$ and the remainder is R.

Then $x^3 - 5x^2 + 11x - 6 \equiv (x - 2)Q(x) + R.$

If we now take the value $x = 2$,

$$2^3 - 5 \times 2^2 + 11 \times 2 - 6 = 0 + R$$
$$\Rightarrow \qquad\qquad R = 8 - 20 + 22 - 6$$
$$\Rightarrow \qquad\qquad R = 4$$

and we have obtained the value of the remainder merely by calculating P(2).

More generally, suppose that when P(x) is divided by $x - \alpha$, the quotient is Q(x) and the remainder is R.

Then $P(x) \equiv (x - \alpha)Q(x) + R$

If we now put $x = \alpha$, $P(\alpha) = 0 + R \Rightarrow R = P(\alpha)$

> This is the *remainder theorem*, that when a polynomial P(x) is divided by $x - \alpha$, the remainder is P(α). In particular,
>
> $\quad$ P(α) = 0 $\Leftrightarrow$ remainder is zero
> so $\quad$ P(α) = 0 $\Leftrightarrow$ $x - \alpha$ is a factor of P(x)

and α is then called a *zero* of P(x), or a *root* of the equation P(x) = 0.

So finding the root of P(x) = 0, or *solving* this equation, is the same problem as factorising P(x).

Example 3

Find the zeros and factors of $x^3 - 2x^2 - 5x + 6$ and so sketch its graph.

We can see at a glance that one of the zeros is $x = 1$, since this value makes the given expression zero. So one of its factors is $x - 1$.

By division we obtain

$$x^3 - 2x^2 - 5x + 6 \equiv (x - 1)(x^2 - x - 6)$$
$$\equiv (x - 1)(x - 3)(x + 2)$$

So there are three zeros, $x = 1, 3$ and -2.

We can also see the sign of the polynomial for other values of x is given by

x	-3	-2	-1	0	1	2	3	4
P(x)	$----$	$-0++$	$++++$	$++++$	$+0--$	$----$	$-0++$	$+++$

Also, as $x \to +\infty$, $x^3 - 2x^2 - 5x + 6 \to +\infty$

and as $x \to -\infty$, $x^3 - 2x^2 - 5x + 6 \to -\infty$

So the graph of the function is

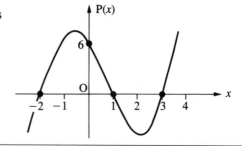

Example 4

Find the factors and zeros of $x^4 - 3x^2 - 2x$ and so sketch its graph.

Two of its zeros are clearly $x = 0$ and $x = -1$. So two factors are x and $x + 1$.

By division, we find that

$$x^4 - 3x^2 - 2x \equiv x(x + 1)(x^2 - x - 2)$$

But $x^2 - x - 2 \equiv (x + 1)(x - 2)$

So $x^4 - 3x^2 - 2x \equiv x(x + 1)^2(x - 2)$

and has zeros 0, -1, and $+2$.

Also as $x \to +\infty$, $x^4 - 3x^2 - 2x \to +\infty$

and as $x \to -\infty$, $x^4 - 3x^2 - 2x \to +\infty$

Hence the values of $x^4 - 3x^2 - 2x$ are given by

x	-3	-2	-1	0	1	2	3	4

P(x) $+ + + + + + + + 0 + + + 0 - - - - - - 0 + + + + + + + +$

and its graph is

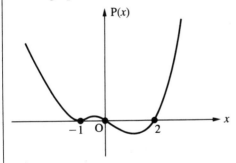

Exercise 2.3b

1   Find the remainder when
   a) $x^3 + 3x^2 - 4x + 5$ is divided by $x - 2$
   b) $2x^3 + 8x^2 - 3$ is divided by $x - 1$
   c) $x^3 + 8$ is divided by $x + 1$
   d) $x^4 + 3x^3 + 2x - 3$ is divided by $x + 3$
   e) $x^3 + a^3$ is divided by $x + a$
   f) $8x^3 - 12x^2 + 7$ is divided by $2x - 1$

2 Factorise the following polynomials, find their zeros and sketch their graphs:

a) $x^3 - 6x^2 + 11x - 6$

b) $x^3 - x^2$

c) $x^3 + x^2 - x - 1$

d) $2x^3 - x^2 - 8x + 4$

e) $-x^4 + 4x^2$

f) $x^4 - 8x^3 + 14x^2 + 8x - 15$

3 a) Show that $3x + 1$ is a factor of the expression $3x^3 + 4x^2 + 7x + 2$.

b) Find the values of a and b if $x - 2$ and $x + 1$ are both factors of $ax^3 + 3x^2 - 9x + b$, and state the third factor of the expression.   (oc)

4 When $x^3 - x^2 + ax + b$ is divided by $x - 1$ the remainder is -8, and when the same expression is divided by $x + 1$ the remainder is 0.

Calculate the values of a and b, and factorise the expression completely.

(oc)

5 Factorise   **a)** $x^3 - a^3$ **b)** $x^3 + a^3$

Polynomial equations

This last exercise will have provided some practice in the solution of simple polynomial equations. Can we always obtain such solutions?

In the case of the quadratic we were successful in discovering the formula

$$x = \frac{-b \pm \sqrt{(b^2 - 4ac)}}{2a}$$

Are there similar formulae for the roots of cubic, quartic and polynomial equations of higher degree?

This problem has had a strange history. The Arabs knew the formula for the solution of the quadratic equation as early as AD 830, and similar—but much more complicated—formulae for cubic and quartic equations were discovered by the Italians Tartaglia and Ferraro in the middle of the sixteenth century.

But for equations of higher degree, like the quintic (of degree 5), no such solution could be found. For centuries the problem remained unsolved and was only finally settled by the young Norwegian mathematician, Abel (1802–29). He discovered, astonishingly, not a solution of such equations, but a proof that a general solution giving the roots in terms of the coefficients is impossible. So their investigation (and in practice that of cubic and quartic equations too) has usually to be undertaken by graphical or numerical methods.

There remains the simpler question, whether we can decide on the number of roots of a polynomial equation. Firstly, is there a maximum number of roots? A little thought shows us that there is, and that a polynomial function of degree n cannot possibly have more than n zeros (unless it is identically zero: $0x^n + 0x^{n-1} + \cdots 0x + 0$).

Suppose for instance, that the cubic function

$$ax^3 + bx^2 + cx + d \quad (a \neq 0)$$

has zeros α, β, and γ.

Then $x - \alpha$, $x - \beta$, and $x - \gamma$ are all factors.

$$\Rightarrow \quad a(x - \gamma)(x - \beta)(x - \gamma) \equiv ax^3 + bx^2 + cx + d$$

So if δ were also a zero,

$$a(\delta - \alpha)(\delta - \beta)(\delta - \gamma) \equiv 0$$

$$\Rightarrow \quad \delta = \alpha, \quad \beta \quad \text{or} \quad \gamma$$

Hence a cubic polynomial cannot have more than three zeros, nor a cubic equation more than three roots. Similarly a polynomial equation of degree n cannot possibly have more than n roots.

To the other question, whether there is a *minimum* number of zeros of a polynomial, we have already seen the answer. For there are no real values of x for which

$$x^2 + 1 = 0 \quad \text{or} \quad x^4 + 3x^2 + 2 = 0$$

On the other hand, we can immediately see that any polynomial of odd degree must have at least one real zero. Why?

Furthermore, it is possible to show that, if we make use of the symbol $i\,(= \sqrt{-1})$, every polynomial has a zero; and thence that we can always factorise a polynomial of degree n into n linear factors. But for the present we confine ourselves to zeros and roots which are real.

Lastly, we can ask whether there are *any* connections between the zeros of a polynomial and its coefficients, and to this question we can provide some easy answers.

Symmetric functions of roots

Suppose that the quadratic equation $ax^2 + bx + c = 0$ has roots $x = \alpha$ and $x = \beta$. Then $ax^2 + bx + c$ has factors $x - \alpha$ and $x - \beta$.

So $ax^2 + bx + c \equiv k(x - \alpha)(x - \beta)$

From the x^2 term, we see that

$$ax^2 \equiv kx^2 \quad \Rightarrow \quad a = k$$

Hence $ax^2 + bx + c \equiv a(x - \alpha)(x - \beta)$

$\Rightarrow$ $ax^2 + bx + c \equiv ax^2 - a(\alpha + \beta)x + a\alpha\beta$

So $-a(\alpha + \beta) = b$

and $a\alpha\beta = c$ $\Rightarrow$ $\boxed{\begin{aligned} \alpha + \beta &= -b/a \\ \alpha\beta &= +c/a \end{aligned}}$

Similarly if $ax^3 + bx^2 + cx + d$ has zeros α, β, γ

then $ax^3 + bx^2 + cx + d \equiv a(x - \alpha)(x - \beta)(x - \gamma)$

$\equiv ax^3 - a(\alpha + \beta + \gamma)x^2$

$+ a(\beta\gamma + \gamma\alpha + \alpha\beta)x - a\alpha\beta\gamma$

$\Rightarrow$ $-a(\alpha + \beta + \gamma) = b$ $\Rightarrow$ $\boxed{\begin{aligned} \alpha + \beta + \gamma &= -b/a \\ \beta\gamma + \gamma\alpha + \alpha\beta &= +c/a \\ \alpha\beta\gamma &= -d/a \end{aligned}}$

$a(\beta\gamma + \gamma\alpha + \alpha\beta) = c$

$-a\alpha\beta\gamma = d$

These relationships are useful in a variety of ways. In example 1, for instance, we could have checked that $x^3 - 2x^2 - 5x + 6 = 0$ has roots, 1, 3 and -2 by noting that

$$1 + 3 + (-2) = \quad 2 = -(-2)$$

$$3 \times -2 + -2 \times 1 + 1 \times 3 = -5 = +(-5)$$

and $$1 \times 3 \times -2 = -6 = -(6)$$

Example 5

If α, β are the roots of $2x^2 + 3x - 5 = 0$, find

a) $\alpha^2 + \beta^2$ b) $\dfrac{1}{\alpha} + \dfrac{1}{\beta}$

We know that $\alpha + \beta = -\frac{3}{2}$ and $\alpha\beta = -\frac{5}{2}$

So $\alpha^2 + \beta^2 = (\alpha + \beta)^2 - 2\alpha\beta$

$= (-\frac{3}{2})^2 - 2 \times (-\frac{5}{2}) = \frac{9}{4} + 5 = \frac{29}{4}$

and $\dfrac{1}{\alpha} + \dfrac{1}{\beta} = \dfrac{\alpha + \beta}{\alpha\beta} = \dfrac{-\frac{3}{2}}{-\frac{5}{2}} \quad = +\frac{3}{5}$

Example 6

If $x^3 + 4x^2 + x - 6 = 0$ has roots α, β, γ, find the equation whose roots are $\alpha + 1, \beta + 1, \gamma + 1$.

Method 1

We know that

$$\alpha + \beta + \gamma = -4$$
$$\beta\gamma + \gamma\alpha + \alpha\beta = +1$$
$$\alpha\beta\gamma = +6$$

Hence the corresponding expressions for the equation with roots $\alpha + 1$, $\beta + 1, \gamma + 1$ are

$$(\alpha + 1) + (\beta + 1) + (\gamma + 1) = \alpha + \beta + \gamma + 3 = -1$$
$$(\beta + 1)(\gamma + 1) + (\gamma + 1)(\alpha + 1) + (\alpha + 1)(\beta + 1)$$
$$= (\beta\gamma + \gamma\alpha + \alpha\beta) + 2(\alpha + \beta + \gamma) + 3$$
$$= 1 + 2 \times (-4) + 3$$
$$= -4$$

and $(\alpha + 1)(\beta + 1)(\gamma + 1) = \alpha\beta\gamma + (\beta\gamma + \gamma\alpha + \alpha\beta) + (\alpha + \beta + \gamma) + 1$
$$= 6 + 1 - 4 + 1 = +4$$

So the required equation has coefficients 1, -4 and -4, and is

$$x^3 + x^2 - 4x - 4 = 0$$

Method 2

More simply, we observe that if x satisfies the equation $x^3 + 4x^2 + x - 6 = 0$, it follows that $t = x + 1$ satisfies another cubic equation, obtained by putting $t = x + 1$, i.e. $x = t - 1$

$$(t - 1)^3 + 4(t - 1)^2 + (t - 1) - 6 = 0$$
$$t^3 + t^2 - 4t - 4 = 0$$

This equation, therefore, has the required roots and so also does

$$x^3 + x^2 - 4x - 4 = 0$$

Exercise 2.3c

1 Find the sums and products of the roots of the following equations:
  a) $3x^2 + 2x + 5 = 0$ **b)** $x^2 + 4x - 3 = 0$
  c) $x^2 + px + q = 0$ **d)** $x^2 - 3 = 2x$

2 Find quadratic equations whose roots have the following sums and products:
  a) $4, \quad 5$ **b)** $-5, \quad +4$
  c) $-\frac{1}{2}, \quad -\frac{7}{2}$ **d)** $p, \quad 1/p$

3 $3x^2 - 2x - 5 = 0$ has roots α, β.
  a) Calculate $\alpha^2 + \beta^2$ and $\dfrac{1}{\alpha} + \dfrac{1}{\beta}$. Hence find the equations whose roots are

b) α^2, β^2 **c)** $\dfrac{1}{\alpha}, \dfrac{1}{\beta}$ **d)** $\dfrac{\alpha}{\beta}, \dfrac{\beta}{\alpha}$

4   Find $\alpha + \beta + \gamma$, $\beta\gamma + \gamma\alpha + \alpha\beta$ and $\alpha\beta\gamma$ where α, β, γ are the roots of
   a) $x^3 - 6x^2 + 11x - 6 = 0$
   b) $x^3 - x = 0$
   c) $x^3 + x^2 - x - 1 = 0$
   d) $2x^3 - x^2 - 8x + 4 = 0$
   Check your results by using the answers to exercise 2.3b, no. **2**.

5   The equation $x^3 - 9x^2 + 26x - 24 = 0$ has roots α, β, γ.
   a) Find $\alpha + \beta + \gamma$, $\beta\gamma + \gamma\alpha + \alpha\beta$, $\alpha\beta\gamma$.
   b) Find the equation whose roots are $\alpha - 1, \beta - 1, \gamma - 1$.
   c) Try to check your answers by solving the given equation.

2.4 Indices and power functions

Indices

So far we have concentrated upon linear functions (like $2x + 1$, $6 - \frac{1}{2}x$), quadratic functions (like $x^2, 3 - x + x^2$), higher degree polynomials and their sums, differences and products. Now we shall investigate functions where the independent variable occurs as an index or power, and we shall call them *power functions*.†

The simplest such functions are $f(x) = 2^x, 10^x$, etc. and we already know the values of these functions when the *index x* is a positive integer.

$2^5 = 32$	$10^5 = 100000$
$2^4 = 16$	$10^4 = 10000$
$2^3 = 8$	$10^3 = 1000$
$2^2 = 4$	$10^2 = 100$
$2^1 = 2$	$10^1 = 10$

As for such expressions as $2^0, 2^{-3}$, and 10^{-4}, we are at liberty to define them however we like. But there is an obvious convenience if we maintain the pattern of this table, so we continue with

$2^0 = 1$ $10^0 = 1$

$2^{-1} = \dfrac{1}{2}$ $10^{-1} = \dfrac{1}{10}$

---

† Also commonly called exponential functions.

$$2^{-2} = \frac{1}{4} = \frac{1}{2^2} \qquad\qquad 10^{-2} = \frac{1}{100} = \frac{1}{10^2}$$

$$2^{-3} = \frac{1}{8} = \frac{1}{2^3} \qquad\qquad 10^{-3} = \frac{1}{1000} = \frac{1}{10^3}, \quad \text{etc.}$$

More generally, if n is a positive integer,

$$\text{we } \textit{define} \quad 2^{-n} = \frac{1}{2^n} \qquad\qquad 10^{-n} = \frac{1}{10^n}$$

and (provided $a \neq 0$), $a^{-n} = \dfrac{1}{a^n}$ and $a^0 = 1$.

Exercise 2.4a

1 Calculate:

   a) 3^{-1} **b)** 4^{-2} **c)** $\left(\dfrac{3}{2}\right)^{-3}$ **d)** $(-5)^{-1}$ **e)** $\left(\dfrac{1}{4}\right)^{-3}$ **f)** 2^{-6}

   g) $\left(\dfrac{1}{3}\right)^{-2}$ **h)** 1^0 **i)** $\left(-\dfrac{1}{2}\right)^0$ **j)** 0^1

2 Calculate:
   a) $3^4 \times 3^2$ and 3^{4+2}
   b) $2^3 \times 2^{-5}$ and 2^{3-5}
   c) $4^{-2} \times 4^5$ and 4^{-2+5}
   d) $10^{-3} \times 10^{-2}$ and 10^{-3-2}
   e) $5^0 \times 5^{-3}$ and 5^{0-3}

3 Calculate:
   a) $(3^4)^2$ and $3^{4 \times 2}$
   b) $(2^{-3})^2$ and $2^{-3 \times 2}$
   c) $(10^2)^3$ and $10^{2 \times 3}$
   d) $(4^{-2})^{-1}$ and $4^{-2 \times -1}$
   e) $(5^0)^3$ and $5^{0 \times 3}$

From this exercise it appears that if $a \neq 0$ and m, n are integers, the above definition leads to

$$\text{and} \quad \left.\begin{array}{l} a^m \times a^n = a^{m+n} \\ (a^m)^n = a^{mn} \end{array}\right\} \text{even if } m \text{ or } n \text{ is negative or zero}$$

Fractional indices

We are now in a position to ask whether any useful meaning can be given to such expressions as $2^{1/2}$, $5^{2/3}$, and $4^{-3/2}$.

If, as before, we aim to maintain the existing pattern, it would be convenient if

$$(2^{1/2})^2 = 2^{(1/2) \times 2} = 2^1 = 2, \quad \text{so we } \textit{define } 2^{1/2} \text{ as } \sqrt{2}$$

and $(5^{2/3})^3 = 5^{(2/3) \times 3} = 5^2,$ so we *define* $5^{2/3}$ as $\sqrt[3]{5^2}$

More generally, if $a \neq 0$ and p, q are positive integers

$$(a^{p/q})^q = a^{(p/q) \times q} = a^p, \quad \text{so we } \textit{define } a^{p/q} \text{ as } \sqrt[q]{a^p}$$

Lastly, the definition is extended to powers that are *negative* fractions, provided that

$$a^{-p/q} = \frac{1}{a^{p/q}} = \frac{1}{\sqrt[q]{a^p}}$$

So $4^{-3/2} = \dfrac{1}{4^{3/2}} = \dfrac{1}{\sqrt[2]{4^3}} = \dfrac{1}{\sqrt[2]{64}} = \dfrac{1}{8}$

The relationships

$$a^m \times a^n = a^{m+n} \quad \text{and} \quad (a^m)^n = a^{mn}$$

are now true for all rational values of m and n, whether positive or negative, integral or fractional.

Exercise 2.4b

1 Calculate:

  a) $25^{1/2}$ **b)** $16^{1/4}$ **c)** $27^{1/3}$ **d)** $8^{2/3}$

  e) $9^{3/2}$ **f)** $8^{5/3}$ **g)** 5^0 **h)** $4^{-1/2}$

  i) $8^{-1/3}$ **j)** $9^{-3/2}$ **k)** $16^{-5/4}$ **l)** $32^{-3/5}$

2 If $2^{x+y} = 1$ and $10^{3x-y} = 100$, find the values of 5^{y-x} and x^y.   (MEI)

The power function, 10^x

Suppose that a colony of insects is breeding so fast that each day its size is multiplied by 10. Then after x days it will have increased x times by a factor 10, i.e. by a factor 10^x.

In order to plot the growth of this *power function*, we first form a table of values of 10^x:

x	0	1	2	3
10^x	1	10	100	1000

To obtain intermediate values (remembering that this first had to be done without use of calculators or logarithms) we first calculate $10^{1/2}$.

Now $10^{1/2} = \sqrt{10} \approx 3.162$ (by the method of section 1.5)

so that $10^{3/2} \approx 31.62, \quad 10^{5/2} \approx 316.2$

and $10^{-1/2} \approx 0.3162$

Similarly $10^{1/4} = (10^{1/2})^{1/2} \approx \sqrt{3.162} \approx 1.778$
$$\Rightarrow \quad 10^{3/4} = 10^{1/2} \times 10^{1/4} \approx 3.162 \times 1.778 \approx 5.623$$

Using these results, we obtain the extended table:

x	−0.50	−0.25	0	0.25	0.50	0.75	1.00	1.25	1.50	1.75	2.00
10^x	0.3162	0.5623	1.0	1.778	3.162	5.623	10	17.78	31.62	56.23	100

and the graph:

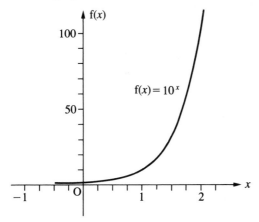

$f(x) = 10^x$

Exercise 2.4c

(Start by plotting $f(x) = 10^x$ on graph paper, from $x = -1$ to $x = +2$.)

1 By what factor has the above population of insects increased after
   a) 0.8 days **b)** 1.3 days **c)** 1.9 days?

2 How long does it take for the population to increase by a factor of
   a) 5 **b)** 27 **c)** 120?

3 Is there any meaning for the values of 10^x when x is negative?

4 By what factor does the population increase
   a) between $x = 1.0$ and $x = 1.2$;
   b) between $x = 1.5$ and $x = 1.7$;
   c) between $x = 1.8$ and $x = 2.0$?

5   Draw similar graphs for the functions 2^x and 3^x, taking values of x from $x = 0$ to 3 at intervals of 0.25, and using the same axes.

Increasing and decreasing power functions

The last section shows the importance of power functions whenever growth is being considered. Two familiar examples will provide further illustration:

Example 1

One of the commonest examples of growth is that of interest on a sum of money placed on deposit in a bank. If £1 is deposited and the interest rate is 5% per year, then usually one of two things happens:

Either **a)** the annual interest £0.05 is withdrawn each year. In this case the total invested remains £1 and the amount earned in n years is the *simple interest*, £0.05n.

Or **b)** the interest is credited to the account each year, so that the future interest will be calculated upon the *new* amount on deposit. In this case the amount on deposit will be multiplied each year by a factor 1.05 and the total interest earned is called the *compound interest*.

The amount on deposit initially is £1

so   after 1 year is £(1.05)

after 2 years is £$(1.05)^2$

and after n years is £$(1.05)^n$

If we suppose that the sums invested are gaining interest continuously rather than at the end of each year, the difference between simple and compound interest is illustrated by the difference between the graphs of a linear function and a power function:

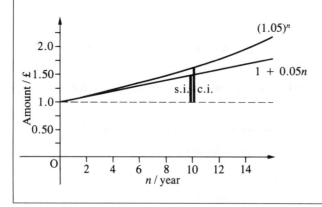

Example 2

Just as the growth of a quantity frequently involves a power function, so does decay; and the strength of these functions is just as marked with decay as with growth.

If, for instance, a car is bought for £1000 and depreciates at 30% per year, then its value is changed each year by a factor $\frac{7}{10}$.

Initial value = £1000

so   after 1 year, value = £1000 $\times \dfrac{7}{10}$ = £700

after 2 years, value = £1000 $\times \left(\dfrac{7}{10}\right)^2$ = £490

and after n years, value = £1000 $\times \left(\dfrac{7}{10}\right)^n$

This too is a power function, though here one of powerful decline. Such functions are extremely common, not only for measuring depreciations of value, but throughout nature, from the decay of radioactive elements to the cooling of one's bath-water. With an increasing power function, the factor of increase was always the same for a given time, whether the actual quantity was large or small: in just the same way, this factor remains constant for a decreasing power function, and the car in our example (for instance) is losing 30% of its value each year, no matter whether it is new and valuable or old and relatively worthless.

If, as with compound interest, we regard the value of the car as depreciating *continuously*, we can represent it by the curve

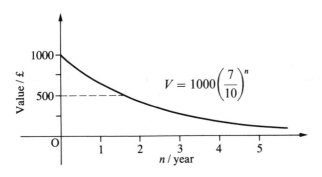

When a quantity is decreasing in this way, we sometimes speak of its 'half life', the time it would take to diminish by 50%. We can see from the graph that our car's 'half-life' is just under 2 years.

We are now in a position to sketch graphs of a number of power functions:

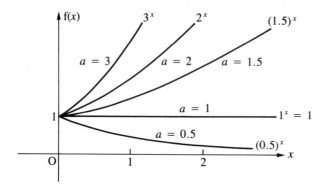

Readers will probably wonder how the increasing power functions compare with, say, x^2 and x^{10}. A little investigation should produce values of x for which 2^x overtakes both of these, and they can then ask themselves whether a more impressive function like x^{1000} would be similarly overtaken.

Exercise 2.4d

1   The cost of living in a certain country is increasing by 5% each year. What will be the total percentage increase after   **a)** 2 years   **b)** 3 years   **c)** 4 years?

2   In the early stages of a measles epidemic there were 100 people infected and each day the number was rising by 10%. How many would be infected   **a)** after 2 days   **b)** after 3 days   **c)** after 4 days   **d)** after 1 week?

3   An absent-minded mathematician makes a pot of tea and immediately puts it in her refrigerator at $0°$ C. If each minute it loses a quarter of its excess temperature, how hot is it when discovered 5 minutes later?

4   If each year the purchasing value of £1 falls by 4%, what is the total percentage fall in the course of   **a)** 2 years   **b)** 3 years?

2.5 Logarithms and logarithmic functions

Logarithms

It was noticed in the last section that indices can be used for the multiplication of two numbers.

For instance, 1000×100 can be written as

$$10^3 \times 10^2 = 10^{3+2} = 10^5 = 100000$$

At first this seems a very unpractical method, but the graph of the power function can be used to extend it a little further, to perform the sum 4×5.

From our graph we see that

$$4 \approx 10^{0.6} \quad \text{and} \quad 5 \approx 10^{0.7}$$

So $4 \times 5 \approx 10^{0.6} \times 10^{0.7} = 10^{0.6+0.7} = 10^{1.3} \approx 20$

This still seems a very perverse and inaccurate way of calculating 4×5, but it will quickly be seen that the replacement of multiplication by addition can have enormous advantages, once there is a quick means of expressing any number as a power of 10.

The power itself is called the *logarithm* (to the base 10) of the number, written $\lg x$ (or $\log_{10} x$).

So $\lg 4 \approx 0.6$ and $\lg 5 \approx 0.7$

More generally, the logarithm (to the base 10) of any number is the power to which 10 has to be raised to give the number:

$$x = 10^y \quad \Leftrightarrow \quad y = \lg x$$

So we see that $\lg x$ is the inverse function of 10^x (and vice versa); and as 10^x is always positive, we can only speak of the logarithms of positive numbers.

Use of logarithms

For many years before the advent of electronic calculators, logarithms and the associated slide-rule provided the standard method of undertaking calculations, and even now it is instructive to use logarithms for this purpose.

Example 1

Calculate $4.163 \times 516.4 \times 0.7296$

By use of logarithms

$$4.163 \times 516.4 \times 0.7296 \approx 10^{0.6194} \times 10^{2.7129} \times 10^{-0.1369}$$
$$\approx 10^{3.1954}$$
$$= 1570, \quad \text{to 3 s.f.}$$

Example 2

Calculate $(2.639)^6$

$$\text{Now} \quad 2.639 \approx 10^{0.4215}$$
$$\Rightarrow \quad (2.639)^6 \approx (10^{0.4215})^6$$
$$\approx 10^{6 \times 0.4215}$$
$$\approx 10^{2.529}$$
$$= 338, \quad \text{to 3 s.f.}$$

Example 3

Calculate $\sqrt[3]{0.02}$

$$\text{Now} \quad 0.02 \approx 10^{-1.6990}$$
$$\text{so} \quad \sqrt[3]{0.02} \approx (10^{-1.6990})^{1/3}$$
$$\approx 10^{-0.5663}$$
$$= 0.271, \quad \text{to 3 s.f.}$$

Exercise 2.5a

1   Express the following as powers of 10. Hence, without using tables, find their logarithms to the base 10.

a) 100000 b) $\sqrt{1000}$ c) 1 d) $\dfrac{1}{100}$ e) $\sqrt[3]{10}$ f) $\dfrac{1}{\sqrt{10}}$

2   It is known that, to five places of decimals

$$2 = 10^{0.30103} \quad \text{and} \quad 3 = 10^{0.47712}$$

Hence express as powers of 10:

a) 20 b) 300 c) 0.03 d) 0.002 e) 6 f) 4 g) $\tfrac{1}{2}$ h) 1.5

3 Using logarithms to express numbers as powers of 10, calculate the following:

a) 5.314×1.297

b) $6.243 \div 2.109$

c) 31.26×29.47

d) $(3.296)^4$

e) $\sqrt{41.26}$

f) $3^{1.5}$

g) 0.026×0.4791

h) $0.2971 \div 4.26$

i) $1 \div 0.02671$

j) $\sqrt[3]{0.0145}$

Properties of logarithmic functions

To investigate the properties of logarithms, let $\lg x = p$ and $\lg y = q$.

Then $x = 10^p$ and $y = 10^q$

So $xy = 10^{p+q}$ $\dfrac{x}{y} = 10^{p-q}$ $x^n = 10^{np}$

$\Rightarrow$ $\lg xy = p + q$ $\lg\dfrac{x}{y} = p - q$ $\lg x^n = np$

So $\lg xy = \lg x + \lg y$

$\lg\dfrac{x}{y} = \lg x - \lg y$

$\lg x^n = n \lg x$

These, of course, are the properties which underlie the use of logarithms in calculations. By adding the logarithms of two numbers we obtain the logarithm of their product, and so are able to find the product itself; and similarly for quotients and powers.

The construction of a slide rule was based on the same principle, the length to a certain number being proportional to the logarithm of the number. So two numbers can be multiplied simply by placing these lengths—or logarithms—end to end.

So far we have considered logarithms to the base 10. But it is clear that we can also define logarithms to other bases. The logarithm to any base of a number is simply the power to which the base must be raised in order to give the number.

For example,

$9 = 3^2$, so $\log_3 9 = 2$

and $8 = 2^3$, so $\log_2 8 = 3$

More generally,

If $a > 0$, $x = a^p \;\Leftrightarrow\; \log_a x = p$ †

To investigate the properties of such logarithms,

let $\log_a x = p \;\Leftrightarrow\; x = a^p$

and $\log_a y = q \;\Leftrightarrow\; y = a^q$

Hence $xy = a^{p+q}$ $\dfrac{x}{y} = a^{p-q}$ $x^n = a^{np}$

and $\log_a xy = p + q$ $\log_a \dfrac{x}{y} = p - q$ $\log_a x^n = np$

so

$$\log_a xy = \log_a x + \log_a y$$
$$\log_a \frac{x}{y} = \log_a x - \log_a y$$
$$\log_a x^n = n \log_a x$$

and the three basic laws of logarithms are therefore true whatever base is being used.

In order to calculate logarithms to a base other than 10 it is usually most convenient to proceed as follows:

To find $\log_3 7$, let $\log_3 7 = x$

$\Rightarrow$ $3^x = 7$

$\Rightarrow$ $\lg 3^x = \lg 7$

$\Rightarrow$ $x \lg 3 = \lg 7$

$\Rightarrow$ $x = \dfrac{\lg 7}{\lg 3}$

Hence $\log_3 7 = \dfrac{\lg 7}{\lg 3} = \dfrac{0.8451}{0.4771} = 1.77$,   to   3 s.f.

Exercise 2.5b

1   Given that $\lg 2 = 0.301030$ and $\lg 3 = 0.477121$, find
    a) $\lg 4$ **b)** $\lg 5$ **c)** $\lg 8$
    d) $\lg 9$ **e)** $\lg 10$ **f)** $\lg 12$
    g) $\lg 15$ **h)** $\lg 30$ **i)** $\lg 180$

† And $x = a^p = $ antilog p.

j) $\lg 0.25$ **k)** $\lg \dfrac{3}{2}$ **l)** $\lg \dfrac{2}{3}$

m) $\lg \sqrt{24}$ **n)** $\lg \sqrt[3]{16}$ **o)** $\lg \dfrac{1}{\sqrt{3}}$

2 Use logarithms to solve the equations:
   a) $3^x = 100$ **b)** $2^x = 1\,000\,000$
   c) $(1.293)^x = 10$ **d)** $10^x = 3$

3 Use logarithms to evaluate:
   a) $\log_4 7$ **b)** $\log_5 10$
   c) $\log_2 40$ **d)** $\log_3 2$

4 Find, correct to three significant figures, the values of x if
   a) $3^{x+1} = 4^{2x-1}$ **b)** $5^x \times 2^{-2x} = 4$ (OC)

5 Find, in surd form, the value of x, given that

$$\log_3 (x^2) + \log_3 x = \log_9 27 \qquad \text{(C)}$$

6 What is the value after 10 years of £100 invested at compound interest if the rate of interest is **a)** 5% **b)** 10%?

7 How long does a sum of money take to double itself when invested at compound interest of **a)** 5% **b)** 10%?

8 A country's population at the end of each year is 2% greater than at the start of the year. How long will it take for its population to increase by 50%?

9 A car bought for £1000 depreciates in value by 15% each year.
   a) What is its value after 6 years?
   b) What is the half-life of its value?

10 Archaeologists can sometimes estimate the age of their discoveries by finding the concentration of the radioactive isotope carbon 14 in them. The half-life of this isotope is 5600 years. How long does it take for 30% to be transmuted?

11 2% of light is absorbed when passing through a glass screen of thickness 1 mm. How much would be absorbed by a screen of thickness 1 cm?

12 Atmospheric pressure diminishes approximately as a power function of height. If at a height of 5000 m it is roughly halved, by what percentage does it diminish for each 1000 m?

Graphs of logarithmic functions

The graph of $y = \lg x$ is the same as that of $x = 10^y$. But $x = 10^y$ is obtained from $y = 10^x$ simply by interchanging x and y.

Such an interchange of x and y corresponds geometrically to a reflection in the line $y = x$.

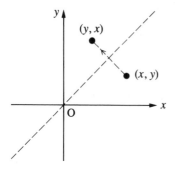

So $y = \lg x$ and $y = 10^x$ are reflections of each other in the line $y = x$.

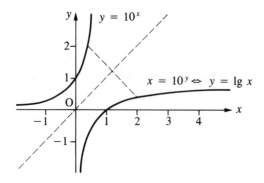

Exercise 2.5c

1 What is the domain and the range of the logarithmic function $\lg x$? Use its graph to illustrate your answer.

2 Draw accurately, in one diagram, graphs of the power functions 2^x, 3^x, 5^x, 10^x and of the logarithmic functions $\log_2 x$, $\log_3 x$, $\log_5 x$, $\log_{10} x$.

Logarithmic scales

Logarithms are frequently used in graphical work when indices are involved and often enable us to replace curved graphs by straight lines.

For example, if $y = 3 \times 2^x$

we can take logarithms (base 10) and obtain

$$\lg y = \lg(3 \times 2^x)$$

$$\Rightarrow \quad \lg y = \lg 3 + x \lg 2$$

So if we plot $\lg y$ against x, we obtain a straight line graph with gradient $\lg 2 \approx 0.3$ and

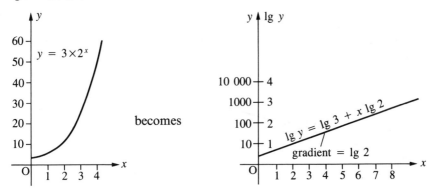

Such representation of power functions by a straight line graph is so common that logarithmic graph paper is available to make the taking of logarithms automatic:

The rise and fall in value of £100 investments
at various rates of interest and depreciation

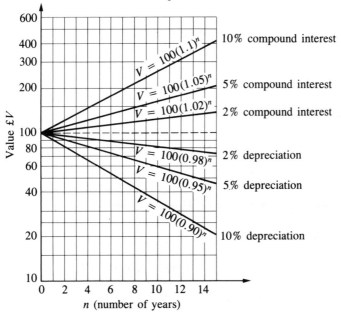

It sometimes happens that y is a function of x of the form

$$y = bx^n$$

Again taking logarithms

$$\lg y = \lg bx^n$$

$$\Rightarrow \quad \lg y = \lg b + n \lg x$$

If this time we plot $\lg y$ against $\lg x$ (and graph paper is also available with a logarithmic scale on *each* axis), we again obtain a straight line graph with gradient n.

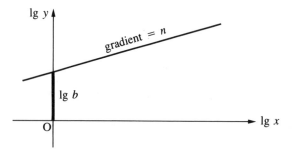

Example 4

It is suspected that the period T of a planet's year is related to r, its mean distance from the sun, by an equation of the form

$$T = kr^n$$

where k and n are constants.

If T is measured in Earth-years and r in astronomical units of distance, their values for the six inner planets are

	Mercury	Venus	Earth	Mars	Jupiter	Saturn
r	0.3871	0.7233	1.000	1.524	5.203	9.539
T	0.2408	0.6152	1.000	1.881	11.86	29.46

From these we obtain the values

$\lg r$	-0.4122	-0.1407	0	0.1829	0.7162	0.9795
$\lg T$	-0.6184	-0.2110	0	0.2744	1.0742	1.4692

These can be plotted on a graph:

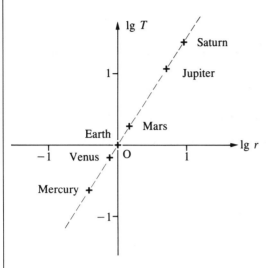

The gradient of the line is 1.5

So $\lg T = 1.5 \lg r = \lg r^{1.5}$

   $\Rightarrow$ $T = r^{1.5}$

and our suspicions are confirmed.

Hence $T^2 = r^3$

a result known as Kepler's third law of planetary motion, after the remarkable German mathematician, Johann Kepler, who discovered it in 1618.

Exercise 2.5d

(To be worked accurately on graph paper.)

1 The population of a rapidly developing country at successive ten-yearly intervals from 1940 was estimated to be:

year	1940	1950	1960	1970
number of years x	0	10	20	30
population P (millions)	2.72	4.14	6.29	9.54

It is suspected that this growth can be expressed by a power function

$$P = ab^n$$

where a and b are constants.
a) Use logarithms to plot these values as an approximate straight line.
b) Estimate values for a and b.
c) What is the percentage growth rate per annum?
d) Estimate the population in 1980.
e) How many years is the population taking to double?

2   The times of oscillation, T second, of heavy weights on the ends of wires of length l metre are given by

l	2	4	6	8
T	2.81	4.01	4.98	5.63

It is suspected that these are related by an equation $T = kl^n$, where k and n are constants.
a) Use logarithms to plot these readings as an approximate straight line.
b) Estimate values for k and n.
c) From your graph find the time of oscillation of a 5-metre pendulum.
d) What is the length of a pendulum whose time of oscillation is 1 second?

3   A car costing £2000 depreciates at a rate of 20% each year.
a) Express its value £P as a function of its age n years.
b) Use a logarithmic scale to plot its depreciation as a straight-line graph.
c) With the same axes plot the depreciation lines when the annual rate is 10%, 30%, 40%.

4   Find y in terms of x if it is known to be proportional to a power of x, with $y = 7.27$ when $x = 2$ and $y = 13.92$ when $x = 3$.

* 2.6 Continuity and limits

You will have noticed, when functions and their graphs were first introduced in section 2.1, a marked difference between graphs which were unbroken, or *continuous*, and those which included sudden jumps. The concept of *continuity* is extremely important in mathematics and needs very careful definition. For the present we shall simply say that a function is *continuous at a particular point* if we can depart from the point along the graph both to the left and to the right without having to make a jump.
So in the examples of section 2.1.

1   $f(h) = 3.6\sqrt{h}$ is continuous if $h > 0$.

2 $f(x) = |x|$ is continuous everywhere.

3 $g(x) = \begin{cases} x - 1, & x < 0 \\ 0, & x = 0 \\ x + 1, & x > 0 \end{cases}$

is continuous except at $x = 0$ (since at this point we have to jump whether we move to left or to right).

4 $h(x) = [x]$

is continuous except when x is an integer (since at these points we have to jump when we move to the left, though not if we move to the right).

Exercise 2.6a

For what values of x are the functions defined in exercise 2.1c nos **1–10**
a) continuous   **b)** discontinuous?

Limits

We shall now reverse the process of the last section and look at the behaviour of a function not as we *leave* a particular point, but as we *approach* it.

This behaviour may be very erratic. Consider, for instance, the function

$$f(x) = \begin{cases} 1 \text{ when } x = \pm 1, \ \pm\frac{1}{2}, \ \pm\frac{1}{4}, \ \pm\frac{1}{8} \cdots \\ 0 \text{ otherwise} \end{cases}$$

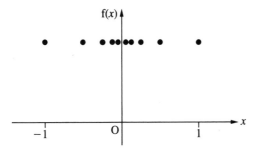

From the graph, we see that although the function is generally zero, it takes the value 1 with increasing frequency as x approaches 0, whether from above or below; so we cannot reasonably say that the function is tending towards any particular limit.

By contrast, the function

$$f(x) = \begin{cases} x^2 + 1 & \text{if } x > 0 \\ 0 & x = 0 \\ x^2 - 1 & x < 0 \end{cases}$$

approaches the value $+1$ as x tends to 0 through positive values, and -1 as x tends to 0 through negative values.

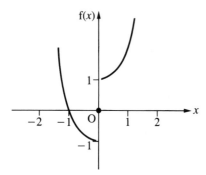

These two values, $+1$ and -1, are sometimes called the *right-hand limit* and *left-hand limit* of the function as x tends to 0, and we can write

$$\lim_{x \to 0+} f(x) = 1 \quad \text{and} \quad \lim_{x \to 0-} f(x) = -1$$

Now consider the function

$$f(x) = \begin{cases} x^2 + 1 & \text{if } x \neq 0 \\ 0 & \text{if } x = 0 \end{cases}$$

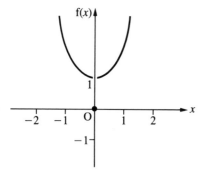

Here the right-hand limit and the left-hand limit have become identical, and when this happens we call their common value simply *the limit of* $f(x)$ *as* x *approaches* 0, and write:

As $x \to 0$, $f(x) \to 1$ or $\lim_{x \to 0} f(x) = 1$

It must be emphasised very strongly that this limit is an indication of the behaviour of the function *near* $x = 0$, not *at* $x = 0$. In this case,

$$\lim_{x \to 0} f(x) = 1, \quad but \quad f(0) = 0$$

So the limit (if it exists) of a function as we approach a point is not necessarily the same as the value of the function at the point.

As with continuity, so with limits, a very precise definition is necessary. For the present, however, we shall content ourselves with the following statement, even though the word *approach* has not been carefully defined:

If when x approaches a particular value $x = a$ (whether from above or below), the values of $f(x)$ approach a certain value l, we call this *the limit of* $f(x)$ *as x tends to a*, and write:

$$\text{As} \quad x \to a, \quad f(x) \to l \quad \text{or} \quad \lim_{x \to a} f(x) = l$$

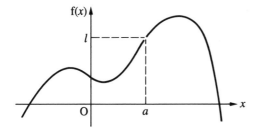

The gap left in the curve is simply to emphasize that the value of the function *at* $x = a$ is irrelevant to the existence of the limit. Now, at last, we are able to come full circle and observe that the function is *continuous* at this point if (and only if) its value is equal to its limit as the point is approached:

$$f(x) \text{ is continuous at } x = a \quad \Leftrightarrow \quad \lim_{x \to a} f(x) = f(a)$$

Exercise 2.6b

Investigate **a)** the right-hand limit **b)** the left-hand limit **c)** the limit of the following functions as x approaches the specified points. In each case state whether the function is continuous at the point.

1 $f(x) = \begin{cases} x \text{ if } x \geqslant 0 \\ 0 \text{ if } x < 0 \end{cases} \quad x = 0$

2 $f(x) = \begin{cases} |x| \text{ if } x \neq 0 \\ 1 \text{ if } x = 0 \end{cases} \quad x = 0$

3 $f(x) = [x]$, $x = 1$

4 $f(x) = x - [x]$, $x = 1$

5 $f(x) = [x]$, $x = 1\frac{1}{2}$

There is another case where, in a rather different sense, we speak of the limit of f(x) as $x \rightarrow a$, and that is when f(x) becomes *unlimitedly* large (either positively or negatively).

For example:

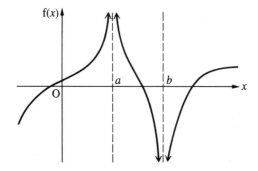

Here we write:

   as $x \rightarrow a$, $f(x) \rightarrow +\infty$ or $\lim_{x \rightarrow a} f(x) = +\infty$

and   as $x \rightarrow b$, $f(x) \rightarrow -\infty$ or $\lim_{x \rightarrow b} f(x) = -\infty$

and the two dotted lines are called *asymptotes*.

Lastly, when x becomes larger and larger without limitation, rather than approach a particular value, we say that x 'tends to infinity', and we use the notation of limits in the following way:

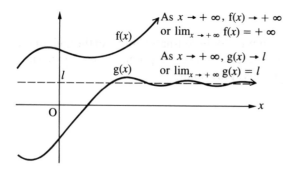

As $x \rightarrow +\infty$, $f(x) \rightarrow +\infty$
or $\lim_{x \rightarrow +\infty} f(x) = +\infty$

As $x \rightarrow +\infty$, $g(x) \rightarrow l$
or $\lim_{x \rightarrow +\infty} g(x) = l$

Similar notation is used when x becomes very large and negative $(x \to -\infty)$; and if we wish to speak of x becoming large without reference to whether it is positive or negative, we simply write $x \to \infty$, without a sign.

So, as $x \to \infty$, $x^2 + 3x \to +\infty$ or $\lim_{x \to \infty} (x^2 + 3x) = +\infty$

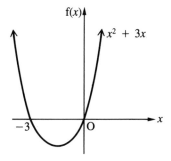

Example

If $f(x) = \dfrac{x+1}{x-1}$, investigate **a)** $\lim_{x \to 1} f(x)$ **b)** $\lim_{x \to \infty} f(x)$

a) As $x \to 1$, $x - 1$ becomes very small.

So $f(x) = \dfrac{x+1}{x-1}$ becomes very large.

For example, $f(1.001) = \dfrac{2.001}{0.001} = 2001$

$f(0.999) = \dfrac{1.999}{-0.001} = -1999$

So the right-hand and left-hand limits are

$\lim_{x \to 1+} f(x) = +\infty$

and $\lim_{x \to 1-} f(x) = -\infty$;

and as these are different, $\lim_{x \to 1} f(x)$ does not exist.

b) As $x \to \infty$, $f(x) = \dfrac{x+1}{x-1} = \dfrac{1 + \dfrac{1}{x}}{1 - \dfrac{1}{x}} \quad \to \quad \dfrac{1}{1} = 1.$

So $\lim_{x \to \infty} f(x) = +1$.

These results are illustrated by the graph of f(x):

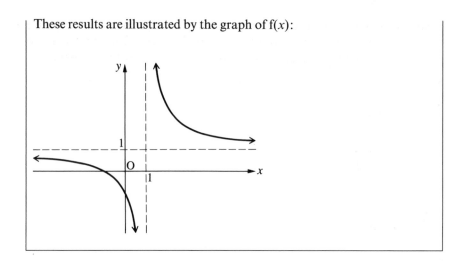

Exercise 2.6c

Investigate the following limits, and illustrate by sketch graphs:

1 $\lim_{x \to +\infty} (2x - x^2)$ $\lim_{x \to -\infty} (2x - x^2)$

2 $\lim_{x \to +\infty} \sqrt{x}$ $\lim_{x \to -\infty} \sqrt{x}$

3 $\lim_{x \to 0} \dfrac{1}{x^2}$ $\lim_{x \to \infty} \dfrac{1}{x^2}$

4 $\lim_{x \to 1} \dfrac{x}{x - 1}$ $\lim_{x \to \infty} \dfrac{x}{x - 1}$

5 $\lim_{x \to 0} \dfrac{x^2 + 1}{x}$ $\lim_{x \to \infty} \dfrac{x^2 + 1}{x}$

6 $\lim_{x \to 0} \dfrac{x + 1}{x^2}$ $\lim_{x \to \infty} \dfrac{x + 1}{x^2}$

7 $\lim_{x \to \infty} [x]$ $\lim_{x \to \infty} x - [x]$

* 2.7 Rational functions

If $P(x)$ and $Q(x)$ are polynomial functions of x, their quotient $P(x)/Q(x)$ is called a *rational function* of x.

Example 1

Investigate the function $f(x) = \dfrac{2x}{x-1}$

We first note that

$$f(0) = 0$$

and $f(1) = \dfrac{2}{1-1} = \dfrac{2}{0},$ which is undefined

The sign of $f(x)$ is indicated by

x	-2	-1	0	1	2	3
$f(x)$	$+++++++++++$		$0 \;---$		$+++++++++++$	

As $x \to 1$, $f(x)$ becomes very large, and by considering the sign of $f(x)$ we see that

as $x \to 1-, \quad f(x) \to -\infty$

as $x \to 1+, \quad f(x) \to +\infty$

Finally, we can see how $f(x)$ behaves as $x \to \pm\infty$ by writing

$$f(x) = \frac{2x}{x-1} = \frac{2}{1 - 1/x} \to 2 \quad \text{as } x \to \pm\infty$$

These can now all be assembled in the graph of the function, where it is seen that there are two asymptotes, one parallel to each of the axes, which the curve approaches but never meets:

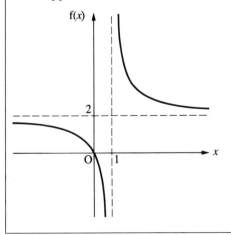

Example 2

Investigate the function $f(x) = \dfrac{x^2}{2 - x}$

In this case $f(0) = 0$ and $f(2)$ is undefined.

The sign of $f(x)$ is given by

x	-2	-1	0	1	2	3
$f(x)$	$+ + + + + + + + + +$		$0 + + + + + + +$		$\vert$ $- - - - - -$	

and we see that

as $x \to 2-$, $f(x) \to + \infty$

$x \to 2+$, $f(x) \to - \infty$

Finally, we can see how $f(x)$ behaves as $x \to \pm \infty$ by writing

$$f(x) = \frac{x^2}{2 - x} = \frac{x}{2/x - 1} \approx \frac{x}{-1} = -x$$

So the graph of $f(x)$ is

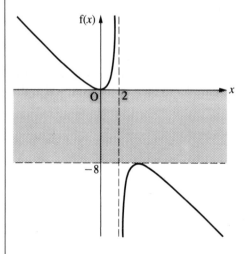

The range of values of $f(x)$ can be found by letting $f(x) = y$.

$\Rightarrow \quad \dfrac{x^2}{2 - x} = y$

$\Rightarrow \quad x^2 + yx - 2y = 0$

So the values of x which produce a given value of y are the roots of this quadratic equation, and these are real if and only if

$$y^2 - 4(-2y) \geqslant 0$$

$$\Leftrightarrow \quad y(y + 8) \qquad \geqslant 0$$

$$\Leftrightarrow \quad y \geqslant 0 \quad \text{or} \quad y \leqslant -8$$

So f(x) can take all values except those between 0 and -8, corresponding to the shaded area of the graph.

Exercise 2.7

Sketch graphs of the following functions, stating their range and clearly marking their asymptotes:

1 $\dfrac{1}{1 - x}$ **2** $\dfrac{1}{x + 1}$ **3** $\dfrac{x^2 + 1}{x}$ **4** $\dfrac{x^2 - 1}{x}$

5 $\dfrac{1}{(x - 1)(x - 2)}$ **6** $\dfrac{x^2}{x^2 + 4}$ **7** $\dfrac{x^2}{x^2 - 4}$ **8** $\dfrac{x}{(x + 1)(x + 2)}$

9 $\dfrac{(x + 1)^2}{x - 1}$ **10** $\dfrac{x}{(x - 1)^2}$

Miscellaneous problems

1 Is it true that all the squares of integers are either divisible by 4 or leave remainder 1 when divided by 8? If so, why?

2 If n is a positive integer, is $n^5 - n$ always divisible by 30? If so, why?

3 When a polynomial P(x) is divided by $x - 1$, x, $x + 1$, the remainders are 1, 2, 3 respectively. Find the remainder when P(x) is divided by $x(x^2 - 1)$.

4 The polynomial P(x) is divided by $(x - a)(x - b)$, where $a \neq b$. Express the remainder in terms of P(a) and P(b).

5 The equation $4x^3 - 3x^2 - 10x - 49 = 0$ has a root which is approximately 3. By substituting $x = 3 + h$ and neglecting squares and higher powers of h, find a closer approximation to the value of this root.   (MEI)

6 The equation

$$px^3 + qx^2 + rx + s = 0$$

has roots α, $1/\alpha$ and β. Prove that

$$p^2 - s^2 = pr - qs$$

Solve the equation

$$6x^3 + 11x^2 - 24x - 9 = 0 \qquad \text{(L)}$$

7   Artists have sometimes paid special attention to the subdivision of a line AB by a point P, called its *golden section*, where AP/PB = PB/AB. Calculate this ratio.

A   P   B

8    $f(x) = x^x$, $g(x) = (x^x)^x$, $h(x) = x^{(x^x)}$
Calculate these functions when $x = 1, 2$ and 3.

9   A fine drizzle is falling steadily and after 1 minute 10% of a pavement is wet. What proportion will be wet after 2 minutes; 3 minutes; 4 minutes? How long will it take for half of the pavement to become wet?

10   The musical notes middle C and top C are taken to have frequencies 256 and 512 respectively. In an equal-tempered scale the octave between them is divided into 12 intervals by the notes

$$C^\#, D, D^\#, E, F, F^\#, G, G^\#, A, A^\#, B$$

and the frequencies of successive notes are in constant ratio. Find this ratio and the frequencies of the notes E and G.

11   Prove that $\log_a b = \dfrac{\lg b}{\lg a}$

12   If £1 gains 100% interest in 1 year, it becomes £2. If, however, 50% interest is added each half-year, it becomes firstly £1.5 and finally £2.25. What would it become
a) if 25% were added quarterly;
b) if 10% were added at the end of 10 equal intervals;
c) if 1% were added 100 times;
d) if 0.1% were added 1000 times?
To what do these amounts seem to be tending?

3 Rates of change: differentiation

3.1 Average velocity and instantaneous velocity

'Everything is in movement', wrote Heraclitus in about 500 BC. Yet there was little attention to the mathematics of movement and variation until the sixteenth and seventeenth centuries. Two thousand years earlier, Archimedes had calculated areas by the 'method of exhaustion' (p. 139), which is at the root of the *integral calculus*. But the discovery (or invention?) of the *differential calculus* had to wait for Newton and Leibniz in the second half of the seventeenth century.

Taken together, the differential and integral calculus form the two edges of a tool which, first used in astronomy, has developed outstanding versatility and power. We shall begin to shape it by a simple example.

If a stone is dropped over the edge of a cliff, the distance y m in which it has fallen after t s, is given approximately by

$$y = 5t^2$$

This can be represented by the table

t	0	1	2	3	4	5	6
y	0	5	20	45	80	125	180

or by the graph

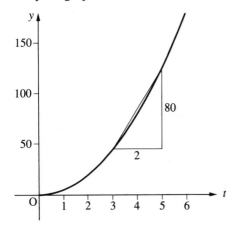

The *average velocity*† of this stone during an interval of time is defined as

$$\frac{\text{displacement during the interval}}{\text{duration of the interval}}$$

So in the 2-second interval between $t = 3$ and $t = 5$,

$$\text{average velocity} = \frac{125 - 45 \,\text{m}}{2 \,\text{s}} = 40 \,\text{m s}^{-1}$$

This is, of course, also the gradient‡ of the line joining corresponding points of the graph.

Exercise 3.1a

When a stone is catapulted with velocity $40 \,\text{m s}^{-1}$ vertically upwards, its height y m after t s is given by $y = 40t - 5t^2$.

1   Construct a table showing the height of the stone at second intervals from $t = 0$ to $t = 10$.

2   Draw accurately the corresponding graph, with clear labels on the axes.

3   What is the maximum height reached by the stone? After how many seconds?

4   How long does it take for the stone to return to ground level?

5   Could there possibly be any meaning for values of y calculated from the above formula when   **a)** $t > 8$ **b)** $t < 0$?

6   What is the average velocity of the stone during the following intervals (velocities being counted positive if upwards and negative if downwards)?
   a) $t = 0$ to $t = 2$
   b) $t = 1$ to $t = 4$
   c) $t = 4$ to $t = 7$
   d) $t = 1$ to $t = 7$

Instantaneous velocity

Although we often wish to know the average velocity over a given interval, we more frequently need the velocity *at a certain instant*. We might, for example, want to know how fast the stone dropped over the cliff was falling after 3 seconds, i.e. at the instant when $t = 3$.

---

† We shall use the word velocity rather than speed whenever we are concerned about the direction of motion.
‡ Throughout this and the next section we shall regard all gradients as being expressed in the appropriate units, usually m s^{-1}. The actual slope of lines drawn on the graph paper will, of course, depend on the scales chosen for distance and time.

For this purpose, we first examine its behaviour in some short intervals near
$t = 3$. A few calculations will show that

t	3	3.1	3.01	3.001
y	45	48.05	45.3005	45.030005

So, between $t = 3$ and $t = 3.1$,

$$\text{average velocity} = \frac{3.05\,\text{m}}{0.1\,\text{s}} = 30.5\,\text{m s}^{-1}$$

Between $t = 3$ and $t = 3.01$,

$$\text{average velocity} = \frac{0.3005\,\text{m}}{0.01\,\text{s}} = 30.05\,\text{m s}^{-1}$$

Between $t = 3$ and $t = 3.001$,

$$\text{average velocity} = \frac{0.030005\,\text{m}}{0.001\,\text{s}} = 30.005\,\text{m s}^{-1}$$

It appears that the smaller the interval we take, the nearer the average velocity
approaches to $30\,\text{m s}^{-1}$, so we are tempted to say that this is the velocity *at the
instant* when $t = 3$.

This becomes still more evident if we consider the interval between

$$t = 3, \qquad \text{when } y = 45$$
$$\text{and} \quad t = 3 + h, \quad \text{when } y = 5(3 + h)^2$$
$$= 45 + 30h + 5h^2$$

The average velocity over this interval

$$= \frac{30h + 5h^2\,\text{m}}{h\,\text{s}}$$

$$= 30 + 5h \ \text{m s}^{-1}$$

So, as $h \to 0$, average velocity $\to 30\,\text{m s}^{-1}$.

We therefore say that the velocity *at the instant when* $t = 3$ is $30\,\text{m s}^{-1}$.

Looking at this on the graph, we see that as $h \to 0$, the line whose gradient
we are calculating is gradually approaching the tangent to the curve at the
point where $t = 3$:

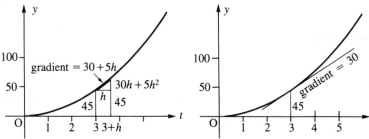

The velocity is therefore represented by the *gradient of the tangent* to the distance–time graph.

Velocity at any instant

Suppose that we need to know the velocity v of the falling stone at any instant.

After time t the stone has fallen a distance $5t^2$

After time $t + h$ the stone has fallen a distance $5(t + h)^2 = 5t^2 + 10th + 5h^2$

Average velocity during this interval

$$= \frac{10ht + 5h^2}{h} = 10t + 5h$$

As $h \to 0$, average velocity $\to 10t$

So $v = 10t$

This enables us to calculate the velocity at *any* instant:

t (s)	0	1	2	3	4	5
$v = 10t$ (m s^{-1})	0	10	20	30	40	50

So we can represent v, as well as y, graphically as a function of t:

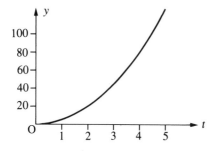 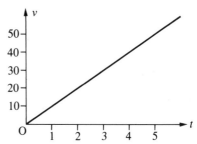

Exercise 3.1b

1 If a stone is catapulted upwards at 40 m s^{-1}, its height y m after time t s is given by $y = 40t - 5t^2$.
   a) Calculate its average velocity during the intervals i) from $t = 2$ to $t = 2.1$, and ii) from $t = 2$ to $t = 2.01$,   and so estimate its velocity when $t = 2$.

b) Calculate its average velocity over the interval from $t = 3$ to $t = 3 + h$, and so find its velocity when $t = 3$.

c) Use the same method to calculate its velocity when $t = 7$.

d) More generally, find its velocity after time t, checking your results for $t = 2, 3, 7$.

e) Sketch the curves of y and v against t.

2 **a)** Repeat no. **1a)** by considering the intervals i) from $t = 1.9$ to $t = 2$, and ii) $t = 1.99$ to $t = 2$.

b) Could you obtain a closer approximation by averaging the values obtained from the intervals $(1.99, 2)$ and $(2, 2.01)$?

c) How does this compare with the average velocity over the interval $(1.99, 2.01)$?

3 If, instead of a stone, a ball is dropped over the edge of a cliff, the distance y m through which it has fallen in t s is found to be

$$y = 5t^2 - \tfrac{1}{300}t^3$$

the term $\tfrac{1}{300}t^3$ arising from air-resistance and the formula being valid only for the first four seconds of fall. (Why, even for an infinitely high cliff, can this formula not always be valid?) Calculate:

a) the effect of air-resistance on the distance fallen in the first three seconds;

b) the velocity when $t = 3$ and hence the effect of air-resistance on the velocity at this instant;

c) the velocity v at time $t(t < 4)$.

3.2 Differentiation

It is clear from these examples that the process of finding the rate of change of a function, or the gradient of a graph, is going to be very important. Sometimes the task is impossible at a particular point or set of points, perhaps because the function has a discontinuity or has different gradients on the two sides of the point; sometimes, and frequently in practical work, it can only be tackled numerically, or by drawing a tangent and estimating its gradient.

But if a function $f(t)$ has a definite rate of change (i.e. its graph has a gradient) this rate of change will itself be a function, which we call the *derivative*, of $f(t)$, and write as $f'(t)$.

For example, in the last section

$$f(t) = 5t^2 \Rightarrow f'(t) = 10t$$

The process of finding such derivatives is called *differentiation*.

Lastly, we see that

when   f'(t) > 0,   f(t) is increasing
          f'(t) < 0,   f(t) is decreasing
if   f'(t) = 0,   f(t) is said to be *stationary*, or to have a *stationary value*.

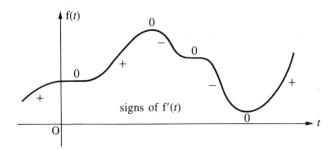

Example 1

If f(x) = x³, find f'(2).

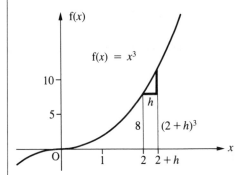

Gradient of chord

$$= \frac{(2 + h)^3 - 8}{h}$$

$$= \frac{(8 + 12h + 6h^2 + h^3) - 8}{h}$$

$$= 12 + 6h + h^2$$

As $h \rightarrow 0$, $12 + 6h + h^2 \rightarrow 12$

So f'(2) = 12

Example 2

Differentiate $f(x) = \dfrac{1}{x}$

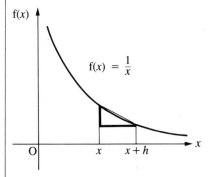

Gradient of chord

$$= \frac{f(x + h) - f(x)}{h}$$

$$= \frac{1/(x + h) - 1/x}{h} = \frac{-h/\{x(x + h)\}}{h}$$

$$= -\frac{1}{x(x + h)}$$

As $h \to 0$, $x + h \to x$

So gradient $\to -\dfrac{1}{x^2}$ and $f'(x) = -\dfrac{1}{x^2}$

Alternatively, we can use the notation of limits:

Example 1

$$f'(2) = \lim_{h \to 0} \frac{(2 + h)^3 - 8}{h}$$

$$= \lim_{h \to 0} \frac{(8 + 12h + 6h^2 + h^3) - 8}{h}$$

$$= \lim_{h \to 0} (12 + 6h + h^2) = 12$$

So $f'(2) = 12$

Example 2

$$f'(x) = \lim_{h \to 0} \frac{\dfrac{1}{x+h} - \dfrac{1}{x}}{h}$$

$$= \lim_{h \to 0} \frac{-h/\{x(x+h)\}}{h}$$

$$= \lim_{h \to 0} -\frac{1}{x(x+h)} = -\frac{1}{x^2}$$

So $f'(x) = -\dfrac{1}{x^2}$

More generally, given a function $f(x)$, we define its derivative as

$$f'(x) = \lim_{h \to 0} \frac{f(x+h) - f(x)}{h}$$

wherever this limit exists and is finite.
  Two further examples will show the force of the words in italics:

Example 3

  $f(x) = |x|$

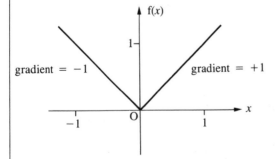

At $x = 0$, we see that, although there is a 'right-hand gradient' $+1$ and a 'left-hand gradient' -1, these are different and the graph has no genuine slope when $x = 0$. So $f'(0)$ does not exist.

Example 4

$$f(x) = \begin{array}{l} +1 \text{ if } x > 0 \\ 0 \text{ if } x = 0 \\ -1 \text{ if } x < 0 \end{array}$$

Here it is clear that at $x = 0$ the right-hand gradient and the left-hand gradient both tend to $+\infty$ as the origin is approached. So here again the graph has no genuine slope, and $f'(0)$ does not exist.

It is readily seen that $f'(x)$ cannot exist at any point where $f(x)$ has a discontinuity.

Exercise 3.2

1 Using the definition

$$f'(x) = \lim_{h \to 0} \frac{f(x + h) - f(x)}{h}$$

a) if $f(x) = x^2$, calculate $f'(3)$;
b) if $f(x) = 2x^3$, calculate $f'(0)$;
c) if $f(x) = 4x^2 + x + 7$, calculate $f'(-1)$;
d) if $f(x) = 3x^4 + 12x$, calculate $f'(1)$;
e) if $f(x) = 1$, calculate $f'(-2)$;
f) if $f(x) = 3/x$, calculate $f'(2)$;
g) if $f(x) = 1/x^2$, calculate $f'(1)$;

2 For each of the functions in no. 1, find the derivative $f'(x)$ and mention particularly any point at which it does not exist.
   For what range of values of x is $f(x)$ a) increasing   b) decreasing   c) stationary?   Illustrate your answers with graphs.

3.3 Standard methods of differentiation

Throughout the above sections, we have discovered each derivative separately 'from first principles'. But it is easily seen that the results of some of the examples can be put in the form of a table

f(x)	x^4	x^3	x^2	x	1	$\dfrac{1}{x}$	$\dfrac{1}{x^2}$
f'(x)	$4x^3$	$3x^2$	$2x$	1	0	$-\dfrac{1}{x^2}$	$-\dfrac{2}{x^3}$

This can be rewritten

f(x)	x^4	x^3	x^2	x	1	x^{-1}	x^{-2}
f'(x)	$4x^3$	$3x^2$	$2x$	1	0	$-x^{-2}$	$-2x^{-3}$

So it appears that

$$f(x) = x^n$$
$$\Rightarrow\ f'(x) = nx^{n-1}$$

This has, of course, not been proved, and in any case n has been restricted to integral values. We shall prove it later (pp.120 and 233) for all $n \in \mathbb{Q}$, but in the meantime we shall assume it true for all values of n.

Example

If $f(x) = \sqrt{x}$, find $f'(x)$.

$$f(x) = \sqrt{x} = x^{1/2}$$

$$\Rightarrow\ f'(x) = \frac{1}{2}x^{-1/2} = \frac{1}{2\sqrt{x}}$$

Exercise 3.3a

Find $f'(x)$ if $f(x)$ is

1 x^5	**2** x^8	**3** $x^{4/3}$	**4** $\sqrt[3]{x}$	**5** $x\sqrt{x}$
6 $\dfrac{1}{x^3}$	**7** $\dfrac{1}{x^4}$	**8** $x^{-1/2}$	**9** $\dfrac{1}{x\sqrt{x}}$	**10** $\dfrac{1}{\sqrt[3]{x}}$

Multiples, sums and differences

Whilst we are now able to differentiate at sight any function of the form x^n, we shall frequently need to find the derivatives of multiples and combinations of such functions.

Here again, the rules of procedure should be apparent from the above exercises. For instance, in exercise 3.2 no. **2** we found (or should have found!) that

a) $f(x) = 2x^3$
$\Rightarrow f'(x) = 6x^2$

b) $f(x) = 3x^4 + 12x$
$\Rightarrow f'(x) = 12x^3 + 12$

So it appears that
a) the derivative of a constant multiple of a function is this multiple of its derivative; and
b) the derivative of the sum of two functions is this sum of its derivatives; or, more briefly, that
a) $(kf)' = kf'$ (if k is a constant)
b) $(f + g)' = f' + g'$
These can very easily be shown as follows:

a) $\{kf(x)\}'$ $\quad = \lim_{h \to 0} \dfrac{kf(x + h) - kf(x)}{h}$

$\quad = k \lim_{h \to 0} \dfrac{f(x + h) - f(x)}{h}$

$\quad = kf'(x)$

So $\quad\quad\quad (kf)' = kf'$

b) $\{f(x) + g(x)\}' = \lim_{h \to 0} \dfrac{\{f(x + h) + g(x + h)\} - \{f(x) + g(x)\}}{h}$

$\quad = \lim_{h \to 0} \dfrac{\{f(x + h) - f(x)\} + \{g(x + h) - g(x)\}}{h}$

$\quad = \lim_{h \to 0} \dfrac{f(x + h) - f(x)}{h} + \lim_{h \to 0} \dfrac{g(x + h) - g(x)}{h}$

$\quad = f'(x) + g'(x)$

So $\quad\quad\quad (f + g)' = f' + g'$

These may seem to the reader very laborious proofs of results which are self-evident. If so, what about products and quotients? Is it always true, only sometimes true, or never true, that

$\{f(x)g(x)\}' = f'(x)g'(x) \quad \text{and} \quad \left\{\dfrac{f(x)}{g(x)}\right\}' = \dfrac{f'(x)}{g'(x)}?$

You could start by investigating particular cases, e.g. $f(x) = x^3$, $g(x) = x^2$.

But perhaps you have chosen an exceptional case. Try some more.

The results $(kf)' = kf'$ (1)

$(f + g)' = f' + g'$ (2)

can now be used. When, for instance, we differentiate

$f(x) = 2x^3 - 3x$ to obtain $f'(x) = 6x^2 - 3$

we use both the above results (and the first one twice).

Exercise 3.3b

Using the above rules, find $f'(x)$ where $f(x)$ is

1   $x^4 + x^3$ 2 $x^5 - x^2$ 3 $x^6 + x^2$ 4 $x^4 - x$

5   $3x^4 + 5x^2$ 6 $2x^6 - 4x^3$ 7 $3x^2 + 4x^3$ 8 $6x^2 - 3x$

9   $2x + \dfrac{1}{x}$ 10 $3x^2 + \dfrac{2}{x^2}$ 11 $5x^2 - \dfrac{2}{x}$ 12 $\dfrac{2 - 3x}{x^2}$

13   $2\sqrt{x}$ 14 $\dfrac{4}{\sqrt{x}}$ 15 $\dfrac{3 + 2x}{\sqrt{x}}$ 16 $\dfrac{2 + x^2}{\sqrt{x}}$

3.4 Higher derivatives

Just as the derivative of f is called f', so the derivative of f' is called f" (the *second derivative* of f); and so on for higher derivatives.

For example $f(x) = 2x^3 - 3x$
$\Rightarrow$ $f'(x) = 6x^2 - 3$
$\Rightarrow$ $f''(x) = 12x$
$\Rightarrow$ $f'''(x) = 12$
$\Rightarrow$ $f''''(x) = 0$, etc.

and the original curve, with its first, second, third and fourth *derived curves* are

$f(x) = 2x^3 - 3x$

whose gradient is

$f'(x) = 6x^2 - 3$

whose gradient is

$$f''(x) = 12x$$

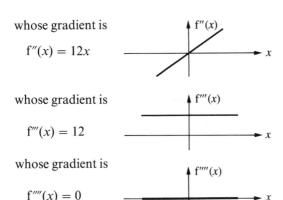

whose gradient is

$$f'''(x) = 12$$

whose gradient is

$$f''''(x) = 0$$

Exercise 3.4a

Find $f'(x)$ and $f''(x)$ where $f(x)$ is

1 **a)** $5x^3$ **b)** $x^4 + x$ **c)** $x^4 - x^2 + 2$ **d)** $\frac{1}{2}x^{10}$

2 **a)** x^{-6} **b)** $14x^{-3}$ **c)** $\dfrac{1}{x^2}$ **d)** $\dfrac{5}{x^3}$

3 **a)** $x^{1/3}$ **b)** $\sqrt{x}$ **c)** $2x\sqrt{x}$ **d)** $\dfrac{1}{\sqrt{x}}$

4 **a)** $(x^3 + 1)^2$ **b)** $\dfrac{2x^2 + 3}{x}$ **c)** $\dfrac{x + 1}{2\sqrt{x}}$ **d)** $\{\sqrt{x+1}\}^2$

* Concavity and convexity: points of inflection

It is readily seen that the sign of $f''(x)$ determines whether $f'(x)$ is increasing or decreasing, i.e. whether the curve of $f(x)$ is concave upwards or downwards (i.e. convex downwards or upwards).

Alternatively, we can think of x increasing and the first stretch as a bend taken with a 'right-hand lock', followed by the second stretch as a bend taken with a left-hand lock. The point where the lock changes (and the curve crosses its tangent, and $f''(x)$ changes sign) is called a *point of inflection*.

$f''(x) < 0 \Rightarrow f'(x)$ is decreasing

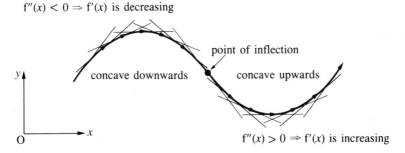

Example

For what values of x is the function $f(x) = x^3 - 3x^2 + 4$
a) increasing, decreasing, stationary;
b) concave upwards, concave downwards, at a point of inflection?

$$f(x) = x^3 - 3x^2 + 4$$

$$\Rightarrow \quad f'(x) = 3x^2 - 6x = 3x(x - 2)$$

and the sign of $f'(x)$ is given by

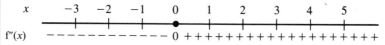

x	-3	-2	-1	0	1	2	3	4	5

$f'(x)$ $+ + + + + + + + + + + +$ 0 $- - - -$ 0 $+ + + + + + + + + + + +$

Hence $f(x)$ is increasing   if $x < 0$ or $x > 2$
          decreasing   if $0 < x < 2$
          stationary   if $x = 0$ or 2

Also $f''(x) = 6x - 6 = 6(x - 1)$ and the sign of $f''(x)$ is given by

x	-3	-2	-1	0	1	2	3	4	5

$f''(x)$ $- - - - - - - - - - - - -$ 0 $+ + + + + + + + + + + + + + + + + + +$

Hence $f(x)$ is concave upwards   when $x > 1$
          concave downwards   when $x < 1$
          at a point of inflection when $x = 1$

These can all be illustrated in the graph of $f(x)$:

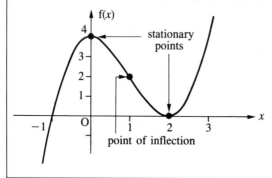

Exercise 3.4b

In each of the following cases
a) find $f'(x)$ and $f''(x)$;
and state the values of x for which the function is

b) increasing, decreasing, stationary;
c) concave up, concave down, at a point of inflection;
d) sketch the graph of f(x).

1 $x^2 - 2x$ **2** $x^3 - 3x$ **3** $3x^2 - 2x^3$

4 $x^4 - 8x^2$ **5** $x^3 - 3x^2 - 9x + 11$ **6** $x + \dfrac{1}{x}$

3.5 Alternative notation: tangents and normals

Newton and Leibniz, when they made their independent discoveries of the calculus in the 1660s, used different notations. It could hardly have been otherwise for men who were saying something for the first time, but ever since their day the history of the subject has been one of varied notation. Sometimes this has been unfortunate, as in the eighteenth century when English mathematics stagnated because Newton's followers became hide-bound to his language and symbols. But more usually it has been a strength: different situations call for different ways of saying the same thing. It was perhaps felt in section 3.3 when we were proving $(f + g)' = f' + g'$ that our language was becoming rather laborious, and we shall soon see how, with a different notation, such proofs can be expressed more simply. But first we shall return to our previous example:

Suppose that we wish to find the gradient at any point of the curve $y = x^3$.

As before, we must find the effect of a small increase in x. We write δx (pronounced 'delta x' and simply meaning 'a bit more x') to indicate this small increase in x, and δy ('a bit more y') to indicate the corresponding increase in y.

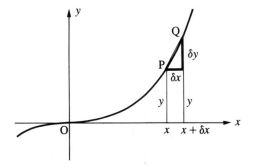

To find the gradient of the tangent at P, we first notice that at P

$$y = x^3$$

and at Q $y + \delta y = (x + \delta x)^3$

$$= x^3 + 3x^2 \, \delta x + 3x(\delta x)^2 + (\delta x)^3$$

$\Rightarrow$ $\delta y = 3x^2 \delta x + 3x(\delta x)^2 + (\delta x)^3$

$\Rightarrow$ gradient of PQ $= \dfrac{\delta y}{\delta x} = 3x^2 + 3x \, \delta x + (\delta x)^2$

So as $\delta x \to 0$, $\dfrac{\delta y}{\delta x} \to 3x^2$, which is clearly the derivative of y and the gradient of

the tangent. As this derivative is the limit of $\dfrac{\delta y}{\delta x}$, it is convenient to denote it by

the symbol $\dfrac{dy}{dx}$.

So $\dfrac{dy}{dx} = \lim_{\delta x \to 0} \dfrac{\delta y}{\delta x} = 3x^2$

Two points must be emphasised:
a) that δ is *not* a quantity, but an adjective meaning 'a bit more', and $\dfrac{\delta y}{\delta x}$ is

simply the ratio $\dfrac{\text{'a bit more } y\text{'}}{\text{'a bit more } x\text{'}}$;

b) that, unlike δx and δy, dx and dy do not (yet) have any meaning by themselves, any more than the two dots in the symbol $\div$.

$\dfrac{dy}{dx}$ is simply a shorthand way of writing $\lim_{\delta x \to 0} \dfrac{\delta y}{\delta x}$

Comparison of notations

To make the position quite clear we shall carry out a differentiation from first principles using the two notations in parallel.

Given $f(x) = 3x^2$, find $f'(x)$ Given $y = 3x^2$, find $\dfrac{dy}{dx}$

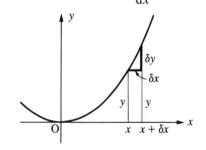

$$f(x) = 3x^2 \qquad\qquad y = 3x^2$$

$$\Rightarrow \quad f(x + h) = 3(x + h)^2 \qquad\qquad \Rightarrow y + \delta y = 3(x + \delta x)^2$$

$$\Rightarrow \quad f(x + h) - f(x) = 6hx + 3h^2 \qquad\qquad \Rightarrow \quad \delta y = 6x\delta x + 3(\delta x)^2$$

$$\Rightarrow \quad \frac{f(x + h) - f(x)}{h} = 6x + 3h \qquad\qquad \Rightarrow \quad \frac{\delta y}{\delta x} = 6x + 3\delta x$$

$$\text{But} \quad f'(x) = \lim_{h \to 0} \frac{f(x + h) - f(x)}{h} \qquad\qquad \text{But} \quad \frac{dy}{dx} = \lim_{\delta x \to 0} \frac{\delta y}{\delta x}$$

$$\Rightarrow \quad f'(x) = 6x \qquad\qquad \Rightarrow \quad \frac{dy}{dx} = 6x$$

It is also convenient to have a symbol to denote the *operation of differentiation*. Just as we use $\sqrt{}$ for the operation of taking the square root of a number, so we use the symbol $\dfrac{d}{dx}$ to denote differentiation, and write

$$\frac{d}{dx}(x^4 + 3x^2) = 4x^3 + 6x$$

and

$$\frac{d}{dx}(y) = \frac{dy}{dx}$$

Further, we can use this notation to denote higher derivatives:

$$\frac{d}{dx}\left(\frac{dy}{dx}\right) = \frac{d^2}{dx^2}(y) \quad \text{or} \quad \frac{d^2y}{dx^2}$$

and

$$\frac{d}{dx}\left(\frac{d^2y}{dx^2}\right) = \frac{d^3}{dx^3}(y) \quad \text{or} \quad \frac{d^3y}{dx^3}$$

So

$$y = x^4 + 3x^2$$

$$\Rightarrow \quad \frac{dy}{dx} = \frac{d}{dx}(x^4 + 3x^2) = 4x^3 + 6x$$

$$\frac{d^2y}{dx^2} = \frac{d^2}{dx^2}(x^4 + 3x^2) = 12x^2 + 6$$

$$\frac{d^3y}{dx^3} = \frac{d^3}{dx^3}(x^4 + 3x^2) = 24x$$

Tangents and normals

Now that we are able to find the gradient of the tangent to a curve, it is possible also to find its equation. Furthermore, we can similarly find the equation of the *normal*, i.e. the line through a point of the curve which is perpendicular to the tangent at this point.

Example

Find the equation of the tangent and normal to the curve $y = x^2 + 2$ at the point $(1, 3)$.

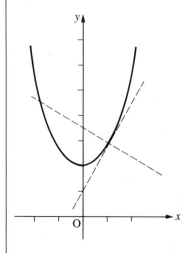

$$y = x^2 + 2$$

$$\Rightarrow \quad \frac{dy}{dx} = 2x = 2, \quad \text{when } x = 1$$

So gradient of tangent at $(1, 3)$ is $+2$ and the gradient of the normal is $-\frac{1}{2}$. Hence the equation of the tangent is

$$\frac{y - 3}{x - 1} = 2 \quad \Rightarrow \quad y = 2x + 1$$

and the equation of the normal is

$$\frac{y - 3}{x - 1} = -\frac{1}{2} \quad \Rightarrow \quad x + 2y - 7 = 0$$

Exercise 3.5

1   Use the notation δx, δy to differentiate from first principles:

  a) $y = 2x^3$ **b)** $y = \dfrac{2}{x}$

2   Use standard methods to find $\dfrac{dy}{dx}$ and $\dfrac{d^2y}{dx^2}$ if

a) $y = x^3 + 3x^2$ b) $y = 4x$ c) $y = \dfrac{1}{x^2}$ d) $y = \dfrac{1}{\sqrt{x}}$

3   Use the operators $\dfrac{d}{dx}$ and $\dfrac{d^2}{dx^2}$ to state the first and second derivatives of

a) x^4 b) $3x^2$ c) $2\sqrt{x}$ d) $\dfrac{1}{x}$

4   Prove that the following curves go through the given points and in each case find:
   i) the gradient of the tangent at the point;
   ii) the angle it makes with the positive x-axis;
   iii the equation of the tangent;
   iv) the equation of the normal.

a) $y = x^2$, $(2, 4)$
b) $y = x^3 - 2x$, $(1, -1)$
c) $y = x - x^2$, $(1, 0)$
d) $y = \dfrac{2}{x}$, $(-2, -1)$

5   Find the equation of the tangent to the curve $y = x^3 - 11x$ at the point $(1, -10)$ and also the equations of the tangents which are parallel to the line $x - y = 0$.   (OC)

6   Find the equation of the tangent at the point $(2, 2)$ on the curve $y = x^3 - 3x$ and the coordinates of the point at which this tangent meets the curve again.   (OC)

7   A woman throws a stone horizontally with velocity $25\,\mathrm{m\,s^{-1}}$ from a point on top of a cliff $500\,\mathrm{m}$ above sea-level.
    The equation of the path of the stone, referred to axes through its point of projection, is $y = -\frac{1}{125}x^2$.
    How far out to sea does it reach and at what angle does it enter the water?

8   The muzzle-velocity of a shell is $300\,\mathrm{m\,s^{-1}}$ and when fired at an angle $45°$ with the horizontal its path is given by the equation

$$y = x - \frac{x^2}{9000}$$

Find:
a) the range of the shell;
b) the angles which its direction makes with the horizontal when it has travelled horizontal distances $3000\,\mathrm{m}$, $4500\,\mathrm{m}$, and $6000\,\mathrm{m}$.

9   Sketch together the curves $y = x^2$ and $y = \dfrac{8}{x}$. Find:

   a) the coordinates of their point of intersection;
   b) their angle of intersection.

3.6 Kinematics

In section 3.1 we introduced the process of differentiation by considering the velocity of a falling stone. The study of the position of a moving body, and how it changes, is called *kinematics*. This forms part of *dynamics*, which also includes study of the forces acting upon the body and their relationship with its movements.

For the present, we shall ignore the *causes* of motion and confine ourselves to its *description* in the case of a small body, or *particle*, which is moving in a straight line.

Displacement, velocity, and acceleration

Suppose that O is a fixed point of a straight line and that a point P is moving so that at time t it has a displacement x (positive or negative) from O:

Its *velocity* v is the rate of change of its displacement, and its *acceleration* a is the rate of change of its velocity.

So $v = \dfrac{dx}{dt}$

and $a = \dfrac{dv}{dt}\left(= \dfrac{d^2x}{dt^2}\right)$

At a particular time t, x tells us where the particle is,
v how it is moving,
and a how its movement is varying.

Units

As the unit of distance is 1 m and the unit of time is 1 s, the unit of velocity is a rate of change of displacement of 1 m in 1 s = 1 m s^{-1}, and the unit of acceleration is a rate of change of velocity of 1 m s^{-1} in 1 s = 1 m s^{-2}.

At the surface of the Earth, and ignoring air resistance, the acceleration of an object falling under gravity is approximately 9.81 m s^{-2}; on the Moon it is 1.62 m s^{-2}. For simplicity, however, the acceleration due to gravity at the Earth's surface will normally be taken as 10 m s^{-2}.

Example

In the simplest case of a stone dropped over the edge of a cliff, let us indicate displacement, velocity and acceleration (in a downwards direction) after time t by the letters y, v and a.

$y = 5t^2$

$\Rightarrow \quad v = \dfrac{\mathrm{d}y}{\mathrm{d}t} = 10t$

$\Rightarrow \quad a = \dfrac{\mathrm{d}v}{\mathrm{d}t} = 10$

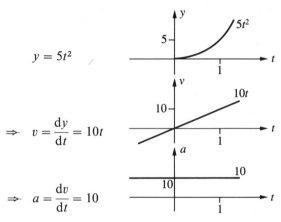

Exercise 3.6

1 The coordinate x m of a moving point A at time t s is given by the formula

$$x = t^3 - 2t^2 + 3t - 4$$

Find the velocity and the acceleration of A at the instant $t = 4$.   (OC)

2 A particle P is travelling along a straight line so that its distance from the starting point, x m, after time t s is given by the equation $x = 8t - \frac{1}{2}t^2$. Calculate the velocity and acceleration of P after 3 s, and find the distance travelled by P when it first comes to rest.

3 A particle is moving in a straight line and its distance s m from a fixed point in the line after t s is given by the equation $s = 12t - 15t^2 + 4t^3$. Find:
a) the velocity and acceleration of the particle after 3 s;
b) the distance travelled between the two times when the velocity is instantaneously zero.   (OC)

4 The velocity of a particle travelling in a straight line is given by $v = t^2 - t - 12$, when v is measured in m s^{-1} and t is the time in seconds measured from a definite instant of the motion.

Find the time that elapses before the particle is instantaneously at rest, and find its acceleration when $t = 6$.   (SMP)

5   A particle moves along the x-axis, its position when t s have elapsed being given by $x = 27 - 36t + 12t^2 - t^3$. Show that for the first 2 s the particle moves in the negative direction, and for the next 4 s it moves in the positive direction.

Find also the accelerations at the instants when the particle is at rest.   (OC)

6   A point moves on the x-axis and at time t its position is given by

$$x = t(t^2 - 6t + 12)$$

Show that its velocity at the origin is 12 and that the velocity decreases to zero at the points where $x = 8$ and thereafter continues to increase.   (OC)

7   If an astronaut on the Moon catapults a stone vertically upwards at a speed of $24 \, \text{m s}^{-1}$, its height y m after t s is given by

$$y = 24t - \frac{4}{5}t^2$$

a) How high does it go?
b) How long does it take to reach its highest point?
c) What are its velocity v and acceleration a after time t?
d) Sketch the graphs of y, v, and a plotted against t.
e) Is there any meaning to the sections of i) the v–t curve, and ii) the y–t curve, when they drop below the t-axis?

3.7 Reverse process

We now ask the reverse question: given $f'(x)$, can we find $f(x)$?

Example 1

Given $f'(x) = 12x - 3x^2$, find $f(x)$.

In this case it is immediately clear that

$$f(x) = 6x^2 - x^3$$

is a possible answer.

But a moment's reflection reminds us that, as the derivative of any constant is zero,

$$6x^2 - x^3 + 15, \quad 6x^2 - x^3 - 10, \quad \text{etc.}$$

are equally possible. Indeed $6x^2 - x^3 + c$, where c is any constant (and therefore sometimes called an 'arbitrary constant') is also such a function.

Looking at this graphically, we readily see the significance of the arbitrary constant c: the original instruction was simply to find a curve whose gradient is $12x - 3x^2$. The first such curve to be found was $6x^2 - x^3$ and the relationship between these two curves can be represented:

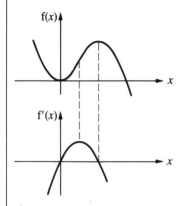

$6x^2 - x^3$

is a function
whose gradient is

$12x - 3x^2$

But the gradient of $6x^2 - x^3$ is unaltered by any vertical displacement of the curves, so *any* function

$$f(x) = 6x^2 - x^3 + c$$

is equally possible.

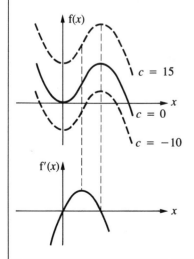

$6x^2 - x^3 + c$

are all functions
whose gradients
are

$12x - 3x^2$

In the last section we learnt to discover the behaviour of velocity and acceleration from that of displacement. We can now see how to reverse this process.

Example 2

A particle leaves the point O with a velocity $4\,\text{m s}^{-1}$ and its acceleration a after time t is given by

$$a = 5 - t$$

Find:
a) its velocity v after time t;
b) its displacement x after time t;
c) the distance travelled in its third second of motion.

$$\frac{\mathrm{d}v}{\mathrm{d}t} = a = 5 - t$$

$$v = 5t - \tfrac{1}{2}t^2 + c, \quad \text{where } c \text{ is a constant}$$

When $t = 0, v = 4$

So $\qquad\qquad 4 = 0 - 0 + c \ \Rightarrow \ c = 4$

$\Rightarrow \qquad\qquad v = 5t - \tfrac{1}{2}t^2 + 4$

So $\qquad\qquad \dfrac{\mathrm{d}x}{\mathrm{d}t} = 5t - \tfrac{1}{2}t^2 + 4$

$\Rightarrow \qquad\qquad x = \tfrac{5}{2}t^2 - \tfrac{1}{6}t^3 + 4t + d, \quad \text{where } d \text{ is a constant}$

When $t = 0, x = 0$

So $\qquad\qquad 0 = 0 - 0 + 0 + d \ \Rightarrow \ d = 0$

$\Rightarrow \qquad\qquad x = \tfrac{5}{2}t^2 - \tfrac{1}{6}t^3 + 4t$

When $t = 2, x = \tfrac{5}{2} \times 2^2 - \tfrac{1}{6} \times 2^3 + 4 \times 2$

$\qquad\qquad\qquad = 10 - \tfrac{4}{3} + 8 = 16\tfrac{2}{3}$

When $t = 3, x = \tfrac{5}{2} \times 3^2 - \tfrac{1}{6} \times 3^3 + 4 \times 3$

$\qquad\qquad\qquad = 22\tfrac{1}{2} - 4\tfrac{1}{2} + 12 = 30$

So in its third second the particle travels

$$30 - 16\tfrac{2}{3} = 13\tfrac{1}{3}\,\text{m}$$

Exercise 3.7

1 Find $f(x)$ if it is known that
  a) $f'(x) = 2x^3$ **b)** $f'(x) = 1 - x$
  c) $f'(x) = 3x^2 + 2$ and $f(0) = 4$ **d)** $f'(x) = \dfrac{1}{x^2}$ and $f(2) = 1$

2 Find the equation of a curve
  a) whose gradient is $4x - 5$;
  b) whose gradient is $4x - 5$, and which goes through the point $(0, 4)$;
  c) whose gradient is 3 and which goes through the point $(1, 2)$;
  d) whose gradient is $-x$ and which goes through the point $(1, 1)$.

3 Find the displacement x m of a particle at time t s if we know that its velocity v m s^{-1} is given by
  a) $v = 10t$
  b) $v = 8t^3 - 6t^2$
  c) $v = 3t^2 + 4t$, and $x = 2$ when $t = 0$
  d) $v = 1 - 2t$, and $x = 3$ when $t = 1$

4 Find the velocity v m s^{-1} and displacement x m of a particle at time t s if we know that its acceleration a m s^{-2} is given by
  a) $a = 6t - 8$, and when $t = 0$, $x = 4$ and $v = 6$
  b) $a = -10$, and when $t = 1$, $x = 5$ and $v = 2$

5 A particle starts from rest at O and moves along a straight line. After t s its velocity is v m s^{-1}, where $v = 3t^2 - t^3$. Show that the particle is momentarily at rest after 3 s and find its distance from O at this time.
  Find also the maximum velocity of the particle during the first 3 s of the motion.   (OC)

6 The velocity of a particle travelling in a straight line is given by $v = t^2 - t - 6$, where v is measured in m s^{-1} and t is the time in s, reckoned from a definite instant of the motion. Find the time that elapses before the particle is instantaneously at rest, and find the acceleration when $t = 5$.
  Find also the distance between its position when $t = 3$ and its position when $t = 9$.   (OC)

7 A train, while travelling from its start at a point A to its stop at a point B, is moving, t hours after it leaves A, with a speed of $(240t - 150t^2)$ km/h. Find:
  a) the time taken in going from A to B;
  b) the distance from A to B;
  c) the greatest speed reached during the journey.   (OC)

8 A particle starts from rest at a point 6 m from O, and moves in a straight line away from O with a velocity v m s^{-1} at time t s given by $v = t - \frac{1}{18}t^2$.

Find:
a) its acceleration and distance from O, each in terms of t;
b) the time at which it begins to return, and the time at which it again reaches its starting-point. (OC)

9 A particle travels in a straight line, being at a point A at zero time. Its velocity in $\mathrm{m\,s}^{-1}$ at time t s is given by the expression $6t^2 - 48t + 42$. Find expressions at time t s for
a) its acceleration;
b) its distance from A.

Find the times at which its velocity is zero. Show also that it passes through A twice after zero time, and find (to one place of decimals) the length of time before its first return. (OC)

10 In the first 20 seconds of its motion down a runway, an aircraft is subject to an acceleration given by $(10 - \frac{1}{2}t)\,\mathrm{m\,s}^{-2}$. Find its velocity at the end of this time, and how far it has travelled, to the nearest metre. (MEI)

11 The acceleration of a particle t s after starting from rest is $(2t - 1)\,\mathrm{m\,s}^{-2}$. Prove that the particle returns to the starting-point after $1\frac{1}{2}$ s, and find the distance of the particle from the starting-point after a further $1\frac{1}{2}$ s.
    Find also at what time after starting the particle attains a velocity of $20\,\mathrm{m\,s}^{-1}$. (OC)

12 The acceleration of a car t s after starting from rest is

$$\frac{75 + 10t - t^2}{20}\,\mathrm{m\,s}^{-2}$$

until the instant when this expression vanishes. After this instant the speed of the car remains constant. Find:
a) the maximum acceleration;
b) the time taken to attain the greatest speed;
c) the greatest speed attained. (OC)

3.8 Stationary values: maxima and minima

In section 3.2 we said that f(x) was stationary whenever f$'(x) = 0$. We now investigate the various types of stationary values.
    It is seen from the graph of f(x) that stationary values are of three types.
a) when the function is *less on both* sides of its stationary value. This is called a *maximum* (as at A, A$'$) and it is clear that f$'(x)$ must be decreasing from positive values through zero to negative values.

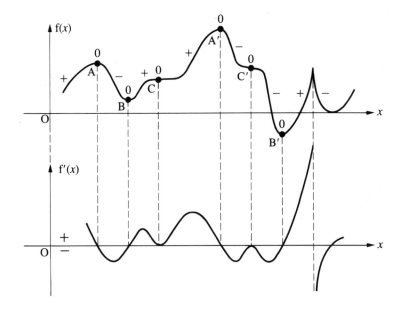

b) when the function is *greater* on *both* sides of its stationary value. This is called a *minimum* (as at **B, B′**) and it is clear that $f'(x)$ must be increasing from negative values through zero to positive values.

c) when the function is greater on one side of its stationary value, but less on the other side. This is a *point of inflection* (as at **C, C′**) and it is clear that $f'(x)$ must have the same sign (whether positive or negative) on each side of the point where it is zero.

Example 1

Investigate the stationary values of $f(x) = x^3 - 3x$

$$f(x) = x^3 - 3x$$

$$\Rightarrow f'(x) = 3x^2 - 3 = 3(x^2 - 1)$$

At stationary values, $f'(x) = 0 \Rightarrow x^2 - 1 = 0 \Rightarrow x = \pm 1$

So $f(x)$ is stationary when $x = \pm 1$ and its two stationary values are

$$f(+1) = 1 - 3 = -2 \quad \text{and} \quad f(-1) = -1 + 3 = +2$$

So the graph of $f(x)$ is stationary at $(+1, -2)$ and $(-1, +2)$.

   We can now investigate in more detail by finding how $f'(x)$ varies as x increases through the values $+1$ and -1.

   As x increases through $+1$,

$f'(x) = 3(x^2 - 1)$ changes from $-$ to 0 to $+$

$\Rightarrow f(+1) = -2$ is a *minimum*.

As x increases through -1,

$f'(x) = 3(x^2 - 1)$ changes from $+$ to 0 to $-$

$\Rightarrow f(-1) = +2$ is a *maximum*.

So the graph of $f(x) = x^3 - 3x$ is:

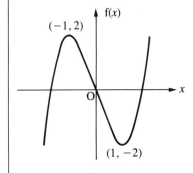

Use of f″(x) at stationary values

If $f'(a) = 0$, i.e. $f(a)$ is a stationary value of $f(x)$, we can often decide whether it is a maximum or a minimum value simply by finding the value of $f''(a)$.

Case 1 $f''(a) < 0$

$f''(a) < 0 \Rightarrow f'(x)$ is *decreasing* at $x = a$.

But $f'(a) = 0$. So, as x increases through a, $f'(x)$ must change from $+$ to 0 to $-$. So $f''(a) < 0 \Rightarrow f(a)$ is a *maximum*.

Case 2 $f''(a) > 0$

$f''(a) > 0 \Rightarrow f'(x)$ is increasing at $x = a$.

But $f'(a) = 0$. So, as x increases through a, $f'(x)$ must change from $-$ to 0 to $+$. So $f''(a) > 0 \Rightarrow f(a)$ is a *minimum*.

In example 1

$$f(x) = x^3 - 3x$$

$$f'(x) = 3x^2 - 3$$

$$f''(x) = 6x \quad \Rightarrow \quad f''(+1) = +6 \quad \text{and} \quad f''(-1) = -6$$

Now $f(x)$ is stationary when $x = \pm 1$,

$$f''(+1) = +6 \quad \Rightarrow \quad f(+1) = -2 \text{ is a minimum}$$

and $\quad f''(-1) = -6 \quad \Rightarrow \quad f(-1) = +2 \text{ is a maximum}$

which confirm what we previously discovered.

Case 3 $f''(a) = 0$

What happens if $f''(a) = 0$?

Sketch the graphs of $f(x)$ and calculate $f'(0)$ and $f''(0)$ when **a)** $f(x) = x^3$ **b)** $f(x) = x^4$ **c)** $f(x) = -x^4$. So if $f''(a) = 0$ it is clear that $f(x)$ might have either a point of inflection, a minimum or a maximum.

Summary $f'(a) = 0$

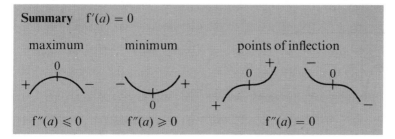

maximum minimum   points of inflection

$f''(a) \leqslant 0$ $f''(a) \geqslant 0$ $f''(a) = 0$

Example 2

Investigate the stationary values of

$$f(x) = x + \frac{4}{x}$$

It follows that $\quad f'(x) = 1 - \dfrac{4}{x^2}$

and $\qquad\qquad f''(x) = \dfrac{8}{x^3}$

At stationary values $\quad f'(x) = 0$

$$\Rightarrow \qquad\qquad 1 - \frac{4}{x^2} = 0$$

$$\Rightarrow \qquad\qquad x = \pm 2$$

Now $f''(-2) = \dfrac{8}{(-2)^3} = -1$

$\Rightarrow$ $f(-2) = -4$ is a maximum value

and $f''(2) = \dfrac{8}{2^3} = +1$

$\Rightarrow$ $f(2) = +4$ is a minimum value

At first it seems surprising that the maximum value should be less than the minimum value. But it is clear that $f(-2) = -4$ is a maximum because it is greater than other values *in the immediate neighbourhood* of $x = -2$; and similarly $f(2) = +4$ is less than other values *in the immediate neighbourhood* of $x = 2$. Their relationship is apparent from the graph of $f(x)$:

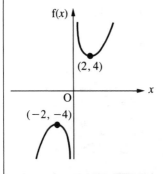

Example 3

Investigate the stationary points of

$$y = 3x^4 - 4x^3$$

and sketch its graph.

$$\dfrac{dy}{dx} = 12x^3 - 12x^2 = 12x^2(x - 1)$$

and $\dfrac{d^2y}{dx^2} = 36x^2 - 24x = 12x(3x - 2)$

Now $\dfrac{dy}{dx} = 0 \;\Rightarrow\; x = 0$ or 1

So y is stationary at $x = 0$, $y = 0$

and at $x = 1$, $y = -1$

a) At $x = 0, \dfrac{d^2y}{dx^2} = 0$, which is indecisive.

But as x increases through $x = 0$,

$$\frac{dy}{dx} = 12x^2(x - 1), \quad \text{and changes from } - \text{ to } 0 \text{ to } -$$

So at $(0, 0)$ the curve has a point of inflection.

b) At $x = 1, \dfrac{d^2y}{dx^2} = 12 \times 1(3 \times 1 - 2) = 12$

So at $(1, -1)$ the curve has a minimum.

Also $\quad y = 0 \quad \Rightarrow \quad x = 0 \quad$ or $\quad \dfrac{4}{3}$, and as $x \to \pm \infty$, $\quad y \to + \infty$

Summarising these, we obtain the graph

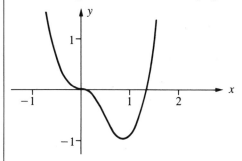

Example 4

When a ship is travelling at a speed of v km h^{-1} its rate of consumption of fuel in tonne h^{-1} is approximately $20 + 0.001\, v^3$.

Find an expression for the total fuel used on a voyage of 5000 km at speed v, and proceed to find the speed for greatest fuel economy and the amount of fuel then used.

Time taken over journey $= \dfrac{5000}{v}$ h

So if total amount of fuel used is A tonne

$$A = (20 + 0.001\, v^3) \times \frac{5000}{v}$$

$$= \frac{100\,000}{v} + 5v^2$$

$$\Rightarrow \quad \frac{\mathrm{d}A}{\mathrm{d}v} = -\frac{100\,000}{v^2} + 10\,v$$

$$\text{and} \quad \frac{\mathrm{d}^2 A}{\mathrm{d}v^2} = +\frac{200\,000}{v^3} + 10$$

For A to be a minimum, $\dfrac{\mathrm{d}A}{\mathrm{d}v} = 0$

$$\Rightarrow \qquad 10v = \frac{100\,000}{v^2}$$

$$\Rightarrow \qquad v^3 = 10\,000$$

$$\Rightarrow \qquad v \approx 21.54$$

Also $\dfrac{\mathrm{d}^2 A}{\mathrm{d}v^2} > 0$, so A is definitely a minimum.

When $v = 21.54, \quad A \approx \dfrac{100\,000}{21.54} + 5(21.54)^2$

$$\approx 4640 + 2320 = 6960$$

So the most economical speed for the voyage is approximately $21.5 \, \mathrm{km \, h^{-1}}$ when the total fuel consumption will be 6960 tonne.

Example 5

If the length, cross-sectional radius and volume of a cylindrical block are l, r, V respectively, prove that its surface area is

$$S = \frac{2V}{r} + 2\pi r^2$$

Hence find the ratio of length to radius if a cylindrical block of given volume has minimum surface area.

Area of curved surface $= 2\pi r l$

Area of two ends $= 2\pi r^2$

$$\Rightarrow \qquad S = 2\pi r l + 2\pi r^2$$

This expression for S is a function of two variables r and l, so we cannot yet find its derivative.

But r and l are connected by the fact that $\pi r^2 l = V$, which is a constant.

So $S = \dfrac{2V}{r} + 2\pi r^2$, which is a function of a single variable r.

Hence $\dfrac{dS}{dr} = -\dfrac{2V}{r^2} + 4\pi r$

and $\dfrac{d^2S}{dr^2} = +\dfrac{4V}{r^3} + 4\pi$

Now S is stationary $\Rightarrow$ $\dfrac{dS}{dr} = 0$

$\Rightarrow \quad 4\pi r = \dfrac{2V}{r^2}$

$\Rightarrow \quad 2\pi r^3 = V$

Also $\dfrac{d^2S}{dr^2} > 0$

So S is a minimum when $2\pi r^3 = V$

$\Rightarrow \quad 2\pi r^3 = \pi r^2 l$

$\Rightarrow \quad \dfrac{r}{l} = \dfrac{1}{2}$

So for minimum surface area we require the diameter to be equal to the length.

Exercise 3.8

1 Find the stationary values of the following functions. In each case find whether the function has a maximum value, a minimum value or a point of inflection, and sketch the graph of the function.
  a) $f(x) = 4x^2 - 8x - 5$ **b)** $f(x) = 3 + 2x - x^2$
  c) $f(x) = x^3 - 3x$ **d)** $f(x) = x^3 + 3x$

2 Investigate the stationary points of the following curves, deciding in each case whether the point is a maximum, a minimum or a point of inflection. Find also any points of oblique inflection and sketch the graph of the function.
  a) $y = x + \dfrac{4}{x}$ **b)** $y = x^2 + \dfrac{2}{x}$
  c) $y = 4x^5 - 5x^4$ **d)** $y = 6x^4 - 4x^6$

3 **a)** Find the coordinates of points on the curve
  $$y = 2x^3 + 3x^2 - 12x + 6$$

at which y has a stationary value and state in each case whether the value of y is a maximum or a minimum.

b) Find the maximum and minimum values of

$$f(x) = 2x^3 - 3x^2 - 12x + 21$$

Hence, or otherwise, show that the equation $f(x) = 0$ has only one real root. Sketch the curve $y = f(x)$.

c) Given that the function $y = ax^3 + bx^2$, where a and b are constants, has a stationary value -4 when $x = 2$, find the values of a and b.   (OC)

4 At all points on a certain curve it is known that $\dfrac{d^2y}{dx^2} = 6x - 2$. If

$\dfrac{dy}{dx} = -1$ and $y = 1$ when $x = 0$,

a) find the equation of the curve;
b) find its stationary points, stating whether y has a maximum or a minimum value.   (OC)

5 A farmer erects a fence along three sides of a rectangle in order to make a sheepfold; the fourth side of the rectangle is provided by a hedge already in existence. Find the maximum area of the enclosure thus made if the total length of the fence is to be 80 m.   (OC)

6 A square sheet of metal of side 12 cm has four equal square portions removed at the corners and the sides are then turned up to form an open rectangular box. Prove that, when the box has maximum volume, its depth is 2 cm.   (OC)

7 The bottom of a rectangular tank is a square of side x m and the tank is open at the top. It is designed to hold 4 m³ of liquid. Express in terms of x the total area of the bottom and the four sides of the tank. Find the value of x for which this area is a minimum.   (OC)

8 A closed rectangular box is made of very thin sheet metal, and its length is three times its width. If the volume of the box is 288 cm³, show that its surface area is equal to

$$\left(\frac{768}{x} + 6x^2\right) cm^2$$

where x cm is the width of the box.

  Find by differentiation the dimensions of the box of least surface area.
    (OC)

9 A rectangular box without a lid is made of cardboard of negligible thickness. The sides of the base are $2x$ cm and $3x$ cm, and the height is y cm. If the total area of the cardboard is 200 cm², prove that

$$y = \frac{20}{x} - \frac{3x}{5}.$$

Find the dimensions of the box when its volume is a maximum.   (oc)

10 The despatch department of a certain firm limits parcels to those whose length and girth are together less than 2 m. If the length of a parcel is x m, show that its volume V m^3, is given by

$$16V = x(2 - x)^2$$

Find the maximum volume of a parcel of this shape, and prove that your result is, in fact, a maximum.

[The girth of the parcel is the perimeter of its square cross-section.]

11 An open tank is to be constructed with a square horizontal base and vertical sides. The capacity of the tank is to be 2000 m^3. The cost of the material for the sides is £4 per square metre and for the base is £2 per square metre. Find the minimum cost of the material, and give the corresponding dimensions of the tank.   (oc)

12 A piece of wire, 150 cm long, is cut into two parts; one is bent to form a square and the other to form a circle. Find the lengths of the two parts when the sum of the two areas is least.

13 A beam of rectangular cross-section is to be cut from a cylindrical log of radius a. The stiffness of the beam is proportional to xy^3 where x is the breadth and y the depth of the section. Find the cross-sectional area of the beam
a) of greatest volume;
b) of greatest stiffness;
that can be cut from the log.   (oc)

14 The cost £S and the time t min of manufacture of a certain article are connected by the formula

$$S = \frac{16}{t^3} + \frac{3t^2}{4}$$

Find:
a) the rate of change of cost when $t = 4$;
b) the minimum cost.   (JMB)

15 ABCD is a square field and the length of a diagonal is $2a$ km. A man starts to walk from A straight across to the opposite corner C at a speed of 4 km/h. At the same instant a second man starts to walk from B straight across to D at a speed of 3 km/h. Show that after t h ($t \leqslant \frac{1}{2}a$) their distance apart is

$$\{(a - 3t)^2 + (a - 4t)^2\}^{1/2}$$

Prove that their least distance apart is $\frac{1}{5}a$. Find how far each man is from the centre of the square at this moment. (OC)

16 A cylinder of height h fits exactly into a sphere of radius a. Show that its volume V is given by

$$V = \tfrac{1}{4}\pi h(4a^2 - h^2)$$

and hence find the maximum fraction of the sphere's volume which can be occupied by the cylinder.

* 3.9 Further differentiation

Function of a function: the chain rule

So far we have confined our attention to functions which are powers of x, and to multiples, sums and differences of such functions. But a function is fundamentally a process which converts an input into an output; and, as with industrial processes, many functions are the succession of a series of such processes. Where the output of one function machine is used as the input of the next, the combined process is known as a 'function of a function'.

Example 1

If $y = \sqrt{(x^2 + 1)}$, we can calculate y when $x = 2$ by three successive processes, each different:

a) using 2 as input, we square it to obtain 4; then
b) using 4 as input, we add 1 to obtain 5; and finally
c) using 5 as input, we take its square root to obtain $\sqrt{5} \approx 2.236$.

If we regard x as passing through intermediate stages u and v in the course of production of y, we can write

input output
$$x \mapsto x^2 = u$$
$$u \mapsto u + 1 = v$$
$$v \mapsto \sqrt{v} = y$$

So $y = \sqrt{v} = \sqrt{(u + 1)} = \sqrt{(x^2 + 1)}$

We now meet the obvious question, whether we can differentiate y when it is known as the final output of a number of functions.

For convenience, let us suppose that we have a two-stage process:

input   intermediate   output

$x \;\longmapsto\; u \;\longmapsto\; y$

Now if x is increased by δx,
     u will increase by δu,
and so y will increase by δy.

Moreover, $\dfrac{\delta y}{\delta x} = \dfrac{\delta y}{\delta u} \times \dfrac{\delta u}{\delta x}$

Now suppose that $\delta x \to 0$ and $\delta u \to 0$, $\delta v \to 0$;

and also that $\dfrac{\delta u}{\delta x}, \dfrac{\delta y}{\delta u}, \dfrac{\delta y}{\delta x} \to \dfrac{du}{dx}, \dfrac{dy}{du}, \dfrac{dy}{dx}$ respectively.

Then $\dfrac{dy}{dx} = \dfrac{dy}{du} \times \dfrac{du}{dx}$

If there are more intermediate stages, say $x \longmapsto u \longmapsto v \longmapsto w \longmapsto y$, this result is easily extended to

$$\frac{dy}{dx} = \frac{dy}{dw} \times \frac{dw}{dv} \times \frac{dv}{du} \times \frac{du}{dx}$$

and for this reason it is often called the *chain rule*.

Though it is one of the advantages of the notation that the result *looks* obvious, it is important to realise that it cannot be proved simply by the cancellation

$$\frac{dy}{du} \times \frac{du}{dx} = \frac{dy}{dx}$$

for the symbol du has, on its own, not been given any meaning and so cannot be cancelled.

Example 2

$y = (x^3 + 1)^{10}$

   Let $y = u^{10}$,   where $u = x^3 + 1$

Now $\dfrac{dy}{du} = 10u^9$ and $\dfrac{du}{dx} = 3x^2$

So $\dfrac{dy}{dx} = \dfrac{dy}{du} \times \dfrac{du}{dx} = 10(x^3 + 1)^9 \times 3x^2 = 30x^2(x^3 + 1)^9$

Exercise 3.9a

1 Find $\dfrac{dy}{dx}$ if y is

    a) $(3x^2 + 1)^4$ **b)** $(9x^2 + 4)^{10}$ **c)** $(1 - x^2)^6$

    d) $(x^2 - x)^5$ **e)** $(3x^2 - 1)^{-1}$ **f)** $\dfrac{1}{x^2 - 1}$

    g) $\dfrac{1}{(x^2 + 2)^3}$ **h)** $\sqrt{(1 + x^2)}$ **i)** $\sqrt[3]{(x^2 - 1)}$

    j) $\dfrac{1}{\sqrt{(1 - x)}}$ **k)** $(\sqrt{x} + 1)^{10}$ **l)** $(ax^2 + b)^n$

2 Find $f'(t)$ when $f(t)$ is

    a) $(t^3 + 1)^4$ **b)** $(1 - 3t^2)^5$ **c)** $\sqrt{(1 - t)}$ **d)** $\dfrac{1}{\sqrt{(t^2 + 1)}}$

3 **a)** If $y = x^2$ (where $x > 0$),

    i) express x as a function of y;

    ii) examine whether $\dfrac{dx}{dy} = 1 \div \dfrac{dy}{dx}$;

    iii) examine whether $\dfrac{d^2x}{dy^2} = 1 \div \dfrac{d^2y}{dx^2}$.

    b) Repeat these questions if $y = (x - 1)^3$.

4 Find the equations of the tangent and normal at the points where $x = 1$ to the curves

    a) $y = \dfrac{1}{1 + x}$ **b)** $y = \sqrt{(1 + 3x^2)}$

5 A man wishes to cross a square ploughed field from one corner to the opposite corner in the shortest possible time. He can walk at 10 km/h when he keeps to the edge of the field and at 6 km/h when he walks on the ploughed land. Prove that he should walk one-quarter of the way along one side and then aim direct for the opposite corner.   (oc)

Differentiation of products

In section 3.3 we investigated the derivatives of multiples and sums of given functions. If u and v were functions of x, we found that

$$\frac{d}{dx}(ku) = k\frac{du}{dx} \quad \text{and} \quad \frac{d}{dx}(u + v) = \frac{du}{dx} + \frac{dv}{dx}$$

We were not, however, able to find an obvious way of differentiating their product uv or their quotient u/v, so we we shall now investigate these.

Let us first consider the product $y = uv$ where u, v are functions of x.
Let x increase by δx. Then u, v increase by δu, δv, and so y increases by δy.

$$y + \delta y = (u + \delta u)(v + \delta v)$$

$$= uv + u\delta v + v\delta u + \delta u \delta v$$

But $y = uv$

Subtracting, $\delta y = u\delta v + v\delta u + \delta u \delta v$

$$\Rightarrow \qquad \frac{\delta y}{\delta x} = u\frac{\delta v}{\delta x} + v\frac{\delta u}{\delta x} + \delta u\frac{\delta v}{\delta x}$$

As $\delta x \to 0, \delta u \to 0$

and $\quad \dfrac{\delta u}{\delta x}, \dfrac{\delta v}{\delta x}, \dfrac{\delta y}{\delta x} \to \dfrac{du}{dx}, \dfrac{dv}{dx}, \dfrac{dy}{dx}$ respectively

So $\qquad \dfrac{dy}{dx} = u\dfrac{dv}{dx} + v\dfrac{du}{dx}$

Example 3

$$y = x^2(x + 1)^5$$

$$\Rightarrow \quad \frac{dy}{dx} = x^2 5(x + 1)^4 + 2x(x + 1)^5$$

$$= x(x + 1)^4\{5x + 2(x + 1)\}$$

$$= x(7x + 2)(x + 1)^4$$

Differentiation of quotients

Let us now consider the quotient, $y = u/v$, where u, v are functions of x.

Again, let x increase by δx. Then u, v increase by δu, δv, and so y increases by δy

$$y + \delta y = \frac{u + \delta u}{v + \delta v}$$

But $y = \dfrac{u}{v}$

Subtracting, $\delta y = \dfrac{u + \delta u}{v + \delta v} - \dfrac{u}{v}$

$$= \frac{v(u + \delta u) - u(v + \delta v)}{v(v + \delta v)}$$

$$= \frac{v\delta u - u\delta v}{v(v + \delta v)}$$

$\Rightarrow$ $\dfrac{\delta y}{\delta x} = \dfrac{v\dfrac{\delta u}{\delta x} - u\dfrac{\delta v}{\delta x}}{v(v + \delta v)}$

Suppose that $\delta x \to 0$ and $\delta v \to 0$;

and $\dfrac{\delta u}{\delta x}, \dfrac{\delta v}{\delta x}, \dfrac{\delta y}{\delta x} \to \dfrac{du}{dx}, \dfrac{dv}{dx}, \dfrac{dy}{dx}$ respectively

Then $\dfrac{dy}{dx} = \dfrac{v\dfrac{du}{dx} - u\dfrac{dv}{dx}}{v^2}$

Example 4

$$y = \frac{x}{1 + x^2}$$

$\Rightarrow$ $\dfrac{dy}{dx} = \dfrac{(1 + x^2) \times 1 - x \times 2x}{(1 + x^2)^2} = \dfrac{1 - x^2}{(1 + x^2)^2}$

These two general results can equally be written:

$$\frac{d}{dx}(uv) = u\frac{dv}{dx} + v\frac{du}{dx}$$

$$\frac{d}{dx}\left(\frac{u}{v}\right) = \frac{v\dfrac{du}{dx} - u\dfrac{dv}{dx}}{v^2}$$

Exercise 3.9b

1   Differentiate:

a) $x(1 + x)^5$ **b)** $x^2(3x + 1)^4$

c) $(x + 1)^2(x + 2)^3$

d) $x(2x^3 + 1)^5$

e) $\dfrac{x - 1}{x + 1}$

f) $\dfrac{x}{2x^2 + 1}$

g) $\dfrac{1 - x^2}{1 + x^2}$

h) $\dfrac{x}{(3x + 1)^2}$

i) $\dfrac{1 + \sqrt{x}}{1 - \sqrt{x}}$

j) $\sqrt{\left(\dfrac{1 - x}{1 + x}\right)}$

2 Investigate the stationary values of **a)** $\dfrac{x^2}{x - 1}$ **b)** $\dfrac{x^3}{1 - x^2}$

3 Find the equations of the tangent and normal at the point where $x = 2$ to the curves

a) $y = \dfrac{x}{x - 1}$ **b)** $y = x\sqrt{(x + 2)}$

4 Prove the rule for the differentiation of the quotient u/v by expressing it as the product uv^{-1}.

* 3.10 Implicit functions and parameters

Implicit functions

The chain rule is frequently useful when y is known only as an *implicit* function of x (i.e. is not given explicitly by a formula in terms of x).

Example 1

Show that the curve

$$x^3 + y^3 = 2xy$$

passes through the point $(1, 1)$, and find its tangent and normal at this point.

The first part follows from the fact that

$$1^3 + 1^3 = 2 \times 1 \times 1$$

Also $\qquad x^3 + y^3 = 2xy$

$$\Rightarrow \quad \frac{d}{dx}(x^3) + \frac{d}{dx}(y^3) = \frac{d}{dx}(2xy)$$

$$\Rightarrow \qquad 3x^2 + 3y^2\frac{dy}{dx} = 2x\frac{dy}{dx} + 2y$$

$$\Rightarrow \qquad (3y^2 - 2x)\frac{dy}{dx} = 2y - 3x^2$$

$$\Rightarrow \qquad \frac{dy}{dx} = \frac{2y - 3x^2}{3y^2 - 2x}$$

So at $(1, 1)$ $\qquad \dfrac{dy}{dx} = \dfrac{2 - 3}{3 - 2} = -1$

Hence the equation of the tangent at $(1, 1)$ is

$$\frac{y - 1}{x - 1} = -1 \quad \Rightarrow \quad x + y - 2 = 0$$

and the equation of the normal is

$$\frac{y - 1}{x - 1} = +1 \quad \Rightarrow \quad y = x$$

Example 2

Given that $\dfrac{d}{dx}(x^n) = nx^{n-1}$ when $n \in \mathbb{Z}^+$, prove that this is also true whenever $n \in \mathbb{Q}$ (i.e. whenever n is rational).

Case 1 $n \in \mathbb{Z}^-$ (i.e. n a negative integer)

Let $n = -m$.

Then $\qquad \dfrac{d}{dx}(x^n) = \dfrac{d}{dx}(x^{-m}) = \dfrac{d}{dx}(x^{-1})^m$

$$= m(x^{-1})^{m-1} \times -x^{-2} \text{ (see p. 86, example 2)}$$

$$= -mx^{-m-1} = nx^{n-1}$$

Case 2 $n = 0$

$$\frac{d}{dx}(x^0) = \frac{d}{dx}(1) = 0 = 0x^{0-1}$$

Case 3 $n \in \mathbb{Q}$

Then $y = x^n = x^{p/q}$, where $p, q \in \mathbb{Z}$

So $\qquad y^q = x^p$

$\Rightarrow \quad qy^{q-1}\dfrac{dy}{dx} = px^{p-1}$

$\Rightarrow \qquad \dfrac{dy}{dx} = \dfrac{p\,x^{p-1}}{q\,y^{q-1}}$

$\qquad\qquad = \dfrac{p}{q}\dfrac{x^{p-1}}{(x^{p/q})^{q-1}}$

$\qquad\qquad = \dfrac{p}{q}\dfrac{x^{p-1}}{x^{p-(p/q)}}$

$\qquad\qquad = \dfrac{p}{q}x^{(p/q)-1} = nx^{n-1}$

So $\dfrac{d}{dx}(x^n) = nx^{n-1}$ whenever $x \in \mathbb{Q}$.

Parameters

Sometimes the equation of a curve is given by expressing the coordinates x and y as functions of a third variable t, called a *parameter*.

Example 3

If a stone is thrown out to sea with velocity $20\ \mathrm{m\,s^{-1}}$ horizontally from the top of a cliff, then its coordinates x, y (referred to axes through its point of projection) can be expressed as functions of time t:

$x = 20t$

$y = -5t^2$

It is easy to see, by eliminating t, that the equation of the stone's path is

$y = -\dfrac{1}{80}x^2$

and its directions of motion could readily be calculated from this equation. Alternatively we can use a version of the chain rule.

For $\dfrac{dy}{dx} = \lim_{\delta x \to 0}\dfrac{\delta y}{\delta x} = \lim_{\delta t \to 0}\dfrac{\delta y/\delta t}{\delta x/\delta t} = \dfrac{dy/dt}{dx/dt}$

So $$\frac{dy}{dx} = \frac{dy/dt}{dx/dt}$$

In our particular case, $\dfrac{dy}{dx} = \dfrac{-10t}{20} = -\dfrac{t}{2}$.

These can be summarised in the figure:

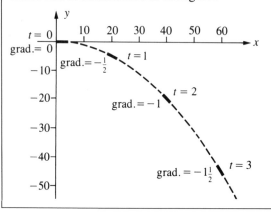

Example 4

Find the equation of the tangent and the normal at the point whose parameter is t to the curve given by $x = 2t^3,\ y = 3t^2$.

Taking different values of t we obtain the following table:

t	-3	-2	-1	0	1	2	3
$x = 2t^3$	-54	-16	-2	0	2	16	54
$y = 3t^2$	27	12	3	0	3	12	27

and so can plot the curve as follows:

Since $x = 2t^3$ and $y = 3t^2$

$$\frac{dy}{dx} = \frac{dy/dt}{dx/dt} = \frac{6t}{6t^2} = \frac{1}{t}$$

Hence the equation of the tangent at the point $(2t^3, 3t^2)$ is

$$\frac{y - 3t^2}{x - 2t^3} = \frac{1}{t} \quad \Rightarrow \quad x - ty + t^3 = 0$$

Furthermore, the gradient of the normal is $-1 \div \dfrac{1}{t} = -t$ and so its equation is

$$\frac{y - 3t^2}{x - 2t^3} = -t \quad \Rightarrow \quad tx + y = 2t^4 + 3t^2$$

Exercise 3.10

1 Find expressions for $\dfrac{dy}{dx}$ for the following curves and calculate their gradients at the given points. Hence find the equations of the tangents and normals at these points.

a) $x^2 - y^2 = 8$, $(3, 1)$
b) $x^2 + 4xy + y^2 = 6$, $(1, 1)$
c) $x^3 + y^3 = 7$, $(-1, 2)$

2 Plot the following curves for values of t from -3 to $+3$, and find $\dfrac{dy}{dx}$ in terms of t. Hence find the equations of the tangents and normals at the point with parameter t.

a) $x = 2t^3$, $y = t^2$
b) $x = 4t$, $y = \dfrac{1}{t}$
c) $x = t^3 - t$, $y = t^2$

3 At high tide a stone is thrown horizontally with speed $10 \, \text{m s}^{-1}$ from a cliff of height 80 m. The horizontal and vertical components of the distance it has travelled after t second are given by $x = 10t$, $y = -5t^2$.
    Find when, where and at what angle it enters the sea.

4 A curve is given by the parametric equations

$$x = t^2 - 3, \quad y = t(t^2 - 3)$$

a) Find its x, y equation, in a form clear of surds and fractions.
b) Prove that it is symmetrical about the x-axis.
c) Show that there are no points on the curve for which $x < -3$.

d) Find $\dfrac{dy}{dx}$ in terms of t, and derive the coordinates of the points on the curve for which $\dfrac{dy}{dx} = 0$.

e) Sketch the form of the curve. (c)

5   Prove that if the velocity of a particle moving in a straight line is always proportional to its displacement from a fixed point, so is its acceleration.

*3.11 Rates of change; approximations and errors

Related rates of change

The chain rule is also useful when we are considering rates of change. Take, for instance, the following case of an expanding sphere whose radius r, surface area S, and volume V are all increasing with time t.

Example 1

A pump is inflating a spherical balloon whose radius at a certain instant is $2\,\mathrm{m}$ and is increasing at a rate of $1\,\mathrm{cm\,s^{-1}}$.
a) At what rate is the pump working?
b) If air continues to be pumped into the balloon at this rate, at what rate will the radius be increasing when it is $5\,\mathrm{m}$?
c) At what rate is the surface area of the balloon increasing when $r = 5$?

a) Here $V = \dfrac{4}{3}\pi r^3,$ where r is a function of t.

So $\dfrac{dV}{dt} = \dfrac{dV}{dr} \times \dfrac{dr}{dt}$

$\dfrac{dV}{dr} = 4\pi r^2 = 4\pi \times 2^2 = 16\pi\,\mathrm{m^2}$

$\dfrac{dr}{dt} = 0.01\,\mathrm{m\,s^{-1}}$

So $\dfrac{dV}{dt} = 16\pi \times 0.01\,\mathrm{m^3\,s^{-1}} \approx 0.50\,\mathrm{m^3\,s^{-1}}$

So the pump is delivering air at the rate of $0.50\,\mathrm{m^3\,s^{-1}}$, or roughly $1\,\mathrm{m^3}$ every $2\,\mathrm{s}$.

b) $\dfrac{\mathrm{d}V}{\mathrm{d}t} = \dfrac{\mathrm{d}V}{\mathrm{d}r} \times \dfrac{\mathrm{d}r}{\mathrm{d}t}$

$\dfrac{\mathrm{d}V}{\mathrm{d}t} = 4\pi r^2 \times \dfrac{\mathrm{d}r}{\mathrm{d}t}$

But $\dfrac{\mathrm{d}V}{\mathrm{d}t} = 0.50$ and $r = 5$

So $0.50 = 4\pi \times 5^2 \dfrac{\mathrm{d}r}{\mathrm{d}t}$

$\dfrac{\mathrm{d}r}{\mathrm{d}t} = \dfrac{0.50}{100\pi} \approx 0.0016 \,\mathrm{m\,s}^{-1}$

So the rate of increase of radius has now been reduced to $0.0016\,\mathrm{ms}^{-1}$.

c) $S = 4\pi r^2$

$\Rightarrow \quad \dfrac{\mathrm{d}S}{\mathrm{d}t} = \dfrac{\mathrm{d}S}{\mathrm{d}r} \times \dfrac{\mathrm{d}r}{\mathrm{d}t} = 8\pi r \times \dfrac{\mathrm{d}r}{\mathrm{d}t}$

$\Rightarrow \quad \dfrac{\mathrm{d}S}{\mathrm{d}t} = 8\pi \times 5 \times 0.0016 = 0.20$

So when the radius of the balloon is 5 m, its membrane is being stretched at the rate of $0.20\,\mathrm{m}^2\,\mathrm{s}^{-1}$.

Example 2

A particle moves from a point O on a straight line and its velocity v is plotted on a graph as a function of its displacement x.

a) Show that its acceleration $a = v\dfrac{\mathrm{d}v}{\mathrm{d}x}$.

b) If its acceleration is constant and it starts with velocity u, show that

$$v^2 = u^2 + 2ax$$

a) $a = \dfrac{\mathrm{d}v}{\mathrm{d}t} = \dfrac{\mathrm{d}v}{\mathrm{d}x} \times \dfrac{\mathrm{d}x}{\mathrm{d}t} = \dfrac{\mathrm{d}v}{\mathrm{d}x} \times v$

So $a = v\dfrac{\mathrm{d}v}{\mathrm{d}x}$

b) Hence $\dfrac{\mathrm{d}x}{\mathrm{d}v} = \dfrac{v}{a}$

If a is constant, $x = \dfrac{v^2}{2a} + A$

When $x = 0$, $v = u$ $\Rightarrow$ $0 = \dfrac{u^2}{2a} + A$

So $x = \dfrac{v^2}{2a} - \dfrac{u^2}{2a}$

$\Rightarrow$ $v^2 = u^2 + 2ax$

Exercise 3.11a

1   If the side of a cube is 5 m and is expanding at a rate of 0.1 m/s, find the rate of increase of   **a)** its total surface area   **b)** its volume.

2   The area of a square is increasing at a rate of 10 cm²/s. If the lengths of its sides are 4 cm, at what rate are they increasing?

3   The radius of a circle is 6 m.
a) If it is increasing at 0.1 m s⁻¹, at what rate is its area increasing?
b) If its area is increasing at the rate of 20 m² s⁻¹, what is the rate of increase of its radius?

4   A spherical balloon is losing air at a rate of 10 m³ s⁻¹. When its radius is 3 m at what rate is it diminishing?

5   A street lamp is at a height 5 m and a man of height 2 m is running away from it at 3 m s⁻¹. How long is his shadow when he is x m away from its base, and at what rate is this length increasing?

6   The top of a 10 m ladder is at a height 8 m up a vertical wall and is slipping down the wall at a rate of 1 m s⁻¹. At what rate is the foot of the ladder sliding along the ground?

7   Liquid is dripping through a conical funnel at a rate of 2 cm³ s⁻¹. When the depth of liquid in the funnel is x cm, its volume is $\frac{1}{3}\pi x^3$. Find the rate at which the level of the liquid is falling when $x = 5$.

8   When the gas in a balloon is kept at constant temperature its volume V and pressure p are connected by Boyle's law, $pV = $ constant.
   If its volume 100 m³ is increasing at a rate 10 m³ s⁻¹, find the rate at which the pressure is decreasing from its present value of 5 atmospheres.

Small changes; approximations and errors

Differentiation was first introduced by considering the effects of small changes. We can now reverse the process and use differentiation to calculate such effects.

Example 3

Calculate $\sqrt[3]{8.01}$.

$$y = x^{1/3} \quad \Rightarrow \quad \frac{dy}{dx} = \tfrac{1}{3}x^{-2/3}$$

$$\Rightarrow \quad \frac{\delta y}{\delta x} \approx \tfrac{1}{3}x^{-2/3}$$

$$\Rightarrow \quad \delta y \approx \tfrac{1}{3}x^{-2/3}\,\delta x$$

If we now let $x = 8$ and $\delta x = 0.01$

then $\qquad y = \sqrt[3]{8} = 2$

and $\qquad \delta y \approx \tfrac{1}{3} \times 8^{-2/3} \times 0.01$

$$\approx \tfrac{1}{3} \times \tfrac{1}{4} \times 0.01$$

$$\approx 0.00083$$

So $\quad y + \delta y \approx 2.00083$

Hence $\quad \sqrt[3]{8.01} \approx 2.00083$

Example 4

The time T taken by a planet to revolve round the Sun and its mean distance r from the Sun are such that $T = kr^{3/2}$.

If the Earth's distance from the Sun were to be increased by 1%, how much longer would a year become?

$$T = kr^{3/2}$$

$$\Rightarrow \qquad \frac{dT}{dr} = \frac{3}{2}kr^{1/2}$$

$$\Rightarrow \qquad \delta T \approx \frac{3}{2}kr^{1/2}\,\delta r$$

Hence $\dfrac{\delta T}{T} \approx \dfrac{3}{2}\dfrac{\delta r}{r}$

But $\dfrac{\delta r}{r} = \dfrac{1}{100}$,   so $\dfrac{\delta T}{T} \approx \dfrac{3}{200} = 1.5\%$

So a year would be increased by $1.5\% \approx 5.5$ days.

Example 5

Show that a small fractional error of $p\%$ in measuring the radius of a sphere causes a fractional error of $2p\%$ in the calculation of its surface area and of $3p\%$ in the calculation of its volume.

If the radius, surface area and volume of the sphere are r, S, V, respectively,

$$S = 4\pi r^2 \qquad \text{and} \qquad V = \tfrac{4}{3}\pi r^3$$

$$\Rightarrow \quad \frac{dS}{dr} = 8\pi r \qquad \text{and} \qquad \frac{dV}{dr} = 4\pi r^2$$

$$\Rightarrow \quad \frac{\delta S}{\delta r} \approx 8\pi r \qquad \text{and} \qquad \frac{\delta V}{\delta r} \approx 4\pi r^2$$

$$\Rightarrow \quad \delta S \approx 8\pi r \delta r \qquad \text{and} \qquad \delta V \approx 4\pi r^2 \delta r$$

$$\text{But} \quad S = 4\pi r^2 \qquad \text{and} \qquad V = \tfrac{4}{3}\pi r^3$$

$$\text{So} \quad \frac{\delta S}{S} \approx 2\frac{\delta r}{r} \qquad \text{and} \qquad \frac{\delta V}{V} \approx 3\frac{\delta r}{r}$$

But the fractional error in the measurement of r is $p\%$

$$\Rightarrow \quad \frac{\delta r}{r} = \frac{p}{100}$$

$$\Rightarrow \quad \frac{\delta S}{S} \approx \frac{2p}{100} \qquad \text{and} \qquad \frac{\delta V}{V} \approx \frac{3p}{100}$$

Hence the fractional errors in the calculation of S and V must be $2p\%$ and $3p\%$, respectively.

Exercise 3.11b

1   If x increases from 4 to 4.001, use differentiation to find the corresponding change in y if
   a) $y = x^3 - 4$

b) $y = \dfrac{1}{x^2}$

c) $y = \sqrt{x}$

2 Find, without use of calculator or long multiplication, approximate values for
a) $(20.1)^3$ **b)** $\sqrt[3]{27.2}$ **c)** $\sqrt[2]{15.9}$

3 The lanes of a circular running track are 1 m wide. What is the difference in length of adjacent lanes?

4 When a hemispherical bowl of radius r contains liquid to a depth x, its volume V is $\frac{1}{3}\pi(3rx^2 - x^3)$.

Find a relationship between slight changes δx and δV in depth and volume respectively.

If a liquid is originally of depth 1 m in a bowl of radius 2 m, find the decrease in depth when 0.1 m³ has evaporated.

5 A cube of metal is heated so that its temperature is raised by 1° C and each side expands by $p\%$. What are the percentage expansions of its surface area and volume?

6 When a wire is slightly stretched, its length l and diameter x vary in such a way that its volume remains constant. Find the percentage decrease in diameter in terms of the percentage increase in length.

* 3.12 Further kinematics

Area under velocity–time curve

When we considered velocity–time curves, we were usually given the velocity as an algebraic function of time, and were able to find displacements simply by reversing the process of differentiation. Frequently, however, we are less fortunate and only know the velocity as a set of readings for different values of t. In such cases we must find a new procedure for the calculation of distances.

As a simple example, let us consider the case of a car which is in the hands of a learner drive and accelerates by a series of jolts:

for 2 s it is moving at $3 \, \text{m s}^{-1}$,

then   for 4 s it is moving at $5 \, \text{m s}^{-1}$,

and   for 3 s it is moving at $6 \, \text{m s}^{-1}$.

The total distance travelled $= 3 \times 2 + 5 \times 4 + 6 \times 3 = 44$ m, and this can be represented by the area beneath the velocity–time graph (overleaf)

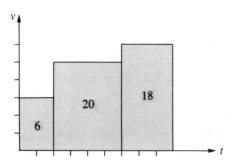

Clearly the distance travelled by a body which moves rather less jerkily can be found in just the same way, by dividing the area beneath its velocity–time graph into narrow strips and estimating their total area.

For instance, the velocity–time graph of an accelerating sports car is:

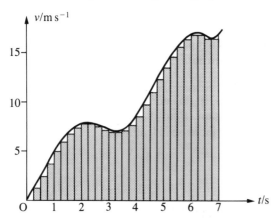

and the distance travelled in the first 7 seconds can be estimated by calculating the area of the rectangles shown. This will be a slight underestimate but, if greater accuracy is required, more rectangles can be drawn. Or, alternatively, a corresponding overestimate can be made by taking rectangles above, rather than below, the curves.

In any event, it is apparent that *the distance travelled in a certain interval of time is represented by the area beneath the corresponding section of the velocity–time curve.*

Moreover, just as distances (or changes of displacement) can be calculated from areas beneath velocity–time curves, so changes of velocity can be calculated from areas beneath acceleration–time curves.

Motion with constant acceleration

It is now instructive to use both methods for the particular case when the acceleration of a body moving in a straight line is constant.

A particle moves from the origin with velocity u, and has *constant* acceleration a. If its displacement and velocity after time t are s and v respectively, we shall find the relationships between s, u, v, a, and t.

Method 1

$$\frac{dv}{dt} = a, \quad \text{which is a constant}$$

$\Rightarrow \quad v = at + c, \quad \text{where } c \text{ is a constant}$

When $t = 0$, $v = u$,

so $\quad u = a0 + c \quad \Rightarrow \quad c = u$

$\Rightarrow \quad v = u + at$

$\Rightarrow \quad \dfrac{ds}{dt} = u + at$

$\Rightarrow \quad s = ut + \frac{1}{2}at^2 + d, \quad \text{where } d \text{ is a constant}$

When $\quad t = 0$, $s = 0$,

so $\quad 0 = 0 + 0 + d \quad \Rightarrow \quad d = 0$

$\Rightarrow \quad s = ut + \frac{1}{2}at^2$

Eliminating a $\quad s = \frac{1}{2}(u + v)t$

Eliminating t $\quad v^2 = u^2 + 2as$

Eliminating u $\quad s = vt - \frac{1}{2}at^2$

Method 2

Since acceleration is constant, the velocity–time curve has constant gradient and so is a straight line:

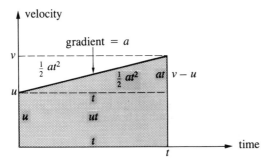

From the figure,

$$\text{gradient of line} = a = \frac{v - u}{t} \quad \Rightarrow \quad v = u + at$$

area under line $= s = \frac{1}{2}(u + v)t = ut + \frac{1}{2}at^2 = vt - \frac{1}{2}at^2$

Eliminating t from any pair of equations

$v^2 = u^2 + 2as$

So both methods lead to the five relationships

$v = u + at$

$s = \frac{1}{2}(u + v)t$

$s = ut + \frac{1}{2}at^2$

$s = vt - \frac{1}{2}at^2$

$v^2 = u^2 + 2as$

It must, however, be emphasised that *these relationships can only be used when it has already been established that the acceleration is constant.* This is not usually so, except in certain special (but important) cases, such as the free fall of a stone under gravity, and even then only under certain limitations.

Example

A stone dropped from a high cliff is falling with speed $20 \, \mathrm{m\,s^{-1}}$ and acceleration $10 \, \mathrm{m\,s^{-2}}$. How long does it take to drop another 160 m and what is then its speed?

In this case, $u = 20$, $a = 10$,   and $s = 160$

So $s = ut + \frac{1}{2}at^2$

$\Rightarrow$ $160 = 20t + 5t^2$

$\Rightarrow$ $t^2 + 4t - 32 = 0$

$\Rightarrow$ $(t + 8)(t - 4) = 0$ $\Rightarrow$ $t = 4 \text{ or } -8$

So the stone takes another 4 s to reach the required level (which it might also have passed 8 s previously had it been catapulted upwards).

Also $v = u + at$

$\Rightarrow$ $v = 20 + 10 \times 4 = 60$

So its speed is then $60 \, \mathrm{m\,s^{-1}}$.

Exercise 3.12

1   A stone is falling at $30 \, \mathrm{m \, s^{-1}}$ with constant acceleration $10 \, \mathrm{m \, s^{-2}}$. How far does it go in the next two seconds, and what is then its speed?

2   A train accelerates uniformly from $12 \, \mathrm{m \, s^{-1}}$ to $36 \, \mathrm{m \, s^{-1}}$ in 2 minutes. What is its acceleration and how far does it go in this time?

3   A cyclist travelling at $20 \, \mathrm{m \, s^{-1}}$ brakes uniformly and stops in 50 m. What is the rate of deceleration and how long does this take?

4   A car travelling at $20 \, \mathrm{m \, s^{-1}}$ accelerates uniformly so that in the next 2 s it covers 42 m. What is its acceleration and its final speed?

5   The driver of a train, travelling at $80 \, \mathrm{km \, h^{-1}}$ on a straight level track, sees a signal against him at a distance of 1 km and, putting on the brakes, comes to rest at the signal. He stops for one minute and then resumes the journey, attaining the original speed in a distance of 4 km. Assuming that retardation and acceleration are uniform, find how much time has been lost owing to the stoppage.   (OC)

6   A car approaches a corner at a speed of $100 \, \mathrm{km \, h^{-1}}$ and has to reduce this speed to $50 \, \mathrm{km \, h^{-1}}$ in a distance of 200 m in order to take the corner. Find the requisite deceleration in $\mathrm{m \, s^{-2}}$.
    After the corner the car regains its former speed in 20 seconds. Find the distance it travels in so doing.
    (Assume that both acceleration and deceleration are uniform.)

7   A train, slowing down with uniform retardation, passes three telegraph posts spaced at intervals of 100 m. It takes 8 seconds between the first and second posts, and 12 seconds between the second and third. Find the retardation of the train, and the distance beyond the third post of the point where the train comes to rest.

8   Given that the maximum acceleration of a lift is $4 \, \mathrm{m \, s^{-2}}$ and the maximum deceleration is $5 \, \mathrm{m \, s^{-2}}$, find the minimum time for a journey of 30 m.
    a) if the maximum speed is $5 \, \mathrm{m \, s^{-1}}$;
    b) if there is no restriction on speed.

Miscellaneous problems

1   Discuss the existence of the derivative at $x = 0$ of the function

   $$f : x \mapsto x|x|$$

   showing that you have considered both positive and negative values as x tends to 0.

State with reasons which of the following are true and which are false:
a) f is an even function;
b) f is continuous at $x = 0$;
c) f is differentiable at $x = 0$.
Sketch a graph of the function f and find the derived function f′.
Is f′ differentiable at $x = 0$? (SMP)

2 Show that the curve $y = x^4 - 6a^2x^2$ has two points of inflection H and K. Prove that the tangents at H and K meet on the y-axis at the point $(0, 3a^4)$. Find the turning points of the curve, and sketch it. (OC)

3 A cylinder of radius x is inscribed in a cone of height h and base radius a, the two figures having a common axis of symmetry. Find an expression for the height of the cylinder in terms of x, a, and h.

Prove that the cylinder of greatest volume that can be so inscribed has a volume $\frac{4}{9}$ of the volume of the cone. (OC)

4 A cylindrical tin of radius r and height h is constructed from a fixed area of sheeting so as to have maximum capacity.

Find the relationship between h and r
a) if the tin has a base but no lid;
b) if it has both base and lid;
c) if it has a slip-on lid with a flange of depth a.

5 At constant temperature the pressure p and volume V of a fixed mass of gas are related by Boyle's Law, $pV = $ constant.

Show that a small fractional increase in volume is equal to the corresponding small fractional decrease in pressure.

6 Plot a graph of the function

$$y = \frac{x(x - 4)}{\sqrt{(x^2 + 3)}}$$

for values of x between -1 and 5 and show that the graph has only one turning point. Find the value of x corresponding to this turning point, correct to two places of decimals. (OC)

7 Sketch the function

$$f(x) = |x|^{-1}$$

and find the radius of the circle which touches both its branches and also the x-axis at the origin. (CS)

8 Without actually plotting to scale, find out all you can about the curves

$$y^2 = \frac{x}{1 + x^2} \quad \text{and} \quad y^2 = \frac{x}{1 - x^2}$$

and makes sketches of them. (OS)

9   A slice of bread is in the shape of a rectangle surmounted by a semicircle. Find the ratio of its total height to width if a given area is to have minimum crust.

10   The despatch department of the firm in exercise 3.8 no. **10** decides that cylindrical parcels be sent. Show that this increases their maximum permissible volume by 27%.

4 Areas: integration

4.1 The area beneath a curve: the definite integral

One of the first uses of geometry was for surveying and the calculation of areas of land. This was clearly simplest with squares and rectangles, but the Greeks were also familiar with means of finding areas of triangles. When, however, an area was wholly or partly bounded by a curve, the problem became more difficult, and they were successful only in a limited number of cases.

The simplest of these was, of course, the circle. The fact that

$$\frac{\text{area}}{(\text{radius})^2}$$

remains constant was well known, and the evaluation of this constant proceeded with increasing accuracy. It is still instructive to carry out such an evaluation of π by drawing a circle of radius 10 cm on graph paper and counting squares. What percentage accuracy would you expect, and what do you achieve? Can it be improved by averaging a number of similar results?

In the first half of the seventeenth century, mathematicians became increasingly concerned with areas bounded by other kinds of curve. Kepler's second law of planetary motion (1609) stated that the speed of a planet in its orbit varies in such a way that a line joining the planet to the Sun sweeps out equal *areas* in equal periods of time. This law and, more simply, the calculation of areas lying beneath parabolas and similar curves were soon to lead Newton to another use for his newly discovered Method of Fluxions, or *calculus* as it was later to be called.

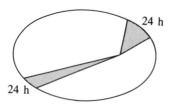

The definite integral

Our first need is for a notation with which to describe an area bounded by curves, and we begin by considering that beneath the graph of a function.

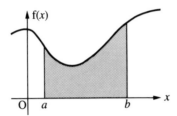

If f(x) is a continuous function which is positive over the domain $a \leqslant x \leqslant b$, we call the area beneath its curve the *definite integral* of f(x) from $x = a$ to $x = b$, and write it

$$\int_a^b f(x)\,dx$$

The symbol $\int$ arose as an elongated form of the letter S (for Sum), simply because splitting the area into approximately rectangular strips, each of small width δx and area f(x)δx, leads us to:

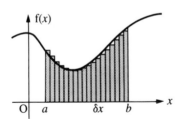
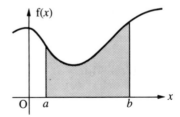

$$\text{area} \approx \operatorname*{sum}_{x=a}^{x=b} f(x)\delta x \qquad \text{and so to the notation:} \qquad \text{area} = \int_a^b f(x)\,dx$$

Example 1

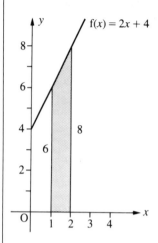

$$\int_{1}^{2} (2x + 4)\,dx$$

indicates the shaded area, which is clearly 7 square units.

So $\int_{1}^{2} (2x + 4)\,dx = 7$

It is apparent that the letter x is simply taken from the name of the axis and has no influence on the result; for this reason it is sometimes called a *dummy*, and it is equally valid to write

$$\int_{1}^{2} (2t + 4)\,dt = 7 \quad \text{or} \quad \int_{1}^{2} (2\theta + 4)\,d\theta = 7$$

Exercise 4.1

Sketch and evaluate the following integrals:

1 $\int_{0}^{5} 7\,dx$ **2** $\int_{0}^{10} 5x\,dx$ **3** $\int_{2}^{4} (3x + 2)\,dx$

4 $\int_{3}^{6} \dfrac{7 - t}{2}\,dt$ **5** $\int_{-1}^{2} (4 - u)\,du$ **6** $\int_{-2}^{2} |v|\,dv$

The two staircases

Each of the above examples and exercises concerned the area beneath a straight line. They were simply to make the symbol ∫ rather more familiar, but we must now meet the challenge of a genuine curve:

Example 2

Find $\displaystyle\int_0^1 x^3 \, dx$

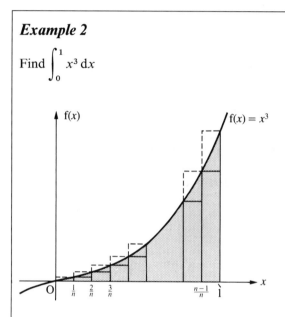

First we set up the two staircases shown, one above the given curve and the other below it, and each consisting of n steps of width $1/n$. For convenience, we shall call the required area A. It is clearly greater than the area beneath the lower staircase and less than the area beneath the upper staircase.

But the heights above the x-axis of the several steps are

$$\left(\frac{1}{n}\right)^3, \left(\frac{2}{n}\right)^3, \left(\frac{3}{n}\right)^3, \ldots \left(\frac{n-1}{n}\right)^3, 1$$

So $\dfrac{1}{n}\left(\dfrac{1}{n}\right)^3 + \dfrac{1}{n}\left(\dfrac{2}{n}\right)^3 + \cdots + \dfrac{1}{n}\left(\dfrac{n-1}{n}\right)^3$

$$< A < \frac{1}{n}\left(\frac{1}{n}\right)^3 + \frac{1}{n}\left(\frac{2}{n}\right)^3 + \cdots + \frac{1}{n}\left(\frac{n-1}{n}\right)^3 + \frac{1}{n}1^3$$

$$\Rightarrow \frac{1}{n^4}\{1^3 + 2^3 + 3^3 + \cdots + (n-1)^3\}$$

$$< A < \frac{1}{n^4}\{1^3 + 2^3 + \cdots + (n-1)^3 + n^3\}$$

Now it can be shown (see section 6.4) that the sum of the cubes of the first n positive integers is $\frac{1}{4}n^2(n+1)^2$. (This is easily verified for particular cases, such as $1^3 + 2^3$, $1^3 + 2^3 + 3^3$, $1^3 + 2^3 + 3^3 + 4^3$, etc.) So the sum of the first $n - 1$ positive integers will be $\frac{1}{4}(n-1)^2 n^2$.

Hence $\dfrac{1}{n^4}\dfrac{1}{4}n^2(n-1)^2 < A < \dfrac{1}{n^4}\dfrac{1}{4}n^2(n+1)^2$

$\Rightarrow$ $\dfrac{(n-1)^2}{4n^2} < A < \dfrac{(n+1)^2}{4n^2}$

$\Rightarrow$ $\dfrac{1}{4}\left(1 - \dfrac{1}{n}\right)^2 < A < \dfrac{1}{4}\left(1 + \dfrac{1}{n}\right)^2$

We now make the two staircases draw closer and closer to the original curve simply by reducing the width of each step and by making many more of them, i.e. by increasing the value of n.

Now as n grows larger and larger, both $\frac{1}{4}(1 - 1/n)^2$ and $\frac{1}{4}(1 + 1/n)^2$ approach $\frac{1}{4}$.

But A always lies between them, so $A = \frac{1}{4}$.

Hence $\displaystyle\int_0^1 x^3 \, dx = \frac{1}{4}$.

4.2 The generation of area: indefinite integrals

The reader may have been irritated by this last section. First, he may suspect our good fortune in having a ready-made formula for $1^3 + 2^3 + 3^3 + \cdots + n^3$ and, though glad to accept it, he might well wonder whether the choice of example was entirely accidental. An attempt to repeat the process for the curve $f(x) = \sqrt{x}$, and to find a corresponding formula for $\sqrt{1} + \sqrt{2} + \sqrt{3} + \cdots + \sqrt{n}$, would probably deepen his suspicion.

Secondly (a more creative irritation), on seeing the simplicity of the final result, he might wonder whether some key can be found to unlock the problem more swiftly.

In the search for such a key, he might recall velocity–time curves in section 3.12, and remember that distances were obtained *both* by finding areas beneath curves *and* by reversing the process of differentiation. Might it not be possible that these are the same process, and that in order to find areas beneath curves we simply have to reverse the process of differentiation? At very least, we can begin by looking at area under a curve in a more dynamic way, not as something presented once and for all as a single snapshot, but rather as a quantity which is gradually growing as our eyes sweep along the curve.

Let us start by looking again at the problem of example 2 in section 4.1.

Example 1

Find the area under the curve $f(x) = x^3$ between $x = 0$ and $x = 1$.

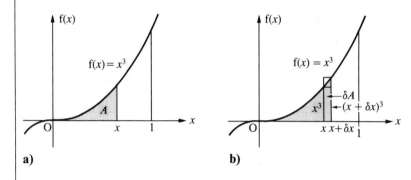

a)   **b)**

Let A be the shaded area shown in Fig. **a)** and let us see how it increases as its boundary gradually sweeps to the right.

If, as in Fig. **b)**, x increases by δx, A will increase by δA and this additional area lies between two rectangles of height x^3 and $(x + \delta x)^3$.

So $x^3 \delta x < \delta A < (x + \delta x)^3 \delta x$

$\Rightarrow \qquad x^3 < \dfrac{\delta A}{\delta x} < (x + \delta x)^3$

Now as $\delta x \to 0$, $(x + \delta x)^3 \to x^3$

$\Rightarrow \qquad \dfrac{\mathrm{d}A}{\mathrm{d}x} = x^3$

$\Rightarrow \qquad A = \tfrac{1}{4}x^4 + c,$ for some value of c

But when $x = 0$, $A = 0$

$\Rightarrow \quad 0 = 0 + c \quad \Rightarrow \quad c = 0$

So $A = \tfrac{1}{4}x^4$

We have, therefore, by this method discovered the area beneath the curve *up to any point*.

In particular, taking $x = 1$, we confirm that the required area between $x = 0$ and $x = 1$ is $\tfrac{1}{4} \times 1^4 = \tfrac{1}{4}$.

Example 2

Let us now consider the problem mentioned earlier of finding the area beneath the curve $f(x) = \sqrt{x}$, and for the sake of variety we shall take this between the limits $x = 1$ and $x = 2$.

Let A be the shaded area of Fig. **a)**.

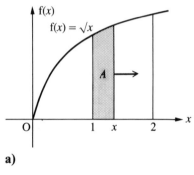
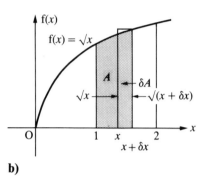

a)   **b)**

As before, we see that

$\quad\quad \delta A$ lies between $\sqrt{x}\,\delta x$ and $\sqrt{(x + \delta x)}\,\delta x$

$\Rightarrow \quad \dfrac{\delta A}{\delta x}$ lies between $\sqrt{x}$ and $\sqrt{(x + \delta x)}$

Again letting $\delta x \to 0$, we see that

$\quad \dfrac{dA}{dx} = \sqrt{x} = x^{1/2}$

$\Rightarrow \quad A = \tfrac{2}{3}x^{3/2} + c, \quad$ for some value of c

But the generation of A begins when $x = 1$

So $0 = \tfrac{2}{3}1^{3/2} + c \quad \Rightarrow \quad c = -\tfrac{2}{3}$

$\Rightarrow \quad A = \tfrac{2}{3}x^{3/2} - \tfrac{2}{3}$

In particular, taking $x = 2$ we see that the required area is

$\quad A = \tfrac{2}{3}2^{3/2} - \tfrac{2}{3} = \tfrac{2}{3}(2\sqrt{2} - 1) \approx 1.22$

So, compared with the method of the last section, our new approach is not only swifter, but also more powerful.

Further, if the area A is denoted by the function $A(x)$, we note that in each case

$\quad A'(x) = f(x), \quad$ and so are led to:

The fundamental theorem of calculus

If f(x) is continuous, and $A(x)$ is the area developed beneath the curve f(x) from a starting line $x = a$, then $A'(x) = f(x)$.

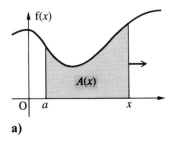 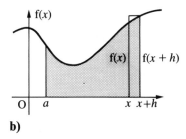

a)   **b)**

When x is increased to $x + h$, the new area added is

$$A(x + h) - A(x)$$

Now this lies between the two rectangles of areas $hf(x)$ and $hf(x + h)$.

$\Rightarrow$ $\qquad \dfrac{A(x + h) - A(x)}{h}$ lies between f(x) and f(x + h).

When $h \to 0$, $\dfrac{A(x + h) - A(x)}{h} \to A'(x)$

and as f(x + h) squeezes closer and closer to f(x), it is clear that

$$A'(x) = f(x)$$

In the language of integrals, the first temptation is to write this result as

$$\frac{d}{dx} \int_a^x f(x)\,dx = f(x)$$

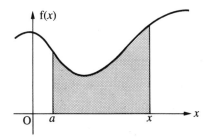

But it will be seen that in this equation, as in the corresponding graph, x is being used in two senses: firstly, as a variable (or name of the x-axis), and secondly, as a particular value of this variable (or particular x-coordinate) at the right-hand boundary of the area. This confusion is very easily avoided by renaming the general variable (or coordinate) as t:

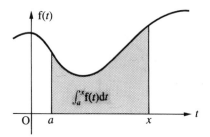

So $$\frac{d}{dx}\int_a^x f(t)\,dt = f(x)$$

which is the usual form of the *fundamental theorem of calculus*.

Indefinite integrals

The significance of this last result is immediately clear: that, as was foreshadowed in the last section, the process of finding the area beneath the curve of a given function $f(x)$ is that of discovering another function which, *when differentiated*, yields the given function, i.e. the reverse process of section 3.7.

Such a function is usually written without any limits, as

$$\int f(x)\,dx$$

So $$\int x^3\,dx = \tfrac{1}{4}x^4, \quad \tfrac{1}{4}x^4 + 3, \quad \tfrac{1}{4}x^4 - 2\tfrac{1}{2}, \quad \text{etc.}$$

More generally

$$\int x^3\,dx = \tfrac{1}{4}x^4 + c, \quad \text{where } c \text{ is a constant}$$

This is usually known as the *indefinite integral* and the constant c, which can take any value, is called an *arbitrary constant*.

Because of the importance of indefinite integrals, we must first gain some familiarity with them.

Example 3

$$\int x\sqrt{x}\,dx = \int x^{3/2}\,dx = \tfrac{2}{5}x^{5/2} + c$$

$$\left[\text{Check:}\quad \frac{d}{dx}(\tfrac{2}{5}x^{5/2} + c) = \tfrac{2}{5}\times\tfrac{5}{2}x^{3/2} + 0 = x^{3/2}\right]$$

Example 4

$$\int \frac{3 + t^2}{t^2}\,dt = \int\left(\frac{3}{t^2} + 1\right)dt = \int (3t^{-2} + 1)\,dt$$

$$= -3t^{-1} + t + c$$

$$= -\frac{3}{t} + t + c$$

$$\left[\text{Check:}\quad \frac{d}{dt}\left(-\frac{3}{t} + t + c\right) = \frac{3}{t^2} + 1\right]$$

Exercise 4.2

Find the following integrals:

1 **a)** $\displaystyle\int x^2\,dx$

   b) $\displaystyle\int (3x^2 - 4x + 2)\,dx$

   c) $\displaystyle\int (x^2 - 2x^3)\,dx$

   d) $\displaystyle\int (x^3 - 2x^2)\,dx$

2 **a)** $\displaystyle\int (t^2 - 2t + 3)\,dt$

   b) $\displaystyle\int t(t + 5)\,dt$

   c) $\displaystyle\int (t + 1)(t + 2)\,dt$

   d) $\displaystyle\int (2t + 1)^2\,dt$

3 **a)** $\displaystyle\int u(u^4 + 1)\,du$

   b) $\displaystyle\int (u^2 + 1)^2\,du$

   c) $\displaystyle\int 2u^2(u^3 - 1)\,du$

   d) $\displaystyle\int (u^4 + 2)^2\,du$

4 a) $\int x^2\sqrt{x}\,dx$ **b)** $\int \sqrt[3]{x}\,dx$

c) $\int (x+1)\sqrt{x}\,dx$ **d)** $\int (\sqrt{x}+3)^2\,dx$

5 a) $\int \dfrac{3}{x^2}\,dx$ **b)** $\int \dfrac{x+3}{x^4}\,dx$

c) $\int \dfrac{x+1}{\sqrt{x}}\,dx$ **d)** $\int \dfrac{\sqrt{x}+5}{x^2}\,dx$

6 a) $\int (ax+b)\,dx$ **b)** $\int (ax^2+bx+c)\,dx$

c) $\int (a\sqrt{x}+b)\,dx$ **d)** $\int \dfrac{ax+b}{\sqrt{x}}\,dx$

7 $\quad \int x^n\,dx$, where n is a constant.

Is your result true, without exception, for all values of n?

8 $\quad \int \sqrt[n]{u}\,du$

4.3 Calculation of definite integrals

Let us now return to the problem of finding $\int_a^b f(x)\,dx$, i.e. the area under the curve $f(x)$ between $x = a$ and $x = b$.

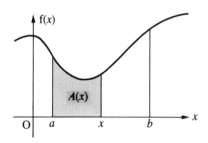

Let $A(x)$ be the area shown.

Then $A'(x) = f(x)$

Now suppose that the indefinite integral of $f(x)$ is $F(x) + c$

Then, for some value of c, $A(x) = F(x) + c$

Putting $x = b$,   required area $= A(b) = F(b) + c$

Putting $x = a$, $0 = A(a) = F(a) + c$

Subtracting, we see that required area $= F(b) - F(a)$, which is usually written

$$\int_a^b f(x)\,dx = \left[F(x) \right]_a^b$$

So, for instance, we can now write

$$\int_1^2 x^4\,dx = \left[\frac{x^5}{5} \right]_1^2 = \tfrac{32}{5} - \tfrac{1}{5} = 6\tfrac{1}{5}$$

Again we see that the value of $\int_a^b f(x)\,dx$ depends only on the function f which gives rise to F, and on the limits of integration a and b. The integral does not depend on x, and could equally well be written

$$\int_a^b f(t)\,dt \quad \text{or} \quad \int_a^b f(\theta)\,d\theta$$

Further, we notice that integrals (as we would expect from their definition as areas beneath curves) are additive

$$\int_a^b f(x)\,dx + \int_b^c f(x)\,dx = \{F(b) - F(a)\} + \{F(c) - F(b)\}$$

$$= F(c) - F(a)$$

So $$\int_a^b f(x)\,dx + \int_b^c f(x)\,dx = \int_a^c f(x)\,dx$$

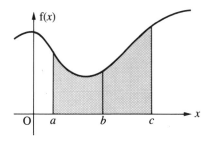

Lastly $$\int_b^a f(x)\,dx = F(a) - F(b)$$

So $$\int_b^a f(x)\,dx = - \int_a^b f(x)\,dx$$

The examples of section 4.2 can now be abbreviated:

Example 1

To find the area beneath $f(x) = x^3$ between $x = 0$ and $x = 1$:

$$\text{Required area} = \int_0^1 x^3 \, dx$$

$$= \left[\tfrac{1}{4} x^4 \right]_0^1$$

$$= \tfrac{1}{4} 1^4 - \tfrac{1}{4} 0^4 = \tfrac{1}{4}$$

Example 2

To find the area beneath $f(x) = \sqrt{x}$ between $x = 1$ and $x = 2$:

$$\text{Required area} = \int_1^2 \sqrt{x} \, dx$$

$$= \int_1^2 x^{1/2} \, dx$$

$$= \left[\tfrac{2}{3} x^{3/2} \right]_1^2$$

$$= \tfrac{2}{3} 2^{3/2} - \tfrac{2}{3} 1^{3/2}$$

$$= \tfrac{2}{3}(2\sqrt{2} - 1)$$

Example 3

To find the area enclosed between $f(t) = 3 + 2t - t^2$ and the t-axis:

$$3 + 2t - t^2 = 0 \quad \Rightarrow \quad t = -1 \text{ or } 3$$

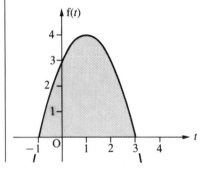

and so the curve cuts the t-axis at $t = -1$ and $t = 3$.

$$\text{Required area} = \int_{-1}^{3} (3 + 2t - t^2)\,dt$$

$$= \left[3t + t^2 - \tfrac{1}{3}t^3 \right]_{-1}^{3}$$

$$= \{3 \times 3 + 3^2 - \tfrac{1}{3} \times 3^3\} - \{3 \times -1 + (-1)^2 - \tfrac{1}{3} \times (-1)^3\}$$

$$= 9 - (-1\tfrac{2}{3}) = 10\tfrac{2}{3}$$

Exercise 4.3a

1 Evaluate:

a) $\displaystyle\int_{1}^{3} x^2\,dx$ **b)** $\displaystyle\int_{2}^{4} (3t^2 - 2t)\,dt$

c) $\displaystyle\int_{-2}^{-1} u^4\,du$ **d)** $\displaystyle\int_{1}^{2} (x^3 - x)\,dx$

2 Evaluate:

a) $\displaystyle\int_{1}^{4} \sqrt{x}\,dx$ **b)** $\displaystyle\int_{1}^{8} \sqrt[3]{t}\,dt$

c) $\displaystyle\int_{1}^{2} \frac{du}{u^2}$ **d)** $\displaystyle\int_{4}^{9} \frac{dx}{x\sqrt{x}}$

3 Calculate the area
a) beneath $f(x) = x^3$ and between $x = 1$ and $x = 2$;
b) beneath $f(x) = \dfrac{1}{x^2}$ and between $x = 2$ and $x = 3$,
c) beneath $f(x) = (x + 1)(x + 2)$ and between $x = 0$ and $x = 1$;
d) beneath $f(x) = x^2 - 2x$ and between $x = -1$ and $x = 0$.

4 Sketch graphs of the following functions and find the areas between them:

a) $y = x^2$ and $y = 4x$
b) $y = 4 - x^2$ and $y = 3$
c) $y = x^2 - 1$ and $y = 1 - x^2$
d) $y = \dfrac{1}{x^2},\ y = x^2$ and $y = \dfrac{x^2}{16}$
e) $y = x^2$ and $y^2 = x$

Negative areas

We have, so far, always considered areas *beneath* curves and have restricted ourselves to curves, or sections of curves, which lie above the x-axis. But it is clearly possible to calculate definite integrals when a curve lies below the x-axis and we must now examine the meaning of such integrals.

Example 4

$$\int_0^1 (-x^3)\,dx = \left[-\frac{1}{4}x^4 \right]_0^1 = -\frac{1}{4}1^4 - 0 = -\frac{1}{4}$$

and it therefore appears that the integral still represents the shaded area:

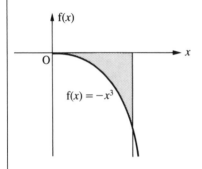

provided that such an area *below* the x-axis is reckoned as *negative*.

It is therefore clear that a definite integral $\int_a^b f(x)\,dx$ represents the *algebraic* sum of areas between a curve and the x-axis, those above the axis being counted positive and those below the axis being counted negative.

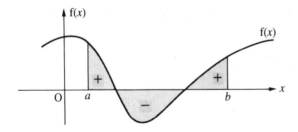

Example 5

What is the area between $f(x) = x^3$, the x-axis and the lines $x = -1$ and $x = 2$?

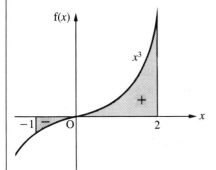

$$\int_{-1}^{2} x^3 \, dx = \left[\tfrac{1}{4}x^4 \right]_{-1}^{2}$$

$$= \tfrac{1}{4}2^4 - \tfrac{1}{4}(-1)^4$$

$$= 4 - \tfrac{1}{4} = 3\tfrac{3}{4}$$

But this is the *algebraic* sum of the shaded areas.

If their *numerical* sum is required, the two parts must be calculated separately:

$$\int_{-1}^{0} x^3 \, dx = \left[\tfrac{1}{4}x^4 \right]_{-1}^{0} = 0 - \tfrac{1}{4} = -\tfrac{1}{4}$$

$$\int_{0}^{2} x^3 \, dx = \left[\tfrac{1}{4}x^4 \right]_{0}^{2} = 4$$

So the total *numerical* area is $4\tfrac{1}{4}$.

Exercise 4.3b

1 Calculate:

a) $\displaystyle\int_{0}^{4} (x^2 - 4x) \, dx$ **b)** $\displaystyle\int_{2}^{6} (x^2 - 4x) \, dx$

and illustrate your answers on a sketch of $y = x^2 - 4x$.

2 **a)** Calculate $\displaystyle\int_{-2}^{2} x(x^2 - 4) \, dx$ and illustrate your answer.

b) What is the numerical value of the area formed between $y = x(x^2 - 4)$ and the x-axis?

3   If a stone is catapulted upwards with velocity $30\,\mathrm{m\,s^{-1}}$, its velocity v upwards after time t is given by $v = 30 - 10t$. Calculate:

a) $\displaystyle\int_0^3 v\,dt$ **b)** $\displaystyle\int_3^6 v\,dt$ **c)** $\displaystyle\int_0^6 v\,dt$ **d)** $\displaystyle\int_0^8 v\,dt$

and show the corresponding areas on the graph of $v = 30 - 10t$.

4   Sketch the graphs of
   a) $y = (x^2 - 1)(x^2 - 4)$ **b)** $y = x(x^2 - 1)(x^2 - 4)$
   and in each case calculate the total numerical area enclosed by the curve beneath the x-axis.

* 4.4 Numerical integration

Faced with the problem of finding the area beneath a curve, we frequently either
a) do not know an algebraic expression $f(x)$ for the curve, or
b) know $f(x)$ but are unable to integrate it, or
c) find it simpler, and sufficient for our purpose, to obtain an approximate numerical value.

We could, of course, simply plot the curve on graph-paper and count the number of squares beneath the required section; or we could (as in section 4.1) regard the curve as squeezed between two staircases. But more frequently we approximate to the curve by a succession of straight lines or a succession of parabolas, and from these approaches we derive two widely-used results, the trapezium rule and Simpson's rule:

The trapezium rule

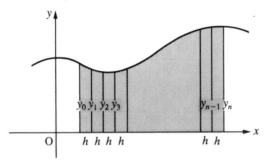

Divide the area into a number of parallel strips, each of width h, and let the ordinates of the curves at successive points be

$$y_0, y_1, y_2, y_3, \ldots y_n$$

Area under curve $\approx$ sum of the areas of trapezia

$$\approx h\frac{y_0 + y_1}{2} + h\frac{y_1 + y_2}{2} + h\frac{y_2 + y_3}{2} + \cdots + \cdots + h\frac{y_{n-1} + y_n}{2}$$

$$\approx h\{\tfrac{1}{2}y_0 + y_1 + y_2 + y_3 + \cdots + y_{n-1} + \tfrac{1}{2}y_n\}$$

$$\approx h\{\tfrac{1}{2}(y_0 + y_n) + (y_1 + y_2 + \cdots + y_{n-1})\}$$

So $\text{area} \approx \text{strip width} \times \begin{cases} \text{average of extreme ordinates} \\ + \text{ sum of remaining ordinates} \end{cases}$

Simpson's rule

Just as any *two* successive points can always be joined by a straight line $y = a + bx$, so any *three* successive points can always be joined by a parabola $y = a + bx + cx^2$. But before we can assemble a collection of areas beneath successive parabolas we first need to know the area beneath a single specimen.

Suppose that a curve has ordinates y_0, y_1, y_2 at points whose successive horizontal separations are h, and we take axes Ox, Oy as follows:

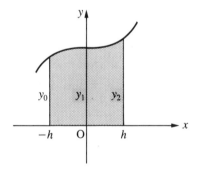

Let the parabolic section joining $(-h, y_0), (0, y_1), (h, y_2)$ be

$$y = a + bx + cx^2$$

$$\text{Area under parabola} = \int_{-h}^{h} (a + bx + cx^2)\,dx$$

$$= \left[ax + \tfrac{1}{2}bx^2 + \tfrac{1}{3}cx^3 \right]_{-h}^{h}$$

$$= 2ah + \tfrac{2}{3}ch^3$$

$$= \tfrac{1}{3}h(6a + 2ch^2)$$

But since the parabola goes through $(-h, y_0), (0, y_1), (h, y_2)$,

$$\left.\begin{array}{ll} a - bh + ch^2 = y_0 \\ a \qquad\quad\; = y_1 \\ a + bh + ch^2 = y_2 \end{array}\right\} \text{from which we see that} \quad 6a + 2ch^2 = y_0 + 4y_1 + y_2$$

So   area under parabola $= \frac{1}{3}h(y_0 + 4y_1 + y_2)$

We can now divide our total area into an *even* number of strips $(2n)$ and let successive heights of the curve be

$$y_0, y_1, y_2, y_3, y_4, \ldots y_{2n}$$

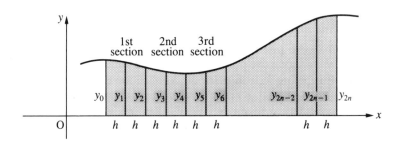

Approximating to the curve by a succession of parabolic sections, we see that

$$\text{area under curve} \approx \tfrac{1}{3}h(y_0 + 4y_1 + y_2) + \tfrac{1}{3}h(y_2 + 4y_3 + y_4) + \cdots$$
$$\cdots + \tfrac{1}{3}h(y_{2n-2} + 4y_{2n-1} + y_{2n})$$
$$\approx \tfrac{1}{3}h\{(y_0 + 4y_1 + y_2) + (y_2 + 4y_3 + y_4) + \cdots$$
$$\cdots + y_{2n-2} + 4y_{2n-1} + y_{2n})\}$$
$$\approx \tfrac{1}{3}h\{y_0 + 4y_1 + 2y_2 + 4y_3 + 2y_4 + \cdots$$
$$\cdots + 2y_{2n-2} + 4y_{2n-1} + y_{2n}\}$$
$$\approx \tfrac{1}{3}h\left\{\begin{array}{l}(y_0 + y_{2n}) \\ + 4(y_1 + y_3 + y_5 + \cdots + y_{2n-1}) \\ + 2(y_2 + y_4 + \cdots + y_{2n-2})\end{array}\right\}$$

This is Simpson's rule, that if a curve be divided into an *even* number of strips of equal width,

$$\text{area beneath curve} \approx \tfrac{1}{3} \times \text{strip width} \times \left\{\begin{array}{l}\text{sum of end ordinates} \\ + 4 \times \text{sum of odd ordinates} \\ + 2 \times \text{sum of remaining} \\ \qquad\qquad\text{even ordinates}\end{array}\right\}$$

Example

Use both the trapezium rule and Simpson's rule to calculate

$$\int_0^6 \sqrt{(1 + x^2)}\, dx$$

The values of $\sqrt{(1 + x^2)}$ at unit intervals can be tabulated, and the calculation set out as follows:

x	$1 + x^2$	$\sqrt{(1 + x^2)}$	trapezium rule		Simpson's rule		
0	1	1.000	1.000		1.000		
1	2	1.414		1.414		1.414	
2	5	2.236		2.236			2.236
3	10	3.162		3.162		3.162	
4	17	4.123		4.123			4.123
5	26	5.099		5.099		5.099	
6	37	6.083	6.083		6.083		

	trapezium		Simpson's		
	2)7.083	16.034	7.083	9.675	6.359
	3.5415		38.700	× 4	× 2
	16.034 ←		12.718 ← 38.700	12.718
	19.5755		3)58.501	
			19.500	

So, to 4 significant figures,

$$\int_0^6 \sqrt{(1 + x^2)}\, dx = \begin{matrix} 19.58 \text{ by the trapezium rule} \\ 19.50 \text{ by Simpson's rule} \end{matrix}$$

[We shall later be able to show that its actual value to 4 significant figures is 19.50.]

Exercise 4.4

1 The depth of a river of width 80 m is measured at 10 m intervals of its cross-section as follows:

distance (m)	0	10	20	30	40	50	60	70	80
depth (m)	0	1.31	3.17	6.24	3.79	1.16	5.25	3.76	2.48

Estimate the area of this cross-section by **a)** the trapezium rule **b)** Simpson's rule.

2 The speed of an accelerating car is recorded at second intervals as follows:

t (s)	0	1	2	3	4	5	6	7	8	9	10
v (m s^{-1})	0	5.3	8.1	10.3	11.9	13.0	14.1	14.8	15.3	15.6	15.8

Estimate the distance it travels in 10 s using **a)** the trapezium rule **b)** Simpson's rule.

3 The acceleration, at half-second intervals, of a stone dropped from the surface of a deep lake is given by

t (s)	0	0.5	1.0	1.5	2.0	2.5	3.0
a (m s^{-2})	3.37	1.54	0.93	0.68	0.53	0.41	0.30

Estimate its speed after 3 seconds using **a)** the trapezium rule **b)** Simpson's rule.

4 Estimate $\int_1^2 \dfrac{dx}{x}$ by means of **a)** the trapezium rule **b)** Simpson's rule

using only two intervals.
　　Given that its correct value, to 4 decimal places, is 0.6931, calculate the percentage errors of each estimate.

5 Repeat the last question, using ten intervals.

6 A quadrant of a circle of radius 10 cm divided into ten parallel strips. Use **a)** the trapezium rule and **b)** Simpson's rule to calculate π.

7 It can be shown that $\displaystyle\int_0^1 \dfrac{dx}{1 + x^2} = \dfrac{\pi}{4}$

Use Simpson's rule (10 strips) to calculate π.

8 Tabulate to 3 decimal places the values of the function $f(x) = \sqrt{(1 + x^2)}$ for values of x from 0 to 0.8 at intervals of 0.1. Use these values to estimate $\displaystyle\int_0^{0.8} f(x)\,dx$ by Simpson's method, **a)** using all the ordinates, and

b) using only ordinates at intervals of 0.2.
Draw any conclusions you can about the accuracy of the results.   (MEI)

* 4.5 Other summations: volumes of revolution

We now return to our observation in section 4.1, that an integral is simply the limit of the sum of a large number of small quantities.
For example,

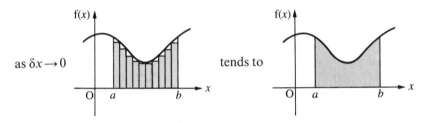

as $\delta x \to 0$ tends to

and $\displaystyle\sum_{x=a}^{x=b} f(x)\,\delta x \to \int_a^b f(x)\,dx$

This is an idea which has wide applications, and we shall usually denote such a sum by $\sum$ (sigma, the Greek capital S).

So $\displaystyle\sum f(x)\,\delta x \to \int_a^b f(x)\,dx$

Volumes of revolution

If $f(x) > 0$ when $a \leqslant x \leqslant b$, find the volume formed when the area between $y = f(x)$, $y = 0$, $x = a$ and $x = b$ is rotated about Ox.

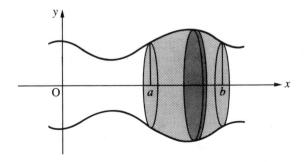

The volume of a typical disc of thickness δx is $\pi y^2 \delta x$.

So the sum of all these volumes is $\displaystyle\sum_{x=a}^{x=b} \pi y^2 \delta x$

$\Rightarrow$ required volume $= \displaystyle\int_a^b \pi y^2 \, dx$

Example 1

Find the volume of the solid formed when the area bounded by $y = x^2$, $y = 0$, $x = 1$, and $x = 2$ is rotated about Ox.

$$\text{Volume} = \int_1^2 \pi y^2 \, dx$$

$$= \int_1^2 \pi x^4 \, dx$$

$$= \left[\frac{1}{5} \pi x^5 \right]_1^2$$

$$= \frac{32\pi}{5} - \frac{\pi}{5} = \frac{31\pi}{5}$$

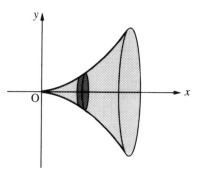

Example 2

Find:
a) the area enclosed by $y = x^2$, $y = 1$, $y = 2$ and $x = 0$.
b) the volume formed when this area is rotated about Oy.

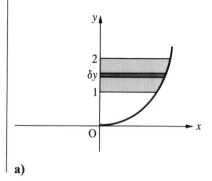

a)

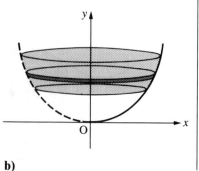

b)

a) Area of a typical rectangular element $= x\delta y$

$$\text{Sum of such areas} = \sum x\delta y$$

$$\text{Required area} = \int_1^2 x\,dy$$

$$= \int_1^2 \sqrt{y}\,dy$$

$$= \left[\frac{2}{3}y^{3/2}\right]_1^2 = \frac{2}{3}(2\sqrt{2}-1)$$

b) Volume of a typical circular disc $= \pi x^2\,\delta y$

$$\text{Sum of such volumes} = \sum \pi x^2\,\delta y$$

$$\text{Required volume} = \int_1^2 \pi x^2\,dy$$

$$= \int_1^2 \pi y\,dy$$

$$= \pi\left[\frac{1}{2}y^2\right]_1^2 = \frac{3\pi}{2}$$

Example 3

Find the volume of water required to fill a hemispherical bowl of radius 5 cm to a depth of 3 cm.

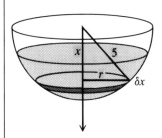

At depth x below centre, radius of cross-section is r, where

$$r^2 = 25 - x^2$$

So volume of a typical disc of thickness δx is

$$\pi r^2\,\delta x = \pi(25 - x^2)\delta x$$

Sum of such volumes $= \sum \pi(25 - x^2)\delta x$

so   total volume $= \displaystyle\int_2^5 \pi(25 - x^2)\,dx$

$$= \pi\left[25x - \tfrac{1}{3}x^3\right]_2^5$$

$$= \pi(25 \times 5 - \tfrac{1}{3} \times 5^3) - \pi(25 \times 2 - \tfrac{1}{3} \times 2^3)$$

$$= \pi(\tfrac{2}{3} \times 125 - \tfrac{142}{3})$$

$$= 36\pi$$

$$= 113\,\text{cm}^3$$

Example 4

A cone has base-area A and height h. Show that its volume is $\tfrac{1}{3}Ah$.

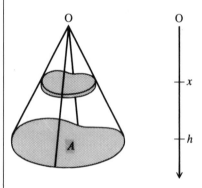

By similarity, the area of a slice parallel to the base at depth x below the vertex $= \dfrac{x^2}{h^2} A$.

If its thickness is δx, it volume is $A\dfrac{x^2}{h^2}\delta x$

So total volume of all such slices is $\sum \dfrac{Ax^2}{h^2}\delta x$

$\Rightarrow$ Volume of cone is $\displaystyle\int_0^h \dfrac{Ax^2}{h^2}\,dx = \left[\dfrac{Ax^3}{3h^2}\right]_0^h = \dfrac{1}{3}Ah$

Exercise 4.5

1 Calculate the volumes formed by rotating about Ox the areas bounded by
   a) $y = x, y = 0, x = 1, x = 2$
   b) $y = x^2 + 1, y = 0, x = 0, x = 1$
   c) $y = \dfrac{1}{x}, y = 0, x = 2, x = 3$
   d) $y = \pm \sqrt{x}, x = 4$
   e) $y = x, y = x^2$
   f) $y = x - x^2, y = 0$

2 Calculate the areas bounded by the following curves:
   a) $y = \frac{1}{2}x, y = 1, y = 2, x = 0$
   b) $y = x^2, y = 1, y = 4, x = 0$
   c) $y = x^3, y = 8, x = 0$
   d) $y = \dfrac{1}{x^4}, y = 1, y = 16$
   e) $y = \sqrt{x}, y = 2, x = 0$

3 Calculate the volumes formed when the areas in no. **2** are rotated about Oy.

4 Calculate the volume of a cone of base-radius r and height h by revolving the line $y = \dfrac{r}{h}x$ about Ox.

5 Calculate the volume of a sphere of radius r by revolving $x^2 + y^2 = r^2$ about Ox.

6 Draw a rough sketch of the curve $y^2 = 16x$ between $x = 0$ and $x = 4$. Calculate the area contained between the curve and the line $x = 4$, and the volume obtained by completely rotating that area about the axis of x.
   (OC)

7 Find the points of intersection of the curve $y = x(4 - x)$ and the line $y = 2x$.
   Find the volume generated when the area enclosed between the curve and the line makes one complete revolution about the x-axis.   (OC)

8 The curves $y^2 = 2x, x^3 = 4y$ intersect at the points $(0, 0)$ and $(2, 2)$. Find the area they enclose and the volume generated by revolving this area through four right angles about the axis of x.   (OC)

9 O is the origin and P is the point $(a, 2a)$ on the parabola $y^2 = 4ax$. Prove that the area bounded by the chord OP and the arc of the parabola between O and P is $\frac{1}{3}a^2$.
   Find the volume generated if this area is revolved about the x-axis through four right angles.   (OC)

10   Find the area enclosed between the curves $y^2 = 4x$ and $x^2 = 4y$.
Find the volume generated when this area is rotated through four right angles about the x-axis.

11   A container 9 m high is such that when the depth of liquid in it is x m, the area of the surface of the liquid is $(10x - x^2)$ m². What is its capacity when full?   (OC)

12   A hole of radius 3 cm is bored symmetrically through a sphere of radius 5 cm. Find the volume of the part which remains.

* 4.6 Mean values

Suppose that a particle is moving so that its velocity v (m s⁻¹) after time t (s) is given by $v = 6t^2$

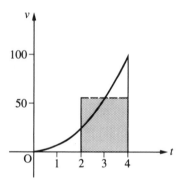

The distance travelled between $t = 2$ and $t = 4$ is

$$\int_2^4 6t^2 \, dt = \left[2t^3 \right]_2^4 = 128 - 16 = 112 \, \text{m}$$

So the average (or *mean*) velocity in this interval is

$$\frac{112}{2} \, \text{m s}^{-1} = 56 \, \text{m s}^{-1}$$

(Is this the same as the average of the velocities when $t = 2$ and $t = 4$?)
More generally, the *mean value of a function* f(x) *between* $x = a$ *and* $x = b$ can be defined as the height of the line beneath which there is the same area:

$$\frac{1}{b - a} \int_a^b f(x) \, dx$$

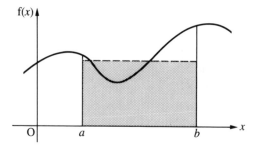

Example

The population of a circular parish of radius 1 km is uniformly distributed, and the church is at its centre. Find the mean distance of a parishioner from the church.

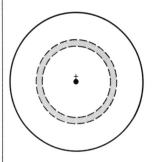

Let the total population be N.

Then   density of population $= \dfrac{N}{\pi}$ parishioners/km²

So number of parishioners at a distance between x and $x + \delta x$ is

$$\frac{N}{\pi}\{\pi(x + \delta x)^2 - \pi x^2\} \approx 2Nx\delta x$$

The sum of their distances from church

$$\approx 2Nx\delta x \times x = 2Nx^2\delta x$$

$$\Rightarrow \quad \text{mean distance} \approx \frac{\sum 2Nx^2\delta x}{N} = 2\sum x^2\delta x$$

Taking limits, their mean distance from church

$$= 2\int_0^1 x^2 \mathrm{d}x = \tfrac{2}{3}$$

So the mean distance is $\tfrac{2}{3}$ km.

Exercise 4.6

1 What is the mean value of
   a) $f(x) = 3x - x^2$ between $x = 0$ and 3;
   b) $f(x) = x^3$ between $x = 1$ and 5;
   c) $f(x) = \sqrt{x}$ between $x = 1$ and 2;
   d) $f(x) = \dfrac{1}{x^2}$ between $x = 1$ and 3?

2 The velocity of a particle after t s is v ms^{-1}, where $v = 6t - 3t^2$.
   What are its mean velocity and mean acceleration during   a) the first
   second   b) the first two seconds?

3 A sphere of radius 1 m has density which increases linearly from 0 at its
   centre to 1000 kg m^{-3} at its surface. Find:
   a) the mass of the sphere;
   b) its mean density.

4 A hive of bees swarms round its queen in a uniform sphere of radius 10 cm.
   What is their mean distance from her?

* 4.7 Integration by substitution

So far, when we have been faced with the problem of integrating a given
function, we have had to guess another function which when differentiated
yields the original. It may be difficult, or even impossible, to find an answer to
the problem; but when we think we have found one it is always possible to
check its truth by differentiation.

   Though integration remains a speculative venture which often calls for
considerable powers of imagination, there are some general methods which
are of assistance. In particular, because it is the reverse of differentiation, we
might expect any general result about differentiation to yield a corresponding
result about integration. In this section we shall look again at the chain rule
and, unlike our usual practice, shall go directly to the general case which we
shall then seek to illustrate by particular examples.

Suppose that $I = \int f(x)\,dx$, so that $\dfrac{dI}{dx} = f(x)$. Let us further suppose that
$x = \phi(t)$, where t is a new variable.

Then $\dfrac{dI}{dt} = \dfrac{dI}{dx} \times \dfrac{dx}{dt}$

                $= f(x)\,\phi'(t)$

$$= f[\phi(t)]\phi'(t)$$

$$\Rightarrow \quad I = \int f[\phi(t)]\phi'(t)\,dt$$

$$\Rightarrow \quad \boxed{\int f(x)\,dx = \int f[\phi(t)]\phi'(t)\,dt}$$

So in the given integral

$$\int f(x)\,dx$$

we can replace x by $\phi(t)$ providing we also replace dx by $\phi'(t)\,dt$.

This is most easily understood from a number of examples.

Example 1

Find $\int x\sqrt{(x+2)}\,dx$

As the integral is not immediately obvious we try a substitution, and as the most disagreeable part of the integral is the square root, we put

$$\sqrt{(x+2)} = t \quad \Rightarrow \quad x = t^2 - 2$$

so that dx is replaceable by $2t\,dt$.

So $\int x\sqrt{(x+2)}\,dx = \int (t^2 - 2)t\,2t\,dt$

$$= \int (2t^4 - 4t^2)\,dt = \tfrac{2}{5}t^5 - \tfrac{4}{3}t^3 + A$$

Hence, returning to the original variable x,

$$\int x\sqrt{(x+2)}\,dx = \tfrac{2}{5}(x+2)^{5/2} - \tfrac{4}{3}(x+2)^{3/2} + A$$

$$= \frac{2}{15}(x+2)^{3/2}\{3(x+2) - 10\} + A$$

$$= \frac{2}{15}(3x - 4)(x+2)^{3/2} + A$$

It is hardly surprising that we were unable to guess this answer, but we can at least check that it is correct:

$$\frac{d}{dx}\left\{\frac{2}{15}(3x-4)(x+2)^{3/2}\right\} = \frac{2}{15}\left\{(3x-4)\frac{3}{2}(x+2)^{1/2} + 3(x+2)^{3/2}\right\}$$

$$= \frac{2}{15}(x+2)^{1/2}\left\{\frac{9x}{2} - 6 + 3(x+2)\right\}$$

$$= \frac{2}{15}(x+2)^{1/2}\frac{15x}{2} = x(x+2)^{1/2}$$

So $\displaystyle\int x\sqrt{(x+2)}\,dx = \frac{2}{15}(3x-4)(x+2)^{3/2} + A$

Example 2

Calculate $\displaystyle I = \int_1^2 \frac{x\,dx}{\sqrt[3]{(x-1)}}$

Here we try $t = \sqrt[3]{(x-1)}$

$$\Rightarrow\quad x = t^3 + 1 \quad \Rightarrow\quad dx = 3t^2\,dt$$

So $\displaystyle\int \frac{x\,dx}{\sqrt[3]{(x-1)}} = \int \frac{(t^3+1)3t^2\,dt}{t}$

$$= \int (3t^4 + 3t)\,dt$$

Now the limits of integration were $x = 1$ and $x = 2$, and so become
$t = \sqrt[3]{(1-1)} = 0$ and $t = \sqrt[3]{(2-1)} = 1$

Hence, $\displaystyle I = \int_0^1 (3t^4 + 3t)\,dt$

$$= \left[\frac{3}{5}t^5 + \frac{3}{2}t^2\right]_0^1 = \frac{3}{5} + \frac{3}{2} = \frac{21}{10} = 2.1$$

Sometimes it is easier to express the connection between x and t with t as a function of x:

Example 3

$$\int x\sqrt{(1 - x^2)}\,dx$$

Here we put $t = 1 - x^2$

$\Rightarrow$ $\quad\quad\quad dt = -2x\,dx$

Now $\quad\displaystyle\int x\sqrt{(1 - x^2)}\,dx = \int \sqrt{(1 - x^2)}x\,dx$

$$= \int \sqrt{t}\,\frac{-dt}{2}$$

$$= \int -\tfrac{1}{2}t^{1/2}\,dt = -\tfrac{1}{3}t^{3/2} + A$$

So $\quad\displaystyle\int x\sqrt{(1 - x^2)}\,dx = -\tfrac{1}{3}(1 - x^2)^{3/2} + A$

Example 4

Calculate $\quad I = \displaystyle\int_0^1 \frac{x^2\,dx}{(x^3 + 1)^2}$

If $u = x^3 + 1$, then $du = 3x^2\,dx$. Also $x = 0 \Rightarrow u = 1$ and $x = 1 \Rightarrow u = 2$.

So $\quad\quad I = \displaystyle\int_1^2 \frac{\tfrac{1}{3}du}{u^2} = \int_1^2 \frac{1}{3u^2}\,du$

$$= \left[-\frac{1}{3u} \right]_1^2 = \left(-\frac{1}{6} \right) - \left(-\frac{1}{3} \right) = \frac{1}{6}$$

Hence $\quad\displaystyle\int_0^1 \frac{x^2}{(x^3 + 1)^2}\,dx = \frac{1}{6}$

Exercise 4.7

1   Find:

a) $\displaystyle\int (x + 1)^5\,dx$ (put $x = t - 1$)   **b)** $\displaystyle\int \frac{dx}{(1 - 2x)^2}$ $\left(\text{put } x = \frac{1 - t}{2} \right)$

c) $\int \sqrt{(2-x)}\,dx$ **d)** $\int (2x+1)^{3/2}\,dx$

e) $\int \dfrac{x\,dx}{\sqrt[3]{(1+x)}}$

2 Find:

a) $\int x(x^2+1)^4\,dx$ (put $t = x^2 + 1$)

b) $\int x^2(2x^3+3)^4\,dx$ (put $t = 2x^3 + 3$) **c)** $\int x\sqrt{(1-x^2)}\,dx$

d) $\int \dfrac{1+2x}{\sqrt{(x+x^2)}}\,dx$ **e)** $\int \dfrac{x}{\sqrt[3]{(x^2-1)}}\,dx$

3 Evaluate:

a) $\displaystyle\int_0^1 (1-x)^4\,dx$ **b)** $\displaystyle\int_2^3 \dfrac{dx}{(2x+1)^2}$

c) $\displaystyle\int_0^3 x\sqrt{(x^2+1)}\,dx$ **d)** $\displaystyle\int_1^2 \dfrac{x^2\,dx}{\sqrt{(x^3-1)}}$

e) $\displaystyle\int_1^4 \sqrt{\dfrac{1+\sqrt{x}}{x}}\,dx$ (put $x = (u-1)^2$)

Miscellaneous problems

1 It is known that the area beneath the curve $y = 1/x$ from $x = 1$ to $x = 10^6$ is approximately 13.8. Use the staircase method to calculate the sum of the reciprocals of the integers from 1 to 1 000 000 as accurately as you can.

2 Find the approximate value of the sum of the square roots of the first million integers.

3 Show that $\displaystyle\int_{-a}^a f(x)\,dx$ is $2\displaystyle\int_0^a f(x)\,dx$ when $f(x)$ is an even function, and 0 when $f(x)$ is an odd function.
   Hence, calculate:

a) $\displaystyle\int_{-1}^1 \dfrac{x^3\,dx}{(1+x^2)^2}$ **b)** $\displaystyle\int_{-2}^2 \dfrac{1+x^3+x^4}{1+x^4}\,dx$

4 Draw a rough sketch of the graph of $y^2 = x(1 - x)^2$, and find the area enclosed by the loop.

Find also the volume of the solid formed by rotating this area about the x-axis.   (OC)

5 Find the coordinates of the points of intersection of the two parabolas $y^2 = 4ax$ and $x^2 = 4ay$ $(a > 0)$. Show that the area between these parabolas is $16a^2/3$.

This area is rotated through four right angles about the x-axis. Show that the volume generated is $96\pi a^3/5$.

6 Sketch the parabolas

$$y^2 = 4x \quad \text{and} \quad y^2 = 5x - 4$$

and find their points of intersection.

A fruit-bowl is made by rotating completely about the x-axis the area enclosed by the curves. Find the volume of the material required to make the bowl.   (SMP)

7 A napkin-ring is formed by boring a cylindrical hole symmetrically through a sphere of radius a. If the height of the ring is h when resting on one of its circular ends, find its volume.

8 An ellipse of length $2a$ and breadth $2b$ has the equation

$$\frac{x^2}{a^2} + \frac{y^2}{b^2} = 1$$

Find the volume of the solid formed when it is rotated about the x-axis.

9 An oil-tanker has a water-line length of 200 m and at a point x m from its bow its underwater cross-sectional area is

$$\frac{3}{1000} x^2 (200 - x)\, \text{m}^2$$

Find its displacement in tonnes.

10 The figure shown is the cross-section of a tyre lying on the ground.

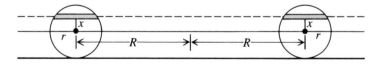

Each circular cross-section has radius r and a centre which is at distance R from the centre of the tyre. The dotted plane is at a height x above the horizontal plane of symmetry. Show that it intersects the tyre in a ring whose outer and inner radii are

$$R + \sqrt{(r^2 - x^2)} \quad \text{and} \quad R - \sqrt{(r^2 - x^2)},$$

and which has area

$$4\pi R\sqrt{(r^2 - x^2)}$$

Hence show that the volume of the tyre is $2\pi^2 Rr^2$.

5 Trigonometric functions

5.1 Sine, cosine and tangent

Sine and cosine

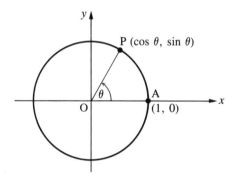

Suppose that a circle of unit radius is drawn with centre at (0, 0) and that A is the point (1, 0). Suppose further that the line OA is rotated anticlockwise through angle θ so that A moves to P. Then the coordinates of P are called the *cosine* and *sine*, respectively, of the angle θ

$$\cos \theta = x \qquad \sin \theta = y$$

It is clear from this definition that a negative rotation will be effectively clockwise, and also that any two angles which differ by a multiple of 360° must have the same sine and cosine. We therefore say that $\sin \theta$ and $\cos \theta$ are *periodic functions*, with period 360°, and it is this feature of $\sin \theta$ and $\cos \theta$ which is fundamental to their wide importance in the theory of waves, alternating currents, oscillations and vibrations of all kinds.

The *periodicity* of $\sin \theta$ and $\cos \theta$ is displayed by plotting their graphs:

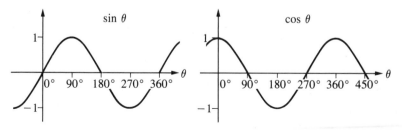

We can obtain a number of useful results simply by looking at the symmetries of the following figure:

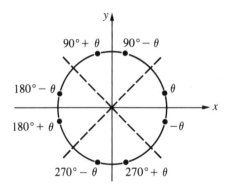

In particular

$$\sin(-\theta) = -\sin\theta \qquad \cos(-\theta) = +\cos\theta$$

i.e. $\sin\theta$ is an odd function and $\cos\theta$ is an even function

$$\sin(90° - \theta) = +\cos\theta \qquad \cos(90° - \theta) = +\sin\theta$$

i.e. the sine of an angle θ is the *co*sine of its *co*mplementary angle $90° - \theta$

$$\sin(90° + \theta) = +\cos\theta \qquad \cos(90° + \theta) = -\sin\theta$$

Each of these has an interpretation for the graphs of the functions, e.g. $\sin(90° + \theta) = \cos\theta$ shows that the graph of $\sin\theta$ is simply the graph of $\cos\theta$ displaced 90° to the right:

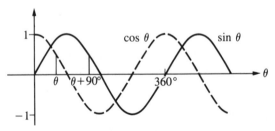

Tangent

The other most frequently used trigonometric function is $\tan\theta$, defined by

$$\tan\theta = \frac{\sin\theta}{\cos\theta}$$

(and so undefined when $\cos\theta = 0$).

In the original diagram, $\tan\theta$ is represented by $\dfrac{\sin\theta}{\cos\theta} = \dfrac{y}{x}$, i.e. by the gradient of the radius.

It, too, can be plotted from the above table and (as would be expected of a gradient) is periodic with period 180°.

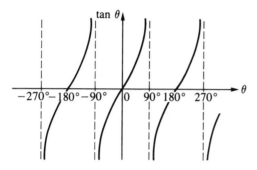

Exercise 5.1a

1 Use the diagram below to find the approximate values of $\sin\theta$, $\cos\theta$ and $\tan\theta$, where θ is
    a) 35° **b)** 167° **c)** 214° **d)** 304° **e)** 400° **f)** −32°

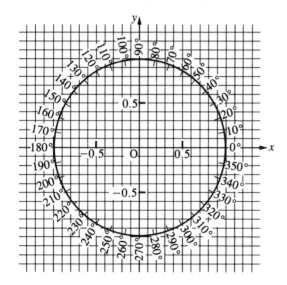

2 Use your calculator to check all your answers to no. **1**.

3 Express in terms of $\sin\theta$, $\cos\theta$, and $\tan\theta$:
    a) $\sin(180° - \theta)$, $\cos(180° - \theta)$, $\tan(180° - \theta)$

b) $\sin(180° + \theta)$, $\cos(180° + \theta)$, $\tan(180° + \theta)$
c) $\sin(270° + \theta)$, $\cos(270° + \theta)$, $\tan(270° + \theta)$

'Set-square' angles: $45°$, $30°$, $60°$

We can easily calculate the trigonometric functions of these angles from the dimensions of two set-squares, each of height 1 unit.

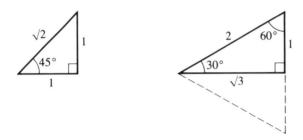

If the second set square is reflected in its base, it forms an equilateral triangle, so that the hypotenuse of the set square must be 2 and its base $\sqrt{3}$.

So
$$\sin 45° = \frac{1}{\sqrt{2}} \qquad \sin 30° = \frac{1}{2} \qquad \sin 60° = \frac{\sqrt{3}}{2}$$

$$\cos 45° = \frac{1}{\sqrt{2}} \qquad \cos 30° = \frac{\sqrt{3}}{2} \qquad \cos 60° = \frac{1}{2}$$

$$\tan 45° = 1 \qquad \tan 30° = \frac{1}{\sqrt{3}} \qquad \tan 60° = \sqrt{3}$$

Exercise 5.1b

What are the exact values of $\sin\theta$, $\cos\theta$, and $\tan\theta$ (in terms of square roots where necessary) if θ is
a) $120°$ **b)** $225°$ **c)** $330°$ **d)** $-60°$ **e)** $-150°$ **f)** $-315°$?

5.2 Simple equations

We are now able to find the value of the sine, cosine and tangent of any angle. But very frequently we need to ask the inverse question: for what value (or values) of θ is $\sin\theta = \quad 0.27$?
$$\text{or} \quad \cos\theta = -0.73?$$
$$\text{or} \quad \tan\theta = -2?$$

Example 1

$\sin \theta = 0.27$

By calculator, we find that $\theta = 15.7°$ is a solution.

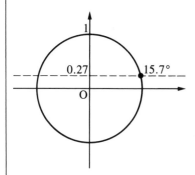

But it is also clear from the figure that another possible solution is $\theta = 164.3°$, and that we can obtain further solutions from these two by adding or subtracting multiples of $360°$.

So $\quad \theta = \ldots -344.3°, 15.7°, 375.7° \ldots$
and $\quad \theta = \ldots -195.7°, 164.3°, 524.3° \ldots$

are all solutions of the given equation.
This can also be seen from the graph of $\sin \theta$:

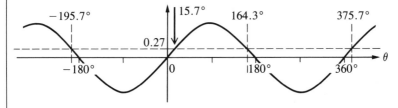

Example 2

$\cos \theta = -0.73$
By calculator, we find that $\cos \theta = +0.73$ has a solution $\theta = 43.1°$ and, as is confirmed by the figure, $\cos \theta = -0.73$ has solutions $\theta = \pm 136.9°$. Finally, by adding multiples of $360°$, we obtain the full set of solutions $\theta = \ldots -223.1°, -136.9°, 136.9°, 223.1° \ldots$

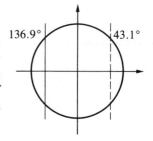

Example 3

$\tan \theta = -2$

In this case we look for the line whose gradient is -2:

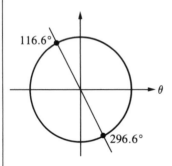

The first solution of $\tan \theta = -2$ is clearly $\theta = 116.6°$, and further solutions can be obtained by half-revolutions from this position. So the full list of solutions is

$$\theta = \ldots -243.4°,\ -63{,}4°,\ 116.6°,\ 296.6°,\ 476.6°,\ 656.6° \ldots,$$

as can also be seen from the graph of $\tan \theta$:

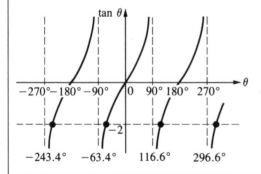

More generally, we can express the solutions to simple equations as follows:

a) $\sin x = \sin \alpha$

$\Rightarrow \qquad x = \alpha + 360n°$

or $(180° - \alpha) + 360n°$

where n is an integer (or $n \in \mathbb{Z}$)

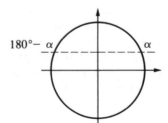

b) $\cos x = \cos \alpha$
$\Rightarrow$ $x = \pm\alpha + 360n°$ $(n \in \mathbb{Z})$

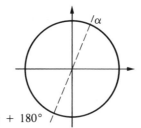

c) $\tan x = \tan \alpha$
$\Rightarrow$ $x = \alpha + 180n°$ $(n \in \mathbb{Z})$

$\alpha + 180°$

Exercise 5.2

Find the values of θ between $0°$ and $360°$ for which

1 **a)** $\sin \theta = 0.37$ **b)** $\sin \theta = 1$ **c)** $\sin \theta = -0.74$

2 **a)** $\cos \theta = 0.45$ **b)** $\cos \theta = -0.61$ **c)** $\cos \theta = -1$

3 **a)** $\tan \theta = -1$ **b)** $\tan \theta = 0$ **c)** $\tan \theta = 2.72$

4 **a)** $\sin^2 \theta + \sin \theta = 0$
      b) $2\cos^2 \theta - 3\cos \theta + 1 = 0$
      c) $3\tan^2 \theta = 1$

5 **a)** $\sin \frac{3}{2}\theta = -0.4266$
      b) $\cos \frac{1}{2}\theta = -0.5230$
      c) $\tan 2\theta = +0.4550$

5.3 Cotangent, secant, cosecant, and the use of Pythagoras' theorem

Three further trigonometric functions are defined by

$$\cot \theta = \frac{1}{\tan \theta} \qquad \sec \theta = \frac{1}{\cos \theta} \qquad \operatorname{cosec} \theta = \frac{1}{\sin \theta}$$

so their properties can be deduced from those of the original functions.

For example, $\cos \theta = \sin (90° - \theta)$

$\Rightarrow \qquad \dfrac{1}{\cos \theta} = \dfrac{1}{\sin (90° - \theta)}$

$\Rightarrow \qquad \sec \theta = \operatorname{cosec} (90° - \theta)$

$(\cos \theta, \sin \theta)$

So the secant of an angle is the *co*secant of its *complementary* angle.

It is clear from Pythagoras' theorem that, for any angle θ,

$$(\sin \theta)^2 + (\cos \theta)^2 = 1$$

We usually write $(\sin \theta)^2$ and $(\cos \theta)^2$ more simply as $\sin^2 \theta$ and $\cos^2 \theta$, so that the above result is written

$$\sin^2 \theta + \cos^2 \theta = 1$$

Dividing by $\cos^2 \theta$ and $\sin^2 \theta$, we obtain

$$\frac{\sin^2 \theta}{\cos^2 \theta} + \frac{\cos^2 \theta}{\cos^2 \theta} = \frac{1}{\cos^2 \theta} \quad \text{and} \quad \frac{\sin^2 \theta}{\sin^2 \theta} + \frac{\cos^2 \theta}{\sin^2 \theta} = \frac{1}{\sin^2 \theta}$$

$\Rightarrow \qquad \tan^2 \theta + 1 = \sec^2 \theta \quad \text{and} \qquad 1 + \cot^2 \theta = \operatorname{cosec}^2 \theta$

Summarising:

$$\sin^2 \theta + \cos^2 \theta = 1$$
$$\tan^2 \theta + 1 = \sec^2 \theta$$
$$\cot^2 \theta + 1 = \operatorname{cosec}^2 \theta$$

Example 1

Simplify $\dfrac{1}{\sec A - \tan A}$

$$\frac{1}{\sec A - \tan A} = \frac{\sec^2 A - \tan^2 A}{\sec A - \tan A}$$

$$= \frac{(\sec A + \tan A)(\sec A - \tan A)}{\sec A - \tan A}$$

$$= \sec A + \tan A$$

Example 2

Solve the equation

$$3 \sin^2 x = 3 \cos^2 x - \cos x + 1$$

for values of x between $0°$ and $360°$.

$$3 \sin^2 x = 3 \cos^2 x - \cos x + 1$$

$\Rightarrow \quad 3 - 3 \cos^2 x = 3 \cos^2 x - \cos x + 1$

$\Rightarrow \quad 6 \cos^2 x - \cos x - 2 = 0$

$\Rightarrow \quad (3 \cos x - 2)(2 \cos x + 1) = 0$

$\Rightarrow \quad \cos x = \frac{2}{3} \quad \text{or} \quad -\frac{1}{2}$

$\Rightarrow \quad x = 48.2°, \quad 311.8°, \quad \text{or} \quad 120°, \quad 240°$

So $x = 48.2°, \quad 120°, \quad 240°, \quad \text{or} \quad 311.8°$

Example 3

Eliminate θ from the equations

$$x = 2 + 3 \cos \theta, \quad y = 1 + 3 \sin \theta$$

As $3 \cos \theta = x - 2$ and $3 \sin \theta = y - 1$,

$$(x - 2)^2 + (y - 1)^2 = 9 \cos^2 \theta + 9 \sin^2 \theta$$

$\Rightarrow \quad (x - 2)^2 + (y - 1)^2 = 9$

So the original equations define a point (x, y) which always lies on the circle

$$(x - 2)^2 + (y - 1)^2 = 9$$

with centre $(2, 1)$ and radius 3.

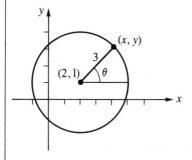

Exercise 5.3

1 Sketch the graph of
a) $\cot x$ **b)** $\sec x$ **c)** $\operatorname{cosec} x$ for values of x from $-180°$ to $+540°$.

2 Simplify:
a) $\cos\theta\tan\theta$
b) $\sin\theta\sec\theta$
c) $\operatorname{cosec}\theta\tan\theta$
d) $\cot\theta\sec\theta$

3 Simplify:
a) $\dfrac{1-\sin^2\theta}{1-\cos^2\theta}$

b) $\dfrac{\cos^2\theta}{1+\sin\theta}+\dfrac{\cos^2\theta}{1-\sin\theta}$

c) $\dfrac{\tan^2\theta}{\sec\theta+1}$

d) $\dfrac{1}{\cot\theta+\tan\theta}$

e) $(\operatorname{cosec}\theta+1)(\operatorname{cosec}\theta-1)$ **f)** $\dfrac{\operatorname{cosec}\theta}{\operatorname{cosec}\theta-\sin\theta}$

g) $\dfrac{\sec^2\theta+2\tan\theta}{(\cos\theta+\sin\theta)^2}$

h) $\dfrac{1+2\sin\theta\cos\theta}{\sin\theta+\cos\theta}$

4 Solve the following equations for values of θ between $0°$ and $360°$:
a) $\sec\theta=2$
b) $\cot\theta=-1$
c) $2\sin^2\theta+\cos\theta=1$
d) $\sec^2\theta=1+\tan\theta$
e) $\tan\theta+4\cot\theta=4\sec\theta$
f) $3\tan^2\theta-5\sec\theta+1=0$

5 **a)** If $\sec x-\tan x=\frac{1}{3}$, find the values of $\sec x$ and $\tan x$.
b) Given that $x=\tan\theta-\sin\theta$, $y=\tan\theta+\sin\theta$, prove that
$(x^2-y^2)^2=16xy$.

* 5.4 Sine and cosine rules

If we know the sizes a, b, c of the three sides of a triangle ABC, then we can construct it, and the angles A, B, C are automatically determined. We therefore expect that there will be some relationships between a, b, c and A, B, C.

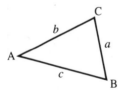

To investigate these, we set the triangle on a framework of axes so that A lies at $(0,0)$ and B at $(c,0)$, with C above the x-axis. (If the original triangle is lettered clockwise rather than anti-clockwise this can be achieved by first turning the triangle upside-down.)

The triangle is then in the following position, and in every possible case the coordinates of C are $(b \cos A, b \sin A)$, even when A is obtuse.

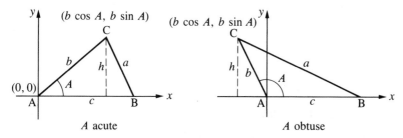

A acute   A obtuse

Sine rule

If the height of C above AB is h, then in each case

$$h = b \sin A$$

Similarly $\qquad h = a \sin B$

So $\qquad a \sin B = b \sin A$

$\Rightarrow \qquad \dfrac{a}{\sin A} = \dfrac{b}{\sin B}$

Similarly $\qquad \dfrac{a}{\sin A} = \dfrac{c}{\sin C}$

So $\qquad \boxed{\dfrac{a}{\sin A} = \dfrac{b}{\sin B} = \dfrac{c}{\sin C}}$

Cosine rules

It is also clear in the above triangle that

$$a^2 = BC^2 = (b \cos A - c)^2 + (b \sin A - 0)^2$$

$\Rightarrow \qquad a^2 = b^2 \cos^2 A - 2bc \cos A + c^2 + b^2 \sin^2 A$

$\Rightarrow \qquad \boxed{a^2 = b^2 + c^2 - 2bc \cos A}$

Similarly, $\qquad b^2 = c^2 + a^2 - 2ca \cos B$

and $\qquad c^2 = a^2 + b^2 - 2ab \cos C$

Example

In $\triangle ABC$, $a = 4$, $b = 3$, and $c = 2$. Find the angles A, B, C.

Using the cosine rule $4^2 = 3^2 + 2^2 - 2 \times 3 \times 2 \cos A$

$$\Rightarrow \qquad\qquad \cos A = -\frac{3}{12} = -0.25 \quad \Rightarrow \quad A = 104.5°$$

By the sine rule $\dfrac{4}{\sin 104.5°} = \dfrac{3}{\sin B} = \dfrac{2}{\sin C}$

$\Rightarrow \quad \sin B = \frac{3}{4}\sin 104.5° = 0.726$

$\Rightarrow \quad B = 46.6°$ (since B cannot be obtuse)

and $\sin C = \frac{1}{2}\sin 104.5° = 0.484$

$\Rightarrow \quad C = 29°$

So $A = 104.5°$, $B = 46.6°$, $C = 28.9°$ (and we check that $A + B + C = 180°$).

Exercise 5.4

In nos. **1–4**, calculate the unknown sides and angles of $\triangle ABC$, where

1   **a)** $a = 5$, $A = 80°$, $B = 30°$
    b) $b = 4.17$, $A = 106.1°$, $B = 36.7°$

2   **a)** $a = 4$, $b = 5$, $c = 6$
    b) $a = 3.49$, $b = 4.62$, $c = 6.93$

3   **a)** $b = 3$, $c = 4$, $A = 50°$
    b) $a = 4.71$, $b = 3.62$, $C = 103.7°$

4   $a = 5$, $b = 3$, $B = 30°$
    (In this question, begin by drawing $\triangle ABC$ to scale; and state if there is more than one possible set of answers.)

5   A destroyer and a cruiser leave harbour at 0900 hours, the destroyer at 24 knots on course 037° and the cruiser at 15 knots on course 139°. Find the bearing, and the distance in nautical miles, of the destroyer from the cruiser at 1400 hours. (1 knot is a speed of 1 nautical mile per hour.)   (OC)

6   **a)** $\triangle ABC$ has BC $= 26$ cm, CA $= 14$ cm, AB $= 30$ cm. Calculate the angle BAC.
    b) $\triangle ABC$ has AB $= 30$ cm, AC $= 14$ cm, $\hat{BAC} = 60°$. BX and CY are

perpendiculars drawn to CA and AB respectively. Calculate the length of XY. (oc)

7 AM is a median of $\triangle ABC$ (i.e. M is the mid-point of BC). Use the cosine rule to prove Apollonius' theorem, that

$$AB^2 + AC^2 = 2(AM^2 + BM^2)$$

8 A triangle has sides 5 cm, 8 cm, 9 cm. Calculate:
   a) the size of its smallest angle;
   b) the length of its shortest median (using Apollonius' theorem).   (oc)

5.5 Compound angles: $a\cos\theta + b\sin\theta$

Suppose we are told that

$$\sin 10° = 0.1736 \quad \text{and} \quad \cos 10° = 0.9848$$
$$\sin 1° \; = 0.0175 \quad \text{and} \quad \cos 1° \; = 0.9998$$

Can we use these values in order to calculate the sines and cosines of $11°$ and $9°$?

The reader will probably find that the answers are not obvious. But similar questions will recur very frequently and it will soon become extremely important to know how

$$\sin(\theta \pm \phi) \quad \text{and} \quad \cos(\theta \pm \phi)$$

are connected with

$$\sin\theta, \cos\theta, \sin\phi, \text{ and } \cos\phi$$

We shall begin by investigating the rotation of a point $P(x, y)$ about O through an angle θ.

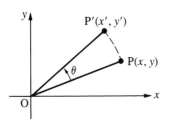
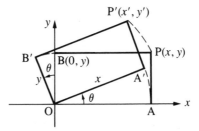

This can be achieved by rotating the rectangle OAPB through an angle θ, so that it takes the position OA'P'B'.

If (as in section 1.7) we denote P by $\begin{pmatrix} x \\ y \end{pmatrix}$, then $\quad A\begin{pmatrix} x \\ 0 \end{pmatrix}\quad$ moves to $\quad A'\begin{pmatrix} x\cos\theta \\ x\sin\theta \end{pmatrix}$

and $\quad B\begin{pmatrix} 0 \\ y \end{pmatrix}\quad$ moves to $\quad B'\begin{pmatrix} -y\sin\theta \\ y\cos\theta \end{pmatrix}$

But P′ is given by

$$x' = x_{A'} + x_{B'} = x\cos\theta - y\sin\theta$$

$$y' = y_{A'} + y_{B'} = x\sin\theta + y\cos\theta$$

Hence $\quad \begin{pmatrix} x' \\ y' \end{pmatrix} = \begin{pmatrix} x\cos\theta - y\sin\theta \\ x\sin\theta + y\cos\theta \end{pmatrix} = \begin{pmatrix} \cos\theta & -\sin\theta \\ \sin\theta & \cos\theta \end{pmatrix}\begin{pmatrix} x \\ y \end{pmatrix}$

and the rotation θ is a linear transformation with matrix

$$\mathbf{T}(\theta) = \begin{pmatrix} \cos\theta & -\sin\theta \\ \sin\theta & \cos\theta \end{pmatrix}$$

But rotation $\theta + \phi$ is the same as rotation ϕ followed by rotation θ, and so has matrix

$$\mathbf{T}(\theta + \phi) = \mathbf{T}(\theta)\mathbf{T}(\phi)$$

Hence $\quad \begin{pmatrix} \cos(\theta + \phi) & -\sin(\theta + \phi) \\ \sin(\theta + \phi) & \cos(\theta + \phi) \end{pmatrix}$

$$= \begin{pmatrix} \cos\theta & -\sin\theta \\ \sin\theta & \cos\theta \end{pmatrix}\begin{pmatrix} \cos\phi & -\sin\phi \\ \sin\phi & \cos\phi \end{pmatrix}$$

$$= \begin{pmatrix} \cos\theta\cos\phi - \sin\theta\sin\phi & -\sin\theta\cos\phi - \cos\theta\sin\phi \\ \sin\theta\cos\phi + \cos\theta\sin\phi & \cos\theta\cos\phi - \sin\theta\sin\phi \end{pmatrix}$$

So

$$\sin(\theta + \phi) = \sin\theta\cos\phi + \cos\theta\sin\phi$$
$$\cos(\theta + \phi) = \cos\theta\cos\phi - \sin\theta\sin\phi$$

We can now replace ϕ by $-\phi$ and remember that

$$\sin(-\phi) = -\sin\phi \quad \text{and} \quad \cos(-\phi) = \cos\phi$$

Then

$$\sin(\theta - \phi) = \sin\theta\cos\phi - \cos\theta\sin\phi$$
$$\cos(\theta - \phi) = \cos\theta\cos\phi + \sin\theta\sin\phi$$

These, therefore, are the formulae which were sought at the start of this section, and the reader can now use them to calculate the sines and cosines of $11°$ and $9°$ from those of $10°$ and $1°$.

They are all extremely important results which should be thoroughly understood and committed to memory: 'sin sum = sin cos + cos sin', etc.

Lastly $\tan(\theta + \phi) = \dfrac{\sin(\theta + \phi)}{\cos(\theta + \phi)} = \dfrac{\sin\theta\cos\phi + \cos\theta\sin\phi}{\cos\theta\cos\phi - \sin\theta\sin\phi}$

$$= \dfrac{\dfrac{\sin\theta\cos\phi + \cos\theta\sin\phi}{\cos\theta\cos\phi}}{\dfrac{\cos\theta\cos\phi - \sin\theta\sin\phi}{\cos\theta\cos\phi}}$$

So $$\tan(\theta + \phi) = \dfrac{\tan\theta + \tan\phi}{1 - \tan\theta\tan\phi}$$

Similarly $$\tan(\theta - \phi) = \dfrac{\tan\theta - \tan\phi}{1 + \tan\theta\tan\phi}$$

Example 1

Find expressions for the values of $\sin 75°$, $\cos 75°$, and $\tan 75°$.

$\sin 75° = \sin(45° + 30°)$

$\quad\quad = \sin 45° \cos 30° + \cos 45° \sin 30°$

$\quad\quad = \dfrac{1}{\sqrt{2}}\dfrac{\sqrt{3}}{2} + \dfrac{1}{\sqrt{2}}\dfrac{1}{2} = \dfrac{\sqrt{3}+1}{2\sqrt{2}}$

$\cos 75° = \cos(45° + 30°)$

$\quad\quad = \cos 45° \cos 30° - \sin 45° \sin 30°$

$\quad\quad = \dfrac{1}{\sqrt{2}}\dfrac{\sqrt{3}}{2} - \dfrac{1}{\sqrt{2}}\dfrac{1}{2} = \dfrac{\sqrt{3}-1}{2\sqrt{2}}$

$\tan 75° = \tan(45° + 30°)$

$\quad\quad = \dfrac{\tan 45° + \tan 30°}{1 - \tan 45° \tan 30°}$

$$= \frac{1 + \dfrac{1}{\sqrt{3}}}{1 - \dfrac{1}{\sqrt{3}}} = \frac{\sqrt{3} + 1}{\sqrt{3} - 1} = \frac{(\sqrt{3} + 1)^2}{(\sqrt{3} - 1)(\sqrt{3} + 1)}$$

$$= \frac{4 + 2\sqrt{3}}{2} = 2 + \sqrt{3}$$

Exercise 5.5a

1 Use the above formulae to calculate, without using tables:
   a) $\sin(60° + 45°)$, $\cos(60° + 45°)$, $\tan(60° + 45°)$
   b) $\sin(60° - 45°)$, $\cos(60° - 45°)$, $\tan(60° - 45°)$

2 Use the compound angle formulae to simplify:
   a) $\sin(90° + x)$, $\cos(90° + x)$, $\tan(90° + x)$
   b) $\sin(180° - x)$, $\cos(180° - x)$, $\tan(180° - x)$

3 Express in terms of $\sin x$, $\cos x$, $\tan x$:
   a) $\sin(45° + x)$, $\cos(45° + x)$, $\tan(45° + x)$
   b) $\sin(60° - x)$, $\cos(60° - x)$, $\tan(60° - x)$

4 Simplify
   a) $\sin 45° \cos 15° + \cos 45° \sin 15°$
   b) $\cos 35° \cos 15° - \sin 35° \sin 15°$
   c) $\sin 2\alpha \cos \beta - \cos 2\alpha \sin \beta$
   d) $\cos 2\theta \cos 3\phi + \sin 2\theta \sin 3\phi$

   e) $\dfrac{\sqrt{3} + \tan \theta}{1 - \sqrt{3} \tan \theta}$

   f) $\dfrac{\tan \theta - 1}{1 + \tan \theta}$

5 Simplify
   a) $\sin(A + B) + \sin(A - B)$
   b) $\cos(A + B) - \cos(A - B)$

   c) $\dfrac{\cos(A + B) + \cos(A - B)}{\sin(A + B) - \sin(A - B)}$

   d) $\dfrac{\sin(A - B)}{\cos A \cos B}$

   e) $\dfrac{\cos(A + B)}{\cos A \sin B}$

f) $\tan(A + 45°)\tan(A - 45°)$

6 Prove that

a) $\cot(A + B) = \dfrac{\cot A\cot B - 1}{\cot A + \cot B}$

b) $\sin(45° + A) - \sin(45° - A) = \sqrt{2}\sin A$

c) $\dfrac{\sin(A + B)}{\cos A\cos B} = \tan A + \tan B$

d) $\dfrac{\cos(A - B)}{\sin A\sin B} = \cot A\cot B + 1$

7 Express $\tan(A + B + C)$ in terms of $\tan A$, $\tan B$, $\tan C$. Hence
a) if A, B, C are acute and $\tan A = \frac{1}{2}$, $\tan B = \frac{1}{5}$, $\tan C = \frac{1}{8}$, calculate $A + B + C$.
b) If A, B, C are the angles of a triangle, show that the sum of their tangents is equal to their product.

The combination of waves

The sound-curve of a tuning-fork which has been lightly tapped is a simple sine-wave (see Fig. 1). But it frequently happens that a wave does not have such a simple curve. If, for example, the tuning-fork were struck with a piece of metal, the sound-wave of the resulting clang would be as shown in Fig. 2,

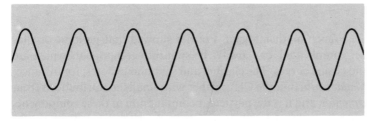

Fig. 1

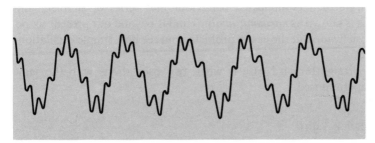

Fig. 2

whilst the waves from a clarinet and a saxophone sounding the same note are as shown in Figs. 3 and 4.

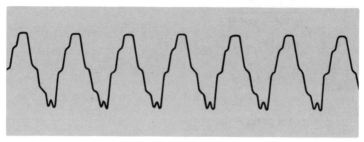

Fig. 3

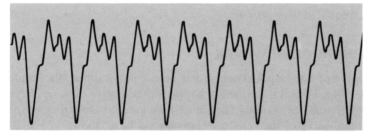

Fig. 4

In 1822, the French mathematician Fourier showed that periodic oscillations, however complicated, can always be split into components which are simple sine and cosine curves. The clarinet and saxophone notes, for instance, consist of a *fundamental* (middle C), together with smaller contributions from a series of *harmonics*, and it is the particular combination of these components which gives each instrument its distinctive tone. In a similar way, engineers studying the vibrations of a gas-turbine blade or the frame of an aircraft are able to separate their complicated behaviour into series of simpler components. This is known as *harmonic analysis* and is beyond our present scope. But we can briefly look at the easier problem, and see how simple oscillations are combined.

The most straightforward case is when two oscillations with the same period are superimposed.

For example

$$f(\theta) = 4\cos\theta + 3\sin\theta$$

Here the two components have the same period (360°), but different *amplitudes* (3 and 4) and a *phase difference* (90°) between successive peaks:

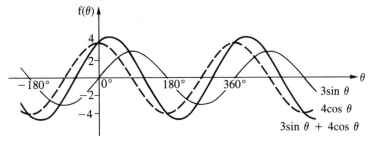

The reader should plot these curves accurately from $\theta = 0°$ to $360°$ and then investigate the behaviour of their sum. It is also instructive to repeat this with the same waves, but separated by a different phase difference, say $60°$.

Alternatively, we see that $4\cos\theta + 3\sin\theta$ can be written as

$$R\cos(\theta - \alpha) \equiv R\cos\alpha\cos\theta + R\sin\alpha\sin\theta$$

provided that $R\cos\alpha = 4$

and $R\sin\alpha = 3$

$\Rightarrow$ $R = \sqrt{(3^2 + 4^2)} = 5$

and $\tan\alpha = \dfrac{3}{4} \;\Rightarrow\; \alpha \approx 37°$

So $4\cos\theta + 3\sin\theta \equiv 5\cos(\theta - 37°)$

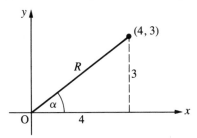

$a\cos\theta + b\sin\theta$

More generally, $a\cos\theta + b\sin\theta$ can be written as $R\cos(\theta - \alpha)$ provided that
$a\cos\theta + b\sin\theta \equiv R(\cos\theta\cos\alpha + \sin\theta\sin\alpha)$

$\Leftrightarrow$ $R\cos\alpha = a$ and $R\sin\alpha = b$

$\Rightarrow$ $R = \sqrt{(a^2 + b^2)}$

        $\tan\alpha = \dfrac{b}{a}$

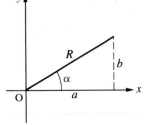

So in every case the sum of two such oscillations with the same period is another oscillation with this period: or, in musical terms, two pure notes of the same frequency always combine perfectly even though their oscillations have different amplitudes and are out of phase.

Example 2

Express $2\cos\theta + \sin\theta$ in the form $R\cos(\theta - \alpha)$ and so
a) find its extreme values;
b) solve the equation $2\cos\theta + \sin\theta = 1.5$.

Suppose that

$$2\cos\theta + \sin\theta = R\cos(\theta - \alpha)$$
$$= R(\cos\theta\cos\alpha + \sin\theta\sin\alpha)$$

Then $R\cos\alpha = 2$ and $R\sin\alpha = 1$

$\Rightarrow \qquad R = \sqrt{5}$

and $\tan\alpha = \frac{1}{2} \Rightarrow \alpha = 26.57°$

So $2\cos\theta + \sin\theta = \sqrt{5}\cos(\theta - 26.6°)$

Hence **a)** the extreme values of

$2\cos\theta + \sin\theta$ are $\pm\sqrt{5}$ (when $\theta \approx 26.6°, 206.6°$, etc.)

$\qquad$ **b)** the equation $2\cos\theta + \sin\theta = 1.5$

$\Rightarrow \qquad\qquad \sqrt{5}\cos(\theta - 26.6°) = 1.5$

$\Rightarrow \qquad\qquad \cos(\theta - 26.6°) = \cos 47.9°$

$\Rightarrow \qquad\qquad \theta - 26.6° = 47.9°, 312.1°$, etc.

$\Rightarrow \qquad\qquad \theta \approx 74.5°, 338.7°$, etc.

So far we have concentrated our attention to the combination of functions which have the same period. But more interesting wave-forms arise from functions with different periods, as in the sound from a musical instrument. For a simple example, see no. **3** of the following exercise.

Exercise 5.5b

1 Express the following functions in the form $R\cos(\theta - \alpha)$, and so find their maximum and minimum values:

a) $\cos\theta + \sin\theta$ **b)** $3\cos\theta + 2\sin\theta$

c) $-3\cos\theta + 4\sin\theta$ **d)** $\cos\theta - 3\sin\theta$

2 Use a similar method to solve the following equations for values between $0°$ and $360°$:

a) $\sin\theta + \cos\theta = 1$ **b)** $2\sin\theta + 3\cos\theta = -1$

c) $4\sin\theta - 3\cos\theta = 2$ **d)** $\cos\theta - 3\sin\theta = -2$

3 Use your calculator to evaluate $f(\theta) = \sin\theta + \frac{1}{2}\sin 2\theta + \frac{1}{4}\sin 4\theta$ at intervals of $10°$ between $0°$ and $370°$, and then plot on one graph the three component oscillations together with the curve of $f(\theta)$.

5.6 Multiple angles

If we put $\phi = \theta$ in the formulae of the last section, we obtain

$$\sin 2\theta = 2\sin\theta\cos\theta$$

$$\cos 2\theta = \cos^2\theta - \sin^2\theta$$

$$= 2\cos^2\theta - 1 = 1 - 2\sin^2\theta$$

$$\tan 2\theta = \frac{2\tan\theta}{1 - \tan^2\theta}$$

This last formula can be used to express $\tan\theta$ in terms of $\tan\frac{1}{2}\theta$. If we let $t = \tan\frac{1}{2}\theta$, then

$$\tan\theta = \frac{2t}{1 - t^2}$$

and it is also easy to express $\sin\theta$ and $\cos\theta$ in terms of t.

For $\sin\theta = 2\sin\frac{1}{2}\theta\cos\frac{1}{2}\theta = \dfrac{2\tan\frac{1}{2}\theta}{\sec^2\frac{1}{2}\theta} = \dfrac{2t}{1 + t^2}$

Similarly $\cos\theta = \cos^2\frac{1}{2}\theta - \sin^2\frac{1}{2}\theta$

$$= \frac{1 - \tan^2\frac{1}{2}\theta}{\sec^2\frac{1}{2}\theta} = \frac{1 - t^2}{1 + t^2}$$

So $\quad \sin\theta = \dfrac{2t}{1+t^2}, \quad \cos\theta = \dfrac{1-t^2}{1+t^2}, \quad \tan\theta = \dfrac{2t}{1-t^2}$

which are easily memorable from the triangle

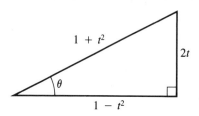

Example 1

Simplify $\quad \dfrac{1-\cos A}{1+\cos A}$

$$1 - \cos A = 1 - (1 - 2\sin^2 \tfrac{1}{2}A) \ = 2\sin^2 \tfrac{1}{2}A$$

and $\quad 1 + \cos A = 1 + (2\cos^2 \tfrac{1}{2}A - 1) \ = 2\cos^2 \tfrac{1}{2}A$

So $\quad \dfrac{1-\cos A}{1+\cos A} = \dfrac{2\sin^2 \tfrac{1}{2}A}{2\cos^2 \tfrac{1}{2}A} = \tan^2 \tfrac{1}{2}A$

Example 2

Express $\tan 3A$ in terms of $\tan A$

$$\tan 3A = \tan(2A + A) = \frac{\tan 2A + \tan A}{1 - \tan 2A \tan A}$$

$$= \frac{\dfrac{2\tan A}{1-\tan^2 A} + \tan A}{1 - \dfrac{2\tan A}{1-\tan^2 A}\tan A}$$

$$= \frac{3\tan A - \tan^3 A}{1 - 3\tan^2 A}$$

Example 3

Find the angles between $0°$ and $360°$ which satisfy $\cos 2\theta = 7\cos\theta + 3$

Now $\qquad\qquad\qquad \cos 2\theta = 7\cos\theta + 3$

$\Rightarrow \qquad\qquad 2\cos^2\theta - 1 = 7\cos\theta + 3$

$\Rightarrow \qquad 2\cos^2\theta - 7\cos\theta - 4 = 0$

$\Rightarrow \quad (2\cos\theta + 1)(\cos\theta - 4) = 0$

$\Rightarrow \qquad\qquad\qquad \cos\theta = -\tfrac{1}{2}$ or $\quad +4$ (which is impossible)

$\Rightarrow \qquad\qquad\qquad \theta = 120°$ or $\quad 240°$

Example 4

$\triangle ABC$ has sides a, b, c and semi-perimeter s (so that $a + b + c = 2s$). Show that its area is

$$\triangle = \sqrt{s(s-a)(s-b)(s-c)}$$

We know that
$$\triangle = \tfrac{1}{2}bc\sin A$$
$$= \tfrac{1}{2}bc\sqrt{(1 - \cos^2 A)}$$
$$= \tfrac{1}{2}bc\sqrt{\left\{1 - \left(\frac{b^2 + c^2 - a^2}{2bc}\right)^2\right\}}$$
$$= \tfrac{1}{4}\sqrt{\{4b^2c^2 - (b^2 + c^2 - a^2)^2\}}$$
$$= \tfrac{1}{4}\sqrt{[\{2bc + (b^2 + c^2 - a^2)\}\{2bc - (b^2 + c^2 - a^2)\}]}$$
$$= \tfrac{1}{4}\sqrt{[\{(b + c)^2 - a^2\}\{a^2 - (b - c)^2\}]}$$
$$= \tfrac{1}{4}\sqrt{\{(a + b + c)(b + c - a)(a - b + c)(a + b - c)\}}$$
$$= \tfrac{1}{4}\sqrt{\{2s(2s - 2a)(2s - 2b)(2s - 2c)\}}$$

$\Rightarrow \quad \triangle = \sqrt{\{s(s-a)(s-b)(s-c)\}} \qquad$ known as Hero's† formula.

† Hero of Alexandria, a Greek mathematician of the first century AD.

Example 5

Use the substitution $t = \tan\dfrac{\theta}{2}$ to solve the equation

$$2\cos\theta + \sin\theta = 1.5$$

Since $\cos\theta = \dfrac{1 - t^2}{1 + t^2}$ and $\sin\theta = \dfrac{2t}{1 + t^2}$, it follows that

$$\dfrac{2(1 - t^2)}{1 + t^2} + \dfrac{2t}{1 + t^2} = 1.5$$

$$\Rightarrow \qquad 2 - 2t^2 + 2t = 1.5(1 + t^2)$$

$$\Rightarrow \qquad 3.5t^2 - 2t - 0.5 = 0$$

$$\Rightarrow \qquad 7t^2 - 4t - 1 = 0$$

$$\Rightarrow \qquad t = \dfrac{2 \pm \sqrt{11}}{7} = 0.760 \text{ or } -0.188$$

$$\Rightarrow \qquad \tan\dfrac{\theta}{2} = 0.760 \text{ or } -0.188$$

$$\Rightarrow \qquad \dfrac{\theta}{2} = 37.23° \text{ or } -10.65°$$

$$\Rightarrow \qquad \theta \approx 74.5° \text{ or } -21.3° \quad \text{(and so } 338.7°, \text{ etc.)}$$

Exercise 5.6

1 Simplify:
   a) $2\sin 15° \cos 15°$
   b) $2\cos^2 15° - 1$
   c) $1 - 2\sin^2 15°$
   d) $\cos^2 15° - \sin^2 15°$

   e) $\cos^2 15° + \sin^2 15°$
   f) $\dfrac{2\tan 15°}{1 - \tan^2 15°}$

2 If $\tan\theta = \frac{3}{4}$, evaluate (without using tables):
   a) $\sin 2\theta$ **b)** $\cos 2\theta$ **c)** $\tan 2\theta$

3 Use double-angle formulae to find expressions for the values of
   a) $\sin 22\frac{1}{2}°$ **b)** $\cos 22\frac{1}{2}°$ **c)** $\tan 22\frac{1}{2}°$

4 **a)** Express $\sin 3A$ in terms of $\sin A$.
   b) Express $\cos 3A$ in terms of $\cos A$.

5 Simplify:

a) $\dfrac{\sin A}{1 + \cos A}$

b) $\operatorname{cosec} 2x - \cot 2x$

c) $\cos^4 \theta - \sin^4 \theta$

d) $\dfrac{\sin 2A + \sin 4A}{1 + \cos 2A + \cos 4A}$

e) $\cot \theta - 2 \cot 2\theta$

f) $\tan 2A(\cot A - \tan A)$

6 a) Express $\cos 4A$ in terms of $\cos A$.
   b) Express $\tan 4A$ in terms of $\tan A$.

7 Find an expression for the value of $\tan 7\frac{1}{2}°$.

8 Find the values of x between $0°$ and $360°$ such that
   a) $4 \cos^3 x + 2 \cos x = 5 \sin 2x$
   b) $\cos 3x - 3 \cos x = \cos 2x + 1$
   c) $\cot 2x = 2 + \cot x$
   d) $\sin 3x = \sin^2 x$

9 By expressing each side in terms of $t = \tan \dfrac{x}{2}$, prove that

$$\sec x + \tan x = \tan \left(45° + \frac{x}{2} \right)$$

10 Use the substitution $t = \tan \dfrac{\theta}{2}$ to solve the equations of exercise 5.5b, no. **2**.

5.7 Factor formulae

There is another group of important formulae which are obtained from the equations of section 5.5, simply by addition and subtraction.

$$\sin (\theta + \phi) + \sin (\theta - \phi) = \quad 2 \sin \theta \cos \phi$$
$$\sin (\theta + \phi) - \sin (\theta - \phi) = \quad 2 \cos \theta \sin \phi$$
$$\cos (\theta + \phi) + \cos (\theta - \phi) = \quad 2 \cos \theta \cos \phi$$
$$\cos (\theta + \phi) - \cos (\theta - \phi) = -2 \sin \theta \sin \phi$$

If in these identities we put

$$\left. \begin{array}{l} \theta + \phi = A \\ \text{and} \quad \theta - \phi = B \end{array} \right\} \quad \text{then} \quad \begin{array}{l} \theta = \frac{1}{2}(A + B) \\ \phi = \frac{1}{2}(A - B) \end{array}$$

and we obtain

$$\sin A + \sin B = \quad 2 \sin \tfrac{1}{2}(A + B) \cos \tfrac{1}{2}(A - B)$$
$$\sin A - \sin B = \quad 2 \cos \tfrac{1}{2}(A + B) \sin \tfrac{1}{2}(A - B)$$

$$\cos A + \cos B = \quad 2\cos\tfrac{1}{2}(A + B)\cos\tfrac{1}{2}(A - B)$$
$$\cos A - \cos B = -2\sin\tfrac{1}{2}(A + B)\sin\tfrac{1}{2}(A - B)$$
$$\qquad\qquad = \quad 2\sin\tfrac{1}{2}(A + B)\sin\tfrac{1}{2}(B - A)$$

As these enable us to factorise $\sin A \pm \sin B$ and $\cos A \pm \cos B$, they are sometimes called the *factor formulae*.

Example 1

Simplify $\dfrac{\sin A + \sin B}{\cos A - \cos B}$

$$\frac{\sin A + \sin B}{\cos A - \cos B} = \frac{2\sin\tfrac{1}{2}(A + B)\cos\tfrac{1}{2}(A - B)}{-2\sin\tfrac{1}{2}(A + B)\sin\tfrac{1}{2}(A - B)} = -\cot\tfrac{1}{2}(A - B)$$

Example 2

If A, B, C are the angles of a triangle, prove that

$$\sin A + \sin B + \sin C = 4\cos\frac{A}{2}\cos\frac{B}{2}\cos\frac{C}{2}$$

$$\sin A + \sin B + \sin C = \sin A + 2\sin\frac{B + C}{2}\cos\frac{B - C}{2}$$

$$= 2\sin\frac{A}{2}\cos\frac{A}{2} + 2\sin\left(90° - \frac{A}{2}\right)\cos\frac{B - C}{2}$$

$$= 2\sin\frac{A}{2}\cos\frac{A}{2} + 2\cos\frac{A}{2}\cos\frac{B - C}{2}$$

$$= 2\cos\frac{A}{2}\left(\sin\frac{A}{2} + \cos\frac{B - C}{2}\right)$$

$$= 2\cos\frac{A}{2}\left(\cos\frac{B + C}{2} + \cos\frac{B - C}{2}\right)$$

$$= 4\cos\frac{A}{2}\cos\frac{B}{2}\cos\frac{C}{2}$$

Example 3

Solve, for values of x between $0°$ and $360°$, the equation

$$\cos x + \cos 2x = \sin 2x - \sin x$$

Now $\qquad \cos x + \cos 2x = \sin 2x - \sin x$

$$\Rightarrow \qquad 2\cos\frac{3x}{2}\cos\frac{x}{2} = 2\sin\frac{x}{2}\cos\frac{3x}{2}$$

$$\Rightarrow \qquad 2\cos\frac{3x}{2}\left(\sin\frac{x}{2} - \cos\frac{x}{2}\right) = 0$$

Case 1

$$\cos\frac{3x}{2} = 0 \quad \Rightarrow \quad \frac{3x}{2} = 90°, 270°, 450°, \ldots$$

$$\Rightarrow \qquad x = 60°, 180°, 300°, \ldots$$

Case 2

$$\sin\frac{x}{2} - \cos\frac{x}{2} = 0$$

$$\Rightarrow \qquad \tan\frac{x}{2} = 1$$

$$\Rightarrow \qquad \frac{x}{2} = 45°, 225°, \ldots$$

$$\Rightarrow \qquad x = 90°, 450°, \ldots$$

So the required solutions are

$$x = 60°, 90°, 180°, 300°$$

Exercise 5.7

1 Factorise:
  a) $\sin 3\theta + \sin 5\theta$
  b) $\sin 4A - \sin 2A$
  c) $\cos(x + 60°) + \cos(x - 60°)$
  d) $\cos(x + 45°) - \cos(x - 45°)$

2 Simplify:
  a) $\dfrac{\cos A + \cos B}{\sin A - \sin B}$
  b) $\dfrac{\sin A + \sin B}{\cos A + \cos B}$

c) $\dfrac{\sin 3A - \sin A}{\cos 3A - \cos A}$

d) $\dfrac{\sin 2A + \sin 3A}{\cos 2A - \cos 3A}$

3   Prove that
   a) $\sin A + \sin 2A + \sin 3A = \sin 2A(2\cos A + 1)$
   b) $\cos A + 2\cos 3A + \cos 5A = 4\cos^2 A \cos 3A$
   c) $\cos A - 2\cos 3A + \cos 5A = 2\sin A\,(\sin 2A - \sin 4A)$
   d) $\dfrac{\sin A - \sin 2A + \sin 3A}{\cos A - \cos 2A + \cos 3A} = \tan 2A$

4   If A, B, C are the angles of $\triangle ABC$, prove that
   a) $\sin 2A + \sin 2B + \sin 2C = 4\sin A \sin B \sin C$
   b) $\cos A + \cos B + \cos C = 1 + 4\sin\dfrac{A}{2}\sin\dfrac{B}{2}\sin\dfrac{C}{2}$
   c) $\sin A - \sin B + \sin C = 4\sin\dfrac{A}{2}\cos\dfrac{B}{2}\sin\dfrac{C}{2}$
   d) $\sin^2 A + \sin^2 B + \sin^2 C = 2 + 2\cos A \cos B \cos C$

5   Solve the following equations, giving all solutions between $0°$ and $180°$:
   a) $\sin x + \sin 5x = \sin 3x$
   b) $\cos x + \cos 2x + \cos 3x = 0$
   c) $\sin x - 2\sin 2x + \sin 3x = 0$
   d) $\cos x - \sin 2x = \cos 5x$

5.8 Radians, general solutions and small angles

So far we have measured angles in degrees, using (like the Babylonians) a sub-division of the circle into 360 parts. There is, however, frequently great advantage if another unit is used, called the *radian*: this is the angle subtended at the centre of a circle by an arc whose length is equal to the radius.

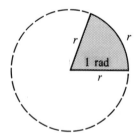

Now the total circumference of the circle is $2\pi \times$ radius.
So the total angle at the centre is $2\pi \times 1\,\text{rad} = 2\pi\,\text{rad}$.

Hence $2\pi \, \text{rad} = 360°$

$$\Rightarrow \qquad 1 \, \text{rad} = \frac{360°}{2\pi} = \frac{180°}{\pi} \approx 57.3°$$

Conversely, $1° = \dfrac{\pi}{180} \approx 0.0175 \, \text{rad}$

From now onwards all angles will be given in radians unless otherwise stated, so we simply say that

$180° = \pi, \quad 90° = \pi/2, \quad 60° = \pi/3, \quad 45° = \pi/4, \quad 30° = \pi/6, \quad \text{etc.}$

Length of arc and area of sector

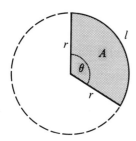

Suppose that a circular sector of radius r and angle θ has arc-length l and area A.

The whole angle 2π generates the complete circle, which has arc-length $2\pi r$ and area πr^2.

So the angle θ will generate a fraction $\dfrac{\theta}{2\pi}$ of the complete circle.

This sector has arc-length $= \dfrac{\theta}{2\pi} \times 2\pi r = r\theta$

and area $\qquad\qquad = \dfrac{\theta}{2\pi} \times \pi r^2 = \tfrac{1}{2}r^2\theta$

$$\Rightarrow \quad \boxed{\begin{array}{l} l = r\theta \\ A = \tfrac{1}{2}r^2\theta \end{array}}$$

Exercise 5.8a

1 Express the following angles in radians:
   a) 10° **b)** 36° **c)** 120° **d)** 180°
   e) 270° **f)** 135° **g)** 1′ **h)** 1″
   (1′ = 1 minute = $\tfrac{1}{60}°$; 1″ = 1 second = $\tfrac{1}{60}′$.)

2   Express in degrees:

a) $\dfrac{\pi}{2}$ rad   b) $\dfrac{3\pi}{2}$ rad   c) $\dfrac{5\pi}{6}$ rad   d) $\dfrac{2\pi}{3}$ rad

e) $\dfrac{3\pi}{4}$ rad   f) $\dfrac{\pi}{12}$ rad   g) $\dfrac{\pi}{15}$ rad   h) $\dfrac{\pi}{10}$ rad

3   What are the values of the following?

a) $\sin\dfrac{\pi}{2}$ b) $\cos\dfrac{\pi}{4}$ c) $\tan\dfrac{\pi}{3}$ d) $\sin\dfrac{\pi}{6}$

e) $\cos\pi$ f) $\sin\dfrac{3\pi}{4}$ g) $\cos\dfrac{2\pi}{3}$ h) $\tan\pi$

i) $\cos\dfrac{3\pi}{2}$ j) $\sin\dfrac{7\pi}{6}$

4   Simplify:

a) $\sin\left(\dfrac{\pi}{2} - x\right)$ b) $\cos(\pi - x)$

c) $\cos(\pi + x)$ d) $\sin(2\pi - x)$

e) $\tan\left(\dfrac{\pi}{2} - x\right)$ f) $\tan(\pi + x)$

5   Solve, in radians (giving values of x between 0 and 2π):

a) $\sin x = \dfrac{1}{2}$ b) $\sin x = \dfrac{1}{\sqrt{2}}$ c) $\cos x = -1$

d) $\cos x = \frac{1}{2}$ e) $\tan x = +1$ f) $\tan x = -\sqrt{3}$

6   Evaluate in radians (to 3 decimal places):
a) $40°$ b) $46.2°$ c) $184.1°$ d) $394.5°$

7   Express in degrees:
a) 0.52 rad   b) 1.73 rad   c) 5.136 rad   d) 0.011 rad

8   Find the length of arc and area of a circular sector whose radius and angle are:
a) 3 m   and   2 rad

b) 2 m   and $\dfrac{\pi}{4}$ rad

c) 4 m   and   25°

9   A chord AB subtends 120° at the centre O of circle whose radius is 10 cm. Find:
a) the length of the minor arc AB;
b) the area of $\triangle$OAB;
c) the area of the major segment cut off by AB.

10   The section of a tunnel consists of the major segment of a circle standing on a chord of length 4 m. If the greatest height of the tunnel is 6 m, calculate the radius of the circle.

Prove that the angle subtended by the chord at the centre of the circle is approximately 1.287 rad, and hence find the area of the cross-section of the tunnel, correct to the nearest square metre.   (oc)

11   A circular cone with base-radius r and slant-height l is unrolled into a circular sector. Find:
   a) the angle of this sector;
   b) the curved surface area of the cone.

12   A nautical mile was defined as the distance between two points of equal longitude whose latitudes differed by 1 minute. Assuming the Earth to be a perfect sphere of radius 6380 km, find the number of kilometres in a nautical mile.

General solutions of trigonometric equations

It will be recalled that in section 5.2 we considered the solution of the trigonometric equations

$$\sin \theta = \sin \alpha \qquad \cos \theta = \cos \alpha \qquad \tan \theta = \tan \alpha$$

where α is a given fixed angle.

These equations can be illustrated as

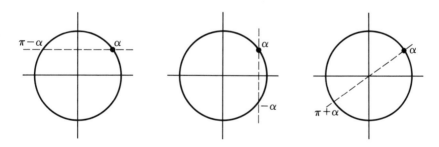

and the corresponding solutions are

$$\theta = \begin{cases} \alpha \\ \pi - \alpha \end{cases} \qquad \theta = \begin{cases} \alpha \\ -\alpha \end{cases} \qquad \theta = \begin{cases} \alpha \\ \pi + \alpha \end{cases}$$

But further solutions are readily obtained by adding any number of complete revolutions, i.e. $2n\pi$, where $n \in \mathbb{Z}$.

So more general solutions are

$$\theta = \begin{cases} 2n\pi + \alpha \\ (2n + 1)\pi - \alpha \end{cases} \qquad \theta = \begin{cases} 2n\pi + \alpha \\ 2n\pi - \alpha \end{cases} \qquad \theta = \begin{cases} 2n\pi + \alpha \\ (2n + 1)\pi + \alpha \end{cases}$$

which can be more economically expressed as

$$\theta = n\pi + (-1)^n\alpha \qquad \theta = 2n\pi \pm \alpha \qquad \theta = n\pi + \alpha$$

Summarising,

$\sin\theta = \sin\alpha$	$\cos\theta = \cos\alpha$	$\tan\theta = \tan\alpha$
$\Rightarrow \theta = n\pi + (-1)^n\alpha$	$\Rightarrow \theta = 2n\pi \pm \alpha$	$\Rightarrow \theta = n\pi + \alpha$

Example

Find the general solutions of
a) $\sin 2\theta = \sin\theta$ **b)** $\tan 3\theta = \cot\theta$

a) $\sin 2\theta = \sin\theta$
$\Rightarrow$ $2\sin\theta\cos\theta = \sin\theta$
$\Rightarrow$ $\sin\theta(2\cos\theta - 1) = 0$

$\Rightarrow$ $\sin\theta = 0$ or $\cos\theta = \dfrac{1}{2} = \cos\dfrac{\pi}{3}$

$\Rightarrow$ $\theta = n\pi$ or $\theta = 2n\pi \pm \dfrac{\pi}{3}$

$\Rightarrow$ $\theta = n\pi$ or $\dfrac{(6n \pm 1)\pi}{3}$

b) $\tan 3\theta = \cot\theta$

$\Rightarrow$ $\tan 3\theta = \tan\left(\dfrac{\pi}{2} - \theta\right)$

$\Rightarrow$ $3\theta = n\pi + \left(\dfrac{\pi}{2} - \theta\right)$

$\Rightarrow$ $4\theta = n\pi + \dfrac{\pi}{2}$

$\Rightarrow$ $\theta = \dfrac{(2n + 1)\pi}{8}$

Exercise 5.8b

Find, in radians, the general solutions of the following equations:

1 $\sin 2\theta = 0$ **2** $\cos\theta = 0.5$

3 $\tan 3\theta = 1$

4 $\sin^2 \theta = 1$

5 $\sec^2 \theta = 2$

6 $\cos 4\theta = \cos 3\theta$

7 $\tan 3\theta = \cot \theta$

8 $\sin 4\theta = \cos 3\theta$

9 $\sin^2 \theta = 1 - \cos \theta$

10 $\sec^2 \theta + 2 \tan \theta = 0$

Small angles

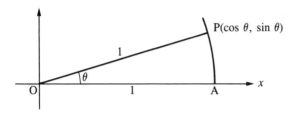

If θ is measured in radians and is small, it is seen from the figure that

$$\sin \theta \approx \widehat{\mathrm{AP}} = \theta$$

and $\quad \cos \theta = \sqrt{(1 - \sin^2 \theta)} \approx (1 - \theta^2)^{\frac{1}{2}} \approx 1 - \tfrac{1}{2}\theta^2$

Hence $\quad \tan \theta = \dfrac{\sin \theta}{\cos \theta} \approx \dfrac{\theta}{1 - \tfrac{1}{2}\theta^2} \approx \theta$

So $\qquad \boxed{\sin \theta \approx \theta \qquad \cos \theta \approx 1 - \tfrac{1}{2}\theta^2 \qquad \tan \theta \approx \theta}$

$$\lim_{\theta \to 0} \frac{\sin \theta}{\theta}$$

Though the above approximations appear to be confirmed by your calculator, the meaning of the symbol $\approx$ has never been properly defined and has simply been taken to mean 'is approximately equal to'. How approximately? And what do we mean by θ being small?

In section 5.9 much is going to depend on a clear understanding of the behaviour of $\sin \theta$ as θ becomes small, so we must now investigate this in the more rigorous language of limits.

We shall consider the same section OAP and let AB be drawn perpendicular to OA.

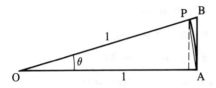

$$\triangle OAP < \text{sector OAP} < \triangle OAB$$

$$\Rightarrow \quad \tfrac{1}{2} \times 1 \times \sin\theta < \tfrac{1}{2} \times 1^2 \times \theta \ < \tfrac{1}{2} \times 1 \times \tan\theta$$

$$\Rightarrow \quad \sin\theta \ < \ \theta \ < \ \tan\theta$$

Dividing by $\sin\theta$, we obtain

$$1 \ < \frac{\theta}{\sin\theta} \ < \ \sec\theta$$

Now as $\theta \to 0$, $\sec\theta \to 1$

$$\Rightarrow \quad \frac{\theta}{\sin\theta} \to 1 \quad \Rightarrow \quad \frac{\sin\theta}{\theta} \to 1$$

So $\qquad \boxed{\lim_{\theta \to 0} \dfrac{\sin\theta}{\theta} = 1}$

5.9 Differentiation of trigonometric functions

Let $\qquad\qquad f(x) = \sin x$

Then $\quad \dfrac{f(x+h) - f(x)}{h} = \dfrac{\sin(x+h) - \sin x}{h}$

$$= \frac{2\sin h/2 \cos(x + h/2)}{h}$$

$$= \cos\left(x + \frac{h}{2}\right)\frac{\sin h/2}{h/2}$$

As $h \to 0$, $\quad \cos\left(x + \dfrac{h}{2}\right) \to \cos x$ and $\dfrac{\sin h/2}{h/2} \to 1$

So $\quad f'(x) = \lim_{h \to 0} \dfrac{f(x+h) - f(x)}{h} = \cos x$

Hence $\dfrac{d}{dx}(\sin x) = \cos x$

This can be illustrated:

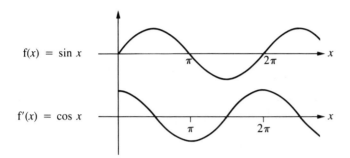

$f(x) = \sin x$

$f'(x) = \cos x$

We see that the derived curve of a sine wave is an exactly similar curve, but with a phase difference of $\pi/2$. (Readers familiar with electrical induction will recognise that this is exactly what is happening in an AC transformer, where the current in the secondary circuit is proportional to the rate of change of that in the primary circuit.) We can therefore predict that the derivative of $\cos x$ will be yet another similar wave with a further phase-difference of $\pi/2$, i.e. be the function $-\sin x$.

For a formal proof, we can proceed as with $\sin x$ and consider the expression

$$\frac{\cos(x + h) - \cos x}{h}$$

Alternatively

$$\frac{d}{dx}(\cos x) = \frac{d}{dx}\left\{\sin\left(x + \frac{\pi}{2}\right)\right\} = \cos\left(x + \frac{\pi}{2}\right) = -\sin x$$

So $\quad \dfrac{d}{dx}(\sin x) = \cos x \quad$ and $\quad \dfrac{d}{dx}(\cos x) = -\sin x$

The reader will have noted the tremendous benefit that has come from the use of radians, enabling us to say that

$$\frac{\sin h/2}{h/2} \to 1 \quad \text{as} \quad h \to 0$$

Example 1

Differentiate:
a) $\sin 3x$ **b)** $\cos^2 x$ **c)** $\sin x°$

a) $\dfrac{d}{dx}(\sin 3x) = (\cos 3x) \times 3 = 3\cos 3x$

b) $\dfrac{d}{dx}(\cos^2 x) = \dfrac{d}{dx}(\cos x)^2 = 2\cos x \times -\sin x = -2\cos x \sin x$

c) $\dfrac{d}{dx}(\sin x°) = \dfrac{d}{dx}\left(\sin \dfrac{\pi x}{180}\right) = \left(\cos \dfrac{\pi x}{180}\right) \times \dfrac{\pi}{180}$

$$= \dfrac{\pi}{180}\cos \dfrac{\pi x}{180} = \dfrac{\pi}{180}\cos x°$$

Example 2

Use differentiation to find the value of $\sin 60.2°$.

Let $y = \sin x$

Then $x = 60° \;\Rightarrow\; y = \sin 60° = \dfrac{\sqrt{3}}{2} = 0.8660$

Also $\delta y \approx (\cos x) \times \delta x$ (x in radians)

So $\delta x = 0.2° = 0.00349 \,\text{rad} \;\Rightarrow\; \delta y \approx \cos 60° \times 0.00349 \approx 0.0017$

$\Rightarrow\quad \sin 60.2° = y + \delta y \approx 0.8660 + 0.0017$

$\Rightarrow\qquad\qquad \sin 60.2° \approx 0.8677$

Exercise 5.9a

1 Prove from first principles that

$$\dfrac{d}{dx}(\cos x) = -\sin x$$

2 Differentiate:
   a) $\sin 2x$ **b)** $\cos 3x$ **c)** $\sin 2\pi x$ **d)** $4\cos \dfrac{3x}{2}$

   e) $\sin^2 x$ **f)** $\cos^3 x$ **g)** $x \sin 2x$ **h)** $\dfrac{\cos x}{x}$

i) $(\sin x + \cos x)^2$ **j)** $\sin^3 4x$ **k)** $\sqrt{\sin x}$ **l)** $\dfrac{x^2}{\sin x}$

3 A curve is defined by the parametric equations

$$x = 3\cos\theta, \quad y = 2\sin\theta$$

a) Sketch the curve by plotting the position of this point for different values of θ.
b) Eliminate θ to find its Cartesian equation.
c) Find dy/dx in terms of θ, and illustrate its values on your sketch.

4 If $x = a\cos\theta, \quad y = b\sin\theta$
a) sketch this curve and finds its Cartesian equation;
b) find dy/dx in terms of θ.

5 A particle is moving along a straight line and its distance x m from a fixed point O in the line at time t s is given by

$$x = 2 + 3\sin t + \cos t$$

Find:
a) the velocity and acceleration at time t;
b) the value of t when the particle first comes to rest;
c) its acceleration and distance from O at this instant.   (OC)

6 A particle is moving on a straight line, and its distance x from a fixed point O on the line at time t is given by

$$x = a(1 + \cos^2 t)$$

Show that the acceleration of the particle is $6a - 4x$.
    Find the values of x at the points where the velocity of the particle is
a) zero   **b)** a maximum.   (OC)

7 The horizontal displacement of a moving spot on a television screen after t milliseconds is x cm, where

$$x = 1 + 2\cos t - 4\cos 2t$$

Show that the spot is momentarily at rest when $t = 0$, and find the next two times when it is momentarily at rest.
    How far does the spot travel between $t = \frac{1}{2}\pi$ and $t = \pi$?   (SMP)

8 The point P moves in a straight line so that after t seconds its distance x m from a fixed point O in the line is given by

$$x = \sin t + 2\cos t$$

Its velocity is then v m s^{-1} and its acceleration a m s^{-2}.
a) Show that $v^2 = 5 - x^2$ and $a = -x$.
b) Find the greatest distance of P from O.

c) Find the velocity of P after 2 seconds, taking 1 radian as 57.3°.

9 Find the values of x for which the following functions have maxima and minima, distinguish between them and sketch the graphs of the functions:
a) $2\sin x - x$　$(-\pi \leqslant x \leqslant \pi)$
b) $2\sin x - \cos 2x$　$(0 \leqslant x \leqslant 2\pi)$

10 Without using your calculator for trigonometric functions, find the values of
a) $\cos 30.5°$　**b)** $\sin 45.3°$
(Take $\sqrt{2} = 1.414$　$\sqrt{3} = 1.732$　$\pi = 3.142$)

11 A girl is standing at the edge of a circular lake of radius 100 m. If she can walk at $2\,\text{m s}^{-1}$ and swim at $1\,\text{m s}^{-1}$, what is the least time in which she can reach the point exactly opposite? Explain your answer carefully.

12 A cylinder is cut from a solid sphere of radius a so that its total surface area is as large as possible. What is the ratio of its height to its diameter?

Differentiation of other trigonometric functions

$$\frac{d}{dx}\left(\frac{\sin x}{\cos x}\right) = \frac{(\cos x)(\cos x) - (\sin x)(-\sin x)}{\cos^2 x}$$

$$= \frac{\cos^2 x + \sin^2 x}{\cos^2 x} = \frac{1}{\cos^2 x} = \sec^2 x$$

So　$\dfrac{d}{dx}(\tan x) = \sec^2 x$

Readers can prove for themselves that

$$\frac{d}{dx}(\cot x) = -\cosec^2 x$$

$$\frac{d}{dx}(\sec x) = \sec x \tan x$$

$$\frac{d}{dx}(\cosec x) = -\cosec x \cot x$$

So the derivatives of the six trigonometric functions can be summarised:

$f(x)$	$\sin x$	$\cos x$	$\tan x$	$\cot x$	$\sec x$	$\cosec x$
$f'(x)$	$\cos x$	$-\sin x$	$\sec^2 x$	$-\cosec^2 x$	$\sec x \tan x$	$-\cosec x \cot x$

Exercise 5.9b

1 Differentiate:
   a) $\tan 3x$ **b)** $\sec 4x$ **c)** $\cot \dfrac{x}{2}$

   d) $\operatorname{cosec} 2x$ **e)** $x \tan x$ **f)** $\tan^2 x$

   g) $\sec^3 x$ **h)** $\dfrac{\sec x}{x}$ **i)** $\cot^2 3x$

   j) $\sqrt{\tan x}$

2 Show that

$$\frac{d}{dx}(\sec^2 x) = \frac{d}{dx}(\tan^2 x)$$

Why is this so?

3 **a)** If $x = \sec \theta$, $y = \tan \theta$, sketch the curve described as θ varies, and find dy/dx in terms of θ. Also find the x, y equation of the curve and so check dy/dx by obtaining it in terms of x, y.
   b) Repeat **a)** if $x = a \operatorname{cosec} \theta$, $y = b \cot \theta$.

4 A boy who can run at $8\,\mathrm{m\,s^{-1}}$ and swim at $2\,\mathrm{m\,s^{-1}}$ wishes to cross a canal of width $40\,\mathrm{m}$ to a point $200\,\mathrm{m}$ along the other bank. If he sets off at an angle θ to the perpendicular width, find:
   a) the total time taken;
   b) the value of θ which makes this a minimum;
   c) the shortest possible time.

5.10 Integration of trigonometric functions

Integration of trigonometric functions is, as always, a speculative process, and depends on a thorough knowledge of differentiation, including the methods of substitution, and of the trigonometric identities. We can, of course, always check indefinite integrals by differentiation. In particular, we see that

$$\int \sin x \, dx = -\cos x + c \quad \text{and} \quad \int \cos x \, dx = \sin x + c$$

Example 1

$$\int \sin 2x \, dx$$

We first recall that $\dfrac{d}{dx}(\cos 2x) = -2 \sin 2x$

So $\displaystyle\int \sin 2x \, dx = -\tfrac{1}{2} \cos 2x + A$

Or we could say that

$$\int \sin 2x \, dx = \int 2 \sin x \cos x \, dx$$

$$= \sin^2 x + B$$

or $\qquad - \cos^2 x + C$

(Why are these three answers all correct?)

Example 2

$$\int \cos^2 x \, dx$$

Our first thought is probably $\tfrac{1}{3} \cos^3 x$.

But $\dfrac{d}{dx}(\tfrac{1}{3} \cos^3 x) = \tfrac{1}{3} 3 \cos^2 x (-\sin x) = -\cos^2 x \sin x$

so we must think again.

Now $\cos 2x = 2 \cos^2 x - 1$

$\Rightarrow \qquad \cos^2 x = \tfrac{1}{2}(\cos 2x + 1)$

So $\displaystyle\int \cos^2 x \, dx = \int \tfrac{1}{2}(\cos 2x + 1) \, dx$

$$= \tfrac{1}{4} \sin 2x + \frac{x}{2} + C$$

$$= \tfrac{1}{2}(\sin x \cos x + x) + C$$

Example 3

$$\int_0^{\pi/2} \sin^3 x \cos x \, dx$$

Here we can put $u = \sin x$

$$\Rightarrow \quad du = \cos x \, dx$$

So $\displaystyle\int_0^{\pi/2} \sin^3 x \cos x \, dx = \int_0^1 u^3 du = \left[\tfrac{1}{4} u^4 \right]_0^1 = \tfrac{1}{4}$

Example 4

$$\int \sin x \cos 2x \, dx$$

Here we recall that

$$\sin 3x - \sin x = 2 \sin x \cos 2x$$

So $\displaystyle\int \sin x \cos 2x \, dx = \tfrac{1}{2} \int (\sin 3x - \sin x) \, dx$

$$= -\tfrac{1}{6} \cos 3x + \tfrac{1}{2} \cos x + A$$

Exercise 5.10

1 a) $\displaystyle\int \sin 2x \, dx$ **b)** $\displaystyle\int \cos \frac{x}{2} \, dx$

   c) $\displaystyle\int_0^{\pi/2} \cos 3x \, dx$ **d)** $\displaystyle\int_0^{\pi/2} \sin \frac{x}{3} \, dx$

2 Use the double-angle formulae to find:

   a) $\displaystyle\int \sin^2 x \, dx$ **b)** $\displaystyle\int \cos^2 \frac{x}{2} \, dx$

   c) $\displaystyle\int_0^{\pi} \sin^2 2x \, dx$ **d)** $\displaystyle\int_0^{\pi/2} \sin x \cos x \, dx$

3 Use the factor formulae to find:

   a) $\displaystyle\int \sin 2x \cos x \, dx$ **b)** $\displaystyle\int \cos 2x \cos x \, dx$ **c)** $\displaystyle\int \sin 2x \sin x \, dx$

4 Use the method of substitution to find the following integrals:

a) $\int x \cos x^2 \, dx$ **b)** $\int x^2 \sin x^3 \, dx$

c) $\int \sin^2 x \cos x \, dx$ **d)** $\int \cos^3 x \sin x \, dx$

5 a) $\int \sec^2 x \, dx$ **b)** $\int \sec x \tan x \, dx$

c) $\int \tan^2 x \, dx$ **d)** $\int \cot^2 x \, dx$

6 Sketch the graph of the curve $y = 2 + \sin x$ for values of x between 0 and 2π. Calculate:
a) the area bounded by the curve, the x-axis and the lines $x = 0$, $x = 2\pi$;
b) the volume generated when this area makes a complete revolution about the x-axis.   (L)

7 A particle moves on the x-axis having an acceleration of $36 \cos 3t$, where t is the time. It is at the origin when $t = 0$ and its velocity then is 12. Prove that its position at time t is given by $x = 4(1 - \cos 3t) + 12t$.
How far from the origin is the particle when it first comes to rest? (OC)

8 What is the average value of $\sin^2 x$ between $x = 0$ and $x = \pi$?

9 Use the substitution $x = \tan \theta$ to find

$$\int_0^1 \frac{dx}{(x^2 + 1)^2}$$

10 a) $\int_3^6 \frac{dx}{\sqrt{\{(x - 3)(6 - x)\}}}$ (put $x = 3 \cos^2 \theta + 6 \sin^2 \theta$)

b) $\int_2^3 \sqrt{\left(\frac{x - 2}{4 - x}\right)} \, dx$ (put $x = 2 \cos^2 \theta + 4 \sin^2 \theta$)

11 On the same diagram, for $-\pi \leqslant x \leqslant \pi$, sketch the curves

$$y = 1 - \frac{8x^2}{\pi^2} \quad \text{and} \quad y = \sin^2 x$$

and verify that the curves intersect when $y = \frac{1}{2}$.
Calculate the area of the finite region enclosed by the two curves.   (L)

*5.11 Inverse trigonometric functions

If we are asked to find the angle whose sine is $\frac{1}{2}$, we have seen that there are any number of answers: $\pi/6$, $5\pi/6$, $13\pi/6$, $17\pi/6$, ... (as well as $-7\pi/6$, $-11\pi/6$ ...).

More generally, the angle whose sine is x does not define a function. Clearly the domain would have to be the set of numbers between -1 and $+1$, but even so there are any number of answers to the instruction, as can be seen from the graph:

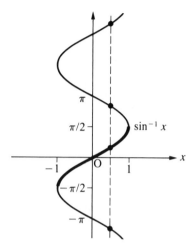

This is not the graph of a function, but it does show that we can obtain a function by making the more definite request for *the angle lying between $-\pi/2$ and $+\pi/2$ whose sine is x.* As we see from the graph, this has a unique answer, so does represent a function, called the inverse sine and written $\sin^{-1}$.

So $\sin^{-1}\left(\frac{1}{2}\right) = \dfrac{\pi}{6}$, $\sin^{-1} 0 = 0$, $\sin^{-1}(-1) = -\dfrac{\pi}{2}$, etc.

Similarly, $\cos^{-1} x$ is defined as *the angle between 0 and π whose cosine is x.*

Hence $\cos^{-1}\left(\frac{1}{2}\right) = \dfrac{\pi}{3}$, $\cos^{-1} 0 = \dfrac{\pi}{2}$, $\cos^{-1}(-1) = \pi$,   etc.

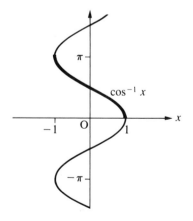

Finally, $\tan^{-1} x$ is defined as *the angle between $-\pi/2$ and $+\pi/2$ whose tangent is x.*

So $\tan^{-1}1 = \dfrac{\pi}{4}$, $\tan^{-1}(-1) = -\dfrac{\pi}{4}$,   etc.

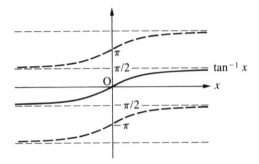

Exercise 5.11a

1 Evaluate:
  a) $\sin^{-1}\frac{1}{2}$, $\cos^{-1}\frac{1}{2}$, $\tan^{-1}\frac{1}{2}$
  b) $\sin^{-1}(-1)$, $\cos^{-1}(-1)$, $\tan^{-1}(-1)$
  c) $\sin^{-1}0.7$, $\cos^{-1}0.7$, $\tan^{-1}0.7$

2 Find, from no. **1**, a relationship between $\sin^{-1}x$ and $\cos^{-1}x$, and illustrate your answer by means of their graphs.

3 Simplify
  a) $\tan^{-1}\frac{1}{2} + \tan^{-1}\frac{1}{3}$
  [Let $\theta = \tan^{-1}\frac{1}{2}$, $\phi = \tan^{-1}\frac{1}{3}$ and calculate $\tan(\theta + \phi)$]
  b) $\tan^{-1}\frac{1}{5} + \tan^{-1}\frac{1}{8}$
  c) $\tan^{-1}x + \tan^{-1}y$,
  where $0 \leqslant x \leqslant 1$ and $0 \leqslant y \leqslant 1$.

Differentiation of inverse trigonometric functions

a) $\dfrac{\mathrm{d}}{\mathrm{d}x}(\sin^{-1}x)$, $\dfrac{\mathrm{d}}{\mathrm{d}x}(\cos^{-1}x)$

$$y = \sin^{-1}x \ \Rightarrow \ x = \sin y$$

$$\Rightarrow \ \frac{\mathrm{d}x}{\mathrm{d}y} = \cos y = \pm\sqrt{(1 - x^2)}$$

$$\Rightarrow \ \frac{\mathrm{d}y}{\mathrm{d}x} = \pm\frac{1}{\sqrt{(1 - x^2)}}$$

But $\sin^{-1}x$ is an increasing function,

so $\dfrac{d}{dx}(\sin^{-1}x) = +\dfrac{1}{\sqrt{(1-x^2)}}$

By similar argument (or alternatively from $\sin^{-1}x + \cos^{-1}x = \tfrac{1}{2}\pi$) we find that

$$\frac{d}{dx}(\cos^{-1}x) = -\frac{1}{\sqrt{(1-x^2)}}$$

b) $\dfrac{d}{dx}(\tan^{-1}x)$

$$y = \tan^{-1}x \quad \Rightarrow \quad x = \tan y$$

$$\Rightarrow \quad \frac{dx}{dy} = \sec^2 y = 1 + x^2$$

$$\Rightarrow \quad \frac{dy}{dx} = \frac{1}{1+x^2}$$

So
$$\frac{d}{dx}(\sin^{-1}x) = \frac{1}{\sqrt{(1-x^2)}} \qquad \int \frac{dx}{\sqrt{(1-x^2)}} = \sin^{-1}x + C$$

$$\frac{d}{dx}(\tan^{-1}x) = \frac{1}{1+x^2} \quad \Rightarrow \quad \int \frac{dx}{1+x^2} = \tan^{-1}x + C$$

The importance of the inverse trigonometric functions largely stems from their value in enabling us to integrate such functions as $\dfrac{1}{\sqrt{(1-x^2)}}$ and $\dfrac{1}{1+x^2}$. These results could equally well have been found using the method of substitution.

More generally

$$\int \frac{dx}{\sqrt{(a^2-x^2)}} = \int \frac{a\cos\theta\,d\theta}{\sqrt{(a^2-a^2\sin^2\theta)}} \qquad (x = a\sin\theta,\, dx = a\cos\theta\,d\theta)$$

$$= \int \frac{a\cos\theta\,d\theta}{a\cos\theta}$$

$$= \int d\theta = \theta + c = \sin^{-1}\frac{x}{a} + c$$

and
$$\int \frac{dx}{a^2+x^2} = \int \frac{a\sec^2\theta\,d\theta}{a^2+a^2\tan^2\theta} \qquad (x = a\tan\theta,\, dx = a\sec^2\theta\,d\theta)$$

$$= \frac{1}{a}\int \frac{\sec^2\theta\,d\theta}{\sec^2\theta}$$

$$= \frac{1}{a} \int d\theta = \frac{\theta}{a} + c = \frac{1}{a} \tan^{-1} \frac{x}{a} + c$$

Exercise 5.11b

1 a) $\displaystyle\int_0^1 \frac{dx}{\sqrt{(4 - x^2)}}$ **b)** $\displaystyle\int_0^3 \frac{dx}{x^2 + 9}$

  c) $\displaystyle\int_0^1 \sqrt{(1 - x^2)}\,dx$ $(x = \sin\theta)$

2 a) $\displaystyle\int \frac{dx}{x^2 + 2x + 2}$ **b)** $\displaystyle\int \frac{dx}{\sqrt{(2x - x^2)}}$

  c) $\displaystyle\int \sqrt{\left(\frac{x}{1 - x}\right)}\,dx$ $(x = \sin^2\theta)$

3 a) $\displaystyle\int \frac{dx}{\sqrt{(9 - 4x^2)}}$ **b)** $\displaystyle\int \frac{dx}{4x^2 + 1}$ **c)** $\displaystyle\int \sqrt{(a^2 - x^2)}\,dx$

4 a) $\displaystyle\int \frac{dx}{x\sqrt{(x^2 - 1)}}$ $(x = \sec\theta)$ **b)** $\displaystyle\int \frac{dx}{(1 + x^2)^2}$ $(x = \tan\theta)$

Miscellaneous problems

1 A wire of length 4 m is shaped into the arc and two radii of a circular sector. Find the maximum area it can enclose.

2 a) Prove, without using tables, that

$$\tan^{-1}\tfrac{1}{2} + \tan^{-1}\tfrac{2}{3} + \tan^{-1}\tfrac{4}{7} = \tfrac{1}{2}\pi$$

  b) Find the values of θ between 0 and 360° which satisfy the equation $\sin 3\theta = \cos 2\theta$, and hence, or otherwise, prove without using tables that $\sin 18° = (\sqrt{5} - 1)/4$.   (OC)

3 Without using tables, prove that

$$\tan\frac{\pi}{8} = \sqrt{2} - 1 \quad \text{and} \quad \tan\frac{\pi}{24} = \frac{1 + \sqrt{3} - \sqrt{6}}{\sqrt{3} + \sqrt{2} - 1} \qquad \text{(OC)}$$

4 Show that $x \tan x = 1$ has an infinite number of real roots, and that if n is a large integer there is a root near $n\pi$. Show that a better approximation is $n\pi + 1/(n\pi)$.

5 A hollow right circular cone is of height h and semi-vertical angle θ. Show that the radius of the largest sphere which can be placed entirely within the cone is

$$\frac{h \sin \theta}{1 + \sin \theta}$$

Prove that the ratio of the volume of the sphere to that of the cone is greatest when $\sin \theta = \frac{1}{3}$, and is then equal to $\frac{1}{2}$.

6   Four towns are situated at the corners of a square of side 100 km. What is the shortest length of road which can be built to connect them?

7   Prove that $\tan^2 \dfrac{A}{2} = \dfrac{1 - \cos A}{1 + \cos A}$, and then use the cosine rule to show that in any $\triangle ABC$

$$\tan \frac{A}{2} = \sqrt{\left\{ \frac{(s - b)(s - c)}{s(s - a)} \right\}}$$

where s is the semi-perimeter.
   Hence find A, B, C in exercise 5.4, no. **2a**.

8   Find all the values of x between 0 and 2π inclusive for which the function $3 \cos x - \cos 3x$ has a turning value, distinguishing between maxima and minima.
   How many inflexions has the function in the range $0 \leqslant x \leqslant 2\pi$?
   Draw a rough graph of the function for this range.   (OC)

9   A circle of radius a is rolling along the x-axis and after it has rotated through an angle θ the point P of the circle which was originally at O has coordinates x and y.
   a) Find x, y in terms of θ.
   b) Find $\dfrac{dx}{d\theta}, \dfrac{dy}{d\theta}$ and $\dfrac{dy}{dx}$ in terms of θ.
   c) Sketch the path of P (known as a *cycloid*).

10   A point P on an astroid has coordinates

$$x = a\cos^3 t, \quad y = a\sin^3 t$$

where t is a parameter. Find:
   a) the gradient of the curve at P;
   b) the equation of the tangent at P;
   c) the coordinates of the points A, B where this tangent meets Ox, Oy;
   d) the length AB.
   Hence show how an astroid can be constructed, and sketch the curve.

                                                                        (C)

11   The coordinates x, y of the spot of light on an oscilloscope are given by

$$x = a \sin mt, \quad y = b \sin (nt + \varepsilon)$$

and so are oscillating independently with amplitudes a, b and frequencies $m/2\pi$, $n/2\pi$ (ε being their initial phase difference). The paths traced by the spot as t varies for different values of a, b, m, n, ε, are known as Lissajou's figures.

If $a = b = 1$, plot these when $\varepsilon = 0$, $\varepsilon = \pi/4$, $\varepsilon = \pi/2$, $\varepsilon = 3\pi/4$ for

a) $m = n = 1$
b) $m = 2, \quad n = 1$
c) $m = 3, \quad n = 1$
d) other values of m and n

How do the figures alter if a and b take other values?

6 Sequences and series

6.1 Sequences and series

a) $1, 2, 3, 4, 5, 6, 7, \ldots$
b) $1, 2, 4, 7, 11, 16, 22, \ldots$
c) $1, 2, 3, 5, 8, 13, 21, \ldots$
d) $1, 2, 4, 8, 16, 32, 64, \ldots$
e) $1, 2, 0, 3, -1, 4, -2, \ldots$
f) $1, 2, 6, 24, 120, 720, 5040, \ldots$

The reader will find little difficulty in spotting how each row (or *sequence*) has been produced, and in writing down the next three members (or *terms*) of each sequence.

The question next arises whether we can state a general formula in each case for the nth term, which for convenience we shall call u_n. In some cases this is obvious. For example in

a) $u_1 = 1, u_2 = 2, u_3 = 3, u_4 = 4$, and $u_n = n$
d) $u_1 = 1, u_2 = 2, u_3 = 4, u_4 = 8$, and $u_n = 2^{n-1}$

Some might be obtained by a certain amount of trial and error. For example in

b) $\quad u_n = \frac{1}{2}(n^2 - n + 2)$

which should be checked for $n = 1, 2, 3, \ldots 9$.

But in other cases, it is doubtful if even the wildest speculation would lead to such a formula. In **c)**, for example,

$$u_n = \frac{1}{\sqrt{5}}\left\{\left(\frac{1 + \sqrt{5}}{2}\right)^{n+1} - \left(\frac{1 - \sqrt{5}}{2}\right)^{n+1}\right\}$$

as can be checked for particular values of n.

Factorials

The sequence **f)**

$\quad 1, 2, 6, 24, 120, 720, \ldots$

presents a special case.

These terms can be written

$$u_1 = 1, \quad u_2 = 2 \times 1, \quad u_3 = 3 \times 2 \times 1, \quad u_4 = 4 \times 3 \times 2 \times 1, \quad \text{etc.}$$

so $u_n = n \times (n-1) \times (n-2) \dots \times 5 \times 4 \times 3 \times 2 \times 1$

This is an expression which occurs so frequently that we abbreviate it to $n!$, called 'n factorial' (or 'n shriek').

So 1! = 1   5! = 120
    2! = 2   6! = 720
    3! = 6   7! = 5040
    4! = 24   etc.

It is clear that the terms of this sequence are growing very rapidly and that the calculation of $n!$ when n is large (e.g. $100! = 100 \times 99 \times 98 \dots$ $\dots 4 \times 3 \times 2 \times 1$) would be very tedious.

There is, in fact, a formula, usually known as *Stirling's formula* (after the Scots mathematician, James Stirling, 1692–1770) which says, rather astonishingly, that

$$n! \approx \sqrt{(2\pi)} n^{n+1/2} e^{-n}, \quad \text{where} \quad e \approx 2.718$$

The percentage error of this approximation diminishes rapidly as n gets larger, and it is interesting to use the formula to compute known numbers like 10!, 12!, etc.

Exercise 6.1a

1 Write down the next three terms of the above sequences **a)**, **b)**, **c)**, **d)**, **e)**, **f)**.

2 Write down the values of u_1, u_2, u_3, u_4, where u_r is

    a) r^3 **b)** $\dfrac{r}{r+1}$ **c)** $1 + (-1)^r$ **d)** $(-1)^r r$

3 Simplify:

    a) $\dfrac{7!}{6!}$ **b)** $\dfrac{12!}{10!}$ **c)** $\dfrac{(n+1)!}{n!}$ **d)** $\dfrac{n!}{(n-2)!}$

Series

We frequently wish to find the sum of a number of terms of a sequence, such as

$$1 + 8 + 27 + 64$$

or $1 + 2 + 4 + 8 + 16 + 32$

or, more generally,

$$u_1 + u_2 + u_3 + u_4 + \cdots + u_n$$

We call each of these a *series* and abbreviate the sum of n terms,

$$s_n = u_1 + u_2 + u_3 + u_4 + \cdots + u_n$$

as $\displaystyle\sum_{r=1}^{n} u_r$ (pronounced 'sigma u_r from $r = 1$ to $r = n$')†

or more briefly as $\displaystyle\sum_{1}^{n} u_r$

So $\displaystyle\sum_{r=1}^{4} r^3 = 1^3 + 2^3 + 3^3 + 4^3 = 100$

and $\displaystyle\sum_{r=0}^{5} 2^r = 2^0 + 2^1 + 2^2 + 2^3 + 2^4 + 2^5 = 63$

Exercise 6.1b

1 Write out fully and find the values of

a) $\displaystyle\sum_{r=1}^{4} r^2$ **b)** $\displaystyle\sum_{r=2}^{5} \frac{1}{r}$ **c)** $\displaystyle\sum_{r=0}^{7} 2^r$ **d)** $\displaystyle\sum_{r=1}^{6} (-1)^r$

e) $\displaystyle\sum_{r=0}^{6} \frac{1}{2^r}$ **f)** $\displaystyle\sum_{r=1}^{3} r(r+1)$ **g)** $\displaystyle\sum_{r=1}^{4} r2^r$ **h)** $\displaystyle\sum_{r=1}^{4} (-1)^{r-1} r^2$

2 Use the $\sum$ notation to abbreviate (but do not evaluate):

a) $1 + 4 + 9 + 16 + \cdots + 625$ **b)** $\frac{1}{2} + \frac{1}{3} + \frac{1}{4} \cdots + \frac{1}{100}$

c) $1 + 8 + 27 + \cdots + 1000$ **d)** $1 + 3 + 9 + 27 + \cdots + 3^{50}$

e) $1 - \frac{1}{2} + \frac{1}{3} \cdots - \frac{1}{100}$ **f)** $1 \times 2 + 2 \times 3 + \cdots + n(n+1)$

6.2 Arithmetic progressions

An *arithmetic progression* (a.p.) is a sequence which proceeds with constant difference, like

$1, 4, 7, 10 \ldots$ or $3, 2\frac{1}{2}, 2, 1\frac{1}{2} \ldots$

Such a sequence is completely defined if we know its first term a and its constant difference d, the above two sequences being given by $a = 1, d = 3$ and by $a = 3, d = -\frac{1}{2}$.

The general a.p. can therefore be expressed as

$a, a + d, a + 2d, a + 3d \ldots$

---

† This is, of course, a refinement of the notation already used in section 4.5.

and it is seen that the nth term is

$$u_n = a + (n-1)d$$

If we now concentrate on n terms of this sequence, stretching from the first term $a\,(=u_1)$ to the last term $l\,(=u_n)$, we see that their sum, S_n, is given by

$$S_n = a + (a+d) + (a+2d) + \cdots + (l-d) + l$$

Writing this backwards,

$$S_n = l + (l-d) + (l-2d) + \cdots + (a+d) + a$$

Adding, we obtain

$$2S_n = (a+l) + (a+l) + \cdots + (a+l)$$
$$= n(a+l)$$

$\Rightarrow \qquad S_n = \tfrac{1}{2}n(a+l) = n\dfrac{a+l}{2}$

Hence the sum of a sequence of terms of an a.p.

= number of terms × average of the first and last terms

Further, since $u_n = a + (n-1)d$

$$S_n = \tfrac{1}{2}n\{a + a + (n-1)d\}$$

$\Rightarrow \qquad S_n = n\dfrac{a+l}{2} = \tfrac{1}{2}n\{2a + (n-1)d\}$

Example

A young woman's initial annual salary was £4500 and increased by £250 a year. How much did she expect to earn in her first six years?

Here we see that the constant increment is causing her salary to rise in an a.p., so we can say either

a) Final term = $4500 + 5 \times 250 = 5750$

$\Rightarrow \qquad$ sum $= 6 \times \dfrac{4500 + 5750}{2} = 30\,750$

or **b)** $S_n = \dfrac{n}{2}\{2a + (n-1)\,d\}$

where $a = 4500,$ $d = 250$ and $n = 6$

$\Rightarrow$ $S_n = \dfrac{6}{2}\{9000 + 5 \times 250\} = 30\,750$

So, by either method, we see that her total income in 6 years was £30 750.

Exercise 6.2

1   Find the nth term and the sum of n terms of each of the following a.p.s:
    a) $2, 6, 10, 14, \ldots$
    b) $1, 2\frac{1}{2}, 4, 5\frac{1}{2}, \ldots$
    c) $3, 1, -1, -3, \ldots$
    Check your results for $n = 5$.

2   Find the number of terms and the sums of each of the following arithmetic series:
    a) $1 + 3 + 5 + \cdots + 99$
    b) $1 + 3\frac{1}{2} + 6 + \cdots + 101$
    c) $10 + 9 + 8 + \cdots - 18 - 19 - 20$

3   Evaluate   a) $\displaystyle\sum_{r=1}^{25} (4r - 1)$ b) $\displaystyle\sum_{1}^{34} (3r - 2)$

4   In a potato race the first and last potatoes are 5 m and 15 m respectively from the starting-line and the rest are equally spaced at intervals of 1 metre. What is the total distance travelled by a runner who brings them one at a time to the starting-line?

5   Find the sum of the numbers divisible by 3 which lie between 1 and 100. Find also the sum of the numbers from 1 to 100 inclusive which are *not* divisible by 3.   (OC)

6   Find how many terms of the progression

    $5 + 9 + 13 + 17 + \cdots$

    have a sum of 2414.   (OC)

7   The first term of an a.p. is 3. Find the common difference if the sum of the first 8 terms is twice the sum of the first 5 terms.   (OC)

8   The fifth term of an arithmetical progression is 24 and the sum of the first five terms is 80. Find the first term, the common difference and the sum of the first fifteen terms of the progression.   (OC)

9   The sum to n terms of a certain a.p. is $2n(n + 5)$. What is
    a) its nth term?
    b) its constant difference?

6.3 Geometric progressions

A *geometric progression* (g.p.) is (by contrast with an a.p.) a sequence which proceeds with constant *ratio*, like $2, 6, 18, 54, \ldots$, or $8, -4, 2, -1, \frac{1}{2}, \ldots$.

Such a sequence is completely defined by its first term a and its constant ratio r, the above two sequences being given by

$$a = 2, r = 3 \quad \text{and} \quad a = 8, r = -\tfrac{1}{2}$$

The general g.p. can therefore be expressed as

$$a, ar, ar^2, ar^3 \ldots$$

so that $u_n = ar^{n-1}$

If we again concentrate on n terms of the sequence, from u_1 to u_n, we see that

$$S_n = a + ar + ar^2 + \cdots + ar^{n-1}$$

Multiplying by r, we obtain

$$rS_n = ar + ar^2 + \cdots + ar^{n-1} + ar^n$$

By subtraction,

$$(1 - r)S_n = a - ar^n = a(1 - r^n)$$

$\Rightarrow$ $$S_n = \frac{a(1 - r^n)}{1 - r} = \frac{a(r^n - 1)}{r - 1}$$

Example 1

On 1 January each year a man puts £50 in a bank which gives 4% interest each year. What will be the value of his investment on 31 December of the tenth year?

In one year £1 would grow to £1.04, and so £50 would grow to £50 × 1.04. Hence, by the end of the 10th year,

his 10th deposit will grow to £50 × 1.04
his 9th deposit will grow to £50 × (1.04)²
his 8th deposit will grow to £50 × (1.04)³
....
his 1st deposit will grow to £50 × (1.04)¹⁰

So their total value (in £) is

$$50\,(1.04) + 50\,(1.04)^2 + \cdots + 50\,(1.04)^{10}$$

$$= 50\,(1.04) \times \frac{(1.04)^{10} - 1}{1.04 - 1}$$

$$= \frac{52}{0.04} \times \{(1.04)^{10} - 1\}$$

$$= 622.6$$

and the final value of his investment is approximately £623.

Infinite geometric series

In the particular case when $a = 1$ and $r = \frac{1}{2}$ the g.p. becomes

$$1, \tfrac{1}{2}, \tfrac{1}{4}, \tfrac{1}{8}, \tfrac{1}{16}, \cdots$$

and the sum of its first n terms is

$$S_n = 1 + \tfrac{1}{2} + \tfrac{1}{4} + \cdots + \frac{1}{2^{n-1}}$$

So $S_1 = 1$, $S_2 = 1\tfrac{1}{2}$, $S_3 = 1\tfrac{3}{4}$, $S_4 = 1\tfrac{7}{8}$

More generally

$$S_n = \frac{1 - (\tfrac{1}{2})^n}{1 - \tfrac{1}{2}} = 2\{1 - (\tfrac{1}{2})^n\} = 2 - (\tfrac{1}{2})^{n-1}$$

As n increases, $(\tfrac{1}{2})^{n-1}$ gets smaller and smaller;

and as $n \to \infty$, $(\tfrac{1}{2})^{n-1} \to 0$ $\Rightarrow$ $S_n \to 2$

In other words, as we take more and more terms of the series, their sum draws nearer and nearer to 2, approaching as closely as we wish.

We therefore say that the *infinite series*

$$1 + \tfrac{1}{2} + \tfrac{1}{4} + \tfrac{1}{8} + \cdots$$

converges to the sum 2.

More generally, consider the geometric series with first term a and common ratio r

$$a + ar + ar^2 + \cdots + ar^n + \cdots$$

Again let $S_n = a + ar + \cdots + ar^{n-1}$

Now if $|r| > 1$ (say $r = 2.5$, -1.5, etc.) successive terms are increasing in size and so the series cannot converge.

If $r = +1$ or -1, successive terms do not diminish in size and so again the series cannot converge.

But if $|r| < 1$ (e.g. $r = 0.8$, -0.9, etc.), $r^n \rightarrow 0$ as $n \rightarrow \infty$.

So $S_n = \dfrac{a(1 - r^n)}{1 - r} = \dfrac{a - ar^n}{1 - r} \rightarrow \dfrac{a}{1 - r}$ as $n \rightarrow \infty$

Hence $a + ar + ar^2 + ar^3 + \cdots$

converges to the sum $\dfrac{a}{1 - r}$, provided that $|r| < 1$

Example 2

A snail moves 2 metres along a straight line in one hour and in each succeeding hour 10% less than in the preceding one. How far does it go?

Here the appropriate g.p. is

$$2, \quad 2 \times \frac{9}{10}, \quad 2 \times \frac{9}{10} \times \frac{9}{10}, \quad 2 \times \frac{9}{10} \times \frac{9}{10} \times \frac{9}{10}, \ldots$$

So the distance travelled in n hours is given by the sum of the first n terms of the series

$$2 + 2\left(\frac{9}{10}\right) + 2\left(\frac{9}{10}\right)^2 + 2\left(\frac{9}{10}\right)^3 + \cdots$$

But this infinite series is convergent to the sum

$$\frac{2}{1 - \frac{9}{10}} = \frac{2}{\frac{1}{10}} = 20$$

So the snail gradually approaches a point on the line which is 20 metres from its starting-point.

Exercise 6.3

1   Find the last term and the sum of the following g.p.s:
   a) $2 + 6 + 18 + \cdots$ (7 terms)
   b) $3 + 1 + \frac{1}{3} + \cdots$ (6 terms)
   c) $2 - 4 + 8 - \cdots$ (8 terms)
   d) $1 + x + x^2 + \cdots$ (10 terms)
   e) $x - x^2 + x^3 - \cdots$ (n terms)

2 Calculate the sum to infinity of the following g.p.s:
  a) $1 + \frac{2}{3} + \frac{4}{9} + \cdots$
  b) $9 - 6 + 4 - \cdots$
  c) $1 + 0.9 + 0.81 + \cdots$
  d) $1 - x + x^2 - \cdots (|x| < 1)$

3 **a)** The recurring decimal $0.\dot{1}\dot{2} = 0.121\,212\ldots$ can be written as

$$\frac{12}{100} + \frac{12}{10000} + \frac{12}{1\,000\,000} + \cdots$$

Hence express $0.\dot{1}\dot{2}$ as a fraction.
  b) Express $0.\dot{1}2\dot{3}$ as a fraction.

4 A man offers to give to a charity by placing 1p on the first square of a chess-board, 2p on the second, 4p on the third, and so on until the board is covered. What is the value of his offer?

5 A woman deposits £100 each year at 5% compound interest. What is the total value of her investment immediately after making her tenth deposit?

6 How much should be deposited at 4% in order to yield £100 in a year's time? How much should be deposited in order to yield £100 at the end of each of the next five years and then leave the account empty?

7 A tennis ball is dropped from a height of 5 m and takes 1 s before hitting the ground. It then takes $1\frac{1}{2}$ s over its first bounce, and then bounces repeatedly so that the intervals between successive impacts are reduced in the ratio $\frac{3}{4}$ and successive heights in the ratio $\frac{9}{16}$.
  Find the total time taken and the total distance travelled before the ball comes to rest.

8 The sum to infinity of a geometric progression is five times its first term. Find its common ratio.
  Can the sum to infinity ever be $\frac{2}{3}$ of the first term?
  Can it be $\frac{1}{3}$ of the first term?

* 6.4 Finite series: the method of differences

In the last two sections we discovered results like

$$\sum_{r=1}^{n} r = 1 + 2 + 3 + \cdots + n = \tfrac{1}{2}n(n + 1)$$

and $\displaystyle\sum_{r=1}^{n} 2^r = 2 + 4 + 8 + \cdots + 2^n = 2\frac{2^n - 1}{2 - 1} = 2^{n+1} - 2$

We now investigate the more general *method of differences*.

Example 1

Calculate

$$\sum_{r=1}^{100} r^3 = 1^3 + 2^3 + 3^3 + \cdots + 100^3$$

The method depends on the fact that

$$r^3 \equiv \tfrac{1}{4}r^2[(r+1)^2 - (r-1)^2] \equiv \tfrac{1}{4}r^2(r+1)^2 - \tfrac{1}{4}(r-1)^2 r^2$$

So $1^3 = \tfrac{1}{4} \times 1^2 \times 2^2 - \tfrac{1}{4} \times 0^2 \times 1^2$

$\qquad\quad 2^3 = \tfrac{1}{4} \times 2^2 \times 3^2 - \tfrac{1}{4} \times 1^2 \times 2^2$

$\qquad\qquad \cdot \quad \cdot \quad \cdot \quad \cdot \quad \cdot \quad \cdot \quad \cdot \quad \cdot \quad \cdot$

$\qquad\quad 99^3 = \tfrac{1}{4} \times 99^2 \times 100^2 - \tfrac{1}{4} \times 98^2 \times 99^2$

$\qquad 100^3 = \tfrac{1}{4} \times 100^2 \times 101^2 - \tfrac{1}{4} \times 99^2 \times 100^2$

When we add these, we notice that all the terms of the right-hand side cancel out, except for

$\qquad \tfrac{1}{4} \times 100^2 \times 101^2 \quad$ and $\quad \tfrac{1}{4} \times 0^2 \times 1^2 \quad$ (which is zero)

So $\displaystyle\sum_{r=1}^{100} r^3 = \tfrac{1}{4} \times 100^2 \times 101^2 = 25\,502\,500$

Example 2

Find the sum of the first n terms of the series

$\qquad 1 \times 2 + 2 \times 3 + 3 \times 4 + \cdots$

i.e. $\displaystyle\sum_{r=1}^{n} r(r+1)$

Now $r(r+1) \equiv \tfrac{1}{3}r(r+1)\{(r+2) - (r-1)\}$

$\qquad\qquad\qquad \equiv \tfrac{1}{3}r(r+1)(r+2) - \tfrac{1}{3}(r-1)r(r+1)$

So $1 \times 2 = \tfrac{1}{3} \times 1 \times 2 \times 3 - \tfrac{1}{3} \times 0 \times 1 \times 2$

$\qquad\quad 2 \times 3 = \tfrac{1}{3} \times 2 \times 3 \times 4 - \tfrac{1}{3} \times 1 \times 2 \times 3$

$\qquad\quad 3 \times 4 = \tfrac{1}{3} \times 3 \times 4 \times 5 - \tfrac{1}{3} \times 2 \times 3 \times 4$

$\qquad\qquad \cdot \quad \cdot \quad \cdot \quad \cdot \quad \cdot \quad \cdot \quad \cdot \quad \cdot \quad \cdot$

and $n(n + 1) = \frac{1}{3}n(n + 1)(n + 2) - \frac{1}{3}(n - 1)n(n + 1)$

Adding, we obtain

$$\sum_{r=1}^{n} r(r + 1) = \frac{1}{3}n(n + 1)(n + 2)$$

Exercise 6.4a

1   Use the identity

$$r(r + 1)(r + 2) \equiv \frac{1}{4}r(r + 1)(r + 2)\{(r + 3) - (r - 1)\}$$

in order to find a formula for

$$\sum_{r=1}^{n} r(r + 1)(r + 2)$$

Check your result when $n = 4$, $n = 5$, $n = 6$.

2   Use the identity

$$\frac{1}{r(r + 1)} \equiv \frac{1}{r} - \frac{1}{r + 1}$$

in order to find a formula for $\displaystyle\sum_{r=1}^{n} \frac{1}{r(r + 1)}$

Check your result when $n = 4$, $n = 5$, $n = 6$. Find the sum to infinity of

$$\frac{1}{1 \times 2} + \frac{1}{2 \times 3} + \frac{1}{3 \times 4} + \cdots$$

Standard series

It is clear that

$$\sum_{1}^{n} 1 = n$$

and we now know that

$$\sum_{1}^{n} r = \frac{1}{2}n(n + 1)$$

and

$$\sum_{1}^{n} r(r + 1) = \frac{1}{3}n(n + 1)(n + 2)$$

and

$$\sum_{1}^{n} r(r + 1)(r + 2) = \frac{1}{4}n(n + 1)(n + 2)(n + 3)$$

These *standard series* can often be used for the calculation of other sums.

Example 3

$$\sum_1^n r^2 = \sum_1^n \{r(r+1) - r\}$$

$$= \sum_1^n r(r+1) - \sum_1^n r$$

$$= \tfrac{1}{3}n(n+1)(n+2) - \tfrac{1}{2}n(n+1)$$

$$= \tfrac{1}{6}n(n+1)\{2(n+2) - 3\}$$

$$= \tfrac{1}{6}n(n+1)(2n+1)$$

Example 4

Find the sum of the first 50 terms of

$$1^2 \times 2 + 2^2 \times 3 + 3^2 \times 4 + \cdots$$

This can be written $\displaystyle\sum_1^{50} r^2(r+1)$

$$\sum_1^n r^2(r+1) = \sum_1^n \{r(r+1)(r+2) - 2r(r+1)\}$$

$$= \tfrac{1}{4}n(n+1)(n+2)(n+3) - \tfrac{2}{3}n(n+1)(n+2)$$

$$= \tfrac{1}{12}n(n+1)(n+2)\{3(n+3) - 8\}$$

$$= \tfrac{1}{12}n(n+1)(n+2)(3n+1)$$

So $\displaystyle\sum_1^{50} r^2(r+1) = \tfrac{1}{12} \times 50 \times 51 \times 52 \times 151 = 1\,668\,550$

Exercise 6.4b

1 Show that $\displaystyle\sum_1^n (2r-1)^2 \equiv 4 \sum_{r=1}^n r(r+1) - 8 \sum_{r=1}^n r + \sum_{r=1}^n 1$

Hence find the sum of the squares of the first n odd integers, and calculate

$$1^2 + 3^2 + 5^2 + \cdots + 99^2$$

2 Find the rth term and use standard series to find the sum to n terms of
   a) $1 \times 3 + 2 \times 4 + 3 \times 5 + \cdots$
   b) $1 \times 2 + 3 \times 4 + 5 \times 6 + \cdots$
   c) $1 \times 3 + 4 \times 6 + 7 \times 9 + \cdots$
   d) $1^3 + 2^3 + 3^3 + \cdots$

*6.5 Mathematical induction

We now come to an exceedingly general method of proof, once a possible result has been suspected. This we shall illustrate by a number of examples.

Example 1

Find the sum of the cubes of the first n integers (see 4.2), i.e. a formula for

$$\sum_1^n r^3 = 1^3 + 2^3 + \cdots + n^3$$

After a certain amount of trial and error with the simplest cases, it may be noticed that

$$1^3 \qquad\qquad = \quad 1 = \quad 1^2 = (\tfrac{1}{2} \times 1 \times 2)^2$$

$$1^3 + 2^3 \qquad\qquad = \quad 9 = \quad 3^2 = (\tfrac{1}{2} \times 2 \times 3)^2$$

$$1^3 + 2^3 + 3^3 \qquad = \quad 36 = \quad 6^2 = (\tfrac{1}{2} \times 3 \times 4)^2$$

$$1^3 + 2^3 + 3^3 + 4^3 = 100 = 10^2 = (\tfrac{1}{2} \times 4 \times 5)^2$$

and we are led to suspect that

$$\sum_1^n r^3 = 1^3 + 2^3 + 3^3 + \cdots + n^3 = [\tfrac{1}{2}n(n + 1)]^2 = \tfrac{1}{4}n^2 (n + 1)^2$$

But the fact that this result has been shown to be true for $n = 1, 2, 3, 4$ and can be verified for any other particular value of n does not constitute a general proof. For, however often our hopes are confirmed, there will remain a fear that for some value of n the result will be untrue. That this is not so can be shown by the method of *mathematical induction*.

To prove that

$$\sum_1^n r^3 = 1^3 + 2^3 + 3^3 + \cdots + n^3 = \tfrac{1}{4}n^2 (n + 1)^2, \quad (n \in \mathbb{Z}^+)$$

first of all let us suppose that this is true for a particular value of n, which we shall call k. We shall try to deduce from this hypothesis that it is then automatically true for the next value, $n = k + 1$.

If the result is true for $n = k$,

then $\qquad \displaystyle\sum_1^k r^3 = \tfrac{1}{4}k^2 (k + 1)^2$

But $\qquad \displaystyle\sum_1^{k+1} r^3 = \sum_1^k r^3 + (k + 1)^3$

$$= \tfrac{1}{4}k^2 (k + 1)^2 + (k + 1)^3$$

$$= \tfrac{1}{4}(k + 1)^2 \{k^2 + 4(k + 1)\}$$

$$= \tfrac{1}{4}(k + 1)^2 (k + 2)^2$$

which has established its truth for the next value, $n = k + 1$.

So *if* it is true for $n = k$, it is also true for $n = k + 1$.

But $1^3 = \tfrac{1}{4} \times 1^2 \times 2^2$

so we know that the result *is* true for $n = 1$.

Hence, from what we have just shown, it is true for the next value, $n = 2$; hence for the next value, $n = 3$; and so on, for all values of n.

So by establishing

a) that its truth for $n = k$ *would imply* its truth for $n = k + 1$; and
b) that it *is* true for the first value $n = 1$,

we have proved that

$$\sum_{1}^{n} r^3 = \tfrac{1}{4}n^2 (n + 1)^2$$

for all positive integral values of n.

Example 2

Prove by induction that

$$\sum_{1}^{n} r(r + 1) = \tfrac{1}{3}n(n + 1)(n + 2)$$

i.e. $1 \times 2 + 2 \times 3 + 3 \times 4 + \cdots + n(n + 1) = \tfrac{1}{3}n (n + 1)(n + 2)$

(We have already proved this by the method of differences.)

Suppose that the result is true for $n = k$,

i.e.   that $\displaystyle\sum_{1}^{k} r(r + 1) = \tfrac{1}{3}k(k + 1)(k + 2)$

Then $\displaystyle\sum_{1}^{k+1} r(r + 1) = \sum_{1}^{k} r(r + 1) + (k + 1)(k + 2)$

$$= \tfrac{1}{3}k(k + 1)(k + 2) + (k + 1)(k + 2)$$

$$= \tfrac{1}{3}(k + 1)(k + 2)(k + 3)$$

So *if* the result is true for $n = k$, it is also true for $n = k + 1$. But we also see that

$$1 \times 2 = \tfrac{1}{3} \times 2 \times 3, \quad \text{so the result } is \text{ true for } n = 1$$

Hence, *by induction*, it is true for all positive integral n that

$$\sum_{1}^{n} r(r + 1) = \tfrac{1}{3}n(n + 1)(n + 2)$$

Example 3

Prove, for all positive integral values of n, that

$$\frac{d}{dx}(x^n) = nx^{n-1}$$

Suppose this is true for $n = k$,

i.e. that $\dfrac{d}{dx}(x^k) = kx^{k-1}$

Then $\dfrac{d}{dx}(x^{k+1}) = \dfrac{d}{dx}(x \times x^k)$

$$= x \times \frac{d}{dx}(x^k) + \frac{d}{dx}(x) \times x^k$$

$$= x \times kx^{k-1} + x^k$$

$$= kx^k + x^k = (k + 1)x^k$$

So *if* the result is true for $n = k$, it will also be true for $n = k + 1$.

But $\dfrac{d}{dx}(x^1) = 1 = 1 \times x^0,$ so the result *is* true for $n = 1$

Hence, by induction, it is true for all positive integral n that

$$\frac{d}{dx}(x^n) = nx^{n-1}$$

Example 4

Show that, for all positive integers n, $u_n = 9^n + 7$ is divisible by 8.

Suppose that the result is true if $n = k$, i.e. that $u_k = 9^k + 7$ is divisible by 8.

Now $u_{k+1} - u_k = (9^{k+1} + 7) - (9^k + 7)$

$$= 8 \times 9^k, \quad \text{which is divisible by 8}$$

So $\qquad u_{k+1} = u_k + 8 \times 9^k$

and if u_k is divisible by 8, so is u_{k+1}.
 So *if* the result is true for $n = k$, it must also be true for $n = k + 1$.

But $u_1 = 9^1 + 7 = 16,$ and so it *is* true for $n = 1$

Hence, by induction, it is true for all positive integral n that $u_n = 9^n + 7$ is divisible by 8.

Exercise 6.5

Check the following conjectures when $n = 1, 2, 3, 4$ and then either prove them for $n \in \mathbb{Z}^+$ by the method of induction or show that they are false:

1 $1 + 2 + 3 + \cdots + n = \frac{1}{2}n(n + 1)$

2 $1^2 + 2^2 + 3^2 \cdots + n^2 = \frac{1}{6}n(n + 1)(2n + 1)$

3 $1 + 3 + 5 + \cdots + (2n - 1) = n^2$

4 $\dfrac{1}{1 \times 2} + \dfrac{1}{2 \times 3} + \cdots + \dfrac{1}{n(n + 1)} = \dfrac{n}{n + 1}$

5 $1 + \dfrac{1}{2^2} + \dfrac{1}{3^2} + \cdots + \dfrac{1}{n^2} = \dfrac{2n + 1}{n + 2}$

6 $1 + x + x^2 + \cdots + x^{n-1} = \dfrac{1 - x^n}{1 - x}(x \neq 1)$

7 $n^3 - n$ is divisible by 6.

8 $8^n + 6$ is divisible by 14.

9 n straight coplanar lines, no two of which are parallel and no three concurrent, meet in $\frac{1}{2}n(n - 1)$ points.

10 The above lines divide the plane into $\frac{1}{2}(n^2 + n + 2)$ regions.

* 6.6 Iteration: recurrence relations

It is frequently very easy, particularly when using a computer, to generate a sequence by a process in which the next term u_{n+1} is produced from its predecessor u_n (or its predecessors). This is usually called *iteration*, simply because it is repeated again and again, and is usually defined by a *recurrence relation* which expresses u_{n+1} in terms of its predecessor(s).

 To produce the sequence a starting point is needed, and so the first term also needs to be given. The recurrence relation, together with this initial term, is sometimes referred to as the *inductive definition* of the sequence.

Possibly the simplest type of sequence to be defined in this way is an arithmetic progression such as 3, 8, 13, 18, ...

where $u_{n+1} = u_n + 5$ and $u_1 = 3$

Similarly for a geometric progression such as 2, 6, 18, 54, ...

$$u_{n+1} = 3u_n \qquad \text{and} \qquad u_1 = 2$$

A slightly more complicated recurrence relation generates the square numbers 1, 4, 9, 16, 25, ... Successive differences between terms are 3, 5, 7, 9, ...

so $u_{n+1} = u_n + (2n + 1),$ with $u_1 = 1$

All such recurrence relations can be neatly represented by simple flow diagrams. For example

3, 8, 13, 18, ...

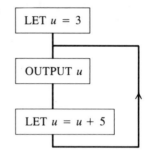

For the sequence 1, 4, 9, 16, 25, ... a rather more involved diagram is required. A decision box can be included if, say, just the first 20 terms of the sequence are required.

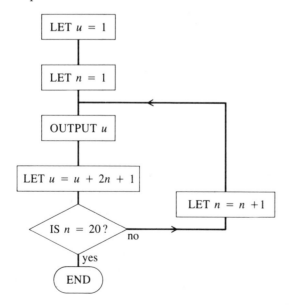

Example 1

Find the sequence defined by the recurrence relation $u_{n+1} = 2u_n + 1$ and initial term $u_1 = 1$. Establish a formula for u_n.

We can see that

$$u_2 = 2u_1 + 1 = 2 \times 1 + 1 = 3$$

$$u_3 = 2u_2 + 1 = 2 \times 3 + 1 = 7$$

$$u_4 = 2u_3 + 1 = 2 \times 7 + 1 = 15$$

etc.

and so obtain the sequence $1, 3, 7, 15, \ldots$.

It is then quickly conjectured that the nth term of the sequence is

$$u_n = 2^n - 1$$

This can easily be proved by the method of induction

for $u_k = 2^k - 1$

$\Rightarrow \quad u_{k+1} = 2u_k + 1$

$$= 2(2^k - 1) + 1$$

$$= 2^{k+1} - 1$$

So if the conjecture is true for $n = k$, it is also true for $n = k + 1$.

But $u_1 = 1 = 2^1 - 1$

So the conjecture is true for $n = 1$.

Hence, by induction, it is true for every positive integer n.

Limits of sequences

Sequences can behave in a number of different ways. For instance,

the sequence $1, 3, 5, 7, \ldots$ (defined by $u_{n+1} = u_n + 2$, $u_1 = 1$) increases without limit;

the sequence $1, 2, 1, 2, \ldots$ (defined by $u_{n+1} = 3 - u_n$, $u_1 = 1$) oscillates finitely,

and the sequence $1, -2, 4, -8, 16, \ldots$ (defined by $u_{n+1} = -2u_n$, $u_1 = 1$) oscillates infinitely.

We shall be particularly interested, however, in sequences which tend to a limit as n increases. One example is the geometric progression

$1, \frac{1}{2}, \frac{1}{4}, \frac{1}{8}, \ldots$ (defined by $u_{n+1} = \frac{1}{2}u_n$, $u_1 = 1$)

whose limiting value is zero.

In most cases it is not so immediately obvious that a sequence tends to a limit, or what that limit is going to be.

Example 2

Evaluate the first eight terms of the sequence defined by $u_{n+1} = \dfrac{u_n + 4}{3}$,

$u_1 = 10$. Investigate whether the sequence tends to a limit.

To calculator accuracy the first eight terms of the sequence are

> 10
> 4.666 666 7
> 2.888 888 9
> 2.296 296 3
> 2.098 765 4
> 2.032 921 8
> 2.010 973 9
> 2.003 658 0

The terms of the sequence *seem* to be tending to a limiting value of 2.

If such a limit, x, does exist, then it must be unaffected by the iteration, i.e. adding 4 and then dividing by 3 will leave its value unchanged.

So $x = \dfrac{x + 4}{3}$

$\Rightarrow$ $3x = x + 4$

$\Rightarrow$ $2x = 4$

$\Rightarrow$ $x = 2$,   as anticipated

Warning: The method given in example 2 for finding the limit of a sequence is valid only if the sequence does converge. The following example illustrates this.

Example 3

Which of the two recurrence relations

a) $u_{n+1} = \dfrac{3}{u_n}$ \qquad\qquad **b)** $u_{n+1} = \dfrac{1}{2}\left(u_n + \dfrac{3}{u_n}\right)$

can be used for finding $\sqrt{3}$ to any required degree of accuracy?

If a limit does exist, then that limit must be $\sqrt{3}$ for each relation since

a) $x = \dfrac{3}{x}$ and   b) $x = \dfrac{1}{2}\left(x + \dfrac{3}{x}\right)$

$\Rightarrow x^2 = 3$ $\Rightarrow 2x = x + \dfrac{3}{x}$

$\Rightarrow x = \dfrac{3}{x}$

$\Rightarrow x^2 = 3$

However, with an initial value $u_1 = 1$ relation a) generates 1, 3, 1, 3, 1, ...
and will in fact give such an oscillating sequence, whatever the value of u_1.
Therefore a limit does not exist, and this relation is of no use for calculating
$\sqrt{3}$.

But with $u_1 = 1$ in relation b),

$u_2 = 2$

$u_3 = 1.75$

$u_4 = 1.7321429$

$u_5 = 1.7320508$ (to calculator accuracy)

This sequence clearly tends to a limit, which must be $\sqrt{3}$.

Exercise 6.6

1 Find the first seven terms of the sequence defined by
  a) $u_{n+1} = u_n + 2$ and $u_1 = 1$
  b) $u_{n+1} = u_n + n$ and $u_1 = 2$
  c) $u_{n+1} = 3u_n + 1$ and $u_1 = 4$
  d) $u_{n+1} = -4u_n$ and $u_1 = -1$
  e) $u_{n+1} = 10/u_n$ and $u_1 = 2$
  f) $u_{n+1} = 2(6 - u_n)$ and $u_1 = 1$

2 Find inductive definitions for the sequences which begin
  a) 3, 7, 11, 15, 19, ...
  b) 1000, 100, 10, 1, 0.1, ...
  c) 3, 4, 6, 10, 18, ...
  d) 1. 0.9, 0.81, 0.729, 0.6561, ...
  e) 2, 4, 16, 256, 65536, ...
  f) $\frac{1}{2}, \frac{1}{3}, \frac{1}{4}, \frac{1}{5}, \frac{1}{6}, \ldots$

3 What are the first seven terms of the sequence u_n defined by

$$u_{n+1} = (n+1)u_n \quad \text{and} \quad u_1 = 1?$$

Find a formula for u_n.

4 Find the first five terms of the sequence defined by

$$u_{n+1} = 5u_n - 6u_{n-1}, \quad u_1 = 5, \quad u_2 = 13$$

and prove that $u_n = 3^n + 2^n$

5 Find the first seven terms of the sequence defined by $u_{n+1} = \dfrac{1}{1 - u_n}$

and $u_1 = 2$. Explain algebraically why $u_{n+3} = u_n$.

6 a) Evaluate the first eight terms of the sequence defined by $u_{n+1} = \dfrac{u_n + 2}{5}$

and $u_1 = 3$. What appears to be the limit to which the sequence tends?
b) Repeat **a)**, but this time with $u_1 = 20$.

7 a) Work out what you think will be the limiting value of the sequence

defined by $u_{n+1} = \dfrac{u_n - 2}{4}$ and $u_1 = 4$.

b) Evaluate the first 8 terms of this sequence, and check your answer to **a)**.

8 Calculate $\sqrt{10}$ to six places of decimals by means of the recurrence relation

$$u_{n+1} = \frac{1}{2}\left(u_n + \frac{10}{u_n}\right)$$

9 $u_{n+1} = \dfrac{1}{2}\left(u_n + \dfrac{10}{u_n^2}\right), \quad u_1 = 2$

If $u_n \to x$, what is the value of x? Find it, correct to 2 decimal places.

* 6.7 Solution of equations: iterative methods

In the last section we saw that if the sequence generated by a recurrence relation $u_{n+1} = f(u_n)$ tended to a limit, then that limit was a solution of the equation $x = f(x)$.

We can now use this process in the solution of equations. If an equation can be written in the form $x = f(x)$, and if the sequence generated by the recurrence relation $u_{n+1} = f(u_n)$ tends to a limit, then the successive terms of the sequence give a closer and closer approximation to a solution of the equation. Therefore, a solution can be obtained to any required degree of accuracy.

Consider the equation $x^3 - 3x - 5 = 0$. The graph of $y = x^3 - 3x - 5$ is illustrated in the diagram, and this shows that the equation has just one solution, which lies between 2 and 3.

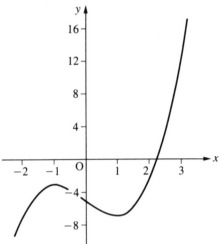

The equation $x^3 - 3x - 5 = 0$ can be rewritten in the form $x = f(x)$ in a number of different ways. Possibly the two most obvious ones are:

a) $x = \sqrt[3]{(3x + 5)}$ and   **b)** $x = \dfrac{x^3 - 5}{3}$

giving rise to the recurrence relations

a) $u_{n+1} = \sqrt[3]{(3u_n + 5)}$ and   **b)** $u_{n+1} = \dfrac{u_n^3 - 5}{3}$

From the graph, 2 seems a reasonable first approximation to the solution of the equation, and so we will take this as the first term in each sequence.

The sequences generated by the two recurrence relations (given to calculator accuracy) are then

a) $u_1 = 2$
$\ u_2 = 2.2239801$
$\ u_3 = 2.2683724$
$\ u_4 = 2.2769672$
$\ u_5 = 2.2786237$
$\ u_6 = 2.2789427$
$\ u_7 = 2.2790041$
$\ u_8 = 2.2790160$

b) $u_1 = 2$
$\ u_2 = 1$
$\ u_3 = -1.3333333$
$\ u_4 = -2.4567901$
$\ u_5 = -6.6095790$

It is already obvious that sequence **b)** does not tend to a limit. However, sequence **a)** is tending to a limit, and to 3 decimal places that limit is 2.279.

Therefore, to 3 decimal places, the solution of the equation

$x^3 - 3x - 5 = 0$ is $x = 2.279$

The following diagrams, on which are marked the line $y = x$ and the curve $y = f(x)$, illustrate geometrically why relation **a)** generates a limiting sequence, whereas relation **b)** does not.

a)

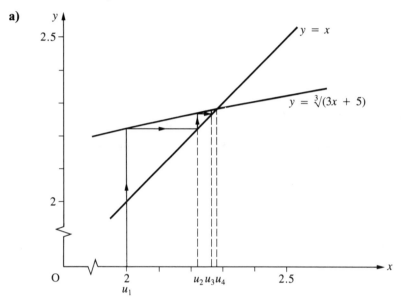

b)

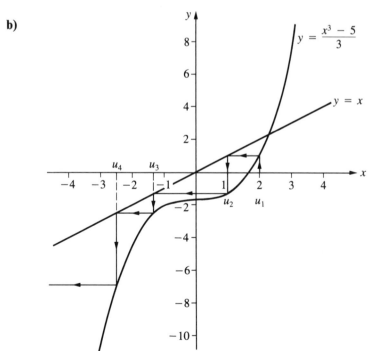

Other less obvious rearrangements of the equation can also be tried.

For example $x^3 - 3x - 5 = 0$

$\Rightarrow$ $x^3 = 3x + 5$

$\Rightarrow$ $x^2 = 3 + \dfrac{5}{x}$

$\Rightarrow$ $x = \sqrt{\left(3 + \dfrac{5}{x}\right)}$

giving the recurrence relation $u_{n+1} = \sqrt{\left(3 + \dfrac{5}{u_n}\right)}$

From the same starting value of 2, this generates the sequence

$u_1 = 2$
$u_2 = 2.3452079$
$u_3 = 2.2653934$
$u_4 = 2.2819120$
$u_5 = 2.2784084$
$u_6 = 2.2791477$
$u_7 = 2.2789916$

The terms of the sequence this time alternate on either side of the limit, and if we look at the iterations geometrically, then we obtain a 'spiral' pattern rather than the previous 'step' pattern.

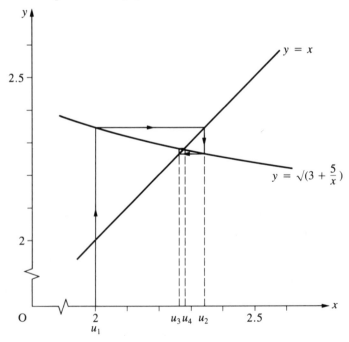

Clearly, in using this type of iterative method for solving equations, the main difficulty will be in deciding which rearrangements of the equation into the form $x = f(x)$ will enable us to generate a sequence converging to a limit. Such a choice is not completely a matter of guesswork, though any guess can be checked very quickly with a calculator.

If we look at the 'step' pattern which leads us to the solution for recurrence relation **a)**, each iteration is represented by a vertical move to the curve, followed by a horizontal move to the straight line $y = x$.

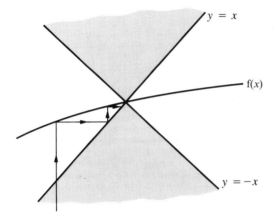

Such a pattern will 'home in' on the intersection of the line and curve only if each vertical move is shorter than the previous horizontal move; for this to be true, the section of the curve near the intersection must lie in the unshaded area of the diagram, i.e. $-1 < f'(x) < 1$ in the vicinity of the intersection.

This value of $f'(x)$ gives other information as well: if $f'(x) > 0$ the sequence 'steps' to the solution, whereas if $f'(x) < 0$ the sequence 'spirals' to the solution; and the closer $f'(x)$ is to zero, the more rapidly the sequence will converge.

Example

Calculate, correct to 2 decimal places, the solution of the equation

$$\sin x = 2 - x$$

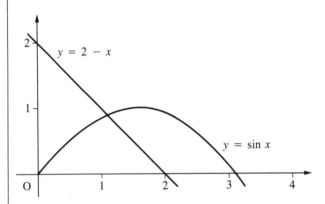

The sketch graph shows that the solution of the equation is slightly greater than 1. So 1 will be taken as the first approximation to the solution.

The equation can be rearranged into the form $x = 2 - \sin x$.

For $f(x) = 2 - \sin x,$ $f'(x) = -\cos x$

and since $|\cos x| < 1$ (for all $x \neq n\pi$), we can be certain that a sequence generated by

$$u_{n+1} = 2 - \sin u_n$$

will tend to the solution of the equation. (Also since $f'(x) < 0$ the terms of the sequence will spiral onto the solution.)

Using the recurrence relation, $u_1 = 1$
$$u_2 = 1.1585290$$
$$u_3 = 1.0837853$$
$$u_4 = 1.1162644$$
$$u_5 = 1.1015334$$
$$u_6 = 1.1080982$$
$$u_7 = 1.1051486$$

and this is sufficient to be able to state that, to 2 decimal places, the solution is $x = 1.11$.

Exercise 6.7

1 With an initial value $u_1 = 3$, find out which of the recurrence relations given below generates a sequence tending to the larger solution of the quadratic equation $x^2 - 3x + 1 = 0$.

   a) $u_{n+1} = 3 - \dfrac{1}{u_n}$ **b)** $u_{n+1} = \dfrac{u_n^2 + 1}{3}$

   Find this solution, correct to 3 decimal places.

2 Repeat no. **1** with an initial value $u_1 = 0.5$ to find the smaller solution of the equation $x^2 - 3x + 1 = 0$.

3 Use the graphs of $y = 3 - \dfrac{1}{x}$ and $y = \dfrac{x^2 + 1}{3}$ together with the graph of $y = x$ in each case, to explain your results in nos. **1** and **2**.

4 One of the recurrence relations below will give the larger solution (with $u_1 = 3$) of the quadratic equation $x^2 - 2x - 4 = 0$, and the other will give the smaller solution (with $u_1 = -1$). Find these two solutions, correct to 3 decimal places.

   a) $u_{n+1} = \sqrt{(2u_n + 4)}$ **b)** $u_{n+1} = \dfrac{4}{u_n - 2}$

5 Sketch the graph of $y = x^3 - 5x + 3$ (for $-3 \leqslant x \leqslant 3$). Use this graph to obtain first approximations to the solutions of the equation $x^3 - 5x + 3 = 0$.

   By using the recurrence relations

   $$u_{n+1} = \sqrt[3]{(5u_n - 3)}, \quad u_{n+1} = \dfrac{u_n^3 + 3}{5}$$

   or other suitable relations, obtain the solutions of the equation, correct to 2 decimal places.

6 Solve the equation $x^2 = \cos x$, giving your answer correct to 3 decimal places. (Use a sketch graph to find a first approximation.)

7 Solve the equation $2^x = 10x$, giving your answer correct to 4 decimal places.

8 By considering the roots of the equation $f'(x) = 0$, or otherwise, prove that the equation $f(x) = 0$, where $f(x) \equiv x^3 + 2x + 4$, has only one real root. Show that this root lies in the interval $-2 < x < -1$.

   Use the iterative procedure

   $$x_{n+1} = -\tfrac{1}{6}(x_n^3 - 4x_n + 4), \quad x_1 = -1$$

   to find two further approximations to the root of the equation, giving your final answer to 2 decimal places.   (L)

9   The solution of the equation

$$x = \sqrt{(9 - x)}$$

is to be found by an iterative method using the relation

$$x_{n+1} = \sqrt{(9 - x_n)}$$

where x_n, x_{n+1} are successive values in the calculation. Taking $x_1 = 2$, find the solution of the equation correct to two decimal places, tabulating your values of x_n as you proceed.

Sketch on one diagram the parts of the graphs of $y = \sqrt{(9 - x)}$ and $y = x$ which lie in the first quadrant. Illustrate the first two stages in the above iterative solution by drawing suitable lines parallel to the axes.

(JMB)

10   Show that the cubic equation $x^3 + 2x - 11 = 0$ has only one real root and further that the root lies between $x = 1$ and $x = 2$.

Two possible iterative schemes for finding the root are

a) $x_{n+1} = (11 - x_n^3)/2$ and

b) $x_{n+1} = (11 - 2x_n)^{1/3}$

Show that only one of these schemes converges from an initial estimate of $x = 2$ and hence find the root correct to 3 d.p., justifying the accuracy of your answer.

(MEI)

11   The sequence x_n is defined by $x_1 = \cos \alpha$, and $x_{n+1} = \cos x_n$, where $0 < \alpha < \frac{1}{2}\pi$.

Explain, graphically or otherwise, why x_n tends to the same limit, X, whatever the value of α, and write down the equation satisfied by X.

Examine whether the bounds given for α are necessary for x_n to tend to X.

Give a reason why the iteration is slow and describe and use a quicker method of finding the value of x rounded to 5 significant figures.

Investigate briefly the sequence obtained if the tangent function were taken instead of cosine in the definition.

(MEI)

12   Show that the equation $x^4 - 32x - 60 = 0$ has a real root between 3 and 4.

It is desired to use an iterative process to find an approximation to the root, starting with initial value $x = 3$. The iterations

$$x_{n+1} = \tfrac{1}{32}(x_n^4 - 60)$$
$$\text{and} \quad x_{n+1} = (32x_n + 60)^{1/4}$$

are both suggested, but one is not suitable. Draw two diagrams, each using a pair of graphs selected from

$$y = \tfrac{1}{32}(x^4 - 60), \quad y = x^4, \quad y = x \quad \text{and} \quad y = 32x + 60$$

to show the first step of each iteration. Hence decide which is the suitable iteration and, for this iteration, correct and complete the flow diagram so that the results of the first five iterations would be printed out.

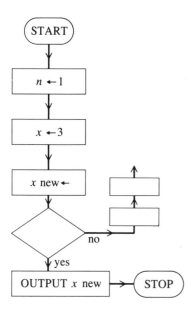

Work out the first two numbers that would be printed in the output, assuming that the computer's arithmetic unit works correct to 3 decimal places.

[$n \leftarrow 1$ is an alternative notation for $n := 1$.]   (SMP)

*6.8 The Newton–Raphson method for solving equations

An alternative, and widely used iterative method for finding the solution of an equation is the Newton–Raphson† method (or more simply just Newton's method). This is based on a quite different principle.

In the Newton–Raphson method the equation has to be written in the form $f(x) = 0$. The solution of the equation is then the x coordinate of the point where $y = f(x)$ cuts the x-axis.

We use the fact that if u_1 is an approximation to the solution of the equation, then the tangent to the curve at $x = u_1$ will usually meet the x-axis at a point which is closer to the solution.

† Joseph Raphson, author of *The History of Fluxions* (1715).

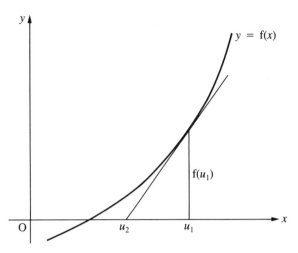

The gradient of the tangent is $f'(u_1)$,

so $f'(u_1) = \dfrac{f(u_1)}{u_1 - u_2}$

$\Rightarrow$ $u_1 - u_2 = \dfrac{f(u_1)}{f'(u_1)}$

$\Rightarrow$ $u_2 = u_1 - \dfrac{f(u_1)}{f'(u_1)}$

This iteration can then be repeated as many times as necessary to obtain the required degree of accuracy, and in general

$$u_{n+1} = u_n - \frac{f(u_n)}{f'(u_n)}$$

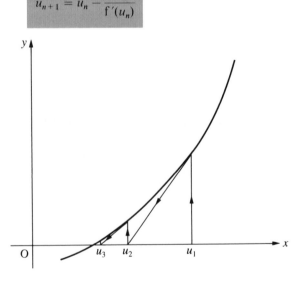

We will use this method to solve once again the equation $x^3 - 3x - 5 = 0$, which occurred in the last section.

$$f(x) = x^3 - 3x - 5 \quad \text{so} \quad f'(x) = 3x^2 - 3$$

Therefore, the recurrence relation to be used is

$$u_{n+1} = u_n - \frac{u_n^3 - 3u_n - 5}{3u_n^2 - 3}$$

Taking the same first approximation as with the previous method, it follows that

$$u_1 = 2$$
$$u_2 = 2.333\,333\,3$$
$$u_3 = 2.280\,5556$$
$$u_4 = 2.2790200$$
$$u_5 = 2.2790187$$

So, to 3 decimal places, the solution of the equation is $x = 2.279$.

If we compare this solution with that in section 6.7, it is clear that, although the recurrence relation is more involved, fewer iterations are required to obtain a solution. Indeed, it is generally true that the Newton–Raphson method gives quite rapid convergence to a limit.

However, there are a few rather unusual situations where the method might break down. The following diagrams illustrate two examples.

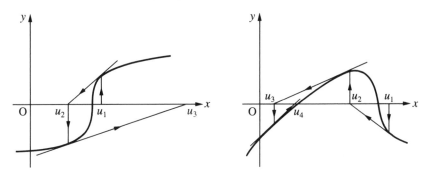

In the first case the sequence does not tend to a limit, and in the second case the sequence homes in on the wrong (or unintended) solution. In both cases the difficulty can be avoided by choosing a better first approximation.

Exercise 6.8

1 Use the Newton–Raphson method to find, correct to 3 decimal places, the solutions of the equation $x^2 - 3x + 1 = 0$. Use 3 and 0.5 as the first approximations. Compare your answers with exercise 6.7, nos. **1** and **2**.

2 Apply the Newton–Raphson method to the equation $x^3 - 18x + 2 = 0$, starting with $x = 4$ as the first approximation. Prove that the second approximation is $x = 4.2$ and that the equation has a root between 4 and 4.2.   (OC)

3 The equation $x^3 - 3x - 1 = 0$ has a root between 1.8 and 2.0. Find the value of this root to 3 decimal places.   (OC)

4 Use the Newton–Raphson method to find the smallest positive solution of the equation $\tan x = 2x$, correct to 3 decimal places. Use $x = 1$ as the first approximation.

5 On the same diagram sketch the graphs of $y = \cos x$ and $y = \sqrt{x}$. Use the diagram to obtain a first approximation to the solution of the equation $\cos x = \sqrt{x}$. Use the Newton–Raphson method to obtain the solution correct to 2 decimal places.

6 Show that the equation

$$x^4 - 3x^2 - 3x + 1 = 0$$

has real roots between 0 and 1 and between 2 and 3. Find the larger of these roots correct to three significant figures.   (OC)

7 Use any method to locate the three roots of the equation

$$x^3 - 5x + 3 = 0$$

and evaluate the numerically smallest root to two significant figures.

(OC)

8 If k is sufficiently large, show that one root of the equation

$$x^{10} - kx + k = 0$$

is approximately $(k - 9)/(k - 10)$.   (L)

9 A first approximation to the positive solution of the equation

$$\frac{1}{x^2} - 6 = 0$$

is denoted by x_1. Show by applying Newton's method to this equation that a second approximation, x_2, is given by

$$x_2 = 1.5x_1 - 3x_1^3$$

Work through the process shown in the flow diagram. Tabulate your values of X, N and Y as you proceed. Work to four decimal places throughout.

Write down the number of iterations which will be needed to solve the given equation correct to two decimal places when x_1 is taken as 0.3. (JMB)

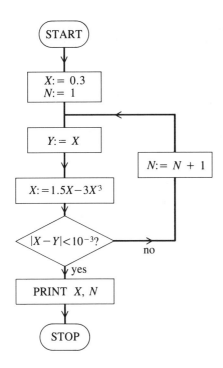

*6.9 Interlude: Fibonacci numbers and prime numbers

Fibonacci numbers

Examine a pineapple. It will be seen that its units are arranged in 5 parallel rows rising gently to the right, 8 steeper rows rising to the left and 13 very steep rows rising to the right (though some pineapples are left-handed!). Why 5 and 8 and 13?

Examine the ancestry of a male bee (or *drone*). It has a mother (a *queen*) but no father, whilst a queen has both a father and a mother. So the genealogy of a drone can be represented:

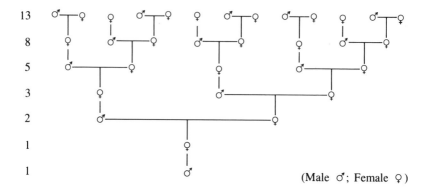

(Male ♂; Female ♀)

Going backwards in time, the numbers in the different generations are

1, 1, 2, 3, 5, 8, 13, ...

and again we notice the numbers 5, 8, and 13.

This sequence of *Fibonacci*† *numbers* is evidently generated by the recurrence relation

$$u_{n+1} = u_n + u_{n-1}$$

Exercise 6.9

1 What are the values of u_8, u_9, u_{10}, u_{11} in the Fibonacci sequence?

2 Show that

a) $\dfrac{u_{n+1}}{u_n} = 1 + \dfrac{u_{n-1}}{u_n}$

b) if $\dfrac{u_{n+1}}{u_n} \to l,$ then $l = 1 + \dfrac{1}{l}$

Hence show that a distant generation of ancestors of a particular drone had $\dfrac{\sqrt{5}+1}{2}$ as many parents. What is the ratio of the sexes in this generation?

† The nickname of Leonardo of Pisa, a thirteenth century mathematician.

3 Check, for $n = 1, 2, 3, 4$, that

$$u_n = \frac{1}{\sqrt{5}}\left\{\left(\frac{1 + \sqrt{5}}{2}\right)^n - \left(\frac{1 - \sqrt{5}}{2}\right)^n\right\}$$

Prime numbers

A prime number is any positive integer larger than 1 which has no factors apart from 1 and itself.

So the sequence of prime numbers is $2, 3, 5, 7, 11, 13, 17, 19, 23 \ldots$.

Unlike Fibonacci numbers, these are highly irregular and there is no iterative method for producing the next of the sequence. Clearly 2 is the only even prime, and as we move through the positive integers there is an increasing number of potential factors and so a diminishing likelihood that a number is prime. Nevertheless, consecutive pairs of prime numbers like 11, 13; 17, 19; 29, 31 keep persisting (the largest known pair at the time of writing is 1 000 000 009 649 and 1 000 000 009 651), though it is not yet certain whether there is an infinite supply of such pairs.†

A simpler question, however, is whether or not there is an infinite supply of the prime numbers themselves? Certainly the further we proceed, the scarcer they become. Are they ever exhausted? This question, 'Is this sequence of prime numbers infinite?', or 'Is there always a next prime number?', was first faced by Euclid and his proof is one of the most famous of mathematics:

Suppose we know that there are n prime numbers, which we shall call $p_1, p_2, p_3 \ldots p_n$ (so $p_1 = 2, p_2 = 3, p_3 = 5$, etc). We now wish to try and establish that there is another prime number larger than p_n.

Let us consider the number

$$(p_1 p_2 p_3 \ldots p_n) + 1$$

which we shall call N. [So if $n = 4$, $N = 2 \times 3 \times 5 \times 7 + 1 = 211$.]

When p_1 is divided into N, there is a remainder 1;
when p_2 is divided into N, there is a remainder 1;
and so on.

So N is not divisible by any of the prime numbers $p_1, p_2 \ldots p_n$.

So *either* N is divisible by a prime number larger than p_n *or* N is itself a prime number, and in either case we have established the existence of a prime number larger than p_n. There is always a *next* prime number and their supply is never exhausted.

---

† Similarly it appears that every even number can be expressed as the sum of two prime numbers, e.g., $8 = 3 + 5, 28 = 11 + 17, 100 = 47 + 53$. But, although there is no known exception, this conjecture also has never been proved.

6.10 Permutations and combinations

Before we leave the subject of sequences and series, we shall investigate a particularly important two-way sequence, known as *Pascal's triangle*, and the closely related *binomial theorem*. But first we must consider the number of ways of *arranging* a number of objects from a given set, called their *permutations*, and also the number of ways of *choosing* them, called their *combinations*. We shall assume, unless otherwise stated, that no two objects are alike.

Permutations

The number of ways of arranging 6 boys in order is found by choosing one of the 6 for the first position (6 ways); then one of the remaining 5 for the next position (5 ways); then one of the remaining 4 for the third position (4 ways), etc. So the total number of permutations is

$$6 \times 5 \times 4 \times 3 \times 2 \times 1 = 6! = 720$$

More generally,

The numbers of permutations of n different objects is

$$n(n-1)(n-2)\ldots 4 \times 3 \times 2 \times 1 = n!$$

If we now ask in how many ways we can arrange (or *permute*) 4 out of a group of 6 boys, which we shall call the number of *permutations of 4 out of 6*, and write $^6P_4$, we see that

$$^6P_4 = 6 \times 5 \times 4 \times 3 = 360$$

and it is also clear that this can be written as

$$\frac{6 \times 5 \times 4 \times 3 \times 2 \times 1}{2 \times 1} = \frac{6!}{2!}$$

So
$$^6P_4 = \frac{6!}{2!}$$

More generally,

$$^nP_r = n(n-1)(n-2)\ldots(n-r+1)$$
$$= \frac{n(n-1)\ldots(n-r+1)(n-r)(n-r-1)\ldots 4 \times 3 \times 2 \times 1}{(n-r)(n-r-1)\ldots 4 \times 3 \times 2 \times 1}$$
$$= \frac{n!}{(n-r)!}$$

When $r = n$, this becomes $\dfrac{n!}{0!}$, which has no meaning as $0!$ is undefined.

But $^nP_n = n!$, so it is highly convenient if we now *define* $0!$ to be 1.

So $\quad ^nP_r = \dfrac{n!}{(n-r)!}$ $\qquad$ (even when $r = n$)

Example 1

In how many ways can three prizes be awarded to a class of ten boys, one for English, one for French, and one for mathematics:
a) if there is a rule that no boy may win more than one prize,
b) with no such rule.

a) There are 10 possible winners of the English prize, then 9 for French, then 8 for mathematics

So $\quad$ number of ways $= 10 \times 9 \times 8 = 720$

b) There are 10 possible winners for English, then 10 for French, then 10 for mathematics

So $\quad$ number of ways $= 10 \times 10 \times 10 = 1000$

Example 2

In how many ways can the letters $\quad$ L E E D S $\quad$ be arranged?

The difficulty here clearly lies with the repeated E. We can, however, mark them with suffixes, $\quad$ L E$_1$ E$_2$ D S, $\quad$ and we then see that rearrangements occur in pairs which are distinguishable only by their suffixes,

e.g. $\quad$ E$_1$ L D E$_2$ S $\quad$ and $\quad$ E$_2$ L D E$_1$ S

So the number of different rearrangements $= \dfrac{5!}{2!} = 60$

Combinations

If we have a set of n different objects, in how many ways can we choose (*irrespective of order*) r out of them?

$\quad$ We shall call this the number of *combinations* of r objects out of n, written $\dbinom{n}{r}$ or ^nC_r.

For particular values of n and r, we can find $\binom{n}{r}$ simply by enumeration. For instance, three letters can be chosen out of the five letters ABCDE in the following ways:

ABC, ABD, ABE, ACD, ACE, ADE, BCD, BCE, BDE, CDE

So $\binom{5}{3} = 10$

But we could also look at the problem of choosing 3 letters out of 5 rather differently, and note that

the number of ways of *arranging* 3 letters out of 5
$$= \text{(the number of ways of } choosing \text{ 3 out of 5)}$$
$$\times \text{(the number of ways of } arranging \text{ those chosen)}$$

So $^5P_3 = \binom{5}{3} \times 3!$

$$\Rightarrow \quad \binom{5}{3} = \frac{^5P_3}{3!} = \frac{5!}{3!2!}$$

More generally,

the number of ways of *arranging* r objects out of n
$$= \text{(the number of ways of } choosing \text{ } r \text{ objects out of } n)$$
$$\times \text{(the number of ways of } arranging \text{ those chosen)}$$

So $^nP_r = \binom{n}{r} \times r!$

$$\Rightarrow \quad \binom{n}{r} = \frac{^nP_r}{r!} = \frac{n!}{r!(n-r)!}$$

We can check this result in particular cases, such as

$$\binom{5}{3} = \frac{5!}{3!2!} = \frac{5 \times 4 \times 3 \times 2 \times 1}{3 \times 2 \times 1 \times 2 \times 1} = 10$$

If, however, $r = 0$, there is no meaning for $\binom{n}{0}$. But $\dfrac{n!}{0!n!} = 1$, so we *define* $\binom{n}{0}$ to be 1.

Hence $$\binom{n}{r} = {}^nC_r = \frac{n!}{r!(n-r)!} \qquad \text{(even when } r = 0)$$

Example 3

If a Council consists of 6 Conservative, 5 Labour and 2 Independent members, in how many ways can a committee be formed with 3 Conservative, 3 Labour, and 1 Independent members?

$$\text{Number of ways} = \binom{6}{3} \times \binom{5}{3} \times \binom{2}{1}$$

$$= 20 \times 10 \times 2 = 400$$

Example 4

Prove that

$$\binom{n+1}{r} = \binom{n}{r-1} + \binom{n}{r}$$

To choose r objects out of $n + 1$, one can
either   choose the first and then the other $r - 1$ from the remaining n,
or   not choose the first, but choose all r from the remaining n.

So $$\binom{n+1}{r} = \binom{n}{r-1} + \binom{n}{r}$$

Exercise 6.10

1   In how many ways can the captain of a cricket team arrange his batting order?

2   In a newspaper competition eight desirable qualities of a good family car (roominess, reliability, etc.) have to be put in order of importance. How many coupons must be completed in order to be sure of obtaining the winning order?

3   If 6 couples go to a party, in how many ways can they pair off to dance?

4   a) If 10 horses run in a race, in how many different ways can the first three places be filled?
    b) If 10 greyhounds race against each other on three successive nights, in how many different ways can the winning places be filled?

5   In how many different orders can colours appear when 5 balls are drawn
    from a bag containing
    a) 1 red ball, 1 yellow, 1 blue, and 2 white;
    b) 1 red ball, 1 yellow, and 3 white;
    c) 2 red balls and 3 white?

6   Repeat the last question, supposing that each ball drawn is immediately
    replaced.

7   A cricket touring party consists of 16 members. In how many ways could
    eleven players be selected for the first match?

8   A man tries to forecast the result (win, lose, or draw) of eleven first-
    division football matches. In how many ways could he be correct
    a) in exactly 9 of the matches;
    b) in at least 9 matches?

9   If n different points are given in space
    a) how many lines are there connecting them in pairs;
    b) how many triangles are there with three of the points as vertices;
    c) how many tetrahedra are there with four of the points as vertices?

10 In how many ways can 12 golfers be divided into three groups of four?

11 By carefully considering their meaning, simplify

    a) $\dbinom{n}{n-r}$ **b)** $\displaystyle\sum_{r=0}^{n}\dbinom{n}{r}$

12 Use the formula for $\dbinom{n}{r}$ in order to simplify

    a) $\dbinom{n}{n-r}$ **b)** $\dbinom{n}{r-1}+\dbinom{n}{r}$

13 Six similar discs are coloured in pairs: red, green, and blue. In how many
    different ways can these discs be placed in a row, and of these how many
    show all three pairs of discs of the same colour lying next to each other?
    (MEI)

14 How many 5-figure numbers can be constructed from the nine digits 1, 2,
    . . . 9
    a) if no digit is repeated;
    b) if repetitions are allowed?

15 **a)** In how many different arrangements, relative to each other, can 6
    people be seated at a circular table?
    b) In how many orders, relative to each other, can 6 different-coloured
    beads be strung on a circular thread?

6.11 Pascal's triangle: the binomial theorem

We now take a different starting point and look at the two-way sequence which arises from the expansions of $(a + b)^n$. For successive values of n, we see that

$$(a + b)^0 = 1$$
$$(a + b)^1 = 1a + 1b$$
$$(a + b)^2 = 1a^2 + 2ab + 1b^2$$
$$(a + b)^3 = 1a^3 + 3a^2b + 3ab^2 + 1b^3$$
$$(a + b)^4 = 1a^4 + 4a^3b + 6a^2b^2 + 4ab^3 + 1b^4$$
$$(a + b)^5 = 1a^5 + 5a^4b + 10a^3b^2 + 10a^2b^3 + 5ab^4 + 1b^5$$

Picking out numerical coefficients, we obtain the array known as Pascal's triangle:†

```
1
1  1
1  2  1
1  3  3  1
1  4  6  4  1
1  5 10 10  5  1
```

and a moment's glance shows how it can be extended for as many rows as we wish. We also notice that it is made up of values of $\binom{n}{r}$ (e.g. $4 = \binom{4}{1}$, $6 = \binom{4}{2}$, $10 = \binom{5}{2}$, $10 = \binom{5}{3}$, etc.), and that if we number its rows and columns from zero, the entry in the nth row and rth column is precisely $\binom{n}{r}$

n \ r	0	1	2	3	4	5
0	1					
1	1	1				
2	1	2	1			
3	1	3	3	1		
4	1	4	6	4	1	
5	1	5	10	10	5	1

If we now look again at the expansion of

$$(a + b)^n = (a + b)(a + b) \ldots (a + b)$$

† Blaise Pascal (1623–62), French philosopher and mathematician.

we see that this product of n brackets has terms:

$a^n,$

$a^{n-1}b,$ $\binom{n}{1}$ terms

$a^{n-2}b^2,$ $\binom{n}{2}$ terms

$\vdots$

$a^{n-r}b^r,$ $\binom{n}{r}$ terms

$\vdots$

$ab^{n-1},$ $\binom{n}{n-1}$ terms

b^n

There are clearly $\binom{n}{r}$ terms of the type $a^{n-r}b^r$, as these are formed by taking b from r of the brackets, which can be done in $\binom{n}{r}$ ways and a from the remaining $n-r$ brackets.

Hence we obtain the *binomial theorem*:

$$(a+b)^n = a^n + \binom{n}{1}a^{n-1}b + \cdots + \binom{n}{r}a^{n-r}b^r + \cdots + \binom{n}{n-1}ab^{n-1} + b^n$$

$$(a+b)^n = \sum_{r=0}^{n}\binom{n}{r}a^{n-r}b^r$$

Example 1

Use the binomial theorem to expand $(3x - 2y)^4$

$$(3x - 2y)^4 = (3x)^4 + \binom{4}{1}(3x)^3(-2y) + \binom{4}{2}(3x)^2(-2y)^2$$
$$+ \binom{4}{3}(3x)(-2y)^3 + (-2y)^4$$

$$= (3x)^4 + 4(3x)^3(-2y) + 6(3x)^2(-2y)^2$$
$$+ 4(3x)(-2y)^3 + (-2y)^4$$

$$= 81x^4 - 216x^3y + 216x^2y^2 - 96xy^3 + 16y^4$$

Example 2

Find the first three terms when $(1 - 2x)^{10}$ is expanded in ascending powers of x, and so evaluate $(0.998)^{10}$.

$$(1 - 2x)^{10} = 1^{10} + \binom{10}{1} 1^9 (-2x) + \binom{10}{2} 1^8 (-2x)^2 \cdots$$

$$= 1 + 10(-2x) + 45(-2x)^2 \cdots$$

$$= 1 - 20x + 180x^2 \cdots$$

Now put $x = 0.001$

$$(0.998)^{10} = 1 - 20(0.001) + 180(0.001)^2 \cdots$$

$$\approx 1 - 0.02 + 0.000\,180$$

$$\approx 0.9802$$

Exercise 6.11

1 Write down the expansions of
  a) $(a + b)^4$ **b)** $(2x + y)^3$ **c)** $(p - 2q)^5$ **d)** $\left(x - \dfrac{1}{x}\right)^4$

  e) $(2x - y)^6$ **f)** $(3x^2 - 2y^2)^4$

2 Write down the first three terms when each of the following is expanded in ascending powers of x:
  a) $(1 + x)^8$ **b)** $(1 + 2x)^6$ **c)** $\left(1 - \dfrac{x}{2}\right)^7$ **d)** $(2 - 3x)^5$

3 Find the given term in the following expansions:
  a) x^2 in expansion of $(1 + x)^8$
  b) x^3 in expansion of $(3 - 2x)^4$

  c) x^4 in expansion of $\left(x^2 + \dfrac{2}{x}\right)^5$

  d) constant term in expansion of $\left(2x + \dfrac{1}{x^2}\right)^3$

  e) constant term in expansion of $\left(x^2 - \dfrac{1}{x}\right)^6$

4 Use the binomial theorem to evaluate:
  a) $(1.01)^{10}$ to 4 significant figures;
  b) $(0.99)^6$ to 5 significant figures;
  c) $(2.001)^8$ to 6 significant figures;
  d) $(0.998)^5$ to 5 significant figures.

5   Expand as far as the term in x^2:
   a) $(1 - x - x^2)^5$
   b) $(1 + 2x + 3x^2)^6$
   c) $(2 - x - x^2)^4$

Miscellaneous problems

1   Investigate whether the sum of the cubes of three consecutive integers is always divisible by 9. Try to prove your conjecture.

2   n points are taken on the circumference of a circle. Investigate u_n, the number of regions into which the chords joining them divide the circle, if no three chords are concurrent. Show that $u_2 = 2$, $u_3 = 4$, $u_4 = 8$ and calculate u_5. Hence try to find a formula for u_n and check when $n = 6$.

3   A magic square of order n consists of the integers 1, 2, 3 ... n^2 arranged in a square in such a way that the numbers in each row, in each column, and in each diagonal have the same sum. What is this sum?

4   Find the sum of all integers less than $10n$ which are not multiples of 2 or 5.
   (cs)

5   Show by induction, or otherwise, that, if n is an integer and $n > 1$, $7^n - 6n - 1$ is divisible by 36, and $5^n - 4n - 1$ is divisible by 16.
   Hence, or otherwise, show that $7^n - 5^n - 2n$ is divisible by 4.
   Find the highest positive integer which will always divide $2 \times 7^n - 3 \times 5^n + 1$ exactly.

6   **a)** A snowflake is made by taking an equilateral triangle of area 1 cm² and adding smaller equilateral triangles based on the middle thirds of the sides. This is repeated on the middle thirds of the segments of the new perimeter, and so on *ad infinitum*.
   What is the area of such a snowflake, and what is its perimeter?
   b) An anti-snowflake is made in a similar way, but by subtracting the equilateral triangles. Calculate its area and perimeter.

7   Find the sum of n terms of the series

   $$x + x^2 + x^3 + \cdots + x^n$$

   Hence, by differentiation, find the sum of the first n terms of

   $$x + 2x^2 + 3x^3 + \cdots$$

   and of $x + 2^2x^2 + 3^2x^3 + \cdots$

8   $f(x) = \sin x + \sin 2x + \cdots + \sin nx$

   By considering the product $f(x) \sin \frac{1}{2}x$ and using the factor formulae, find the sum of this series.

9 Use the expansion of $(1 + x)^n$ to find:

a) $\dbinom{n}{0} + \dbinom{n}{1} + \dbinom{n}{2} + \cdots + \dbinom{n}{n}$

b) $\dbinom{n}{0} - \dbinom{n}{1} + \dbinom{n}{2} + \cdots + (-1)^n\dbinom{n}{n}$

c) $\dbinom{n}{1} + 2\dbinom{n}{2} + 3\dbinom{n}{3} + \cdots + n\dbinom{n}{n}$

d) $\dbinom{n}{0}^2 + \dbinom{n}{1}^2 + \dbinom{n}{2}^2 + \cdots + \dbinom{n}{n}^2$

10 An absent-minded mathematician writes n letters and addresses n envelopes for them. u_n is the number of ways in which the letters could be put in the envelopes so that each is in a wrong envelope.
a) Find u_n for $n = 1, 2, 3, 4, 5$.
b) Try to find a recurrence relation for u_{n+1} in terms of u_n, u_{n-1} and hence calculate u_6, u_7.
c) Try to find u_n in terms of u_{n-1}.

7 Probability and statistics

7.1 Trials, events, and probabilities

If a coin is spun and is just as likely to turn up heads as tails we call it *fair*, or *unbiased*.

Such a spin is usually called a *trial* and the appearance of a head is an *event*. As this event occurs in only one of the two possible *outcomes*, and these are equally likely, we say that the *probability* of a head is $\frac{1}{2}$:

$\text{P (head)} = \frac{1}{2}$

This definition is fortified by our belief that, in a large number of similar trials, the proportion of heads would gradually approach $\frac{1}{2}$.†

Similarly, if a playing card is chosen from a well-shuffled pack, the probability of its being an ace is $\frac{4}{52}$, as the event of choosing an ace takes place in 4 of the 52 equally possible outcomes:

$\text{P (ace)} \quad = \frac{4}{52} = \frac{1}{13}$

and likewise

$\text{P (heart)} = \frac{13}{52} = \frac{1}{4}$

More generally, let us consider the case of a trial that has N equally likely outcomes, in some of which a particular event E takes place.

We shall use $\mathscr{E}$ to denote the set of all N outcomes, and E also to denote the set of outcomes in which E takes place.

Then the probability of E is defined as

$$P(E) = \frac{\text{number of elements of } E}{\text{number of elements of } \mathscr{E}} = \frac{n(E)}{N}$$

Example 1

Trial: the toss of a fair coin

$\mathscr{E} = \{\text{head, tail}\} \quad \text{and} \quad n(\mathscr{E}) = 2$

---

† Indeed, there are many probabilities (e.g. that of a drawing-pin landing point upwards when dropped) which can be assessed only from such a large number of trials.

If E is the event of the coin turning up heads then

$$E = \{\text{head}\} \quad \text{and} \quad n(E) = 1$$

So $\quad P(E) = \dfrac{n(E)}{n(\mathscr{E})} = \dfrac{1}{2}$

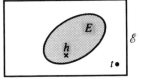

Example 2

Trial: the choice of a card from a well-shuffled pack

$\mathscr{E} = \{\text{the 52 cards of the pack}\}$ and $n(\mathscr{E}) = 52$

If A is the event of choosing an ace, then

$$A = \{\text{the four aces}\} \quad \text{and} \quad n(A) = 4$$

So $\quad P(A) = \dfrac{n(A)}{n(\mathscr{E})} = \dfrac{4}{52} = \dfrac{1}{13}$

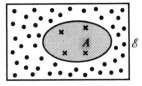

If H is the event of choosing a heart, then

$$H = \{\text{the thirteen hearts}\} \quad \text{and} \quad n(H) = 13$$

So $\quad P(H) = \dfrac{n(H)}{n(\mathscr{E})} = \dfrac{13}{52} = \dfrac{1}{4}$

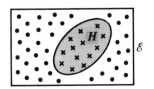

If we take particular cases.

a) when E never happens, i.e. is impossible.

The set E has no members, and so

$$P(E) = \frac{n(E)}{N} = \frac{0}{N} = 0$$

b) when E always happens, i.e. is certain.

The set E is identical with $\mathscr{E}$, so

$$P(E) = \frac{n(E)}{N} = \frac{n(\mathscr{E})}{N} = \frac{N}{N} = 1$$

Furthermore, as every outcome belongs either to E or to E', but never to both:

$$n(E) + n(E') = N$$

$$\Rightarrow \quad \frac{n(E)}{N} + \frac{n(E')}{N} = 1$$

$$\Rightarrow \quad P(E) + P(E') = 1$$

Example 3

A fair six-sided die is thrown and E is the event of it showing a 5 or a 6.

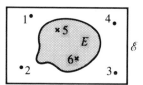

Here $\mathscr{E} = \{1, 2, 3, 4, 5, 6\}$
$E = \{5, 6\}$
$E' = \{1, 2, 3, 4\}$

So $P(E) = \dfrac{n(E)}{n(\mathscr{E})} = \dfrac{2}{6} = \dfrac{1}{3}$

and $P(E') = \dfrac{n(E')}{n(\mathscr{E})} = \dfrac{4}{6} = \dfrac{2}{3}$

Example 4

Two six-sided dice are thrown. What is the chance of scoring more than 10?

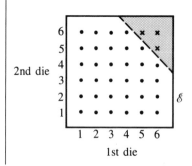

$\mathscr{E} = \{\text{all possible pairs of scores}\}$

$\quad = \{(1, 1), (1, 2), (1, 3) \ldots (6, 4), (6, 5), (6, 6)\}$

So $\quad n(\mathscr{E}) = 36$

If E is the event of scoring more than 10,

then $\qquad E = \{(5, 6), (6, 5), (6, 6)\}$

and $\qquad n(E) = 3$

Hence $\quad P(E) = \dfrac{n(E)}{n(\mathscr{E})} = \dfrac{3}{36} = \dfrac{1}{12}$

It might alternatively be argued that the total score on two dice could be

$\quad 2, 3, 4, 5, \ldots, 12$

and that of these eleven possible totals exactly two are greater than 10. So is it true to say that the probability of scoring more than 10 is $\frac{2}{11}$? Why not? Are the eleven possible outcomes all equally likely?

Exercise 7.1

In the following trials state the universal set $\mathscr{E}$ corresponding to all possible outcomes and the sets corresponding to the stated events. Hence calculate the probabilities of these events.

1 A card is chosen from a complete pack.
   a) A: the choice of an ace
   b) B: the choice of a black card
   c) C: the choice of a court card (ace, king, queen, or jack)

2 A die is thrown.
   a) S: scoring a six
   b) E: scoring an even number
   c) Z: scoring less than 4 or more than 5

3 Two coins are spun.
   a) H: two heads
   b) T: two tails
   c) E: a head and a tail

4 Three coins are spun.
   a) H: three heads
   b) M: more heads than tails
   c) S: the same number of heads and tails

5 Two poker dice are thrown.
   a) E_2 : two aces
   b) E_1 : one ace
   c) E_0 : no aces

6 Two boys each choose a positive integer less than 10.
   a) A : they choose the same number
   b) B : they choose different numbers
   c) C : their total is less than 5

7 I meet a friend who has four children in his family.
   a) A : it will contain 2 boys and 2 girls
   b) B : it will contain more boys than girls
   c) C : the eldest and youngest will be boys

8 Five tulips, of which two are red, are planted in a row.
   a) E : the red tulips are at the two ends
   b) N : the red tulips are next to each other

9 Two boys have their birthdays in the same week.
   a) M : both birthdays are on the Monday
   b) S : both birthdays are on the same day
   c) C : the birthdays are on consecutive days

7.2 Compound events

If A and B are two separate events, then

   $A \cap B$ represents the event of both A and B taking place; and
   $A \cup B$ represents the event of either A or B (or both) taking place.

Suppose for example, that in the trial of drawing a card from a pack,

   A is the event of choosing an ace: $A = \{$the 4 aces$\}$; and
   B is the event of choosing a black card: $B = \{$the 26 spades and clubs$\}$.

Then $A \cap B = \{$ace of spades, ace of clubs$\}$

and $A \cup B = \{$all 4 aces together with all the remaining black cards$\}$

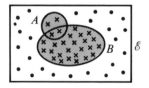

Further, we see that

$$n(A) = 4, \qquad n(B) = 26$$

$$n(A \cap B) = 2, \quad n(A \cup B) = 28$$

These are connected by the equation

$$n(A \cup B) + n(A \cap B) = n(A) + n(B)$$

and inspection of other Venn diagrams will satisfy the reader that this must always be so.

Hence, dividing by N,

$$\frac{n(A \cup B)}{N} + \frac{n(A \cap B)}{N} = \frac{n(A)}{N} + \frac{n(B)}{N}$$

$$\Rightarrow \qquad \boxed{P(A \cup B) + P(A \cap B) = P(A) + P(B)}$$

Exclusive events

If A and B are two events which can never happen together, they are called *exclusive* events.

In such a case, $A \cap B = \varnothing \quad \Rightarrow \quad P(A \cap B) = 0$

$$\Rightarrow \qquad P(A \cup B) = P(A) + P(B)$$

For example, in drawing a card from a pack

A = the choice of an ace⎫

and K = the choice of a king⎭ are exclusive events

In this case,

$$P(A) = \tfrac{1}{13}, \quad P(K) = \tfrac{1}{13}, \quad \text{and} \quad P(A \cup K) = \tfrac{2}{13}$$

Similarly,

C = the choice of a club ⎫

and R = the choice of a red card⎭ are exclusive

and in this case

$$P(C) = \tfrac{1}{4}, \quad P(R) = \tfrac{1}{2}, \quad P(C \cup R) = \tfrac{3}{4}$$

By contrast,

A = the choice of an ace ⎫

and B = the choice of a black card⎭ are *not* exclusive

In this case,

$$P(A) = \tfrac{1}{13}, \quad P(B) = \tfrac{1}{4}, \quad P(A \cup B) = \tfrac{28}{52} = \tfrac{7}{13}$$

and $P(A \cup B) \neq P(A) + P(B)$

Exhaustive events

If A and B are two events at least one of which must happen, they are called *exhaustive*. In such a case, $P(A \cup B) = 1$

$\Rightarrow \quad 1 + P(A \cap B) = P(A) + P(B)$

Exercise 7.2a

In nos **1–6** an ordinary six-sided die is thrown.

A is the event of scoring a 6
B is the event of scoring an even number
C is the event of scoring less than 3

1   Find:
   a) $P(A)$, $P(B)$, $P(C)$
   b) $P(B \cap C)$, $P(C \cap A)$, $P(A \cap B)$
   c) $P(B \cup C)$, $P(C \cup A)$, $P(A \cup B)$

2   Check that

$$P(A \cup B) + P(A \cap B) = P(A) + P(B)$$

   and the two similar results.

3   Which pairs of events are **a)** exclusive **b)** exhaustive?

4–6 Repeat nos. **1–3** for the events A', B', and C'.

7   The probability of a boy being selected for the Cricket XI is $\tfrac{1}{10}$, for the Football XI is $\tfrac{1}{20}$, and for both is $\tfrac{1}{40}$. What is the chance of his being selected for
   a) at least one of the teams;
   b) just one of the teams?

8   An integer is chosen at random. Calculate the probability that it is not divisible by 3 or 7.   (MEI)

Independent events

Suppose that a person is chosen at random from a football crowd, perhaps by selecting the ten thousandth to be admitted to the ground.

Let A be the event of the person having a surname beginning with the
       letter A
and B be the event of the person being bald

Then it may happen that

$P(A) = \frac{1}{10}$ and $P(B) = \frac{1}{20}$

What would be the probability of the person having a surname starting with A
and being bald?
   As there is no likely connection between baldness and the first letter of one's
surname, we should expect this probability to be

$\frac{1}{10} \times \frac{1}{20} = \frac{1}{200}$

so that $P(A \cap B) = P(A)P(B)$.

If, however, A is the event of the person being an adult
and B is the event of being bald

we should expect a very different answer. It is almost certain that a bald person
will be adult, so $P(A \cap B)$ is not very different from $P(B)$ and in this case

$P(A \cap B) \neq P(A)P(B)$

When events A and B are such that

$P(A \cap B) = P(A)P(B)$

we call them *independent*.

So having a surname beginning with A and being bald are independent
events, but being adult and being bald are not independent.
   If A and B are independent, it can easily be shown that A' and B are also
independent.

For A and B independent $\Leftrightarrow$ $P(A \cap B) = P(A)P(B)$

But $P(A' \cap B) = P(B) - P(A \cap B)$
                    $= P(B) - P(A)P(B)$
                    $= (1 - P(A))P(B)$
                    $= P(A')P(B)$

So A' and B are independent.
   Similarly it can be shown that A and B' are independent and that A' and B'
are independent.

Contingency tables

Compound events, whether independent or not, can be illustrated by means of
probability, or *contingency*, tables.

Example 1

A card is drawn from a pack.

A: the choice of an ace
and B: the choice of a black card

These are independent events and the various probabilities are as follows:

	B	B'	
A	$P(A \cap B) = \dfrac{1}{26}$	$P(A \cap B') = \dfrac{1}{26}$	$P(A) = \dfrac{1}{13}$
A'	$P(A' \cap B) = \dfrac{6}{13}$	$P(A' \cap B') = \dfrac{6}{13}$	$P(A') = \dfrac{12}{13}$
	$P(B) = \dfrac{1}{2}$	$P(B') = \dfrac{1}{2}$	1

Example 2

A card is drawn from a pack.

B: the choice of a black card
and C: the choice of a club

As a club must be a black card, these are *not* independent events and the probabilities are:

	B	B'	
C	$\frac{1}{4}$	0	$\frac{1}{4}$
C'	$\frac{1}{4}$	$\frac{1}{2}$	$\frac{3}{4}$
	$\frac{1}{2}$	$\frac{1}{2}$	1

Example 3

10% of a large consignment of apples is known to be bad. If three are chosen at random, what is the chance that:
a) all will be bad;
b) none will be bad;
c) at least one will be bad?

As the consignment is large, the chance that an apple will be bad is not affected by those previously chosen.

Now $\qquad P(B) = \dfrac{1}{10}$

So $\quad P(B, B, B) = \dfrac{1}{10} \times \dfrac{1}{10} \times \dfrac{1}{10} = \dfrac{1}{1000} = 0.001$

Also $\qquad P(B') = \dfrac{9}{10}$

So $\quad P(B', B', B') = \dfrac{9}{10} \times \dfrac{9}{10} \times \dfrac{9}{10} = \dfrac{729}{1000} = 0.729$

$\Rightarrow \quad$ Chance of having at least one bad $= 1 - 0.729 = 0.271$

Exercise 7.2b

In the following trials calculate $P(A), P(B)$, and $P(A \cap B)$. In each case state whether events A and B are independent.

1   Andrew and Barbara each throw a die.
    A: Andrew throws a 6
    B: Barbara throws a 6

2   Andrew and Barbara each has a pack of cards and picks a card.
    A: Andrew picks a heart
    B: Barbara picks a heart

3   Andrew and Barbara each take a card from the same pack at the same time.
    A: Andrew picks a heart
    B: Barbara picks a heart

4   Andrew chooses at random a number between 1 and 200.
    A: It is divisible by 4
    B: It is divisible by 5

5   Barbara chooses at random a number between 1 and 200.
    A: It is divisible by 4
    B: It is divisible by 10

6   A pack of 52 ordinary playing cards is shuffled and one card is withdrawn.
    D denotes the event that the card is a diamond, K that it is a king, R that it
    is red. Prove by calculating the appropriate probabilities that D and K are
    independent, that K and R are independent, but that D and R are not
    independent.
        What is the value of $P[(D' \cap K) \cup (D \cap K')]$?   (MEI)

7   How many times should an unbiassed die be thrown if the probability that
    a six should appear at least once is to be greater than $\frac{9}{10}$?

8   The chance of hitting a target with a single shot is $\frac{1}{3}$. How many shots must
    be fired to make the chance of at least one hit on the target more than 90
    per cent?
        What is the chance that the first hit scored will be with the third shot?
                                                                              (OC)

* 7.3 Conditional probability

We frequently wish to consider the probability of an event A amongst
occurrences of another event B. This we call the *conditional probability of A
upon B*, written $P(A|B)$, or the *probability of A given B*.

    For example, in drawing a card from an ordinary pack,

$$P(\text{ace}|\text{club}) = \text{probability of a card being an ace if it is known to be a club} = \tfrac{1}{13}$$

$$P(\text{black card}|\text{club}) = \text{probability of a card being black if it is known to be a club} = 1$$

and similarly $P(\text{club}|\text{black card}) = \tfrac{1}{2}$
                 $P(\text{heart}|\text{black card}) = 0$

Now $P(A|B)$ is the probability of event A amongst occurrences of B, and so is
the probability of event $A \cap B$ amongst occurrences of B.

So $$P(A|B) = \frac{n(A \cap B)}{n(B)} = \frac{n(A \cap B)/N}{n(B)/N}$$

$\Rightarrow$ $$P(A|B) = \frac{P(A \cap B)}{P(B)}$$

and this can quickly be verified in the above examples.

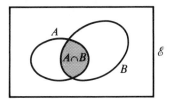

In the particular case when A and B are independent

$$P(B|A) = P(A)P(B)$$

$$\Rightarrow \quad P(A|B) = \frac{P(A \cap B)}{P(B)} = \frac{P(A)P(B)}{P(B)} = P(A)$$

$$\text{and} \quad P(B|A) = \frac{P(B \cap A)}{P(A)} = \frac{P(B)P(A)}{P(A)} = P(B)$$

So if two events are independent the probability of either is (as we would expect) always the same, whether or not the other is known to occur.

Exercise 7.3a

1 When a card is chosen from a pack A is the event of obtaining an ace and B is the event of obtaining a black card. Calculate:
   a) $P(A)$ $P(A')$ $P(B)$ $P(B')$
   b) $P(A \cap B)$ $P(A \cap B')$ $P(A' \cap B)$ $P(A' \cap B')$
   c) $P(A|B)$ $P(A|B')$ $P(A'|B)$ $P(A'|B')$
   d) $P(B|A)$ $P(B|A')$ $P(B'|A)$ $P(B'|A')$

   Which of the following pairs of events are independent: A, B; A, B'; A', B; A', B'?

2 Repeat no. **1** if a die is thrown and A is the event of obtaining a six, B is the event of obtaining an even number.

3 The events A and B are such that

$$P(A) = 0.4, P(B) = 0.45, P(A \cup B) = 0.68.$$

Show that the events A and B are neither mutually exclusive nor independent.   (L)

4 Events A and C are independent. Probabilities relating to events A, B and C are as follows:

$$P(A) = \tfrac{1}{5}, P(B) = \tfrac{1}{6}, P(A \cap C) = \tfrac{1}{20}, P(B \cup C) = \tfrac{3}{8}.$$

Evaluate $P(C)$ and show that events B and C are independent.   (L)

5 Events A, B and C occur so that $P(A) = 3p$, $P(B) = 2p$ and $P(C) = p$, $p \neq 0$. Given also that $P(A|B) = P(B|C) = 0$, $P(A|C) = P(A)$ and $P(A \cup B \cup C) = \frac{11}{12}$, find:
a) $P(A \cap B)$, $P(B \cap C)$ and $P(A \cap B \cap C)$
b) $P(C \cap A)$ in terms of p
c) $P(A)$, $P(B)$ and $P(C)$ (L)

6 Each of three identical boxes A, B and C has two drawers.

   Box A contains a prize in each drawer.
   Box B contains a prize in one drawer only.
   Box C does not contain any prizes.

   A box is chosen at random and a drawer is opened and found to be empty. Find the probability that a prize will be found:
a) if the other drawer in the same box is opened;
b) if one of the other two boxes is chosen at random and a drawer is opened.   (L)

7 Prove that if two events A, B are independent, so are A, B'; A', B; A', B'.

Conditional probabilities are very conveniently illustrated by means of *probability trees*:

Example

The probabilities that a man makes a certain dangerous journey by car (C), motor-bike (M), or on foot (F) are $\frac{1}{2}$, $\frac{1}{6}$, and $\frac{1}{3}$ respectively. If the probabilities of an accident (A) when he uses these means of transport are $\frac{1}{5}$, $\frac{3}{5}$, and $\frac{1}{10}$ respectively, what is the chance of an accident?

Here the situation can best be summed up in a figure:

$P(C \cap A) = \frac{1}{2} \times \frac{1}{5} = \frac{1}{10}$

$P(C \cap A') = \frac{1}{2} \times \frac{4}{5} = \frac{2}{5}$

$P(M \cap A) = \frac{1}{6} \times \frac{3}{5} = \frac{1}{10}$

$P(M \cap A') = \frac{1}{6} \times \frac{2}{5} = \frac{1}{15}$

$P(F \cap A) = \frac{1}{3} \times \frac{1}{10} = \frac{1}{30}$

$P(F \cap A') = \frac{1}{3} \times \frac{9}{10} = \frac{3}{10}$

So $P(A) = P(C \cap A) + P(M \cap A) + P(F \cap A)$
$= P(C)P(A|C) + P(M)P(A|M) + P(F)P(A|F)$
$= \frac{1}{2} \times \frac{1}{5} + \frac{1}{6} \times \frac{3}{5} + \frac{1}{3} \times \frac{1}{10}$
$= \frac{1}{10} + \frac{1}{10} + \frac{1}{30}$
$= \frac{7}{30}$

So the chance of an accident $= \frac{7}{30}$.

It is also interesting to ask the reverse question: if an accident is known to have happened, what is the probability that the man was travelling by car, motor-bike, or on foot?

Using the above notation, these probabilities can be written

$$P(C|A) = \frac{P(C \cap A)}{P(A)} = \frac{\frac{1}{10}}{\frac{7}{30}} = \frac{3}{7}$$

$$P(M|A) = \frac{P(M \cap A)}{P(A)} = \frac{\frac{1}{10}}{\frac{7}{30}} = \frac{3}{7}$$

and $$P(F|A) = \frac{P(F \cap A)}{P(A)} = \frac{\frac{1}{30}}{\frac{7}{30}} = \frac{1}{7}$$

So if it is known that the man had an accident, the probabilities of his having travelled by car, motor-bike, or on foot are $\frac{3}{7}, \frac{3}{7}, \frac{1}{7}$ respectively.

If he is known to have arrived safely, show that the corresponding probabilities are $\frac{12}{23}, \frac{2}{23}, \frac{9}{23}$.

Exercise 7.3b

1 A coin is spun twice. Draw a probability tree showing the various possible combinations of heads and tails.

2 Repeat no. 1 when the coin is spun three times. What are the probabilities of 0, 1, 2, 3 heads?

3 The chance of tomorrow being fine is $\frac{3}{4}$. If it is fine, our football team's chance of victory is $\frac{4}{5}$; but otherwise it is only $\frac{1}{2}$. What is our chance of victory?

4 Ten counters are in a bag, 7 being white and 3 black. They are taken out at random one by one without replacement. Calculate the probabilities that **a)** the first taken is black **b)** the second is black **c)** the first two are black **d)** exactly one of the first two is black **e)** at least one of the first two is black.   (OC)

   [This question should be answered both by use of combinations and by a probability tree.]

5 The probability that a candidate attempts a certain question is $\frac{9}{10}$ and, having done so, the probability of success is $\frac{2}{3}$. Find the probability that the examiner will find at least one correct solution in the first three scripts which she marks.   (SMP)

6 A garden has 3 flower beds. The first bed has 50 flowers of which 10 are red, the second bed has 30 flowers of which 10 are red and the third bed

has 20 flowers of which 10 are red. One of the beds is chosen at random
and a flower is selected at random from that bed and removed from the
garden. Find:
a) the probability that a red flower will be taken from the first bed;
b) the probability that a red flower will be taken from the garden.
If a flower is chosen at random from the 100 flowers in the garden, find the
probability that the flower will be red.   (MEI)

7 John and Peter each have two pennies. John tosses each of his coins once
and gives to Peter those which fall heads. Peter now tosses once each of
the coins in his possession, giving to John those which fall heads. Using a
probability tree, or otherwise, determine the probability that
a) Peter now has all the coins;
b) John now has all the coins;
c) John and Peter each now have two coins.

8 Events A and B are such that

$$P(A) = \tfrac{2}{3} \quad P(B|A) = \tfrac{1}{2} \quad P(B|A') = \tfrac{1}{4}$$

Draw the corresponding probability tree with A before B and all
probabilities marked.
a) Calculate $P(A \cap B)$ $P(A \cap B')$ $P(A' \cap B)$ $P(A' \cap B')$.
b) Calculate $P(B)$ $P(B')$.
c) Calculate $P(A|B)$ $P(A'|B)$ $P(A|B')$ $P(A'|B')$.
d) Draw the probability tree with B before A, again marking all the
probabilities.

9 Box A contains two white balls and Box B a white ball and a black ball. A
man spins a coin in order to decide the box from which to choose a ball. If
we are told that the chosen ball is white, what is the chance of its having
come from Box A?

10 20% of the inhabitants of a city have been inoculated against a certain
disease. In an epidemic the chance of infection amongst those inoculated
is $\tfrac{1}{10}$, but amongst the rest is $\tfrac{3}{4}$.
a) What proportion are infected?
b) If a man is chosen at random and found to be infected, what is the
chance of his having been inoculated?
c) What is the corresponding chance if he is not infected?

11 A bowler delivers off-breaks, leg-breaks and straight balls with pro-
babilities $\tfrac{1}{3}, \tfrac{1}{6}$, and $\tfrac{1}{2}$, and the batsman's chances of being bowled by these
balls are $\tfrac{1}{10}, \tfrac{1}{10}$, and $\tfrac{1}{20}$ respectively.
a) What is the chance of his being bowled by a particular ball?
b) If he is bowled, what is the chance of the ball having been straight?
c) If he is not bowled, what is the chance of it having been a leg-break?

12 A company advertises a professional vacancy in three national daily newspapers A, B, and C which have readerships in the proportions 2, 3, 1 respectively. From a survey of the occupations of the readers of these papers it is thought that the probabilities of an individual reader replying are 0.002, 0.001, 0.005 respectively.

a) If the company receives one reply, what are the probabilities that the applicant is a reader of papers, A, B, and C respectively?

b) If two replies are received, what is the probability that both applicants are readers of paper A?

[You may assume that each reader sees only one paper.]   (MEI)

* 7.4 Probability distributions: three standard types

We shall now investigate three different trials that can be carried out with dice, calculating the probabilities of the various events that can occur, and investigating the way in which these are distributed. The reader may like to carry out these trials a number of times and find the discrepancies between his distributions and the ones we obtain. A further investigation of these discrepancies is one of the major tasks of statistics. If for instance, we find that the discrepancy is considerable and that the chance of such a large discrepancy is exceedingly small, we begin to suspect the basis of our calculation and are forced to doubt that the dice were unbiased. But for the moment we shall concentrate on calculating the probabilities of the various events which are possible.

The rectangular distribution

Suppose that a single die is thrown just once. Then the probabilities of scoring 1, 2, 3, 4, 5, 6 are all $\frac{1}{6}$ and we obtain a *rectangular distribution*:

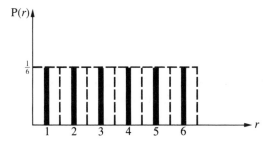

Exercise 7.4a

1 Two dice are thrown and their total score r recorded. Calculate the probabilities $P(r)\,(2 \leqslant r \leqslant 12)$ and sketch the corresponding probability distribution.

2 Repeat no. **1** when r is the numerical difference between their scores.

The binomial distribution

If four dice are now thrown, what are the probabilities of obtaining

0,   1,   2,   3,   4   sixes?

With a single die, the chance

of throwing a six is $\frac{1}{6}$
of not throwing a six is $\frac{5}{6}$

So when four dice are thrown,

$$P(0 \text{ sixes}) = (\tfrac{5}{6})^4$$

If 1 six is thrown, this can be in any one of $\binom{4}{1} = 4$ ways.

So $P(1 \text{ six})\quad = \binom{4}{1}\left(\dfrac{5}{6}\right)^3\left(\dfrac{1}{6}\right)$

Now 2 sixes can occur in any one of $\binom{4}{2} = 6$ ways.

So $P(2 \text{ sixes}) = \binom{4}{2}\left(\dfrac{5}{6}\right)^2\left(\dfrac{1}{6}\right)^2$

Similarly $P(3 \text{ sixes}) = \binom{4}{3}\left(\dfrac{5}{6}\right)\left(\dfrac{1}{6}\right)^3$

and $P(4 \text{ sixes}) = \qquad\left(\dfrac{1}{6}\right)^4$

So we see that the probabilities of 0, 1, 2, 3, 4 sixes are given by the terms of the binomial expansion

$$\left(\frac{5}{6} + \frac{1}{6}\right)^4$$

More generally, if in a single trial the probability of a certain event is p and $q = 1 - p$, then in n independent trials the probabilities of

0,   1,   2,..., n events are given by the terms of $(q + p)^n$

This is therefore called a *binomial distribution*.

The individual probabilities can easily be calculated.

In the above case, for instance, $n = 4$, $p = \frac{5}{6}$, and $q = \frac{1}{6}$ and the chances of 0, 1, 2, 3, 4 sixes were given by the terms of

$$(\tfrac{5}{6} + \tfrac{1}{6})^4$$

and so are

$$(\tfrac{5}{6})^4, \quad 4(\tfrac{5}{6})^3(\tfrac{1}{6}), \quad 6(\tfrac{5}{6})^2(\tfrac{1}{6})^2, \quad 4(\tfrac{5}{6})(\tfrac{1}{6})^3, \quad (\tfrac{1}{6})^4$$

which can easily be shown to be approximately

0.482,   0.386,   0.116,   0.015,   0.001

These can be most conveniently illustrated:

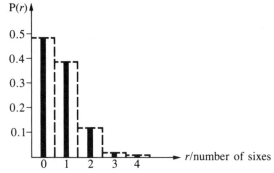

If, instead of concentrating on the number of sixes, we had counted the number of dice which showed either a five or a six, then the probability for a single die would be $\frac{1}{3}$.

So, $n = 4$, $p = \frac{1}{3}$, $q = \frac{2}{3}$

and the corresponding probabilities of a five or a six being shown by 0, 1, 2, 3, 4 dice are given by the terms of

$$(\tfrac{2}{3} + \tfrac{1}{3})^4$$

These can easily be shown to be approximately

0.198,   0.395,   0.296,   0.099,   0.012.

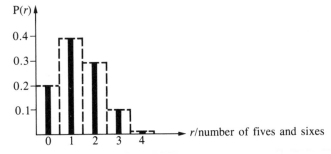

Exercise 7.4b

1 Find the probabilities of the various possible number of heads when a boy spins **a)** 1 coin **b)** 2 coins **c)** 3 coins **d)** 4 coins **e)** 5 coins **f)** 6 coins. Illustrate each answer by a probability diagram.

2 A marksman has an 80% chance of scoring a bull. When he fires five shots at a target calculate the probabilities of his scoring 0, 1, 2, 3, 4, 5 bulls.

3 A machine is faulty and 10% of the articles it produces are defective. In a batch of ten articles, what is the chance of more than one being defective?

4 A question paper contains six questions, whose answers are either 'yes' or 'no'. If a candidate guesses at all the answers, what is the chance that more than half will be correct?

5 The probability of germination of a certain type of seed is $\frac{4}{5}$. If the seeds are planted out in rows of six, find the probability that, in any one row:
a) all the seeds will germinate;
b) five or six seeds will germinate.   (MEI)

6 A test is applied to a certain type of component produced in a factory. If the probability that a component chosen at random fails is $\frac{1}{5}$, calculate the following probabilities for a test on 10 components so chosen **a)** that none will fail **b)** that less than two will fail **c)** that exactly two will fail **d)** that more than two will fail.   (MEI)

7 The probability that a colony of bees will survive a particularly hard winter is 0.35. If I have four colonies what is the probability that exactly one of these colonies will survive? I have a friend who has twice as many colonies (eight). Write down an expression for the probability that exactly one of his colonies will survive. Is the probability greater or less than in my case?   (SMP)

8 A gun is engaging a target and it is desired to make at least two direct hits. It is estimated that the chance of a direct hit with a single round is $\frac{1}{4}$ and that this chance remains constant throughout the firing. A burst of five rounds is fired and if at least two direct hits are scored firing ceases. Otherwise a second burst of five rounds is fired. Find the chance that at least two direct hits will be scored
a) only five rounds being fired;
b) ten rounds having to be fired.   (OC)

9 The probability of a bowler taking a wicket with a particular delivery is $\frac{1}{4}$. What is the chance in a six-ball over of his taking
a) 2 wickets;
b) at least 2 wickets;
c) a hat-trick (and no other wickets)?

10 A coin, which is thought to be unbiassed, is spun 10 times and only once
does it come down heads. What is the probability
a) of obtaining so few heads;
b) of obtaining such an extreme result, either heads or tails?

The geometric distribution

A simple die is thrown repeatedly until a 6 occurs. The chance of this
happening

at the 1st throw is $\frac{1}{6} \approx 0.167$

at the 2nd throw is $\frac{5}{6} \times \frac{1}{6} \approx 0.138$

at the 3rd throw is $(\frac{5}{6})^2 \times \frac{1}{6} \approx 0.116$

and as these form the terms of an infinite g.p., we call it a *geometric
distribution*.

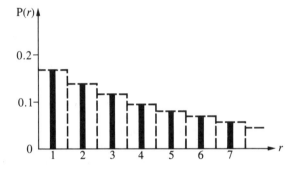

Exercise 7.4c

1 The chance of a girl passing her driving test is $\frac{2}{3}$, and does not alter if she
has to take it again. What is the probability that she will have to take it at
least three times?

2 The last pair of our side is batting and is just about to face a six-ball over.
a) If the chance of a wicket is $\frac{1}{4}$ for each successive ball, what is the chance
that they will not survive the over?
b) If the probability is the same for the other bowler, what is the chance
that the innings will end during the following over?

3 Two guns are firing together at a fortified emplacement. Their chances of
hitting the target are $\frac{1}{3}$ and $\frac{1}{4}$ respectively. What is the chance that it will
survive three simultaneous rounds
a) if simultaneous hits are needed to destroy it;
b) if one hit is sufficient?

4   A rifleman is firing at a distant target and has only a 10% chance of hitting it. How many rounds must he fire in order to have a 50% chance of hitting it at least once?

7.5 Statistics: location and spread

Whilst the calculation of probabilities concerns the uncertainties of the future, the study of statistics appears at first sight to be about the certainties of the past. But our predictions of the future usually depend on the evidence of the past; and our study of the past frequently involves investigation of the probability, on certain assumptions, that events should have happened as they did. So the two subjects are very closely interwoven.

Statistics, therefore, is largely concerned with assessing the strength of numerical evidence. But first must come the humbler task of recording, classifying and describing this evidence as clearly as possible, and this will be our next objective.

Location

A test, marked out of 40, is set to a class of 15 pupils, who obtain the following marks

32   27   30   26   19   20   34   26   23   30   34   26   31   24   29

The statistical term for any such set of numbers is a *population*.

For purposes of comparison with the performance of other classes some measure of the general level or *location* of such marks is needed. There are three measures which are commonly used in statistics:

a) The *mode* is the value which occurs most frequently. In the case of the test marks the mode is 26 (which occurs three times). However the mode does not always take one distinct value. If there had been another mark of 30 in the class then there would have been two modal values, 26 and 30, and the usefulness of the mode as a measure of location is very limited.

b) The *median* is the middle value when the numbers are arranged in order. If this is done for the test marks

19   20   23   24   26   26   26   ㉗   29   30   30   31   32   34   34

then the median is 27. If there had been an even number of marks, then the median would be taken as half-way between the two middle values.

c) The *mean*, or average, is the most important and widely used measure of location. Unlike the mode and median it is calculated from all the members of the population.

The sum of the test marks, $19 + 20 + 23 + \cdots + 34$ is 411, so the mean is $411/15 = 27.4$.

More generally, the mean of n values $x_1, x_2, x_3, \ldots, x_n$ can be written

$$m = \frac{x_1 + x_2 + \cdots + x_n}{n} = \frac{1}{n} \sum_{r=1}^{n} x_r$$

Or, more briefly: $m = \frac{1}{n} \sum x_r$

m is sometimes called the *arithmetic mean A*, to distinguish it from the *geometric mean*,

$$G = \sqrt[n]{(x_1 x_2 x_3 \ldots x_n)}$$

and the *harmonic mean, H*, where

$$\frac{1}{H} = \frac{1}{n}\left(\frac{1}{x_1} + \frac{1}{x_2} + \cdots + \frac{1}{x_n}\right)$$

Both the geometric and (less frequently) the harmonic means are sometimes used in statistics, but we shall restrict ourselves to the overwhelmingly more important arithmetic mean.

Spread

A second class (class B), this time of 20 pupils, obtained the following marks in the same test.

6 13 15 16 21 23 26 27 28 28 29 31 32 32 33 35 37 38 38 40

The total of these marks is 548, and therefore their mean is $548/20 = 27.4$, exactly the same as for the first class (class A).

So by considering just the mean, the performance of the two classes appears to be the same. However, even allowing for the different number of pupils, there is a disparity between the classes which is not accounted for in the calculation of the mean. Clearly, the marks for class B are much more spread out than those for class A. Therefore, besides a measure of location such as the mean, we shall also need a measure of dispersion or *spread*.

The simplest such measure is the *range*, which is the difference between the largest and smallest values in the population. So for class A the range is $34 - 19 = 15$, and for class B the range is $40 - 6 = 34$.

However when a population has an isolated extreme value (such as the mark of 6 for class B), the range can give a distorted impression of the spread. To eliminate this problem the *interquartile range* is often used in preference to the range. The lower and upper quartiles are the values which are one-quarter and three-quarters of the way through the population, when arranged in order. The interquartile range is the difference between these quartiles.

So for class A

19   20   23   (24)   26   26   26   (27)   29   30   30   (31)   32   34   34

       lower   median   upper
      quartile   quartile

Interquartile range $= 31 - 24 = 7$

It is also common to use half this value, known as the *semi-interquartile range* (or *quartile deviation*).

Another way of investigating the spread of a population might be to calculate the average difference, or deviation, from the mean.

Subtracting m from each mark for class A gives values:

$$-8.4, -7.4, -4.4, \ldots, 6.6$$

and their average is found to be zero. This is hardly a surprise, since

$$\frac{1}{n}\sum(x_r - m) = \frac{1}{n}\sum x_r - \frac{1}{n}\sum m$$

$$= m - \frac{1}{n}nm = 0$$

So the average of such deviations is always zero.

Nevertheless, we could use the *numerical* values of such deviations in order to calculate the *mean absolute deviation*:

$$\tfrac{1}{15}(8.4 + 7.4 + 4.4 + \cdots + 6.6) = 3.76$$

More generally, mean absolute deviation $= \dfrac{1}{n}\sum |x_r - m|$.

In practice, however, it is so inconvenient to deal with such numerical values that we almost always avoid this by squaring deviations from the mean, to eliminate negative signs, and call their average value the *variance* (written s^2). To compensate for the squaring of the deviations, we can square-root the variance to give the most useful and important measure of spread, the *standard deviation, s*.

For the marks of class A,

$$s^2 = \tfrac{1}{15}\{(-8.4)^2 + (-7.4)^2 + (-4.4)^2 + \cdots + (6.6)^2\}$$

$$= 19.97$$

$$\Rightarrow \quad s = 4.47$$

and for class B,

$$s = 9.04$$

So the marks of class B are spread out roughly twice as much as those of class A.

More generally,
$$s^2 = \frac{1}{n}\sum(x_r - m)^2$$

Now $m = \frac{1}{n}\sum x_r$

So $s^2 = \frac{1}{n}\sum(x_r - m)^2$

$$= \frac{1}{n}\left\{\sum x_r^2 - 2m\sum x_r + \sum m^2\right\}$$

$$= \frac{1}{n}\left\{\sum x_r^2 - 2mnm + nm^2\right\}$$

$$= \frac{1}{n}\sum x_r^2 - m^2$$

which is usually a more convenient formula.

Summarising
$$m = \frac{1}{n}\sum x_r$$

and
$$s^2 = \frac{1}{n}\sum(x_r - m)^2 = \frac{1}{n}\sum x_r^2 - m^2$$

Change of variable

The teacher of class A wishes to record his test results as percentages. Each mark is therefore multiplied by 2.5. Can we conveniently work out the new mean and standard deviation from the old ones?

It is fairly clear that since each mark is multiplied by 2.5, then the mean will be multiplied by 2.5 also, giving a new value of $m = 2.5 \times 27.4 = 68.5$.

Also, since the spread of marks will be multiplied by 2.5, the new standard deviation will be 2.5 times as large

$$s = 2.5 \times 4.47 \approx 11.2$$

The teacher of class B also wishes to give his marks out of 100. But to avoid half-marks he decides instead to multiply his marks by 2 and then add 20. How are the mean and standard deviation affected this time?

It seems reasonable here again that the mean is changed in the same way as the individual marks, so the new mean is

$$m = 2 \times 27.4 + 20 = 74.8$$

When we come to look at the spread of marks however, the standard deviation is multiplied by 2; but the addition of 20 to each of the marks does not change their positions in relation to one another (as can be seen in the diagram), and therefore has no effect on the standard deviation. The new standard deviation is then

$$s = 2 \times 9.04 = 18.08$$

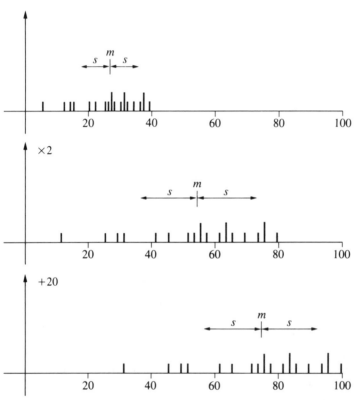

In general, if each member of a population, x_r ($r = 1$ to n), with mean m_x and standard deviation s_x, is multiplied by a factor λ, and is then increased by a, it follows that for the new values y_r, with $y_r = \lambda x_r + a$,

$$m_y = \frac{1}{n}\sum y_r = \frac{1}{n}\sum(\lambda x_r + a)$$

$$= \frac{1}{n}\lambda \sum x_r + \frac{1}{n}\sum a$$

$$= \frac{1}{n}\lambda n m_x + \frac{1}{n}na$$

$$= \lambda m_x + a$$

and $\quad s_y^2 = \dfrac{1}{n}\sum (y_r - m_y)^2$

$$= \dfrac{1}{n}\sum \{(\lambda x_r + a) - (\lambda m_x + a)\}^2$$

$$= \dfrac{1}{n}\sum (\lambda x_r - \lambda m_x)^2$$

$$= \lambda^2 \dfrac{1}{n}\sum (x_r - m_x)^2$$

$$= \lambda^2 s_x^2$$

So $\quad y = \lambda x + a \quad \Rightarrow \quad$
$$m_y = \lambda m_x + a$$
$$s_y = \lambda s_x$$

These results of a change of variable can also be used to ease the burden of computation, particularly if a calculator is not available.

Example 1

The temperatures, in degrees Celsius, of a furnace on 8 successive days were

   1620   1640   1610   1620   1620   1630   1610   1650.

What is the mean and standard deviation of these temperatures?

From each of these temperatures we can subtract 1600, giving

   20   40   10   20   20   30   10   50

and then divide by 10 to give

   2 4 1 2 2 3 1 5

So we have carried out the transformation $y = \dfrac{x - 1600}{10}$.

It is now quite easy to evaluate $m_y = 2.5$ and $s_y \approx 1.32$,

but $\quad x = 10y + 1600$

so $\quad m_x = 10 \times 2.5 + 1600 = 1625$

and $\quad s_x = 10 \times 1.32 = 13.2$ (to 3 s.f.)

Combining populations

We now investigate, by means of a worked example, how the means and standard deviations of two populations can be combined.

Example 2

A greengrocer has two consignments of grapefruit delivered. The first consignment of 200 fruit has a mean weight (in grams) of 240 and a standard deviation of 30. The second consignment of 300 fruit has a mean weight of 210 and standard deviation of 20. When the consignments are mixed together, what is their combined mean and standard deviation?

The formlae for m and s can be rearranged to give:

$$\sum x_r = nm \quad \text{and} \quad \sum x_r^2 = n(s^2 + m^2)$$

and using these results the information can be presented in a table:

	n	m	$\sum x_r$	s^2	$\sum x_r^2$
first consignment	200	240	48 000	900	11 700 000
second consignment	300	210	63 000	400	13 350 000
total consignment	500		111 000		25 050 000

So combined mean $= \dfrac{111\,000}{500} = 222$

and combined standard deviation $= \sqrt{\left(\dfrac{25\,050\,000}{500} - 222^2\right)}$

$$= 28.6$$

Calculation of the combined mean can be simplified by using the ratio of the population sizes, but there is usually no short-cut to the combined standard deviation of the two populations.

Exercise 7.5

For the data given in each of nos. **1** to **6** evaluate   **a)** the mean   **b)** the median   **c)** the mode   **d)** the range   **e)** the standard deviation.

1   7   10   6   11   10

2   13   17   21   12   19   14   19   20   15   18

3   7   8   8   13   15   10   5   14

4 37   9   35   29   40   55   33   37   31

5 370   90   350   290   400   550   330   370   310

6 87   59   85   79   90   105   83   87   81

7 The masses, in kilograms, of the crew members for a recent Oxford and Cambridge boat race are given below. Calculate the mean and standard deviation of these masses for each crew.

Oxford   76   80   86   87   92   91   87   78   48
Cambridge   72   76   86   85   89   83   78   78   57

8 The monthly rainfall figures (in millimetres) for Edinburgh in one year were

   43   41   48   36   51   48   69   79   51   66   53   53

a) Calculate the mean and standard deviation of these figures.
b) In how many months was the rainfall greater than the average for the year?
c) In how many months was the rainfall within one standard deviation of the mean?

9 In an experiment to find the acceleration due to gravity the following results were obtained:

$x(\text{m s}^{-2})$ 9.813   9.803   9.811   9.814   9.809   9.815   9.816   9.815

a) Use the transformation $y = \dfrac{x - 9.810}{0.001}$ to convert them to a set with origin at 9.810, and 0.001 as unit.
b) Find m_y and s_y.
c) Hence find the mean and standard deviation of the original results.

10 Five estimates for the radius of the Sun (in km) are

   695400   695800   695300   695900   695100

Find their mean and standard deviation.

11 Given that the ten numbers $a_1, a_2, a_3, \ldots, a_{10}$ have mean m and standard deviation S, *write down* the means and standard deviations of

a) $3a_1, 3a_2, 3a_3, \ldots, 3a_{10}$
b) $a_1 + 3, a_2 + 3, a_3 + 3, \ldots, a_{10} + 3$

in terms of m and S. (No working need be shown.)   (SMP)

12 A chemistry examination has two sections, theory and practical. For a certain class the marks on the theory section had a mean of 60 and a standard deviation of 20. The marks on the practical paper had a mean of 45 and a standard deviation of 10.

a) Devise a transformation of the practical marks to give them the same mean and standard deviation as the theory marks.

b) Using this transformation, what theory marks are equivalent to practical marks of   i) 50,   ii) 35,   iii) 57?

13 A university admissions officer interviews 30 candidates on Monday and 20 on Tuesday. On each day the mean length of all the interviews is exactly 15 minutes, but the standard deviations are 2 minutes on Monday and $2\frac{1}{2}$ minutes on Tuesday. Find (correct to 1 decimal place) the standard deviation for the two days taken together.   (SMP)

14 A class consists of 15 boys and 10 girls. The mean height of the boys is 153 cm with a standard deviation of 11 cm; the mean height of the girls is 148 cm with a standard deviation of 14 cm. Find the mean and standard deviation of heights for the whole class.

15 Suppose that the values of a random sample taken from some population are $x_1, x_2, \ldots, x_n$. Prove the formula

$$\sum_{i=1}^{n} (x_i - \bar{x})^2 = \sum_{i=1}^{n} x_i^2 - n\bar{x}^2$$

Parplan Opinion Polls Ltd. conducted a nationwide survey into the attitudes of teenage girls. One of the questions asked was "What is the ideal age for a girl to have her first baby?" In reply, the sample of 165 girls from the Northern zone gave a mean of 23.4 years and a standard deviation of 1.6 years. Subsequently, the overall sample of 384 girls (Northern plus Southern zones) gave a mean of 24.8 years and a standard deviation of 2.2 years.

Assuming that no girl was consulted twice, calculate the mean and standard deviation for the 219 girls from the Southern zone.   (AEB, 1981)

16 **a)** Given that the mean and variance of $x_1, x_2, \ldots, x_n$ are μ and σ^2 respectively, state the mean and variance of
i) $3x_1, 3x_2, \ldots, 3x_n$
ii) $1 - 2x_1, 1 - 2x_2, \ldots, 1 - 2x_n$
b) Twenty values of a continuous variable, recorded to the nearest cm, had a mean value of 12.5 cm and a variance of 1.35 cm^2.
i)   Find the two original values which, if combined with the twenty original values, would leave the mean unchanged but increase the variance by 9.9 cm^2.
ii) The twenty original values were combined with a further thirty values whose mean and variance were 11.5 cm and 1.21 cm^2 respectively. Calculate the mean and variance of the combined fifty values.

(AEB, 1983)

7.6 Frequency distributions

When dealing with a large amount of data, it is impracticable to list all the individual values. Instead the number of times each different value occurs is recorded, and this is called its *frequency*. The frequency distribution for the number of siblings (brothers and sisters) of 100 schoolchildren is given below, and is illustrated by a frequency diagram:

number of siblings x_r	frequency f_r
0	19
1	42
2	23
3	9
4	4
5	1
6	2
	100

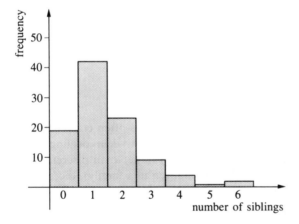

So, for instance, 19 children have no brothers and sisters. Also the sum of the frequencies must give the total number of children:

$$n = \sum f_r$$

In calculating the mean or standard deviation of a frequency distribution, instead of adding all the individual values of x_r, x_r^2, each different value is multiplied by its frequency, and these quantities are then added. So the formulae for mean and standard deviation become (see over):

$$m = \frac{1}{n}\sum f_r x_r$$

$$s^2 = \frac{1}{n}\sum f_r(x_r - m)^2 = \frac{1}{n}\sum f_r x_r^2 - m^2$$

We set out the calculation for the frequency distribution above in tabular form, though in practice the whole calculation can usually be done on a calculator.

x_r	f_r	$f_r x_r$	$f_r x_r^2$
0	19	0	0
1	42	42	42
2	23	46	92
3	9	27	81
4	4	16	64
5	1	5	25
6	2	12	72
	100	148	376

So $\quad m = \dfrac{148}{100} \quad$ and $\quad s = \sqrt{\left(\dfrac{376}{100} - 1.48^2\right)}$

$\qquad = 1.48 \qquad\qquad \approx 1.25$

Grouped data

When presented with a large number of readings, it is often convenient to arrange them in groups according to size. Indeed when dealing with continuous data (such as length, weight or time) rather than discrete data (such as examination marks or number of children in a family), there is no alternative but to present the information in this way.

In the following example, the diameters are given correct to two decimal places; therefore the interval 23.27 − 23.31 will include any bolt whose head diameter lies between 23.265 mm and 23.315 mm, and so will be centred on 23.29 mm. We do not know the individual diameters of the 52 bolts in the interval, so in calculating the mean and standard deviation, we take all of them to be at the mid-point of the interval, 23.29 mm. If we assume an even distribution throughout the interval, this approximation is valid for calculation of the mean, but it gives a slight distortion to the standard deviation. It is possible to offset this distortion by an adjustment known as Sheppard's correction, but for the present we shall analyse grouped data without such adjustment.

Example

The frequency distribution of a random sample of 250 steel bolts according to their head diameter, measured to the nearest 0.01 mm, is shown in the following table:

diameter (mm)	number of bolts
23.07–23.11	10
23.12–23.16	20
23.17–23.21	28
23.22–23.26	36
23.27–23.31	52
23.32–23.36	38
23.37–23.41	32
23.42–23.46	21
23.47–23.51	13

We shall denote the mid-points of the intervals, 23.09, 23.14, ..., 23.49 by x_r. Since all values are slightly greater than 23 mm, we shall subtract 23 from each and then, to save bothering with a decimal point, multiply each by 100. The resulting values will be denoted by y_r.

We are clearly using the transformation

$$y = 100(x - 23)$$

so that $x = \dfrac{y}{100} + 23$

The calculation can then be set out:

x_r	y_r	f_r	$f_r y_r$	$f_r y_r^2$
23.09	9	10	90	810
23.14	14	20	280	3920
23.19	19	28	532	10108
23.24	24	36	864	20736
23.29	29	52	1508	43732
23.34	34	38	1292	43928
23.39	39	32	1248	48672
23.44	44	21	924	40656
23.49	49	13	637	31213
		250	7375	243775

$$m_y = \frac{7375}{250} \qquad \text{and} \quad s_y = \sqrt{\left(\frac{243775}{250} - 29.5^2\right)}$$

$$= 29.5 \qquad\qquad \approx 10.2$$

$$\text{so} \quad m_x = \frac{29.5}{100} + 23 \quad \text{and} \quad s_x \approx \frac{10.2}{100}$$

$$= 23.295 \qquad\qquad \approx 0.102$$

The frequency diagram for the steel bolts is shown below, with the mean and standard deviation illustrated on it.

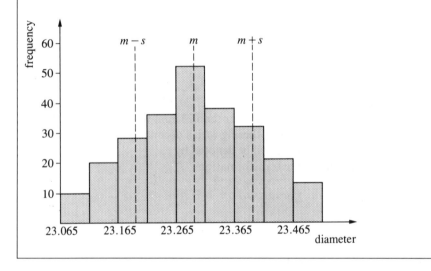

Cumulative frequency

We sometimes need to evaluate the median and interquartile range of a large population, where it is not feasible to arrange the data in order. These measures (and other information) can be obtained quite easily from a cumulative frequency diagram.

The *cumulative frequency* is the total frequency up to the end of a particular interval. The frequency table for the diameters of bolt heads is given on page 297 with the cumulative frequencies added.

The cumulative frequency diagram is then obtained by plotting cumulative frequency against the end-point of the interval to which it applies, and joining the points with a smooth curve.

diameter (mm)	frequency	cumulative frequency
23.07–23.11	10	10
23.12–23.16	20	30 (= 10 + 20)
23.17–23.21	28	58 (= 30 + 28)
23.22–23.26	36	94
23.27–23.31	52	146
23.32–23.36	38	184
23.37–23.41	32	216
23.42–23.46	21	237
23.47–23.51	13	250

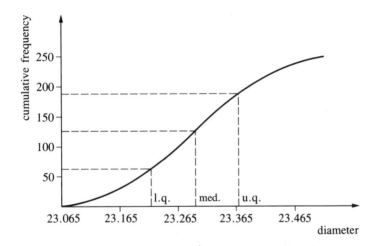

The median is the point on the horizontal axis corresponding to the mid-point value on the vertical (cumulative frequency) axis. Similarly for the lower and upper quartiles.

So   median = 23.30
and   interquartile range = 23.37 − 23.22 = 0.15

The median can be thought of as the 50th *percentile*, and the quartiles as the 25th and 75th percentiles. Other percentiles can be obtained in a similar way.

Histograms

The following data, derived from a traffic survey on a main road, gives the time, in seconds, between successive vehicles passing a given point. (Measurements were taken to the nearest second, so that a measurement registered as one second could be anywhere in the interval 0.5 to 1.5 seconds.)

time interval	number of vehicles f_r
0.5–1.5	11
1.5–2.5	19
2.5–3.5	23
3.5–4.5	14
4.5–5.5	10
5.5–10.5	15
10.5–20.5	8
	100

These result can be displayed in a frequency diagram:

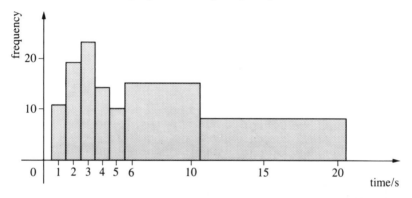

However, it is immediately clear that the diagram gives a very misleading impression. This arises because the time intervals are not all of equal width, and to correct the impression we require a diagram in which the *areas* of rectangles, rather than their heights, represent frequencies. We therefore divide each frequency by the width of the corresponding interval. The resulting quantity is known as the *frequency density* (in this case measured in cars per second), and the corresponding diagram is called a *histogram*:

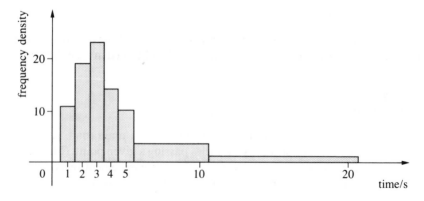

To give diagrams of comparable size when total frequencies may be different, it is often more convenient to use the proportion, or *relative frequency* (f_r/n), rather than frequency itself on the vertical scale of the diagrams. In the case of a histogram, the total area must then be equal to 1.

For the traffic survey histogram the frequency densities of 10 and 20 are equivalent to relative frequency densities of 0.1 and 0.2. The histogram is then directly comparable with the one below, which illustrates the results of an extension of the survey to 5000 intervals between successive vehicles, this time being measured to a greater accuracy of 0.1 s.

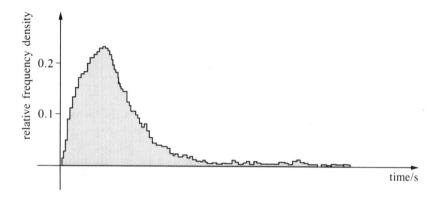

As we would expect, the general form of the histogram is similar to that of its predecessor, but because the time intervals are now much smaller, the outline of the histogram is correspondingly smoother. If the time intervals are measured to greater and greater accuracy, and the survey size increased accordingly, the outline of the histogram will clearly tend closer and closer to a curve.

Exercise 7.6

1 Two dice were thrown together 200 times and for each throw the total score was recorded. The frequencies with which the scores occurred were as follows:

score	12	11	10	9	8	7	6	5	4	3	2
frequency	7	10	20	20	30	34	27	21	14	11	6

Find the mean and the standard deviation of the scores.   (L)

2 The contents of 50 boxes of matches of the same make were found to be as follows:

number of matches	42	43	44	45	46	47	48
number of boxes	10	6	20	6	5	2	1

Find the mean number of matches in a box and the standard deviation.

(L)

In nos. **3**, **4** and **5** calculate the mean and standard deviation of the data. Represent the data on a histogram, and mark the position of the mean, and one standard deviation on either side of the mean.

3   The lifetimes (in completed weeks) of 100 mice of a certain species are given below.

lifetime (weeks)	0–9	10–39	40–49	50–59	60–69	70–79	80–99
number of mice	8	3	7	18	31	22	11

4   Each of a group of 50 schoolboys attempts to throw a javelin for the first time. The distances they achieve (to the nearest metre) are given below.

distance (m)	0–19	20–29	30–34	35–39	40–49
number of throws	5	11	18	13	3

5   120 people set out on a sponsored walk to raise money for charity. The distances they covered are given below.

distance (km)	less than 10	10–15	15–20	20–30	30–50
number of walkers	12	15	25	41	27

6   The table below summarises a car-dealer's sales for one year. Find, correct to 3 significant figures, the mean and standard deviation of the car prices.

price (£)	3001–4000	4001–4500	4501–5000	5001–5500
number of cars	18	37	41	34

price (£)	5501–6000	6001–7000	7001–10000
number of cars	23	16	9

7   The following data give the percentage ash in the bones of chicks which have been fed on a diet containing supplementary vitamin D. Form a frequency tabulation and plot a histogram.

Calculate *from the grouped data* the mean and standard deviation of the percentage.

percentage ash in chick bones

34.68	33.57	36.17	37.53	37.17	38.51	42.13
38.70	37.67	37.35	34.64	39.19	38.88	43.44
38.30	40.23	37.14	37.58	36.39	37.17	39.38
35.74	34.92	35.37	41.85	34.64	36.69	35.57
34.28	36.41	35.98	34.99	34.15	35.30	37.87
34.49	40.44	36.47	30.41	35.00	41.09	41.20
35.17	38.66	41.39	37.27	35.30	33.32	37.41
34.96	40.65	37.19	36.41	39.72	29.11	(MEI)

8 A small factory employs 100 people. It publishes the following statistics, giving the numbers of employees whose basic weekly wage is less than various amounts:

basic weekly wage (£)	40	50	60	80	100	150	200
number of employees getting less than this amount	0	6	31	65	85	95	100

Draw a *histogram* showing the distribution of the employees into various wage-bands; indicate clearly on your diagram the scales to which it is drawn.

It is not possible to calculate the mean wage exactly from these statistics, but it can be stated that the mean weekly wage lies between $£x$ and $£y$, for certain values of x and y. Give the values of x and y which are as close together as the data allow. (SMP)

9 The marks of 492 candidates who entered for an examination are shown:

marks		marks	
0 to 4	1	5 to 9	2
10 to 14	5	15 to 19	6
20 to 24	10	25 to 29	11
30 to 34	21	35 to 39	26
40 to 44	44	45 to 49	53
50 to 54	63	55 to 59	66
60 to 64	58	65 to 69	42
70 to 74	30	75 to 79	21
80 to 84	15	85 to 89	9
90 to 94	5	95 to 99	4

The maximum mark was 99.
a) Draw the cumulative frequency curve and estimate the median and inter-quartile range.

b) What should be the pass mark if at least 80% of the candidates are to pass?

c) If the pass mark is 47, estimate how many candidates fail by no more than 5 marks.   (SMP)

10   The marks of 1000 candidates in an examination were distributed as follows:

mark	0–9	10–19	20–29	30–39	40–49
number of candidates	26	64	112	170	212

mark	50–59	60–69	70–79	80–89	90–99
number of candidates	192	130	62	22	10

Construct the cumulative frequency curve for these marks. It is required to pass 60% of the candidates, to give a grade A to the top 10%, and to give a grade B to the next 20% of the candidates. It being understood that all the marks given are whole numbers, obtain from your cumulative frequency curve an estimate of the least mark which a candidate must obtain to be given **a)** a pass, **b)** a grade A, **c)** a grade B.   (L)

11   The number of bargains marked in certain ordinary shares on the London Stock Exchange on 50 business days were:

```
66  133   89    97  112  104    96  105  117  88
77   89   91   113  109  102    94  107  119  87
71  127  125   109   99  103    96  108  117  88
81   91  124    98  109  101   105   93  119  86
84  131   92    97  100  101   106   93  121  87
```

Draw up a frequency table showing the numbers in the intervals 60 to 69, 70 to 79, etc., and from this table calculate the mean of the distribution.

Draw a histogram to illustrate the data.

Draw a cumulative frequency curve to fit the distribution and deduce the values of the median and the semi-inter-quartile range.   (OC)

12   In 1970 it was estimated that the age-distribution of the population (in thousands) of England and Wales in the year 2000 would be as follows:

age	male	female	total
0–4	2528	2394	4922
5–9	2438	2318	4756
10–14	2345	2238	4583
15–19	2225	2150	4375
20–24	2168	2118	4286
25–29	2072	2022	4094
30–34	2035	1978	4013
35–39	2088	2021	4109
40–44	1794	1741	3535
45–49	1640	1601	3241
50–54	1803	1775	3578
55–59	1434	1443	2877
60–64	1235	1283	2518
65–69	1045	1175	2220
70–74	815	1097	1912
75–79	594	1003	1597
80–84	287	626	913
85–	158	510	668

[Population Projections, 1970–2010: HMSO 1971]

Plot frequency and cumulative frequency diagrams for the total population.
a) Calculate the median age of the population and its lower and upper quartiles.
b) Calculate the mean and standard deviation.

13 Repeat no. 12 for the male population.

14 Repeat no. 12 for the female population.

7.7 The Normal distribution

Frequency diagrams and histograms can take a wide variety of different shapes, depending on the data which is being represented. However, one particular shape (or distribution) occurs in a wide variety of circumstances, which we can illustrate by three quite different sets of data, each with its own histogram.

1 Masses of five-penny pieces

100 five-penny pieces were weighed and their masses were classified into intervals of 0.05 g. The resulting frequency table, with corresponding relative frequencies and relative frequency densities was found to be as follows:

mass (g)	frequency	relative frequency	r.f.d.
5.45–5.50	1	0.01	0.2
5.50–5.55	3	0.03	0.6
5.55–5.60	4	0.04	0.8
5.60–5.65	16	0.16	3.2
5.65–5.70	37	0.37	7.4
5.70–5.75	27	0.27	5.4
5.75–5.80	10	0.10	2.0
5.80–5.85	1	0.01	0.2
5.85–5.90	1	0.01	0.2
total	100	1.00	

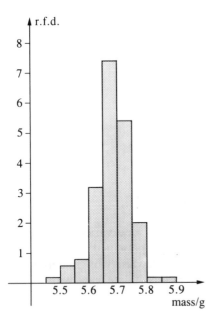

2 Lifetimes of electric light bulbs

The lifetimes of 200 electric light bulbs were classified into intervals of 200 hours, as follows:

lifetime (h)	frequency	relative frequency	r.f.d. $\times 10^3$
800–1000	2	0.010	0.05
1000–1200	7	0.035	0.175
1200–1400	17	0.085	0.425
1400–1600	43	0.215	1.075
1600–1800	78	0.390	1.950
1800–2000	34	0.170	0.850
2000–2200	15	0.075	0.375
2200–2400	4	0.020	0.100
total	200	1.000	

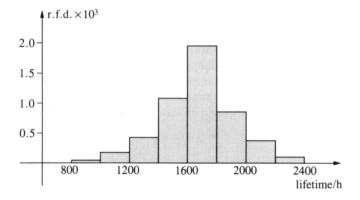

3 Heights of adult males

The heights of 1000 adult males were measured and classified into intervals of 2 cm:

height (cm)	frequency	height (cm)	frequency	height (cm)	frequency
150–152	2	166–168	101	182–184	23
152–154	3	168–170	116	184–186	16
154–156	5	170–172	125	186–188	7
156–158	9	172–174	112	188–190	2
158–160	16	174–176	102	190–192	2
160–162	29	176–178	93	192–194	1
162–164	55	178–180	61	194–196	0
164–166	79	180–182	40	196–198	1

From this table it is again easy to calculate the corresponding relative frequencies and relative frequency densities, which can be plotted as follows:

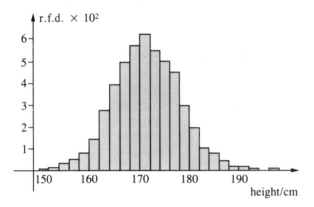

The three histograms appear to have the same general shape; they are approximately symmetrical, with most of their area concentrated in a central region, but tailing off fairly rapidly at each end. In the third example, where the total frequency and the number of class intervals are both greater than in the other two examples, the symmetry is more marked, and it is quite easy to imagine the 'bell-shaped' curve which will approximate to the outline of the histogram as the class intervals become still smaller. The histogram is reproduced below with such a curve superimposed, and similar (though less closely fitting) curves are also superimposed on the other two histograms.

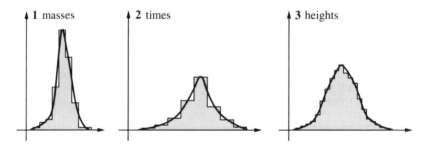

Though there are differences between the curves – one is 'short and wide' whilst another is 'tall and thin' – these differences are only of location and spread; the curve is basically the same shape in each case. Indeed, each of the curves can be defined by a single function, involving two parameters (the mean and standard deviation), and this function is known as the *Normal function*. (The term *Gaussian function* is also used sometimes after one of its earliest investigators†, and Gauss himself called it the *error function* from the way in which it originally arose.) A more detailed analysis of this function and its

† K. F. Gauss (1777–1855), German mathematician and physicist.

equation is given in Book 2, Chapter 15. Data which can be represented in this way is said to be *Normally distributed*.

It must be strongly emphasised however, that in many cases data is only approximately described by the Normal function. In our second example the lifetimes of electric light bulbs, though having an approximately bell-shaped distribution, are not exactly Normal in their distribution. Even so, this should not be regarded as in any way 'abnormal', for although the Normal distribution is extremely important, it is far from being universal. Indeed, perfect 'normality' is highly exceptional.

Since so much data is Normally distributed, we can use the Normal curve to make predictions and assessments of such data.

Properties of the Normal curve

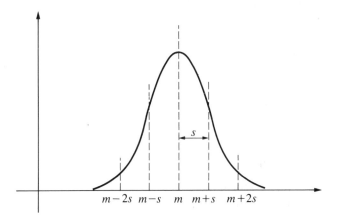

a) The axis of symmetry is at the mean of the distribution.
b) The curve is asymptotic to the x-axis, i.e. $y \to 0$ as $x \to \pm\infty$.
c) The total area beneath the curve is equal to 1.
d) Just over 68% of the area beneath the curve lies between $m - s$ and $m + s$, so that approximately $\frac{2}{3}$ of the data lies within one standard deviation of the mean.
e) Approximately 95% of the area lies within two standard deviations of the mean.

The standard Normal function

We now develop a single method of dealing with all Normal probability distributions, irrespective of the values of their parameters m and s. This is done by standardising the variable, that is transforming it so that its new mean and standard deviation become respectively 0 and 1.

Example 1

The marks in a mathematics examination are found to be approximately Normally distributed with mean 56 and standard deviation 18. Standardise the marks and find the standardised equivalent to a mark of 70.

(Examination marks, of course, are discrete, rather than continuous, variables; nevertheless, as we shall see in later examples, a continuous probability curve can give us a good approximation to a discrete statistical distribution.)

Let us denote the examination marks by the variable x, We first reduce the mean to zero by subtracting 56 from each mark. So x becomes $x - 56$.

As the mean of our new variable, $x - 56$, is now at the origin, and standard deviation measures spread from the mean, we reduce the standard deviation to 1 by dividing each new mark by 18. So $x - 56$ becomes $(x - 56)/18$. Therefore $(x - 56)/18$ is now a standardised Normal variable.

So a mark of 70 becomes $\dfrac{70 - 56}{18} = 0.78$.

If we now look at our Normal curve, in geometrical terms the standardisation process can be regarded as two successive transformations

a) a translation with vector $\begin{pmatrix} -m \\ 0 \end{pmatrix}$

b) a two-way stretch with factor $1/s$ parallel to the x-axis and factor s parallel to the y-axis (so that the total area under the curve remains constant).

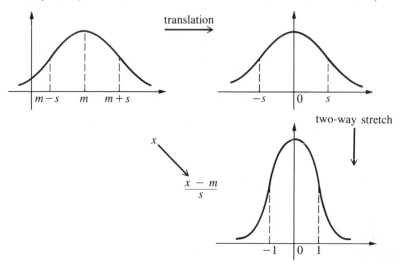

Hence the general standardised variable now becomes $(x - m)/s$.

The most convenient way of handling the standard Normal curve is to use the proportion of its area up to a given point t. This proportion is denoted by $\Phi(t)$, and is shown by the shaded area on the diagram:

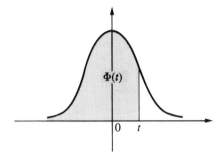

$\Phi(t)$ corresponds to the cumulative frequency introduced in section 7.6. Values of $\Phi(t)$ are tabulated below:

t	$\Phi(t)$	t	$\Phi(t)$	t	$\Phi(t)$	t	$\Phi(t)$
0.0	0.5000	1.0	0.8413	2.0	0.9772	3.0	0.99865
0.1	0.5398	1.1	0.8643	2.1	0.9821	3.1	0.99903
0.2	0.5793	1.2	0.8849	2.2	0.9861	3.2	0.99931
0.3	0.6179	1.3	0.9032	2.3	0.9893	3.3	0.99952
0.4	0.6554	1.4	0.9192	2.4	0.9918	3.4	0.99966
0.5	0.6915	1.5	0.9332	2.5	0.9938	3.5	0.99977
0.6	0.7257	1.6	0.9452	2.6	0.9953	3.6	0.99984
0.7	0.7580	1.7	0.9554	2.7	0.9965	3.7	0.99989
0.8	0.7881	1.8	0.9641	2.8	0.9974	3.8	0.99993
0.9	0.8159	1.9	0.9713	2.9	0.9981	3.9	0.99995
1.0	0.8413	2.0	0.9772	3.0	0.9986	4.0	0.99997

Two points should be noted about the use of such a table:

a) Intermediate values

More accurate statistical tables are generally available, but otherwise linear interpolation must be used to evaluate $\Phi(t)$ for values of t not given in the table. However, the values of $\Phi(t)$ so obtained should be regarded as accurate only to 3 decimal places for values of $t \leqslant 3$, e.g.

$$\Phi(1.37) \approx \Phi(1.3) + \tfrac{7}{10}[\Phi(1.4) - \Phi(1.3)]$$
$$\approx 0.9032 + \tfrac{7}{10} \times 0.0160$$
$$\approx 0.9032 + 0.0112$$
$$\approx 0.914, \quad \text{to 3 d.p.}$$

b) Negative values

No negative values of t are given in the table since $\Phi(-t)$ can be calculated very easily from $\Phi(t)$.

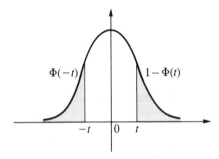

By symmetry, the two shaded areas on the diagram are equal. The areas represent $\Phi(-t)$ and $1 - \Phi(t)$, so it follows that

$$\Phi(-t) = 1 - \Phi(t)$$

So, for instance, $\Phi(-0.6) = 1 - \Phi(0.6)$
$$= 1 - 0.7257$$
$$= 0.2743$$

The following two examples illustrate practical applications of the Normal distribution:

Example 2

Assuming that intelligence quotients are Normally distributed with mean 100 and standard deviation 15, calculate:
a) the proportion of people with an IQ between 88 and 118;
b) the percentage of the population with an IQ greater than 130.

To standardise the variable in this case, we subtract 100 and divide by 15. Therefore:

a) The proportion, denoted by

$P(88 < IQ < 118)$
$\quad = P(IQ < 118) - P(IQ < 88)$

$\quad = \Phi\left(\dfrac{118 - 100}{15}\right) - \Phi\left(\dfrac{88 - 100}{15}\right)$

$\quad = \Phi(1.2) - \Phi(-0.8)$

$\quad = \Phi(1.2) - [1 - \Phi(0.8)]$

$\quad = \Phi(1.2) + \Phi(0.8) - 1$

$\quad = 0.8849 + 0.7881 - 1$

$\quad = 0.673$

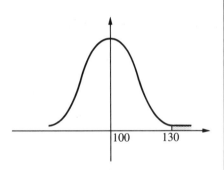

b) $P(IQ > 130)$
$\quad = 1 - P(IQ < 130)$

$\quad = 1 - \Phi\left(\dfrac{130 - 100}{15}\right)$

$\quad = 1 - \Phi(2)$

$\quad = 1 - 0.9772$

$\quad = 0.0228$

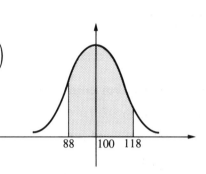

So approximately 2.3% of the population have IQ greater than 130.

Example 3

Machine components have lengths which are Normally distributed with mean 57.2 mm. It is found that 13% of the components have lengths greater than 57.4 mm.
a) What is the standard deviation of the component lengths?
b) What length will be exceeded by 20% of the components?

a) Denote the standard deviation by s
Now $P(\text{length} < 57.4) = 1 - P(\text{length} > 57.4)$

$\Rightarrow \qquad \Phi\left(\dfrac{57.4 - 57.2}{s}\right) = 1 - 0.13$

$\Rightarrow \qquad \Phi\left(\dfrac{0.2}{s}\right) = 0.87$

But from the Normal probability table, $\Phi(1.13) = 0.87$.

So $\dfrac{0.2}{s} = 1.13$

$\Rightarrow$ $s = \dfrac{0.2}{1.13} = 0.18\,\text{mm}$ (correct to 2 d.p.)

b) Let the required length be denoted by X

then $P(\text{length} > X) = 0.2$

$\Rightarrow$ $P(\text{length} < X) = 0.8$

$\Rightarrow$ $\Phi\!\left(\dfrac{X - 57.2}{0.18}\right) = 0.8$

But (correct to 2 d.p.) $\Phi(0.84) = 0.8$

$\Rightarrow$ $\dfrac{X - 57.2}{0.18} = 0.84$

$\Rightarrow$ $X - 57.2 = 0.84 \times 0.18 = 0.15$

$\Rightarrow$ $X = 57.35\,\text{mm}$

Exercise 7.7

1   The mean weight of 100 sixth-formers at a certain school is 63 kg and the standard deviation is 7 kg. Assuming that the weights are approximately Normally distributed, estimate how many sixth-formers weigh   **a)** more than 70 kg   **b)** less than 73 kg   **c)** less than 60 kg   **d)** between 61 and 65 kg.

2   The times taken by a large group of students to complete a project are approximately Normally distributed, with mean 30 hours and standard deviation 4 hours. Estimate the percentage of students who
   a) spend more than 36 hours on the project;
   b) finish the project in less than 25 hours;
   c) finish the project in less than 20 hours.

3   A factory produces ball-bearings whose masses are Normally distributed with mean 45.32 g and standard deviation 0.17 g. Find the proportion of ball-bearings that will have a mass   **a)** more than 45.40 g   **b)** more than 45.50 g   **c)** less than 45.00 g   **d)** between 45.20 and 45.45 g.

4   The daily delivery of mail at a biological research station follows a time

pattern conforming to the Normal distribution, with a mean time of arrival at 8.40 a.m. and with a standard deviation of 20 minutes.

Estimate the number of occasions during the 250 working days in the year when the mail arrives

a) before the main gates open, at 8.00 a.m.

b) after the arrival of the office staff, at 8.20 a.m.

c) during the Director's daily meeting with heads of research sections (9.00 a.m.–9.20 a.m.).

5 The heights in centimetres of a sample of 700 six-month old babies, attending a post-natal clinic, are given in the following frequency table:

height (cm)	62	63	64	65	66	67	68	69	70	71	72	total
frequency	25	35	52	84	120	135	101	61	40	33	14	700

Calculate the mean and standard deviation of the heights.

Assuming that the heights of all such babies are Normally distributed about this mean with this standard deviation, estimate:

a) the percentage of all six-month old babies of height 70 cm or more;

b) the height that will be exceeded by 60% of all babies of this age.

6 The table shows the frequency f with which x α-particles were radiated from a source in a given time, the values of x being integers only.

x	0	1	2	3	4	5	6	7	8	9	10
f	30	37	71	126	130	108	67	5	3	2	1

Calculate the mean value of x and its standard deviation. Assuming that the sample fits a Normal distribution, find the value of x that will be exceeded on 70% of occasions.   (MEI)

7 Packets of soap-powder are filled in such a way that the masses of their contents are Normally distributed with mean 520 g and standard deviation 15 g. The cartons are nominally of 500 g.

a) Calculate (correct to 1 d.p.) the percentage of packets that will be 'under-weight'.

b) Calculate the percentage of packets containing more than 5% above their nominal contents.

c) If the mean mass of soap-powder can be altered, without affecting the standard deviation, what should the new mean be to ensure that only 2% of packets are under-weight?

8 Hens' eggs have mean mass 60 g with standard deviation 15 g, and the distribution may be taken as Normal. Eggs of mass less than 45 g are classified as 'small'. The remainder are divided into 'standard' and 'large',

and it is desired that these should occur with equal frequency. Suggest the mass at which the division should be made (correct to the nearest gram).

(SMP)

9 In a cross-country race the times taken by competitors to complete the course are approximately Normally distributed with mean 47 minutes and standard deviation 9 minutes. The race organisers wish to split the competitors, on the basis of their times, into 5 equal categories, very slow, slow, average, fast and very fast, so that there will be an equal number of competitors (as far as possible) in each category.

a) What is the greatest time in which a competitor must complete the race to be recorded as 'very fast'?

b) Between what times will a competitor be recorded as 'average'?

c) Between what times will a competitor be regarded as slow? (Give answers in minutes, correct to 1 d.p.)

10 A machine produces components to any required length specification with a standard deviation of 1.40 mm. At a certain setting it produces to a mean length of 102.30 mm. Assuming the distribution of lengths to be Normal, calculate:

a) what percentage would be rejected as less than 100 mm long;

b) to what value, to the nearest 0.01 mm, the mean should be adjusted if this rejection rate is to be 1%;

c) whether at the new setting more than 1% of components would exceed 107 mm in length.   (MEI)

11 Machined components are accepted if they pass through a gauge of 1.040 cm and do not pass through a gauge of 0.960 cm. It was found that over a period of production the percentages of components rejected by the larger and smaller gauges were 3.5 and 1.5 respectively. Assuming that the distribution of the dimension tested is Normal, find the mean and the standard deviation.   (OC)

12 A manufacturer hopes, for the sake of his reputation, to produce an article of such quality that not more than 5% of the articles may be expected to last for less than 6 months; and for reasonable economy in manufacturing costs, that not more than 15% may be expected to last for more than 20 months. He can control quality to give (assuming Normal distribution) a pre-determined average m and standard deviation s for the durability of his product. At what values of m and s should he aim?   (C)

13 Skulls may be classified into 3 types according to an index based on their length–breadth ratio

    type A,   under 75
    type B,   between 75 and 80
    type C,   over 80

A large number of skulls are examined and it is found that 58% are of type A, 38% of type B and 4% of type C. Assuming that the length–breadth ratio is Normally distributed, determine the mean and standard deviation of the length–breadth ratio of the skulls. (c)

14 A woman leaves home at 8.00 a.m. every morning in order to arrive at work at 9.00 a.m. She finds that over a long period she is late once in forty times. She then tries leaving home at 7.55 a.m. and finds that over a similar period she is late once in one hundred times. Assuming that the time of her journey has a Normal distribution, before what time should she leave home in order not to be late more than once in two hundred times? (smp)

15 In a particular school 1460 pupils were present on a particular day. By 8.40 a.m. 80 pupils had already arrived, and at 9.00 a.m. 12 pupils had not arrived but were on their way to school. By assuming that the frequency function of arrival times approximates to Normal form, use tables to estimate:
a) the time by which half of those eventually present had arrived;
b) the standard deviation of the times of arrival.
   If registration occurred at 8.55 a.m. how many would not have arrived by then?
   If each school entrance permitted a maximum of 30 pupils per minute to enter, find the minimum number of entrances required to cope with the 'peak' minute of arrival. (smp)

16 The length of a certain mass-produced item is a Normal variable with mean 24.3 mm and standard deviation 0.4 mm. Items are rejected if their lengths are below 23.7 mm or above 25.2 mm. Calculate the proportion of items that are rejected, and the proportion of rejects that are too long. The machine making these items is adjusted with the intention of reducing the proportion rejected; the standard deviation of length cannot be altered, but the mean can. State what value of the mean leads to the minimum proportion of rejects, and give the value of this proportion.
(c)

Miscellaneous problems

1 An electrically heated hot water service is operated by two relay switches, A and B, whose probabilities of failure, at the moment the service is switched on, are

$$P(A) = 0.1, \qquad P(B) = 0.05$$

Assuming the switches operate independently of each other, calculate the probability that at least one of the switches will fail at this moment.

In practice it has been observed that the failure of switch A depends on switch B and that the conditional probability of A failing given that B has not failed is 0.105. Calculate the probability that both of the switches will fail, at the moment the service is switched on, if the individual probabilities of failure are as before.   (MEI)

2   A random-number table consists of a succession of digits each chosen at random and independently from the set $\{0, 1, 2, 3, \ldots, 9\}$. Two successive digits are taken from such a table: show that the probability that their sum is 9 is $\frac{1}{10}$.

   Four successive digits are taken from the table. Calculate the probability that the sum of the first two equals the sum of the third and fourth.   (MEI)

3   Electric light bulbs are chosen at random and tested until a dud is found. It was thought that 10% were faulty, but the thirtieth bulb is the first dud. How surprising is it that so many have been tested successfully?

4   Three boxes A, B, C contain respectively 2 white and 2 black balls, 3 white and 2 black balls, and 4 white and 2 black balls. A ball is drawn unseen from A and placed in B and then a ball is drawn from B and placed in C. If a ball is now drawn from C, find the probability that it will be black.   (C)

5   A large batch of manufactured articles is accepted if either a random sample of 10 articles contains no defective article or, failing this, a second random sample of ten contains no defective article. Otherwise it is rejected.

   If, in fact, 5% of the articles in a batch to be examined are defective, what is the chance of its being accepted?   (OC)

6   Ignoring February 29, what is the probability of at least two members of a class having the same birthday if it contains **a)** 20 members **b)** 25 members? How large must it be in order for this event to be more likely than not?

7   In a game of tennis one point is scored either by A or his opponent B. The winner of the game is the player who first scores four points, unless each player has won three points, when 'deuce' is called and play proceeds until one player is two points ahead of the other and so wins. If A's chance of winning any point is 2/3 and B's chance is 1/3 calculate the chance of
   a) A winning the game without deuce being called;
   b) a similar win by B;
   c) 'deuce' being called.

   If deuce is called, prove that A's subsequent chance of winning the game is 4/5.

   Deduce that A's chance of winning the game is nearly six times that of B.   (C)

8 An absent-minded mathematician writes three letters and addresses three envelopes. What is the chance that they are put

a) all in the right envelopes;

b) all in wrong envelopes?

How would these answers be changed if there were four letters and four envelopes; five letters and five envelopes? (see Miscellaneous problems 6, no. **10**).

9 Three events A, B, C are said to be *pairwise independent* if any two are independent of each other, and *completely independent* when

$$P(A \cap B \cap C) = P(A)P(B)P(C)$$

Consider the spin of two coins, a 5 p piece and a 10 p piece and

let A be the event of 5 p showing head,

B be the event of 10 p showing head,

and C be the event of them being both heads or both tails.

Calculate:

a) $P(A)$, $P(B)$, $P(C)$

b) $P(B \cap C)$, $P(C \cap A)$, $P(A \cap B)$

c) $P(A \cap B \cap C)$

Are three events which are pairwise independent necessarily completely independent?

10 Public houses are scattered randomly along a road at *average* intervals of 1 km.

a) What is the approximate chance that a stretch of length 100 m   i) has a pub, ii) doesn't have a pub?

Hence estimate the probability that a particular kilometre of road (made up of ten such stretches) doesn't have a pub.

b) Repeat with 100 stretches of length 10 m.

c) To what figure do these probabilities appear to be tending? Compare your answer with no. **8** and with Miscellaneous problems 2 no. **12**.

8 Introduction to vectors

8.1 Vectors and vector addition

Many physical quantities have no direction. If we enquire about the temperature of a bath, the pressure of the atmosphere, or the charge of an electron, we expect only to be told a magnitude, i.e., the number of whatever are the appropriate units for the quantity. It would be unusual to speak of one's bath having a temperature 42 °C vertically upwards, or of the atmospheric pressure being on a bearing of 241°, or of an electron having its charge of -1.6×10^{-19} coulomb in a direction North-North-West. Such quantities, which are completely described by their magnitudes, are called *scalars*.

But there are also others known as *vector* quantities, such as the pass of a football from one player to another, the velocity of a car and the intensity of a magnetic field, where information is quite inadequate unless it includes not only a magnitude (or *modulus*) but also a direction.

As an example, we shall begin by considering the possible operation of a helicopter service between five points: Aberdeen, Buckie, Crieff, Dundee, and Edinburgh.

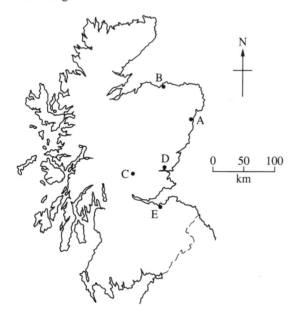

We shall represent a flight from Aberdeen to Buckie, called a *displacement vector*, by **AB** (or, in manuscript, by $\overrightarrow{AB}$), a flight from Buckie to Crieff by **BC** etc.

Then **AB** followed by **BC** achieves the same as **AC**. So we write **AB** + **BC** = **AC**.

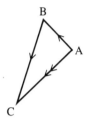

This is the *vector law of addition*, and **AC** is called the *vector sum* of **AB** and **BC**. The symbol + and the words 'addition' and 'sum' have, as yet, no connection with the symbol + and the same words in elementary arithmetic and algebra, but it will soon be clear that they are intimately related.

We are now able to *define* a vector quantity as one which (like displacement) has magnitude and direction and obeys the triangle law of addition; and we shall see later that among such quantities are velocity and acceleration, force and momentum.

Exercise 8.1a

1   Simplify:
   a) **AB** + **BE**   b) **AC** + (**CD** + **DE**)
   c) (**AC** + **CD**) + **DE**   d) (**AB** + **BC**) + (**CD** + **DE**)

2   Use vectors to list ways of flying from Buckie to Crieff
   a) with only one stop between;
   b) with two stops.

Localised vectors and free vectors

Displacements like these which we have just considered are called *localised vectors*: *vectors* because they have magnitude and direction and combine by the triangle law, and *localised* because they also have a particular location. The flight **AB** from Aberdeen to Buckie, for instance, is approximately the same distance and in the same direction as the flight **EC** from Edinburgh to Crieff, but most passengers wishing to make a particular journey would regard the differences between them at least as important as their similarities, and almost certainly they would be expected to have different tickets.

Often, however, we wish to concentrate only on the magnitudes and directions of vectors. When we do so, by considering them dislocated from any particular position, we call them *free* vectors. So **AB** and **EC** are particular examples of the free vector which has (approximately) magnitude 70 km and direction North-West.

Such free vectors are usually represented by small letters in bold type

a, b, c, . . .

(or in manuscript $a, b, c, \ldots$), and their magnitudes (or moduli) are denoted by $a, b, c, \ldots$

Addition of vectors

Suppose we are given two vectors **a** and **b**:

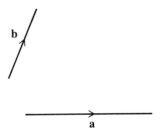

We can localise these in a variety of ways, and in particular both as:

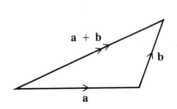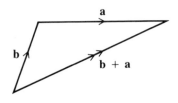

From what has been said about the combination of localised vectors, it is clear that the combined displacements may be called **a** + **b** and **b** + **a**, respectively.

But these have the same magnitude and direction, so are the same free vector.

Hence   **a** + **b** = **b** + **a**

This is the *commutative* law of vector addition.

By combining the two figures, we see that the sum of the two vectors **a** and **b** is represented by the diagonal of the parallelogram formed by **a** and **b**:

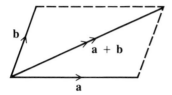

This is sometimes called the *parallelogram law of vector addition*.
    Furthermore, given three vectors **a**, **b**, and **c**, it is clear from the figures that

$$(a + b) + c = a + (b + c)$$

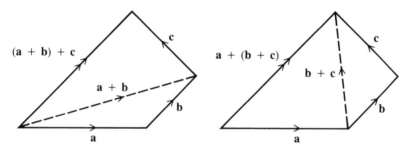

This is the *associative* law of vector addition.

Negation

If we wish to consider the vector which has the same magnitude as **a** but the opposite direction, it is convenient to call it $-\mathbf{a}$:

We can then write the addition of **a** and $-\mathbf{a}$ as

$$a + (-a) = 0$$

where **0** is called the *zero vector*.

Subtraction

We are now in a position to define $\mathbf{a} - \mathbf{b}$ as $\mathbf{a} + (-\mathbf{b})$. It can therefore be represented by either **OC** or **BA**, both of these having the same magnitude and direction:

$$b + (a - b) = a$$

$$a + (-b) = a - b$$

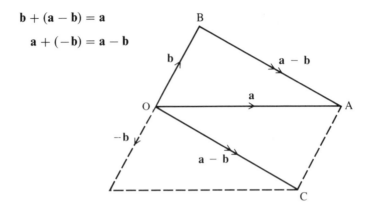

Exercise 8.1b

Draw accurately the following displacement vectors:

a: 6 cm due North (0°);
b: 8 cm North-East (045°);
c: 4 cm due West (270°).

Hence construct the following vectors and state their magnitudes and directions:

1 **a)** a + b   **b)** b + c   **c)** a + c   **d)** (a + b) + c
   e) (b + c) + a   **f)** (a + c) + b

2 **a)** −a   **b)** −b   **c)** −c   **d)** a − b
   e) a − c   **f)** b − c   **g)** b − a   **h)** c − a
   j) b + c − a   **i)** c − b

Multiplication by a scalar

If λ is a scalar, then λa is defined as the vector which has the same direction as **a** and λ times its magnitude:

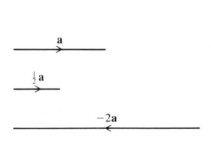

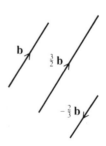

Now if we multiply by λ a vector which has already been multiplied by μ, the result is clearly the same as if we had multiplied the original vector by $\lambda\mu$:

$$\lambda(\mu\mathbf{a}) = (\lambda\mu)\mathbf{a}$$

This is the *associative law* for multiplication by a scalar.

Fortunately, the operation of multiplying by a scalar also combines very conveniently with that of vector addition. For $2\mathbf{a} + 3\mathbf{a}$ is clearly the same as $(2 + 3)\mathbf{a}$; and more generally,

$$(\lambda + \mu)\mathbf{a} = \lambda\mathbf{a} + \mu\mathbf{a}$$

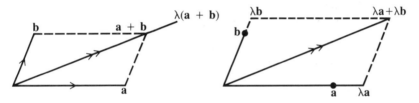

Secondly, multiplying the vector sum $\mathbf{a} + \mathbf{b}$ by λ produces the same result as adding $\lambda\mathbf{a}$ to $\lambda\mathbf{b}$:

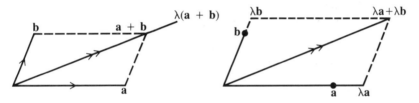

So $\lambda(\mathbf{a} + \mathbf{b}) = \lambda\mathbf{a} + \lambda\mathbf{b}$

These two results,

$$(\lambda + \mu)\mathbf{a} = \lambda\mathbf{a} + \mu\mathbf{a}$$

and $\lambda(\mathbf{a} + \mathbf{b}) = \lambda\mathbf{a} + \lambda\mathbf{b}$

are the *distributive laws* which link multiplication by a number with vector addition.

We have already seen that the addition of vectors is different from the addition of numbers, and this could have been emphasised if we had used another symbol, say $\mathbf{a} \oplus \mathbf{b}$, to represent a vector 'sum'. This operation would then obey the *associative* and *commutative* laws.

$$(\mathbf{a} \oplus \mathbf{b}) \oplus \mathbf{c} = \mathbf{a} \oplus (\mathbf{b} \oplus \mathbf{c})$$

and $\mathbf{a} \oplus \mathbf{b} = \mathbf{b} \oplus \mathbf{a}$

and furthermore, it is now clear that it would also obey the *distributive* laws

$$(\lambda + \mu)\mathbf{a} = \lambda\mathbf{a} \oplus \mu\mathbf{a}$$

$$\lambda(\mathbf{a} \oplus \mathbf{b}) = \lambda\mathbf{a} \oplus \lambda\mathbf{b}$$

The first of this latter pair demonstrates the relationship between the symbols
$\oplus$ and $+$, so that it is only natural to refer to the *addition* of **a** and **b**, and to
a $\oplus$ **b** as their *vector sum*, which for simplicity can be written **a** + **b**.

Exercise 8.1c

If **a** is the vector 4 cm due North and **b** is the vector 3 cm due East, construct
the following vectors and state their magnitudes and directions.

1 a) 2**a** b) 3(2**a**) c) 6**a**

2 a) 3**b** b) 4**b** c) 7**b**

3 a) $\frac{1}{2}$**a** b) $\frac{1}{2}$**b** c) $\frac{1}{2}$**a** + $\frac{1}{2}$**b**
   d) **a** + **b** e) $\frac{1}{2}$(**a** + **b**)

8.2 Position vectors and components

If, in section 8.1, helicopter flights had all been directed from an operational
centre at Oban, then we could have represented the *positions* of

	Aberdeen	Buckie	Crieff	Dundee	Edinburgh
by	**OA**	**OB**	**OC**	**OD**	**OE**

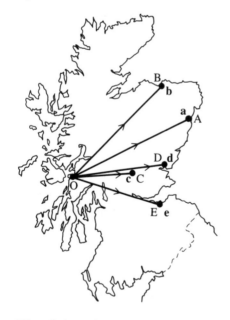

We call these the *position vectors*, referred to O, of A, B, C, D, E, and usually
abbreviate them as **a**, **b**, **c**, **d**, **e**.

Now if we are given the position vectors of two points A and B we can clearly use them to represent other points of their plane:

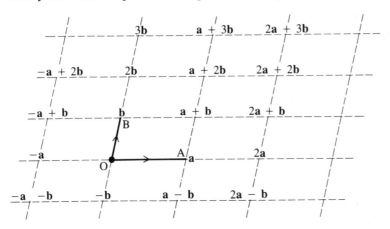

Indeed, providing that **a** and **b** are not parallel, any point P of the plane can be represented in terms of **a** and **b**:

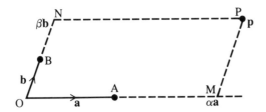

For we can always draw a parallelogram as shown.

Since M is on OA, **OM** = α**a** $\Big\}$ where α, β are numbers
and since N is on OB, **ON** = β**b**

So **OP** = **OM** + **ON**

$\Rightarrow$ **p** = α**a** + β**b**

Similarly, if we are given three non-coplanar vectors **a**, **b**, **c**, then any point in space can be represented by its position vector referred to an origin O, and this vector is a linear combination of **a**, **b**, **c**:

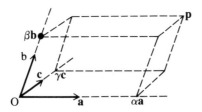

For, as in the figure, we can construct a *cuboid* (or *parallelepiped*), whose opposite faces are parallel, so that

$$\mathbf{p} = \alpha\mathbf{a} + \beta\mathbf{b} + \gamma\mathbf{c}$$

where α, β, γ are numbers.

Exercise 8.2a

1 O is an origin and points with position vectors $\mathbf{a}$, $\mathbf{b}$ are 6 km due East from O and 4 km North-East from O respectively. Sketch these points and mark the points which have position vectors
   a) $2\mathbf{a}$　b) $\frac{1}{3}\mathbf{b}$　c) $-\frac{1}{2}\mathbf{a}$　d) $\mathbf{a} + 2\mathbf{b}$　e) $2\mathbf{a} + \frac{1}{2}\mathbf{b}$　f) $\mathbf{a} - 2\mathbf{b}$

2 $\mathbf{i}$, $\mathbf{j}$ are the position vectors (from O) of points whose Cartesian coordinates are $(1, 0)$ and $(0, 1)$. What are
   a) the position vectors of $(3, 2)$,　$(1, 3)$,　$(0, -2)$,　$(\frac{1}{2}, \frac{1}{2})$?
   b) the Cartesian coordinates of points with position vectors $2\mathbf{i} + \mathbf{j}$,　$\frac{1}{2}\mathbf{i} + \mathbf{j}$,　$-3\mathbf{i}$,　$-\mathbf{i} + \mathbf{j}$?

3 OAPB is a parallelogram. If, from O, A and B have position vectors $\mathbf{a}$ and $\mathbf{b}$, express in terms of $\mathbf{a}$ and $\mathbf{b}$:
   a) $\mathbf{AB}$ and $\mathbf{BA}$;
   b) the position vector of P;
   c) the position vectors of the mid-points of AB and OP. What do you deduce?

4 ABCD is a skew quadrilateral and E, F, G, H are the mid-points of AB, BC, CD, and DA respectively. Denoting position vectors by corresponding letters, find:
   a) $\mathbf{e}$, $\mathbf{f}$, $\mathbf{g}$, $\mathbf{h}$ in terms of $\mathbf{a}$, $\mathbf{b}$, $\mathbf{c}$, $\mathbf{d}$;
   b) $\mathbf{EF}$ and $\mathbf{HG}$ in terms of $\mathbf{a}$, $\mathbf{b}$, $\mathbf{c}$, $\mathbf{d}$;
   c) the position vectors of the mid-points of EG and FH in terms of $\mathbf{a}$, $\mathbf{b}$, $\mathbf{c}$ and $\mathbf{d}$. What do you deduce?

5 The corners A, B, C of this parallelepiped have position vectors $\mathbf{a}$, $\mathbf{b}$, $\mathbf{c}$ from the corner O. Find in terms of $\mathbf{a}$, $\mathbf{b}$, $\mathbf{c}$:

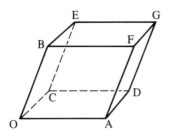

   a) the position vectors of D, E, F, G;

b) AC, FE;

c) the position vectors of the mid-points of AC and EF;

d) the position vectors of the mid-points of OG, AE, BD, CF. What do you deduce?

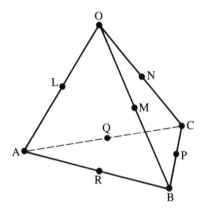

6   The corners A, B, C of a tetrahedron have position vectors **a**, **b**, **c** from the corner O. Find the position vectors of:

a) the mid-points of OA, OB, OC (L, M, N);

b) the mid-points of BC, CA, AB (P, Q, R);

c) the mid-points of LP, MQ, NR. What do you deduce?

Components

So far we have regarded a vector as a single entity, having both magnitude and direction and combining with similar vectors according to a parallelogram law of addition.

If, however, we take a point O as origin, we can refer to the position of another point P either by its position vector **r** or by its coordinates relative to a set of mutually perpendicular axes Ox, Oy, Oz. It is usual, for reasons which will be clear later, to take a right-hand set of axes, i.e., axes for which a rotation from Ox to Oy threads a right-handed screw in the direction Oz.

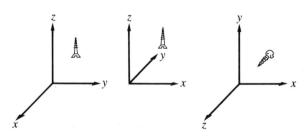

If we let **i**, **j**, and **k** be unit vectors along the axes Ox, Oy, Oz, we obtain the figure

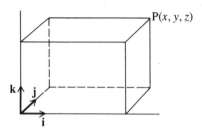

So the point P(x, y, z) can also be described as

r = x**i** + y**j** + z**k**

where x**i**, y**j**, z**k** are said to be the *components* of **r**.†

Using Pythagoras' theorem, we see that the modulus of **r** is given by

$r = \sqrt{(x^2 + y^2 + z^2)}$

In this way, every result which we have so far established can be expressed in terms of such components.

Example 1

If **i**, **j** are unit vectors in the directions East and North respectively, find
a) the magnitude and direction of the vector **a** = 3**i** − 4**j**;
b) the components along **i** and **j** of a displacement **b** which has magnitude 80 units on a bearing of 320°.

a) From the figure,

$$a = \sqrt{(3^2 + 4^2)} = 5$$
and $\tan \alpha = \frac{3}{4} \quad \Rightarrow \quad \alpha \approx 37°$

so the direction of **a** is on bearing 143°.

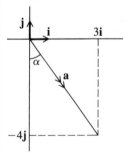

---

† Though the magnitudes x, y, z are also frequently referred to as the components of **r**.

b) Again from the figure, **b** is seen to have components

$$-80 \sin 40° \approx -51.4 \quad \text{along } \mathbf{i}$$
$$\text{and} \quad 80 \cos 40° \approx +61.3 \quad \text{along } \mathbf{j}$$
$$\text{so} \qquad\qquad \mathbf{b} = -51.4\mathbf{i} + 61.3\mathbf{j}$$

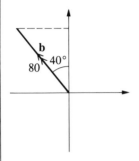

Example 2

We can also use the unit vectors **i**, **j** to investigate the effect of a rotation about O through an angle ϕ.

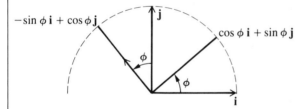

It is clear from the figure that

$$\mathbf{i} \;\mapsto\; \cos\phi\,\mathbf{i} + \sin\phi\,\mathbf{j}$$
$$\text{and} \quad \mathbf{j} \;\mapsto\; -\sin\phi\,\mathbf{i} + \cos\phi\,\mathbf{j}$$

Furthermore, we can investigate the effect of this rotation on the unit vector **p** which makes an angle θ with **i**:

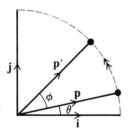

Now $\qquad$ $\mathbf{p} = \cos\theta\,\mathbf{i} + \sin\theta\,\mathbf{j}$

So $\quad \mathbf{p} \;\mapsto\; \mathbf{p}' = \cos\theta\,(\cos\phi\,\mathbf{i} + \sin\phi\,\mathbf{j}) + \sin\theta\,(-\sin\phi\,\mathbf{i} + \cos\phi\,\mathbf{j})$

$\qquad\qquad\quad\; = (\cos\theta\cos\phi - \sin\theta\sin\phi)\mathbf{i} + (\sin\theta\cos\phi + \cos\theta\sin\phi)\mathbf{j}$

But $\mathbf{p}'$ is a unit vector in the direction $\theta + \phi$

So $\quad \mathbf{p}' = \cos(\theta + \phi)\mathbf{i} + \sin(\theta + \phi)\mathbf{j}$

and comparing the two expressions for $\mathbf{p}'$ we see that

$$\cos(\theta + \phi) = \cos\theta\cos\phi - \sin\theta\sin\phi$$
$$\sin(\theta + \phi) = \sin\theta\cos\phi + \cos\theta\sin\phi$$

This is, of course, fundamentally the same proof of the addition formulae as that given in section 5.5, though here expressed in the language of vectors rather than that of matrices.

Exercise 8.2b

1 If $\mathbf{i}, \mathbf{j}$ are unit vectors in the directions East and North respectively,
   a) find the magnitudes and directions of $\mathbf{i} + 2\mathbf{j}, \quad -\mathbf{i} + 3\mathbf{j}, \quad 2\mathbf{i} - 5\mathbf{j}$;
   b) express, in terms of $\mathbf{i}$ and $\mathbf{j}$, vectors
   of magnitude 3 units in the direction North-East;
   of magnitude 5 units in the direction South-East;
   of magnitude 10 units on a bearing of 287°.

2 If $\quad \mathbf{a} = \mathbf{i} + 2\mathbf{j} + 3\mathbf{k} \quad$ and $\quad \mathbf{b} = 2\mathbf{i} + 3\mathbf{j} + 4\mathbf{k}$,
   find the magnitudes of $\quad$ a) $\mathbf{a}$ $\quad$ b) $\mathbf{b}$ $\quad$ c) $\mathbf{a} + \mathbf{b}$ $\quad$ d) $\mathbf{a} - \mathbf{b}$

   Hence verify that the sum of the squares of the diagonals of a parallelogram is equal to the sum of the squares of its four sides.

3 The unit vectors $\mathbf{i}, \mathbf{j}$ are directed along the axes Ox, Oy respectively. Referred to these axes the points A, B, C have coordinates $(1, 2)$, $(4, 3)$, $(3, -1)$ respectively. D is the mid-point of the line BC. Express the vectors $\mathbf{BA}$ and $\mathbf{CA}$ in terms of $\mathbf{i}, \mathbf{j}$. *Verify* that $\mathbf{BA} + \mathbf{CA} = 2\mathbf{DA}$. $\qquad$ (OC)

4 Prove that the points with coordinates $(0, -2, -\frac{1}{2})$, $(1, 1, 1)$, $(5, 3, 4)$, and $(8, 2, 5\frac{1}{2})$ are the vertices of a trapezium in which the non-parallel sides are equal in length. $\qquad$ (SMP)

5 Three points A, B, and C have position vectors $\mathbf{i} + \mathbf{j} + \mathbf{k}$, $\mathbf{i} + 2\mathbf{k}$ and $3\mathbf{i} + 2\mathbf{j} + 3\mathbf{k}$ respectively, relative to a fixed origin O. A particle P starts from B at time $t = 0$, and moves along BC towards C with constant speed 1 unit per s. Find the position vector of P after t s $\quad$ a) relative to O and $\quad$ b) relative to A. $\qquad$ (L)

6 An aircraft takes off from the end of a runway in a southerly direction and climbs at an angle of $\tan^{-1}$ $(\frac{1}{2})$ to the horizontal at a speed of $225\sqrt{5}\,\text{km}\,\text{h}^{-1}$. Show that t s after take-off the position vector $\mathbf{r}$ of the aircraft with respect to the end of the runway is given by $\mathbf{r} = \frac{1}{16}t(2\mathbf{i} + \mathbf{k})$ where $\mathbf{i}, \mathbf{j}, \mathbf{k}$ represent vectors of length 1 km in directions South, East, and vertically upwards.

At time $t = 0$ a second aircraft flying horizontally South-West at $720\sqrt{2}\,\text{km}\,\text{h}^{-1}$ has a position vector $-1.2\mathbf{i} + 3.2\mathbf{j} + \mathbf{k}$. Find its position vector at time t in terms of $\mathbf{i}, \mathbf{j}, \mathbf{k}$ and t. Show that there will be a collision unless courses are changed and state at what time it will occur.   (SMP)

8.3 Scalar products

So far we have confined ourselves to the addition and subtraction of vectors and to their multiplication by a scalar. But there are two other operations which can be performed on a pair of vectors, those of *scalar multiplication* and *vector multiplication*. These are both widely used throughout applied mathematics, especially for describing (respectively) the work done by a force and the moment of a force. In this section we shall restrict ourselves to a consideration of scalar multiplication.

Definition

If $\mathbf{a}$ and $\mathbf{b}$ are two vectors with magnitudes a and b whose directions are separated by an angle θ, we define their *scalar product* (or *dot product*) as

$$\mathbf{a}.\mathbf{b} = ab\cos\theta$$

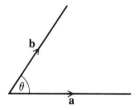

As $\cos(-\theta) = \cos\theta$ we need not be concerned about the direction in which to measure θ, and it is also apparent that

$$\mathbf{a}.\mathbf{b} = a(b\cos\theta) = a \times \text{projection of } \mathbf{b} \text{ on } \mathbf{a}$$
$$= b(a\cos\theta) = b \times \text{projection of } \mathbf{a} \text{ on } \mathbf{b}$$

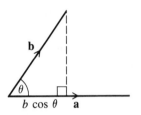 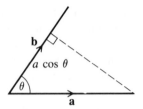

As an illustration, if **F** is a force and **c** is a displacement of its point of application, the scalar product **F.c** is called the *work done by the force* in this displacement.

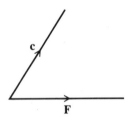

So the work can be regarded as

      (projection of force **F** in direction of displacement **c**)
                          × (magnitude of displacement)

or   (projection of displacement **c** in direction of force **F**)
                             × (magnitude of force)

and if **F** and **c** are perpendicular, then **F.c** = 0, so that no work is done when a force is displaced perpendicular to its line of action.

A number of results follow immediately from the definition of the scalar product:

a) $\mathbf{a.a} = aa\cos 0 = a^2$ **a.a** is usually abbreviated as $\mathbf{a}^2$. So $\mathbf{a}^2 = a^2$.

b) $\mathbf{a.b} = \mathbf{b.a}$

(i.e., scalar product is *commutative*).

c) If **a** and **b** are non-zero vectors,

    $\mathbf{a.b} = 0 \quad \Leftrightarrow \quad ab\cos\theta = 0 \quad \Leftrightarrow \quad \theta = \dfrac{\pi}{2}$

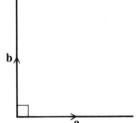

So $\mathbf{a.b} = 0 \quad \Leftrightarrow \quad$ **a** and **b** are perpendicular.

d) $\mathbf{a}.(\lambda \mathbf{b}) = a(\lambda b)\cos\theta = \lambda ab\cos\theta$

   $\lambda(\mathbf{a}.\mathbf{b}) = \lambda(ab\cos\theta) = \lambda ab\cos\theta$

So $\mathbf{a}.(\lambda \mathbf{b}) = \lambda(\mathbf{a}.\mathbf{b})$

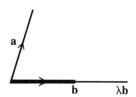

i.e., the scalar product is *distributive with respect to multiplication by a scalar.*

This last result, like some of its predecessors and also the next, at first appears trivial. But it is worthwhile pausing to appreciate its importance, and to be grateful that when we are confronted with $\mathbf{a}.(2\mathbf{b})$, $(2\mathbf{a}).\mathbf{b}$ and $2(\mathbf{a}.\mathbf{b})$, we know without hesitation that they are all equal.

e) $(\mathbf{a} + \mathbf{b}).\mathbf{c} = \mathbf{a}.\mathbf{c} + \mathbf{b}.\mathbf{c}$

i.e., scalar product is *distributive with respect to addition.*
This result, too, is clear from the figure.

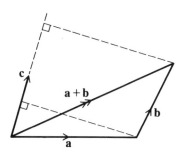

For $(\mathbf{a} + \mathbf{b}).\mathbf{c} = $ (projection of $\mathbf{a} + \mathbf{b}$ on $\mathbf{c}$) $\times c$
$= $ (projection of $\mathbf{a}$ on $\mathbf{c}$ + projection of $\mathbf{b}$ on $\mathbf{c}$) $\times c$
$= $ (projection of $\mathbf{a}$ on $\mathbf{c}$) $\times c$
$+ $ (projection of $\mathbf{b}$ on $\mathbf{c}$) $\times c$

So $(\mathbf{a} + \mathbf{b}).\mathbf{c} = \mathbf{a}.\mathbf{c} + \mathbf{b}.\mathbf{c}$

and, similarly, $\mathbf{a}.(\mathbf{b} + \mathbf{c}) = \mathbf{a}.\mathbf{b} + \mathbf{a}.\mathbf{c}$

This result is particularly important in mechanics. For if $\mathbf{P}$ and $\mathbf{Q}$ are two forces and $\mathbf{c}$ is a certain displacement, then

   $(\mathbf{P} + \mathbf{Q}).\mathbf{c} = \mathbf{P}.\mathbf{c} + \mathbf{Q}.\mathbf{c}$

So, for any displacement, the work done by the resultant of two forces is the sum of the amounts of work done by the forces separately.

Example 1

If two pairs of opposite edges of a tetrahedron are perpendicular, so is the third.

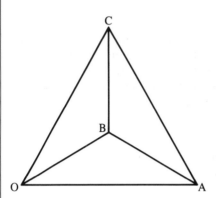

Take one of the vertices as the origin O, and let the position vectors of the others be **a, b, c**.

Suppose that **OB** $\perp$ **CA** and   **OC** $\perp$ **AB**

Then $\mathbf{b}.(\mathbf{a} - \mathbf{c}) = 0$ and $\mathbf{c}.(\mathbf{b} - \mathbf{a}) = 0$

$\Rightarrow$ $\mathbf{b}.\mathbf{a} - \mathbf{b}.\mathbf{c} = 0$ and $\mathbf{c}.\mathbf{b} - \mathbf{c}.\mathbf{a} = 0$

$\Rightarrow$ $\mathbf{a}.\mathbf{b} = \mathbf{b}.\mathbf{c}$ and $\mathbf{b}.\mathbf{c} = \mathbf{a}.\mathbf{c}$

$\Rightarrow$ $\mathbf{a}.\mathbf{b} = \mathbf{a}.\mathbf{c}$

$\Rightarrow$ $\mathbf{a}.(\mathbf{b} - \mathbf{c}) = 0$

$\Rightarrow$ **OA** $\perp$ **BC**

Finally, we see that in the particular case when the four points are coplanar, each is the *orthocentre* of the triangle formed by the other three.

Exercise 8.3a

1   Letting **a** and **b** be the position vectors of two points A and B and expanding the scalar product $(\mathbf{a} - \mathbf{b})^2$, prove the cosine rule for $\triangle$OAB.

2   Taking O as the centre of a circle, points with position vectors **a**, $-\mathbf{a}$ at opposite ends of a diameter, and **r** anywhere else on the circle, prove that the angle in a semi-circle must be a right angle.

3   Let O be the middle point of the base BC of $\triangle$ABC and let A, B, C have

position vectors $\mathbf{a}$, $\mathbf{b}$, $-\mathbf{b}$. By expanding $(\mathbf{a} - \mathbf{b})^2 + (\mathbf{a} + \mathbf{b})^2$, prove Apollonius' theorem, that

$$AB^2 + AC^2 = 2(AO^2 + BO^2)$$

4 AB, CD are two skew (non-coplanar) lines which have their mid-points at P, Q respectively. Prove that

$$AC^2 + AD^2 + BC^2 + BD^2 = AB^2 + CD^2 + 4PQ^2$$

5 A point P lies in the plane of a triangle ABC, and G is the centroid of this triangle. Prove that

$$PA^2 + PB^2 + PC^2 = 3PG^2 + \tfrac{1}{3}(BC^2 + CA^2 + AB^2)$$

What is the least value of $PA^2 + PB^2 + PC^2$ as P varies in the plane? What is the locus of P when P varies so that

$$PA^2 + PB^2 + PC^2 \text{ is constant?} \tag{OC}$$

Scalar product in component form

Suppose that $\mathbf{a}$ and $\mathbf{b}$ are expressed in terms of unit vectors $\mathbf{i}$, $\mathbf{j}$ and $\mathbf{k}$:

$$\mathbf{a} = a_1\mathbf{i} + a_2\mathbf{j} + a_3\mathbf{k}$$
$$\text{and} \quad \mathbf{b} = b_1\mathbf{i} + b_2\mathbf{j} + b_3\mathbf{k}$$

We can readily find $\mathbf{a} . \mathbf{b}$ in terms of these components.

Firstly we note that $\mathbf{i}$, $\mathbf{j}$, $\mathbf{k}$ are unit vectors, so $\mathbf{i}.\mathbf{i} = \mathbf{j}.\mathbf{j} = \mathbf{k}.\mathbf{k} = 1$; and $\mathbf{i}$, $\mathbf{j}$, $\mathbf{k}$ are perpendicular, so $\mathbf{i}.\mathbf{j} = \mathbf{i}.\mathbf{k} = 0$, etc.

Now $\mathbf{a}.\mathbf{b} = (a_1\mathbf{i} + a_2\mathbf{j} + a_3\mathbf{k}).(b_1\mathbf{i} + b_2\mathbf{j} + b_3\mathbf{k})$

$$\Rightarrow \quad \mathbf{a}.\mathbf{b} = a_1 b_1 + a_2 b_2 + a_3 b_3$$

In particular, $\mathbf{a}^2 = \mathbf{a}.\mathbf{a} = a_1^2 + a_2^2 + a_3^2$

and $\mathbf{b}^2 = \mathbf{b}.\mathbf{b} = b_1^2 + b_2^2 + b_3^2$

Also, if $\mathbf{a} = l\mathbf{i} + m\mathbf{j} + n\mathbf{k}$ is a unit vector which makes angles θ, ϕ, ψ with the three axes, then

$\cos\theta = \mathbf{a}.\mathbf{i} = l$

$\cos\phi = \mathbf{a}.\mathbf{j} = m$

$\cos\psi = \mathbf{a}.\mathbf{k} = n$

so that l, m, n are known as the *direction cosines* of $\mathbf{a}$.

These results are frequently of great use in finding the angle between two directions in space.

Example 2

If three corners A, B, C of a crystal are known to be at points $(3, 2, 1)$, $(4, 0, 2)$, $(5, 2, 4)$ respectively, we may wish to find the angle θ between **AB** and **AC**.

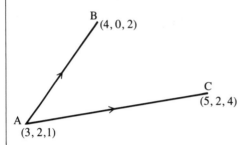

Taking unit vectors $\mathbf{i}, \mathbf{j}, \mathbf{k}$ in the direction of the three axes,

$$\mathbf{a} = \mathbf{AB} = \mathbf{i} - 2\mathbf{j} + \mathbf{k}$$

and $\qquad \mathbf{b} = \mathbf{AC} = 2\mathbf{i} + 3\mathbf{k}$

Then $\quad a^2 = 1^2 + 2^2 + 1^2 = 6 \;\; \Rightarrow \;\; a = \sqrt{6}$

and $\qquad b^2 = 2^2 + 0^2 + 3^2 = 13 \;\; \Rightarrow \;\; b = \sqrt{13}$

Also $\quad \mathbf{a.b} = 1 \times 2 + -2 \times 0 + 1 \times 3 = 5$

Now $\quad \mathbf{a.b} = ab \cos \theta$

so $\qquad 5 = \sqrt{6} \times \sqrt{13} \cos \theta$

$\Rightarrow \quad \cos \theta = \dfrac{5}{\sqrt{6}\sqrt{13}} = \dfrac{5}{\sqrt{78}} = \dfrac{5}{8.832} = 0.5661$

$\Rightarrow \qquad \theta = 55.5°$

So   **AB** and **AC** are inclined at $55.5°$.

Exercise 8.3b

1   Find the scalar products of the following pairs of vectors and hence the angle between them:
   a) $\mathbf{i} + \mathbf{j}$ and $\mathbf{i} + 2\mathbf{j}$
   b) $\mathbf{i} + \mathbf{j} + \mathbf{k}$ and $\mathbf{i} + \mathbf{j} - \mathbf{k}$
   c) $-\mathbf{i} + 2\mathbf{j} - \mathbf{k}$ and $\mathbf{i} - \mathbf{j} + \mathbf{k}$

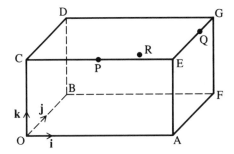

2   In the above rectangular box, OA = 4, OB = 3, OC = 2. Furthermore, P
    is the mid-point of CE; Q divides EG in the ratio 2:1; and R is the centre of
    rectangle BFGD.
       For each of the following pairs of lines, find vectors which represent the
    two lines (in terms of **i**, **j**, **k**) and hence the angle between the lines:
    a) OP and **OD**
    b) AQ and **AB**
    c) ER and **RC**
    d) PR and **QF**

* 8.4 Lines, planes and angles

The equations of a line

Suppose that P_0 is a point whose position vector is $\mathbf{r}_0$ and that a straight line
passes through P_0 in the direction of the vector **u**. Then if P, with position
vector **r**, lies on this line, it follows that

$$\mathbf{r} = \mathbf{r}_0 + \lambda\mathbf{u} \quad \text{(where } \lambda \text{ is a scalar)}$$

Suppose now that the components of

$\quad$ **r** are (x, y, z), $\mathbf{r}_0$ are (x_0, y_0, z_0), and **u** are (l, m, n)

Then $x = x_0 + \lambda l, \quad y = y_0 + \lambda m, \quad z = z_0 + \lambda n$

$$\Rightarrow \qquad \frac{x - x_0}{l} = \frac{y - y_0}{m} = \frac{z - z_0}{n}$$

which are called the *Cartesian equations of the line*.

Example 1

Find the vector and Cartesian equations of the line joining the points $(1, 2, 3)$ and $(2, 3, 5)$.

In this case, $\quad \mathbf{r}_0 = \mathbf{i} + 2\mathbf{j} + 3\mathbf{k}$
$$\mathbf{u} = \mathbf{i} + \mathbf{j} + 2\mathbf{k}$$

So $\quad \mathbf{r} = (\mathbf{i} + 2\mathbf{j} + 3\mathbf{k}) + \lambda(\mathbf{i} + \mathbf{j} + 2\mathbf{k})$
$\Rightarrow \quad \mathbf{r} = (1 + \lambda)\mathbf{i} + (2 + \lambda)\mathbf{j} + (3 + 2\lambda)\mathbf{k}$

and in Cartesian form:

$$x = 1 + \lambda, \quad y = 2 + \lambda, \quad z = 3 + 2\lambda$$

$$\Rightarrow \quad \frac{x - 1}{1} = \frac{y - 2}{1} = \frac{z - 3}{2}$$

Example 2

Find the angle between the lines

$$\frac{x}{1} = \frac{y - 1}{2} = \frac{z - 2}{3} \quad \text{and} \quad \frac{x}{2} = \frac{y + 1}{3} = \frac{z + 2}{4}$$

These are in the directions of the vectors

$$\mathbf{i} + 2\mathbf{j} + 3\mathbf{k} \quad \text{and} \quad 2\mathbf{i} + 3\mathbf{j} + 4\mathbf{k}$$

Now, if the angle between these vectors is θ, then

$$(\mathbf{i} + 2\mathbf{j} + 3\mathbf{k}) \cdot (2\mathbf{i} + 3\mathbf{j} + 4\mathbf{k}) = \sqrt{14} \times \sqrt{29} \cos \theta$$

$$\Rightarrow \qquad 20 = \sqrt{14} \times \sqrt{29} \cos \theta$$

$$\Rightarrow \qquad \cos \theta = \frac{20}{\sqrt{14} \times \sqrt{29}} = 0.9926$$

$$\Rightarrow \qquad \theta = 7°$$

The equation of a plane

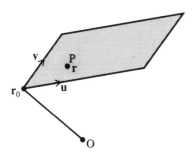

Suppose that P (with position vector **r**) lies in a plane which passes through a fixed point P_0 (whose position vector is r_0), and that **u**, **v** are two non-parallel vectors in the plane.

Then $\mathbf{r} = \mathbf{r}_0 + \lambda\mathbf{u} + \mu\mathbf{v}$ (where λ, μ are scalars)

which is therefore a *parametric equation of the plane*.

Suppose further that the normal to the plane is in the direction **n**.

Then $\mathbf{n}.(\mathbf{r} - \mathbf{r}_0) = 0$

$\Rightarrow$ $\mathbf{n}.\mathbf{r} = \mathbf{n}.\mathbf{r}_0$

which is called the *vector equation of the plane*.

If we now suppose that $\mathbf{n} = a\mathbf{i} + b\mathbf{j} + c\mathbf{k}$,   it follows that

$(a\mathbf{i} + b\mathbf{j} + c\mathbf{k}).(x\mathbf{i} + y\mathbf{j} + z\mathbf{k}) = \mathbf{n}.\mathbf{r}_0$

and if we let $\mathbf{n}.\mathbf{r}_0 = -d$,   we see that

$ax + by + cz + d = 0$

which is the *Cartesian equation of the plane*.

Example 3

Find the Cartesian and vector equations of the plane through the point $(1, 2, 3)$ which is perpendicular to the line

$$\frac{x - 1}{3} = \frac{y - 2}{2} = \frac{z - 3}{1}$$

The direction of this line is given by the vector $3\mathbf{i} + 2\mathbf{j} + \mathbf{k}$, and the vector from $(1, 2, 3)$ to (x, y, z) is

$$(x - 1)\mathbf{i} + (y - 2)\mathbf{j} + (z - 3)\mathbf{k}$$

As these must be perpendicular, it follows that

$$3(x - 1) + 2(y - 2) + (z - 3) = 0$$

$$\Rightarrow \qquad\qquad 3x + 2y + z - 10 = 0$$

is the Cartesian equation of the plane.

Alternatively, in vector form, if $\mathbf{r}$ is a point of the plane, $\mathbf{r} - (\mathbf{i} + 2\mathbf{j} + 3\mathbf{k})$ is perpendicular to $3\mathbf{i} + 2\mathbf{j} + \mathbf{k}$.

$$\Rightarrow \quad \{\mathbf{r} - (\mathbf{i} + 2\mathbf{j} + 3\mathbf{k})\}.\{3\mathbf{i} + 2\mathbf{j} + \mathbf{k}\} = 0$$

$$\Rightarrow \qquad\qquad \mathbf{r}.(3\mathbf{i} + 2\mathbf{j} + \mathbf{k}) = (\mathbf{i} + 2\mathbf{j} + 3\mathbf{k}).(3\mathbf{i} + 2\mathbf{j} + \mathbf{k})$$

$$\Rightarrow \qquad\qquad \mathbf{r}.(3\mathbf{i} + 2\mathbf{j} + \mathbf{k}) = 10$$

is the vector equation of the plane.

Exercise 8.4a

1   Find, in both parametric and Cartesian forms, the equations of
    a) the line joining $(0, 1, 2)$ and $(2, 1, 0)$;
    b) the line joining $(0, 1, 2)$ and $(-2, -1, 0)$.

2   Find the equations of:
    a) the plane which passes through $(1, 2, 3)$ and is perpendicular to the line $(x - 1)/2 = y + 1 = z/3$;
    b) the plane which passes through $(0, 1, 2)$ and is perpendicular to $\mathbf{i} + 2\mathbf{j} + 3\mathbf{k}$;
    c) the line through $(1, 1, 2)$ which is perpendicular to $2x + y + z = 4$.

3   Find the angles between the lines
    a) $\mathbf{r} = (1 + \lambda)\mathbf{i} + (1 - \lambda)\mathbf{j} + (2 + \lambda)\mathbf{k}$
    and $\mathbf{r} = (1 - \mu)\mathbf{i} + (1 - 2\mu)\mathbf{j} + (1 + \mu)\mathbf{k}$

    b) $\dfrac{x - 1}{2} = \dfrac{y - 2}{1} = \dfrac{z}{3}$ and $\dfrac{x}{3} = \dfrac{y + 1}{2} = \dfrac{z + 2}{1}$

4   Find whether the line through $(1, 0, 2)$ and $(4, 3, -1)$ meets the line through $(0, 3, -1)$ and $(5, -2, 4)$. Calculate the (acute) angle between their directions.   (SMP)

5   The rectangular Cartesian coordinates of two points A and B are $(1, 2)$ and $(-1, 4)$ respectively. Show that the column vector $\begin{pmatrix} t \\ t \end{pmatrix}$ is always normal to

AB and hence, or otherwise, find the coordinates of the foot of the perpendicular from the origin to AB. (C)

6 Lines l, m have respectively the parametric equations

$$\begin{pmatrix} x \\ y \\ z \end{pmatrix} = t \begin{pmatrix} 2 \\ 1 \\ -1 \end{pmatrix} + \begin{pmatrix} 0 \\ 1 \\ 3 \end{pmatrix} \qquad \begin{pmatrix} x \\ y \\ z \end{pmatrix} = u \begin{pmatrix} -2 \\ 1 \\ 1 \end{pmatrix} + \begin{pmatrix} 1 \\ 1 \\ -1 \end{pmatrix}$$

A is the point on l with parameter t_1, B the point on m with parameter u_1. Write down an expression for the vector **AB**.

Given that the line AB is perpendicular to both l and m, find the values of t_1 and u_1, and show that the length of AB is $7/\sqrt{5}$ units. (SMP)

Angles between planes and lines

We have already seen how to calculate the angle between two lines in space. We now proceed to calculate the angle between two planes; and, finally, the angle between a line and a plane.

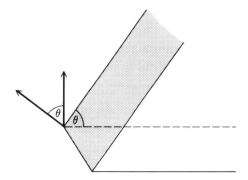

The angle between two planes can be defined as the angle between two lines in the planes both of which are perpendicular to the line of intersection of the planes. Hence it must also be the angle between the normals to the two planes.

Example 4

Find the angle between the planes

$$3x + 2y + z = 0 \quad \text{and} \quad x + 2y + 4 = 0$$

The directions of the normals to these planes are given by

$$3\mathbf{i} + 2\mathbf{j} + \mathbf{k} \quad \text{and} \quad \mathbf{i} + 2\mathbf{j}$$

and the angle between these two vectors is θ, where

$$\cos\theta = \frac{(3\mathbf{i} + 2\mathbf{j} + \mathbf{k}).(\mathbf{i} + 2\mathbf{j})}{\sqrt{14} \times \sqrt{5}}$$

$$= \frac{7}{\sqrt{70}} = \sqrt{0.7} = 0.8367$$

$$\Rightarrow \quad \theta = 33.2°$$

The angle between a line and a plane is defined as the angle (θ) between the line (l) and its projection (p) on the plane (π).

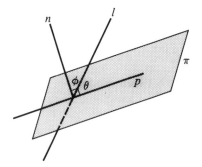

Hence it is conveniently calculated by finding the angle ϕ between the line and the normal n to the plane, and then taking its complement, $\frac{1}{2}\pi - \phi$.

Example 5

Find the angle between the line

$$\frac{x - 1}{2} = \frac{y - 2}{3} = \frac{z - 3}{4}$$

and the plane

$$4x + 3y + 2z + 1 = 0$$

Now the line is in the direction of the vector $2\mathbf{i} + 3\mathbf{j} + 4\mathbf{k}$ and the normal to the plane is in the direction $4\mathbf{i} + 3\mathbf{j} + 2\mathbf{k}$.

So the angle θ between these directions is given by

$$\cos\theta = \frac{(2\mathbf{i} + 3\mathbf{j} + 4\mathbf{k}).(4\mathbf{i} + 3\mathbf{j} + 2\mathbf{k})}{\sqrt{29} \times \sqrt{29}} = \frac{25}{29}$$

$$\Rightarrow \quad \theta = 30.5°$$

Hence the required angle is

$$90° - 30.5° = 59.5°$$

The distance from a point to a plane

Find the perpendicular distance from the point (x_1, y_1, z_1) to the plane

$$ax + by + cz + d = 0$$

Let $\mathbf{r}_1 = x_1\mathbf{i} + y_1\mathbf{j} + z_1\mathbf{k}$
       $\mathbf{r} = x\mathbf{i} + y\mathbf{j} + z\mathbf{k}$ be a point in the plane

and $\mathbf{n} = a\mathbf{i} + b\mathbf{j} + c\mathbf{k}$

Then $\mathbf{n.r} = ax + by + cz = -d = \text{constant}$

so that the projection of $\mathbf{r}$ in the direction of $\mathbf{n}$ is a constant.

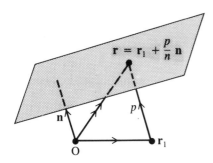

Hence $\mathbf{n}$ is normal to the plane. But the unit vector in this direction is $\mathbf{n}/n$; so if the required perpendicular distance is p, it follows that the foot of the perpendicular from P_1 to the plane is

$$\mathbf{r}_1 \pm p\frac{\mathbf{n}}{n} \quad \text{(the sign depending on the direction of } \mathbf{n})$$

But this point lies in the plane.

So $\mathbf{n.}\left(\mathbf{r}_1 \pm p\dfrac{\mathbf{n}}{n}\right) = -d$

$\Rightarrow$ $\mathbf{n.r}_1 \pm pn + d = 0$

$\Rightarrow$ $|p| = \dfrac{|\mathbf{n.r}_1 + d|}{n}$

Now $\mathbf{n.r}_1 = ax_1 + by_1 + cz_1$
and $n = \sqrt{(a^2 + b^2 + c^2)}$

So $|p| = \dfrac{|ax_1 + by_1 + cz_1 + d|}{\sqrt{(a^2 + b^2 + c^2)}}$

In exactly similar fashion in a plane, the perpendicular distance from a point

$P_1(x_1, y_1)$ to the line $ax + by + c = 0$ is given (see section A.1 on page 478) by

$$\frac{|ax_1 + by_1 + c|}{\sqrt{(a^2 + b^2)}}$$

So, for example, the distance of the point $(1, 2)$ from the line

$3x + 4y + 1 = 0$ is

$$\frac{|3 \times 1 + 4 \times 2 + 1|}{\sqrt{(3^2 + 4^2)}} = \frac{12}{5} = 2.4$$

Exercise 8.4b

1   Find pairs of vectors which are normal to the following planes, and hence the angles between them:
   a) $x + 2y + 3z = 6$ and $x - y - z = 4$
   b) $x - y - 3 = 0$ and $y + z - 1 = 0$

2   Find the angles between the following planes and lines:

   a) $3x + 2y + z = 1$ and $\dfrac{x - 1}{1} = \dfrac{y - 2}{2} = \dfrac{z - 3}{3}$

   b) $x + 2y - z = 3$ and $\dfrac{x - 2}{1} = \dfrac{y - 3}{-2} = \dfrac{z - 1}{4}$

3   **a)** Calculate the perpendicular distances from $(2, 1)$ to the lines (in the plane $z = 0$)

   $$3x + 4y + 5 = 0 \quad \text{and} \quad 5x - 12y + 15 = 0$$

   b) Find the locus of a point $P(x, y)$ which is equidistant from these two lines. Describe this locus.

4   **a)** Calculate the perpendicular distances from $(1, 2, 3)$ to the planes

   $$x + 2y + 2z + 7 = 0 \quad \text{and} \quad 2x + y + 2z - 1 = 0$$

   b) Find the locus of a point $P(x, y, z)$ which is equidistant from these two planes. Describe this locus.

5   A cube $ABCDA'B'C'D'$ has base $ABCD$ and vertical sides AA' etc. The mid-points of the edges AB, AD, AA' are respectively P, Q, R. The cube is cut along the plane PQR.
   a) Find the angle the edge AA' makes with the plane PQR.
   b) Find the angle the plane PQR makes with each face of the cube.

c) Determine the ratio of the volume of the tetrahedron APQR to that of the remainder of the cube.

If the cube is also cut along the plane AB′C, draw a diagram of the triangle PQR and mark in the line of this second cut.   (SMP)

6 Find the perpendicular distance of the point $(3, 0, 1)$ from the line whose Cartesian equation is

$$\frac{x-1}{3} = \frac{y+2}{4} = \frac{z}{12}$$ (MEI)

(First find the general point P of this line and then the condition for the vector from $(3, 0, 1)$ to P to be perpendicular.)

7 Find the perpendicular (i.e., shortest) distance between the two skew lines

$$\mathbf{r} = (\mathbf{i} - 3\mathbf{j} + 3\mathbf{k}) + \lambda(-2\mathbf{i} + \mathbf{j} - 2\mathbf{k})$$
$$\mathbf{r} = (3\mathbf{i} - \mathbf{j} + 2\mathbf{k}) + \mu(2\mathbf{i} - 3\mathbf{j} - 2\mathbf{k})$$

8 Find the equation of the plane through $A = (2, 2, -1)$ perpendicular to OA. Which point on this plane is closest to the point $(2, -1, 2)$?   (SMP)

9 The lines OA, OB and OC are mutually perpendicular, and are each of unit length. M is the mid-point of OA. Calculate:
a) the area of the triangle ABC;
b) the angle between the lines AC and BM;
c) the length of the perpendicular from M to the plane ABC.   (AEB, 1982)

Problems in three dimensions

Example 6

A path up a hillside makes $40°$ with its line of greatest slope. If the hillside is inclined at $20°$ to the horizontal, find the inclination of the path.

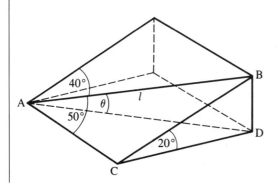

Let AB be a section of the path of length l and suppose that it is inclined at θ to the horizontal.

In the diagram

$$BC = l \sin 50°$$

$\Rightarrow \quad BD = BC \sin 20° = l \sin 50° \sin 20°$

$\Rightarrow \qquad \sin \theta = \dfrac{BD}{AB} = \dfrac{l \sin 50° \sin 20°}{l}$

$\Rightarrow \qquad \sin \theta = \sin 50° \sin 20°$

$\Rightarrow \qquad \theta = 15.2°$

Example 7

If ABCD is a regular tetrahedron, find:
a) the angle between AB and the base BCD;
b) the angle between any two faces.

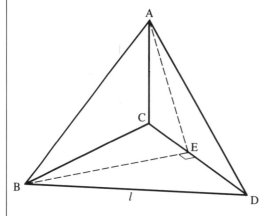

Suppose that each edge has length l and that E is the mid-point of CD

Then $\quad AE = BE = l \sin 60° = \dfrac{l\sqrt{3}}{2}$

Now a) the angle between AB and BCD is, by definition, the angle θ between AB and its projection BE on BCD
b) the angle between ACD and BCD is, by definition, the angle ϕ between AE and BE, since both these are perpendicular to the line of intersection CD.

Hence, the required angles can be calculated from the isosceles $\triangle ABE$:

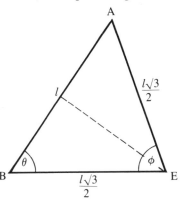

$$\cos \theta = \frac{l/2}{\sqrt{3}l/2} = \frac{1}{\sqrt{3}}$$

$$\Rightarrow \quad \theta = 54.74°$$

$$\Rightarrow \quad \phi = 180° - 2 \times 54.74°$$

$$= 70.52°$$

Exercise 8.4c

1 A man is walking up a hillside which is inclined at 20° to the horizontal, but chooses a path whose inclination to the horizontal is only 15°. What is the angle between this direction and the line of greatest slope?

2 A woman walks down a hillside in a direction which makes 45° with the line of steepest descent, and finds that the inclination of her path to the horizontal is 18°. What is the slope of the hillside?

3 Three points O, A, B lie on level ground. OA = OB = 5 m and AB = 6 m. A pole OT of length 6 m has its base at O and its other end T is held above the ground by two ropes AT and BT each of length 8 m. Find the angle that OT makes with the horizontal.   (AEB, 1981)

4 A tetrahedron has a horizontal equilateral triangular base of side 6 cm and the sloping edges are each of length 4 cm.
a) Find the height of the tetrahedron.
b) Find the inclination of a sloping edge to the horizontal.
c) Show that the sloping faces are inclined to one another at an angle $\cos^{-1}(-\frac{1}{7})$.   (AEB, 1981)

5 A pyramid VABCD stands on a square base ABCD and has four identical sloping faces meeting at the vertex V. Given that VA is 60 cm and $A\hat{V}B = 80°$, calculate:
  a) the length of AB;
  b) the length of VM, where M is the mid-point of AB;
  c) the angle that the face VAB makes with the horizontal;
  d) the perpendicular distance from A to VB;
  e) the angle between the faces VAB and VBC.   (AEB, 1983)

6 The rhombus ABCD of side 17 cm is the horizontal base of a pyramid VABCD. The vertex V is vertically above the point M where the diagonals AC and BD intersect. Given that VA = VC = 17 cm and that VB = VD = $\sqrt{128}$ cm, find
  a) AC and BD and verify that AC:BD = 15:8;
  b) the cosine of the angle between the planes VBA and ABCD.
                                                              (AEB, 1983)

8.5 Differentiation of vectors

Suppose that a moving point goes from position P at time t to position P′ at time $t + \delta t$.

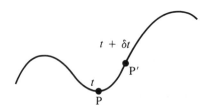

Its displacement in this interval is **PP′**, so we call $(1/\delta t)$**PP′** its *average velocity* during the interval; and as this is a scalar multiple of **PP′**, it also is a vector.

Now if, as $\delta t \to 0$ and P′ approaches P, the vector $(1/\delta t)$**PP′** (which we shall also denote by **PP′**$/\delta t$) tends to a certain limiting vector **v**, we call this limit the *velocity of P at the instant t*:

$$\mathbf{v} = \lim_{\delta t \to 0} \frac{\mathbf{PP'}}{\delta t}$$

We now refer the positions P and P′ to an origin O, and suppose that

at time t,   **OP** = **r**

at time $t + \delta t$,   **OP′** = **r** + δ**r**

Then $\quad \lim_{\delta t \to 0} \dfrac{\mathbf{PP'}}{\delta t} = \lim_{\delta t \to 0} \dfrac{\delta \mathbf{r}}{\delta t}$

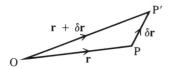

We naturally call this limit $\dfrac{d\mathbf{r}}{dt}$, so that $\mathbf{v} = \dfrac{d\mathbf{r}}{dt}$.

Similarly, if $\dfrac{\delta \mathbf{v}}{\delta t}$ tends to a limit as $\delta t \to 0$, we call this limit $\dfrac{d\mathbf{v}}{dt}$, the *acceleration* $\mathbf{a}$ of the point at time t.

So $\quad \mathbf{v} = \dfrac{d\mathbf{r}}{dt}$

and $\quad \mathbf{a} = \dfrac{d\mathbf{v}}{dt} = \dfrac{d^2\mathbf{r}}{dt^2}$

When any new operation, like differentiation of vectors, is introduced, it is essential to discover how it combines with other operations.

Now $\qquad \delta(\mathbf{r} + \mathbf{s}) = (\mathbf{r} + \delta\mathbf{r} + \mathbf{s} + \delta\mathbf{s}) - (\mathbf{r} + \mathbf{s}) = \delta\mathbf{r} + \delta\mathbf{s}$

So $\qquad \dfrac{\delta(\mathbf{r} + \mathbf{s})}{\delta t} = \dfrac{\delta\mathbf{r}}{\delta t} + \dfrac{\delta\mathbf{s}}{\delta t}$

$\Rightarrow \quad \lim_{\delta t \to 0} \dfrac{\delta(\mathbf{r} + \mathbf{s})}{\delta t} = \lim_{\delta t \to 0} \dfrac{\delta\mathbf{r}}{t} + \lim_{\delta t \to 0} \dfrac{\delta\mathbf{s}}{\delta t}$

$\Rightarrow \qquad \dfrac{d}{dt}(\mathbf{r} + \mathbf{s}) = \dfrac{d\mathbf{r}}{dt} + \dfrac{d\mathbf{s}}{dt}$

So the derivative of a vector sum is the vector sum of its derivatives; and similarly it can be proved that, if λ is a constant,

$$\dfrac{d}{dt}(\lambda\mathbf{r}) = \lambda\dfrac{d\mathbf{r}}{dt}$$

When we are considering such derivatives with respect to time it is frequently convenient to denote them by means of dots placed above their symbols.

So $\quad \mathbf{v} = \dfrac{d\mathbf{r}}{dt}$ is written $\dot{\mathbf{r}}$ and its components as $\dot{x}, \dot{y}, \dot{z}$.

Similarly $\quad \mathbf{a} = \dfrac{d\mathbf{v}}{dt} = \dfrac{d^2\mathbf{r}}{dt^2}$ is written $\ddot{\mathbf{r}}$ and its components as $\ddot{x}, \ddot{y}, \ddot{z}$.

Summarising,

$$\mathbf{r} = x\mathbf{i} + y\mathbf{j} + z\mathbf{k}$$
$$\Rightarrow \quad \mathbf{v} = \dot{\mathbf{r}} = \dot{x}\mathbf{i} + \dot{y}\mathbf{j} + \dot{z}\mathbf{k}$$
$$\Rightarrow \quad \mathbf{a} = \ddot{\mathbf{r}} = \ddot{x}\mathbf{i} + \ddot{y}\mathbf{j} + \ddot{z}\mathbf{k}$$

Example 1

The position vector of a particle after time t is given by

$$\mathbf{r} = 3t\mathbf{i} + 4t\mathbf{j} - 5t^2\mathbf{k}$$

Find its velocity and acceleration after time t and show that its acceleration is constant.

$$\mathbf{r} = 3t\mathbf{i} + 4t\mathbf{j} - 5t^2\mathbf{k}$$
$$\Rightarrow \quad \mathbf{v} = \dot{\mathbf{r}} = 3\mathbf{i} + 4\mathbf{j} - 10t\mathbf{k}$$
$$\Rightarrow \quad \mathbf{a} = \ddot{\mathbf{r}} = -10\mathbf{k}, \quad \text{which is constant}$$

Exercise 8.5

1   The position vector $\mathbf{r}$ of a particle after time t is given by

$$\mathbf{r} = t^2\mathbf{i} - 5t\mathbf{j} + (t^2 - 1)\mathbf{k}$$

Find expressions for its velocity $\mathbf{v}$ and acceleration $\mathbf{a}$ and calculate their magnitudes when $t = 2$.

2   The position vector $\mathbf{r}$ of a particle can be expressed in terms of the constant orthogonal unit vectors $\mathbf{i}, \mathbf{j}$ by means of the relation

$$\mathbf{r} = \mathbf{i} \cos nt + \mathbf{j} \sin nt$$

Find the speed of the particle at time $t = 2$.
Show that its acceleration is the vector $-n^2\mathbf{r}$.　(SMP)

3   A particle moves in a plane so that its coordinates (x, y) at any time t are $(t + \cos t, 1 + \sin t)$. Find the values of t for which it is stationary.
Show that the acceleration of the particle has constant magnitude.　(SMP)

4 The position vector **p** of a passenger in a roundabout at a fairground is given by

$$\mathbf{p} = 9\mathbf{i}\cos\frac{2t}{3} + 9\mathbf{j}\sin\frac{2t}{3} + \frac{9}{2}\mathbf{k}\left(1 - \cos\frac{4t}{3}\right)$$

where **i**, **j**, **k** are mutually perpendicular vectors each of length 1 m, **k** being vertical, t is the time measured in seconds, and the angles are measured in radians.

How long does it take to perform a complete revolution, and what happens in the vertical direction while this is taking place?

Find the acceleration of the passenger at time t. What is the magnitude of the greatest acceleration he experiences?   (SMP)

5 If a particle moving on a smooth horizontal plane is connected to the origin O by means of an elastic spring, it can be shown that its position vector **r** obeys the equation

$$\ddot{\mathbf{r}} + \omega^2\mathbf{r} = 0$$

where ω is a constant.

Verify that a possible solution of this differential equation is

$$\mathbf{r} = \mathbf{a}\sin\omega t + \mathbf{b}\cos\omega t$$

where **a**, **b** are constant vectors.

6 At a certain instant a body has a velocity of $3\,\mathrm{m\,s}^{-1}$ eastward and a constant acceleration of $2.5\,\mathrm{m\,s}^{-2}$ in a direction North 30° East. Find, graphically or otherwise, the magnitude and direction of its velocity two seconds later. When the body is moving in a direction due North-East find the magnitude of its velocity.   (MEI)

7 Two particles A and B start simultaneously from the origin. Particle A moves with constant velocity $2\mathbf{i} + 6\mathbf{j}$, while particle B has initial velocity $4\mathbf{i} - 8\mathbf{j}$ and a constant acceleration $\mathbf{i} + \mathbf{j}$.
   a) Find the position of each particle after t seconds;
   b) Find the time at which the velocities are perpendicular;
   c) If C is the point with position vector $17\mathbf{i} + 25\mathbf{j}$, find the time at which the line joining the positions of A and B passes through C.   (W)

8.6 Motion under gravity: projectiles

A stone is thrown horizontally out to sea from the top of a cliff. If its initial velocity is **u** and thereafter it has an acceleration **g** vertically downwards, find its position after time t.

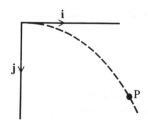

Taking unit vectors **i**, **j**, horizontally and vertically downwards, we see that

$$\ddot{\mathbf{r}} = g\mathbf{j}$$
$$\Rightarrow \quad \dot{\mathbf{r}} = gt\mathbf{j} + \mathbf{c} \quad \text{(where } \mathbf{c} \text{ is an arbitrary constant vector).}$$
But when $t = 0$, $\dot{\mathbf{r}} = u\mathbf{i}$

So $\mathbf{c} = u\mathbf{i}$

Hence $\dot{\mathbf{r}} = u\mathbf{i} + gt\mathbf{j}$

$$\Rightarrow \quad \mathbf{r} = ut\mathbf{i} + \tfrac{1}{2}gt^2\mathbf{j} + \mathbf{d} \quad \text{(where } \mathbf{d} \text{ is another arbitrary constant vector).}$$

But when $t = 0, \mathbf{r} = \mathbf{0}$

So $\mathbf{d} = \mathbf{0}$

Hence $\mathbf{r} = ut\mathbf{i} + \tfrac{1}{2}gt^2\mathbf{j}$

If, for instance, the stone was thrown with speed $20\,\mathrm{m\,s^{-1}}$, then $u = 20$ and $g \approx 10$.

So $\mathbf{r} \approx 20t\mathbf{i} + 5t^2\mathbf{j}$

$\dot{\mathbf{r}} \approx 20\mathbf{i} + 10t\mathbf{j}$

and $\ddot{\mathbf{r}} \approx 10\mathbf{j}$

and the position and velocity of the stone can be found at successive instants:

t	$\mathbf{r}$	$\dot{\mathbf{r}}$
0	**0**	$20\mathbf{i}$
1	$20\mathbf{i} + 5\mathbf{j}$	$20\mathbf{i} + 10\mathbf{j}$
2	$40\mathbf{i} + 20\mathbf{j}$	$20\mathbf{i} + 20\mathbf{j}$
3	$60\mathbf{i} + 45\mathbf{j}$	$20\mathbf{i} + 30\mathbf{j}$
4	$80\mathbf{i} + 80\mathbf{j}$	$20\mathbf{i} + 40\mathbf{j}$

More generally, we now consider the motion of a bullet which has been fired from a point O with velocity $\mathbf{u}$ at an angle α to the horizontal and which has constant acceleration $\mathbf{g}$ vertically downwards, air resistance being negligible.

Suppose that after time t the bullet is at a point P, with position vector $\mathbf{r}$.

Then $\qquad\qquad \ddot{\mathbf{r}} = \mathbf{g}$

$\Rightarrow \qquad\qquad \dot{\mathbf{r}} = t\mathbf{g} + \mathbf{c}$ (where $\mathbf{c}$ is a constant vector)

But when $t = 0$, $\quad \dot{\mathbf{r}} = \mathbf{u}, \quad$ so $\mathbf{c} = \mathbf{u}$

Hence $\qquad\qquad \dot{\mathbf{r}} = \mathbf{u} + t\mathbf{g}$

$\Rightarrow \qquad\qquad \mathbf{r} = t\mathbf{u} + \frac{1}{2}t^2\mathbf{g} + \mathbf{d}$ (where $\mathbf{d}$ is another constant vector).

But when $t = 0$, $\quad \mathbf{r} = \mathbf{0}, \quad$ so $\mathbf{d} = \mathbf{0}$

Hence $\qquad\qquad \mathbf{r} = t\mathbf{u} + \frac{1}{2}t^2\mathbf{g}$

Summarising,

$$\mathbf{r} = t\mathbf{u} + \tfrac{1}{2}t^2\mathbf{g}$$
$$\dot{\mathbf{r}} = \mathbf{u} + t\mathbf{g}$$
$$\ddot{\mathbf{r}} = \mathbf{g}$$

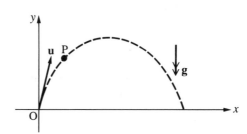

Alternatively, we could take axes Ox, Oy horizontally and vertically upwards, and carry out the above work in components:

$\qquad \ddot{\mathbf{r}} = \mathbf{g}$

$\Rightarrow \quad \ddot{x} = 0 \qquad \ddot{y} = -g$

Integrating with respect to t,

$\qquad \dot{x} = a \qquad \dot{y} = -gt + b$

where a and b are constants.

But when $t = 0$, $\quad \dot{x} = u\cos\alpha \quad$ and $\qquad \dot{y} = u\sin\alpha$

So $\qquad\qquad u\cos\alpha = a \qquad\qquad u\sin\alpha = b$

$\Rightarrow \qquad\qquad \dot{x} = u\cos\alpha \qquad\qquad \dot{y} = u\sin\alpha - gt$

Integrating again, we obtain

$$x = ut\cos\alpha + c \qquad y = ut\sin\alpha - \tfrac{1}{2}gt^2 + d$$

But when $t = 0$, $x = 0$ and $y = 0$

So $c = d = 0$

$\Rightarrow$ $x = ut \cos \alpha$ $y = ut \sin \alpha - \frac{1}{2}gt^2$

Summarising, we see that the components of displacement, velocity, and acceleration are

$$x = ut \cos \alpha \qquad y = ut \sin \alpha - \tfrac{1}{2}gt^2$$
$$\dot{x} = u \cos \alpha \qquad \dot{y} = u \sin \alpha - gt$$
$$\ddot{x} = 0 \qquad \ddot{y} = -g$$

Now the bullet hits the ground when $y = 0$

$\Rightarrow$ $ut \sin \alpha - \frac{1}{2}gt^2 = 0$

$\Rightarrow$ $t = 0$ or $\dfrac{2u \sin \alpha}{g}$

So the *time of flight* is $(2u \sin \alpha)/g$, and when $t = (2u \sin \alpha)/g$,

$$x = u \frac{2u \sin \alpha}{g} \cos \alpha = \frac{u^2 \sin 2\alpha}{g}$$

This is therefore the *horizontal range* of the projectile and, if α is allowed to vary, this range is greatest when

$$\sin 2\alpha = 1 \quad \Rightarrow \quad \alpha = \tfrac{1}{4}\pi$$

So the *maximum range* is u^2/g.

We also see that

$$x = ut \cos \alpha, \qquad y = ut \sin \alpha - \tfrac{1}{2}gt^2$$

provide us with parametric equations for the trajectory, and that we can eliminate t to obtain its x, y equation.

For $t = \dfrac{x}{u \cos \alpha}$

$\Rightarrow$ $y = u \dfrac{x}{u \cos \alpha} \sin \alpha - \tfrac{1}{2}g \dfrac{x^2}{u^2 \cos^2 \alpha}$

$\Rightarrow$ $$y = x \tan \alpha - \frac{gx^2}{2u^2} \sec^2 \alpha$$

which is a parabola.

Exercise 8.6

(Take $g = 10\,\mathrm{m\,s^{-2}}$.)

1 A golfer strikes a ball with velocity $25\mathbf{i} + 15\mathbf{j}$ (using SI units and unit vectors $\mathbf{i}, \mathbf{j}$ horizontally and vertically). Find:
   a) its position vector $\mathbf{r}$ after time t;
   b) its velocity $\mathbf{v}$ at this instant;
   c) its time of flight before pitching on to a horizontal fairway;
   d) its horizontal range;
   e) its maximum height.

2 A ball is thrown with velocity $20\,\mathrm{m\,s^{-1}}$ at an angle of $60°$ to the horizontal. Find in terms of unit vectors $\mathbf{i}, \mathbf{j}$, horizontally and vertically:
   a) its displacement and velocity after $t\,\mathrm{s}$;
   b) its speed after $2\,\mathrm{s}$ and the direction in which it is travelling;
   c) when it is travelling horizontally.

3 If a man could throw a ball at $30\,\mathrm{m\,s^{-1}}$ in any direction, find:
   a) the greatest height it could reach;
   b) its greatest horizontal range.

4 The record hit at cricket is reputedly (and incredibly) $175\,\mathrm{yd}$, or approximately $160\,\mathrm{m}$. What must have been the least initial speed of the ball for such a feat?

5 A man throws a ball with speed u at an angle α to the horizontal. Prove that
   a) its maximum height is $(u^2 \sin^2 \alpha)/2g$;
   b) it greatest horizontal range is four times the maximum height it would reach during such a trajectory.

6 A golfer strikes a ball so that its initial velocity makes an angle $\tan^{-1}\frac{1}{2}$ with the horizontal. If the range of the ball on the horizontal plane through the point of projection is $156.8\,\mathrm{m}$, calculate the greatest height of the ball above the plane and its time of flight.   (OC)

7 A particle is projected from a point O with speed V at an angle of elevation α. Prove that the equation of its trajectory referred to horizontal and vertical axes through O is

$$y = x \tan \alpha - \frac{gx^2}{2V^2} \sec^2 \alpha$$

Taking $g = 10$, find the two values of $\tan \alpha$ so that a particle projected with $V = 70$ will pass through the point $(280, 40)$. Find the ranges, on the horizontal plane through O, corresponding to these two angles of projection.   (JMB)

8 A ball is kicked from a point A on level ground and hits a goalpost at a point 4 m above the ground. The goalpost is at a distance of 32 m from A. Initially the velocity of the ball makes an angle $\tan^{-1}\frac{3}{4}$ with the ground. Show that the initial speed of the ball is 20 m s^{-1}. Find the speed of the ball when it hits the goalpost. (L)

9   A golf ball is struck from a point A, leaving A with speed 30 m s^{-1} at an angle of elevation θ, and lands, without bouncing, in a bunker at a point B, which is at the same horizontal level as A. Before landing in the bunker, the ball just clears the top of a tree which is at a horizontal distance of 72 m from A, the top of the tree being 9 m above the level of AB. Show that one of the possible values of θ is $\tan^{-1}\frac{3}{4}$ and find the other value. Given that θ was in fact $\tan^{-1}\frac{3}{4}$, find the distance AB. (L)

8.7 Motion in a circle

Angular velocity

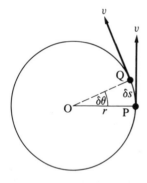

Suppose that a point is moving round a circle of radius r with constant speed v.

Further, suppose that in a brief time δt, the point moves from P to Q through an angle $\delta\theta$ and distance δs.

Then $r\,\delta\theta \approx \delta s \approx v\,\delta t$

$\Rightarrow$ $\dfrac{d\theta}{dt} = \dfrac{v}{r}$

The quantity $\dfrac{d\theta}{dt}$ (or $\dot\theta$) is called the *angular velocity of* P *about* O, and is commonly denoted by ω.

So $\omega = \dot\theta = \dfrac{v}{r}$ and $v = r\dot\theta = r\omega$

Acceleration of a particle moving in a circle with constant speed

Method 1

Although the speed of P is constant, its *velocity* (a vector) is always changing direction. So we can now proceed to find its rate of change, which is the acceleration of P.

Suppose that the velocity at P is $\mathbf{v}$ and at Q is $\mathbf{v} + \delta\mathbf{v}$.

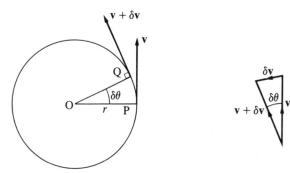

Now the magnitudes of $\mathbf{v}$ and $\mathbf{v} + \delta\mathbf{v}$ are the same, and we see from the right-hand diagram that $\delta\mathbf{v}$ is almost perpendicular to $\mathbf{v}$ and has magnitude $v\,\delta\theta$.

So $\dfrac{\delta\mathbf{v}}{\delta t}$ is almost perpendicular to $\mathbf{v}$ and has magnitude $v\dfrac{\delta\theta}{\delta t}$.

In the limit, $\mathbf{a} = \dfrac{d\mathbf{v}}{dt}$ is perpendicular to $\mathbf{v}$ and has magnitude $v\dfrac{d\theta}{dt} = \dfrac{v^2}{r}$.

So a particle moving in a circle with constant speed v has acceleration $\dfrac{v^2}{r}$ directed towards its centre.

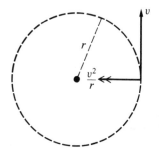

If the angular velocity of the point is ω, then $v = r\omega$, so that this acceleration can also be expressed as $r\omega^2$.

These are such important results that we shall now derive them by another method.

Method 2

Suppose that P is moving round a circle with constant angular velocity ω, and that initially **r** was along **i**.

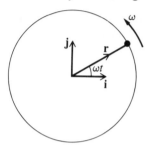

Then, after time t,

$$\mathbf{r} = r\cos\omega t\,\mathbf{i} + r\sin\omega t\,\mathbf{j}$$

$$\Rightarrow \quad \mathbf{v} = \dot{\mathbf{r}} = -r\omega\sin\omega t\,\mathbf{i} + r\omega\cos\omega t\,\mathbf{j}$$

and $\quad v = \sqrt{(r^2\omega^2\sin^2\omega t + r^2\omega^2\cos^2\omega t)} = r\omega$

Furthermore,

$$\mathbf{a} = \ddot{\mathbf{r}} = -r\omega^2\cos\omega t\,\mathbf{i} - r\omega^2\sin\omega t\,\mathbf{j}$$

$$= -\omega^2\mathbf{r}$$

Hence the acceleration of P is directed towards O and has magnitude

$$r\omega^2 = \frac{v^2}{r}$$

Summarising, for motion in a circle of radius r at constant angular velocity ω, the magnitudes of velocity and acceleration are

$$v = r\omega \quad \text{and} \quad a = r\omega^2 = \frac{v^2}{r}$$

Example 1

A motor-cyclist is going round a Wall of Death at a speed of $50\,\mathrm{km\,h^{-1}}$ in a horizontal circle of radius 20 m. What is his acceleration?

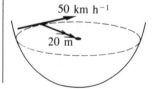

Using SI units

$$\text{speed} = v = 50\,\text{km}\,\text{h}^{-1} = \frac{50 \times 10^3}{3600}\,\text{m}\,\text{s}^{-1} \approx 14\,\text{m}\,\text{s}^{-1}$$

$$\text{radius} = r = 20\,\text{m}$$

Hence $\text{acceleration} = \dfrac{v^2}{r} \approx \dfrac{14^2}{20} = 9.8\,\text{m}\,\text{s}^{-2}$

(and, by sheer coincidence, is the same in magnitude as he would experience if he were falling under gravity).

We shall soon see that any such acceleration necessitates a force on the motor-cycle in the same direction, which is provided by a combination of its weight and the thrust of the wall on its tyres.

Example 2

The Earth is moving round the Sun in an orbit which is approximately a circle of radius 1.50×10^8 km. What is its acceleration?

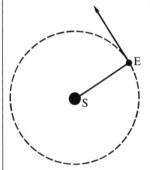

The Earth takes a year (of roughly 365 days) to complete the circumference which is approximately

$$2\pi \times 1.50 \times 10^{11}\,\text{m} = 9.43 \times 10^{11}\,\text{m}$$

So the Earth's speed $= \dfrac{9.43 \times 10^{11}}{365 \times 24 \times 3600}\,\text{m}\,\text{s}^{-1}$

$$= 2.99 \times 10^4\,\text{m}\,\text{s}^{-1}$$

Hence its acceleration $= \dfrac{v^2}{r} = \dfrac{(2.99 \times 10^4)^2}{1.50 \times 10^{11}}\,\text{m}\,\text{s}^{-2}$

$$\approx 0.006\,\text{m}\,\text{s}^{-2}$$

Again, this acceleration necessitates a force towards the Sun, called *gravitational attraction*.

Exercise 8.7

1   What is the acceleration of a particle which moves in a circle
   a) with radius 2 m and constant speed $3\,\mathrm{m\,s^{-1}}$;
   b) with radius 1 km and constant speed $20\,\mathrm{m\,s^{-1}}$;
   c) with radius 4 m and constant angular velocity $2\,\mathrm{rad\,s^{-1}}$;
   d) with radius 1 cm and constant angular velocity $20\,\mathrm{rad\,s^{-1}}$;
   e) with radius 1 m and constant angular velocity $80\,\mathrm{rev\,min^{-1}}$;
   f) with radius 2 cm and constant angular velocity $100\,\mathrm{rev\,min^{-1}}$?

2   A supersonic jet is travelling at a constant speed of $500\,\mathrm{m\,s^{-1}}$ and turning in a horizontal circle of radius 50 km. What is its acceleration? Find the turning radius of a light aircraft whose pilot experiences the same acceleration when travelling at $50\,\mathrm{m\,s^{-1}}$.

3   A pilot can loop the loop without using a safety harness providing that his vertical acceleration at the top of the loop is greater than $g(10\,\mathrm{m\,s^{-2}})$. If the diameter of the loop is $\frac{1}{2}$ km, what is the minimum speed for safety?

4   A woman swings a small bucket of water round and round in a vertical circle of radius 1 m at a constant speed of 1 rev. every 2 s. Show that the magnitude of its acceleration is slightly greater than g, so that the water just stays in the bucket.

* 8.8 Relative motion

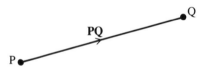

Suppose that P, Q are two moving points. Then **PQ** represents the position vector of Q relative to P, and $\dfrac{\mathrm{d}}{\mathrm{d}t}(\mathbf{PQ})$ represents its rate of change, which we call the *velocity of Q relative to P*, written $\mathbf{v}_P(Q)$.

So $\mathbf{v}_P(Q) = \dfrac{\mathrm{d}}{\mathrm{d}t}(\mathbf{PQ})$

If we now consider three moving points, P, Q, R,

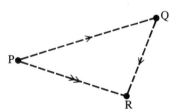

PQ + QR = PR

Hence $\dfrac{d}{dt}(PQ) + \dfrac{d}{dt}(QR) = \dfrac{d}{dt}(PQ + QR) = \dfrac{d}{dt}(PR)$

$\Rightarrow$ $\boxed{\mathbf{v}_P(Q) + \mathbf{v}_Q(R) = \mathbf{v}_P(R)}$

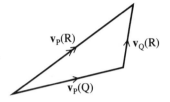

Furthermore, the rate of change of $\mathbf{v}_P(Q)$ can be called the *acceleration of Q relative to P*, written $\mathbf{a}_P(Q)$.

So $\mathbf{a}_P(Q) = \dfrac{d}{dt}\mathbf{v}_P(Q)$

and it immediately follows that

$\boxed{\mathbf{a}_P(Q) + \mathbf{a}_Q(R) = \mathbf{a}_P(R)}$

Example 1

A train is travelling at $40\,\text{m s}^{-1}$ and rain is falling vertically at a speed of $10\,\text{m s}^{-1}$. Find the velocity of the rain relative to the train and hence the direction of its streaks on the carriage windows.

Letting T, R, G denote the train, rain, and ground respectively, and taking unit vectors $\mathbf{i}, \mathbf{j}$ horizontally and vertically,

$$\mathbf{v}_G(T) = 40\mathbf{i}$$
$$\Rightarrow \quad \mathbf{v}_T(G) = -40\mathbf{i}$$
$$\text{and} \quad \mathbf{v}_G(R) = -10\mathbf{j}$$

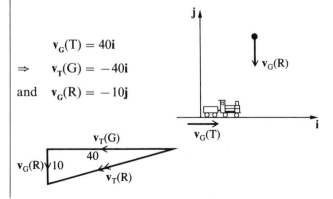

So $v_T(R) = v_T(G) + v_G(R)$ and we see that the streaks of rain make an angle θ with the horizontal, where $\tan \theta = \frac{10}{40} = \frac{1}{4}$

$\Rightarrow \quad \theta = 14°$

Example 2

A ship, which steams at $18 \, \text{km h}^{-1}$, has to travel North-East through water in which there is a current flowing from the West at $4 \, \text{km h}^{-1}$. Find, by drawing or calculation, the direction in which the ship should be headed and the actual speed of the ship towards the North-East.

Letting S, W, L represent the ship, water, and land respectively, we know that

$v_W(S)$ had magnitude $18 \, \text{km h}^{-1}$
$v_L(W)$ has magnitude $4 \, \text{km h}^{-1}$ due East
$v_L(S)$ has to be in the direction North-East

But $v_W(S) = v_W(L) + v_L(S)$

By drawing this to scale, we see that $\theta = 126°$, so that the ship should be set on course $036°$ and $v_L(S) = 20.6$, which is its actual speed towards North-East.

Alternatively, if we let $v_L(S) = v$, we see from the cosine rule that

$18^2 = v^2 + 4^2 - 2 \times v \times 4 \cos 45°$

$\Rightarrow \quad v^2 - 4\sqrt{2}v - 308 = 0$

$\Rightarrow \quad v = 2\sqrt{2} \pm \sqrt{(8 + 308)}$

$\qquad = 2\sqrt{2} \pm \sqrt{316}$

$\qquad = 2.83 \pm 17.77 = 20.6 \quad (\text{or } -14.94)$

Hence the ship should be headed on course $036°$ and her actual speed will be $20.6 \, \text{km h}^{-1}$.

Exercise 8.8

1   A motor-boat is steered due North and has a speed of $12 \, \text{km h}^{-1}$ through the water, in which there is a tide of $5 \, \text{km h}^{-1}$ running from East to West. Obtain by calculation the direction in which the boat travels and the time the boat will take to travel $6\frac{1}{2}$ km in this direction.   (SMP)

2 A canoeist, who can paddle at $2\,\mathrm{m\,s^{-1}}$ in still water, is on a river which is flowing at $1.5\,\mathrm{m\,s^{-1}}$. She wishes to make a 'ferry glide', that is to go straight across the river on a track perpendicular to the banks and current. By calculation or scale drawing find the direction in which she must paddle.

How long will she take to cross if the river is $60\,\mathrm{m}$ wide?   (SMP)

3 On a day when there is a $20\,\mathrm{km\,h^{-1}}$ breeze blowing from the South-West, a cyclist is travelling in a direction $110°$ at $24\,\mathrm{km\,h^{-1}}$. By scale drawing, find the direction from which the wind appears to the cyclist to be coming.   (SMP)

4 A destroyer with speed $60\,\mathrm{km\,h^{-1}}$ sights an enemy ship $10\,\mathrm{km}$ due East proceeding on a course $120°$ at $20\,\mathrm{km\,h^{-1}}$. Find, by drawing, the course the destroyer should take to intercept the enemy.   (SMP)

5 A man travelling due East finds that the wind appears to blow from the North. On doubling his velocity he finds that the wind appears to blow from the North-East. Find the true direction from which the wind is blowing.   (SMP)

6 A ship X, whose maximum speed is $30\,\mathrm{km\,h^{-1}}$, is due South of a ship Y which is sailing at a constant speed of $36\,\mathrm{km\,h^{-1}}$ on a fixed course of $150°$. If X sails immediately to intercept Y as soon as possible, find the required course.

If initially the ships are $50\,\mathrm{km}$ apart, find to the nearest minute how long it takes X to reach Y.   (L)

7 When a boat travels due West at $v\,\mathrm{km\,h^{-1}}$, the wind appears to come from the North-North-West. When the boat travels North-West at the same speed, the wind appears to come from the North. Find the speed and direction of the wind.

When the boat travels due South at $v\,\mathrm{km\,h^{-1}}$, from what direction does the wind appear to come?   (OC)

8 A hawk flying slightly above level ground sees a mouse $100\,\mathrm{m}$ due West. The mouse is running Northwards at a speed of $5\,\mathrm{m\,s^{-1}}$, and there is a $10\,\mathrm{m\,s^{-1}}$ wind blowing *from* the South-West. The hawk flies in a straight line and catches the mouse after $5\,\mathrm{s}$. Find the magnitude and direction of its velocity relative to the air.   (OC)

9 The equation of the path of a particle P is $\mathbf{r} = \mathbf{i}t + \mathbf{k}t^2$, where t is the time. Show that the acceleration of P is constant.

The velocity of another particle Q relative to P is $(\mathbf{i} - \mathbf{j})$ and when $t = 0$, $\mathbf{PQ} = \mathbf{j}$. Find the equation of the path of Q and the time at which Q is nearest to P.   (L)

10 The flight controller at an airport is about to 'talk-down' an aircraft A whose position vector relative to the control tower is $(10\mathbf{i} + 20\mathbf{j} + 5\mathbf{k})$ km and whose constant velocity is $(-210\mathbf{i} - 50\mathbf{j})$ km h^{-1}.

At this instant a second aircraft B appears on his radar screen with position vector $(-20\mathbf{i} - 10\mathbf{j} + 3\mathbf{k})$ km and constant velocity $(150\mathbf{i} + 250\mathbf{j} + 60\mathbf{k})$ km h^{-1}. Find:

a) the velocity of A relative to B;

b) the position vector of A relative to B t min after B first appeared on the radar screen;

c) the time that elapses, to the nearest 10 s, until the two aircraft are nearest to one another. (JMB)

11 A particle P moves in a horizontal circle with fixed centre O and radius 3 m with angular velocity $\frac{1}{4}\pi$ rad s^{-1} in a clockwise sense. A second particle Q moves in the same plane in a concentric circle of radius 2 m with angular velocity $\frac{1}{2}\pi$ rad s^{-1} in a clockwise sense. Initially O, P, and Q are collinear with P and Q due North of O. Find the magnitude and direction of the velocity of P relative to Q **a)** after 2 s and **b)** after 3 s.

(A graphical method may be used.) (C)

12 Two trains T$_1$ and T$_2$ are moving on perpendicular tracks which cross at the point O. Relative to O, the position vectors of T$_1$ and T$_2$ at time t are $\mathbf{r}_1$ and $\mathbf{r}_2$ respectively, where

$$\mathbf{r}_1 = Vt\mathbf{i}, \quad \mathbf{r}_2 = 2V(t - t_0)\mathbf{j}$$

Here $\mathbf{i}$ and $\mathbf{j}$ are unit vectors, and V and t_0 are positive constants. Which train goes through O first, and how much later does the other train go through O?

Show that the trains are closest together when $t = 4t_0/5$, and calculate their distance apart at this time. Draw a diagram to show the positions of the trains at this time; show also the directions in which they are moving. (MEI)

13 At time $t = 0$ a ship A is at the point O and a ship B is at the point with position vector $10\mathbf{j}$ nautical miles referred to O. The velocities of the two ships are constant. Ship A sails at 17 knots, where 1 knot is 1 nautical mile per hour, in the direction of the vector $8\mathbf{i} + 15\mathbf{j}$ and ship B sails at 15 knots in the direction of the vector $3\mathbf{i} + 4\mathbf{j}$. Write down

a) the velocity vector of each ship;

b) the velocity of B relative to A;

c) the position vector of B relative to A at time t hours.

Given that visibility is 5 nautical miles, show that the ships are within sight of each other for $\sqrt{6}$ hours. (L)

14 At time t two points P and Q have position vectors $\mathbf{p}$ and $\mathbf{q}$ respectively, where

$$\mathbf{p} = 2a\mathbf{i} + (a\cos \omega t)\mathbf{j} + (a\sin \omega t)\mathbf{k},$$

$$\mathbf{q} = (a \sin \omega t)\mathbf{i} - (a \cos \omega t)\mathbf{j} + 3a\mathbf{k}$$

and a, ω are constants. Find $\mathbf{r}$, the position vector of P relative to Q, and $\mathbf{v}$, the velocity of P relative to Q. Find also the values of t for which $\mathbf{r}$ and $\mathbf{v}$ are perpendicular.

Determine the smallest and greatest distances between P and Q.   (L)

Miscellaneous problems

1   If P, Q are the mid-points of AB, CD respectively, prove that
   a) $\mathbf{AC} + \mathbf{BD} = 2\mathbf{PQ}$
   b) $\mathbf{AC} + \mathbf{BC} + \mathbf{AD} + \mathbf{BD} = 4\mathbf{PQ}$

2   If the opposite sides of a hexagon are parallel, what can you prove about its diagonals? Need the hexagon be plane?

3

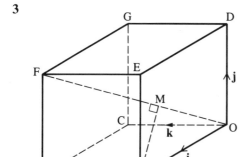

The vertices of a unit cube are labelled as shown in the figure. Show that the perpendiculars from the six vertices A, B, C, D, E, G meet the diagonal OF in just two points, which trisect OF.

Let M be the trisection point that corresponds to the vertex A. Find the position vector (in terms of $\mathbf{i}, \mathbf{j}, \mathbf{k}$) of the point at which the line BM, when extended, leaves the cube.

Suppose that AB = BM and calculate the angles of the triangle ABM.
    (SMP)

4   A particle is moving in a plane in such a way that

$$\mathbf{r} = \mathbf{a} \cos nt + \mathbf{b} \sin nt$$

where n is constant and $\mathbf{a}$ and $\mathbf{b}$ are fixed non-parallel vectors. Show that the particle describes an ellipse, and that the force acting on it is directed towards the centre O of the ellipse and is proportional to the distance of the particle from O.   (OC)

5 The coordinates of a moving point P are

$$(a \cos \omega t, a \sin \omega t)$$

where a and ω are constants. Find the magnitude and direction of the velocity and acceleration of P at time t.

There is a speed limit of 80 km h^{-1} round an unbanked curve on a railway track. The curve is in the shape of an arc of a circle and the angle turned through is $30°$ in 500 m. Find:

a) the radius of the circle;

b) the lateral force on the rails at maximum speed as a fraction of the weight;

c) the least distance before the curve that a warning sign must be placed if the trains cruise at 150 km h^{-1} and a comfortable braking retardation is $\frac{1}{2} \text{m s}^{-2}$.   (MEI)

6 A shell is projected from O with fixed speed v and the initial gradient of its trajectory is m. Show that its path is

$$y = mx - \frac{g(1 + m^2)}{2v^2} x^2$$

Hence show that possible initial directions of a trajectory that goes through (x, y) are given by the roots of the quadratic equation

$$gx^2 m^2 - 2v^2 xm + (gx^2 + 2v^2 y) = 0$$

and that there are two such trajectories provided that

$$y < \frac{v^2}{2g} - \frac{gx^2}{2v^2}$$

What happens if (x, y) is such that this inequality is not satisfied?

Sketch the *parabola of safety*,

$$y = \frac{v^2}{2g} - \frac{gx^2}{2v^2},$$

showing its relationship to the various trajectories for different values of m. Why is it important?

7 A particle starts from a point A whose position vector is **a** with velocity **u** under a constant acceleration **g**. Obtain an expression for the position vector of the particle after time t.

Two particles move freely under gravity; show that their relative velocity remains constant.

Two boys stand on level ground, $b \text{ m}$ apart. One throws a stone vertically upwards with velocity $u \text{ m s}^{-1}$, the other waits until this stone is at its highest point then throws a stone directly at it with speed $v \text{ m s}^{-1}$.

Show that the stones will collide provided v is big enough and find the lower limit for v. Find u to make this lower limit least.

List the assumptions you are making.   (MEI)

8 A river flows between parallel banks which are at a distance $2a$ apart. The speed of the current at any point is proportional to the distance, x, of the point from the nearer bank and has the value U at midstream. A motor boat whose speed is V in still water ($V > U$) crosses the river.

a) If the boat is steered so as always to point in a direction perpendicular to the current, find how far it is carried downstream in making the complete crossing.

b) If the boat is steered so as to cross in a straight line from a point on one bank to a point directly opposite to it on the other bank, find an expression for the retardation of the boat in terms of U, x and a during the first half of the crossing and calculate the time taken to cross the river completely.   (JMB)

9 An aircraft whose speed in still air is $V \, \text{km h}^{-1}$ describes a circuit which may be taken as a horizontal square of side a km. A wind of magnitude $v \, \text{km h}^{-1}$ (where $v < V$) blows parallel to one of the sides of the square. Assuming that no time is lost on corners, find the time taken by the aircraft to complete the circuit.   (OC)

10 An aircraft flies a distance a due North at a constant airspeed V through air which is blowing from the North East with constant speed $v \, (v < V)$.

Find the ratio of the time of the outward flight to that of the return flight under the same wind conditions.   (MEI)

11 An aircraft flying in a straight line is climbing at an angle ϕ to the horizontal. It is observed from a point O on horizontal ground. The aircraft passes through a point A′ vertically above a point A on the ground due east of O and, at a later time, through a point B′ vertically above a point B on the ground due north of O, where OA = OB. When the aircraft is at A′, its angle of elevation from O is α and when at B′, its angle of elevation from O is β. Prove that

$$\tan \phi = \frac{1}{\sqrt{2}} (\tan \beta - \tan \alpha)$$

Show further that, when the aircraft is vertically above the mid-point of AB, its angle of elevation, γ, from O is given by

$$\tan \gamma = \frac{1}{\sqrt{2}} (\tan \alpha + \tan \beta)$$ (C)

9 Introduction to mechanics

9.1 Historical introduction: Newton's first law

Nature, and Nature's laws, lay hid in night.
God said 'Let Newton be!' and all was light.

It is a commonplace to speak of the Scientific Revolution of the seventeenth century, and in many ways this is an oversimplification. Much was known before 1600, much else remained unknown in 1700, and such movements rarely fall neatly into centuries. Nevertheless, and particularly in the science of mechanics, the century witnessed a revolution whose importance can hardly be exaggerated.

In 1600, men's understanding of motion depended on a tradition of nearly two thousand years, since the writings of Aristotle in the fourth century BC. They believed that earthly objects have a natural desire to fall to the ground, but can be diverted from this tendency if an *impetus* is imparted to them. When a stone is thrown it is given an impetus which, until exhausted, causes it to defy its natural destiny. As for the movements of the planets, they are subject to quite different laws. They are celestial objects and so move in orbits composed of systems of perfect circles, and are propelled along their heavenly spheres by supernatural forces.

In 1700, however, man's understanding of motion was utterly different. Thirteen years earlier, Isaac Newton had published his great work, *Philosophiae naturalis principia mathematica*: 'The mathematical foundations of science', or 'Theoretical mechanics', as it might now be called. This was the culmination of twenty years' work, ranging from the discovery of the Calculus to optical investigations, the fundamental laws of motion and the law of universal gravitation. The first volume of *Principia* was devoted to the laws of motion, universal gravitation, and the orbits of the planets. In the second volume Newton discussed the motion of a body in a resisting medium, the motion of fluids and of waves. His third volume began:

In the preceding books I have laid down the principles of philosophy; principles not philosophical but mathematical. But lest they should appear dry and barren, I have illustrated them here and there, giving an account of such things as the density and the resistance of bodies and the motion of light and sounds. It remains that, from the same principles, I now demonstrate the frame of the system of the World.

Newton then set out to investigate the movement of satellites round their planets, to calculate the mass of the Sun and of the planets which have satellites, to explain and calculate the flattened shape of the Earth and the wobble of its axis, to discuss the irregularities of the Moon's motion due to the attraction of the Sun, to inaugurate the study of tides, to explain the movements of comets and to calculate their return. The inscription beneath his statue in Cambridge, '*Genus humanum ingenio superavit*' was no exaggeration.

What then were Newton's laws of motion, why were they so revolutionary, and how were they discovered? The laws themselves, like the rest of *Principia*, were in Latin, and can be translated as:

a) Every object remains in a state of rest or of uniform motion in a straight line, unless a force is acting on it.

b) When an object is accelerating there is a force acting upon it which is proportional to the acceleration and in the same direction.

c) If two objects act upon each other, the forces between them are equal and opposite.

Before considering the significance of these laws, a preliminary remark should be made about the word 'object'. Newton meant by this a quantity of matter which can be regarded as concentrated at a point, and which we shall usually call a *particle*. A particle, therefore, is any object whose dimensions are small compared with all other distances under consideration; in one instance this might be a bullet moving through the air or, in another, the Earth relative to the rest of the solar system.

What, then, was so revolutionary about these laws, and why are they of such immense importance?

Perhaps the two most striking features occur in the first law:

'Every object'. The very first two words marked the end of the ancient distinction between the mechanics of earthly objects and the mechanics of heavenly objects: the motion of every object, terrestrial and celestial, was to be subject to the same laws.

'In a state of rest or of uniform motion in a straight line.' No longer, when confronted with an object moving with uniform velocity, were men to look for a force or 'impetus' sustaining its motion. Forces are to be sought not when there is motion, but only when there is *change* of motion.

This was the major revolution, and its principal agents were the concepts of force and acceleration. So far as the latter is concerned, it is assumed that we always know the acceleration of an object. But a little reflection shows that this is so only relative to a particular frame of reference. Are we to measure the acceleration of a car relative to the road, to axes fixed within the spinning

Earth or fixed within the solar system? This was one of the questions which ultimately, for the consideration of very high velocities, led to the theory of relativity; but mercifully it will not be necessary for us to be detained by such doubts, providing we decide upon a particular frame of reference (usually a set of axes fixed in the Earth) and then use it consistently.

The other concept, force, has not been defined, though we all have intuitive ideas of push and pull, thrust and tension and pressure. Newton's second law can be regarded as a refinement of these ideas, relating them to the acceleration of the object on which they are acting; and Newton's third law states that such forces always occur in pairs, as inter-actions between two objects.

Applying Newton's second law to the case of a falling apple, we see that its acceleration downwards must necessitate a downward force acting upon it, which we call its *weight*. Now it will also be recalled that when a particle is moving with constant speed round a circle, its acceleration is directed towards the centre of the circle. So if we think of the Moon revolving round the Earth, we should expect there to be a force drawing it towards the Earth. It was Newton's insight to recognise that this could be regarded as the same kind of force as the weight of an object on the Earth. So the Earth's pull on both the apple and the Moon was called *gravitation*, which he later understood as a force of attraction between every pair of particles in the universe.

This is a highly condensed account of Newton's revolutionary view of force and motion. How was it achieved?

Firstly, it must be regarded as the climax of nearly a century and a half of patient observation, calculation, and reflection. In 1543 was published, while its author was dying, Copernicus' book *De revolutionibus orbium caelestium*, 'On the revolutions of the heavenly spheres', explaining the simplifications that can be achieved if the Sun, rather than the Earth, is regarded as the centre round which the planets turn. Sixty years later, Kepler was to devote himself to years of patient calculation based on observations of the Danish astronomer Tycho Brahé. In 1618 Kepler published *Harmonice mundi*, 'On the harmony of the world', in which he stated his three laws of planetary motion:

a) Each planet moves in an ellipse with one focus at the Sun.
b) The line joining a planet to the Sun sweeps out equal areas in equal times.
c) If the mean distance of a planet from the Sun is r and its period (or time of revolution) is T, then T^2 is proportional to r^3: $T^2 = kr^3$.

Meanwhile, in Italy, Galileo was combining astronomical discoveries with investigation of the accelerations, as well as the velocities, of moving objects; and only a few months after he died, Newton was born. The final words are his, written when he was an old man about the work he accomplished by the age of twenty three:

In the same year I began to think of gravity extending to the orb of the moon, and from Kepler's third law I deduced that the forces which keep the planets in their orbs must be reciprocally as the squares of their distances from the centres about which they revolve; and thereby compared the force requisite to keep the moon in her orb with the force of gravity at the surface of the earth, and found them answer pretty nearly. All this was in the two plague years of 1665 and 1666, for in those days I was in the prime of my age for invention, and minded mathematics and philosophy more than at any time since.

9.2 Force, mass, and weight: Newton's second law

The force acting on a particle is proportional to its acceleration, and in the same direction.
Newton's second law implies that force, like acceleration, must be a vector quantity. So we call these **F** and **a** respectively, and write

$$\mathbf{F} = m\mathbf{a}$$

where the constant of proportionality m is a scalar quantity, called the *mass* of the particle. This equation is also commonly called *the equation of motion*.

The SI unit of mass is the kilogram (kg), being the mass of a standard cylinder kept in Paris; and the unit of force is the newton (N), the force required to give a mass of 1 kg an acceleration of $1\,\mathrm{m\,s^{-2}}$. (So $1\,\mathrm{N} = 1\,\mathrm{kg\,m\,s^{-2}}$.)

Exercise 9.2a

1 What force must act on a mass of 3 kg to give it an acceleration of
   a) $2\,\mathrm{m\,s^{-2}}$ **b)** $40\,\mathrm{m\,s^{-2}}$ **c)** $0.05\,\mathrm{m\,s^{-2}}$?

2 What would be the acceleration of a mass of 40 kg when acted upon by a force of
   a) 120 N   **b)** 24 N   **c)** 0.1 N?

3 What would be the mass of an object that is given an acceleration $2\,\mathrm{m\,s^{-2}}$ by a force of
   a) 40 N   **b)** $5 \times 10^3\,\mathrm{N}$ **c)** 0.03 N?

4 A 3 kg mass is moving at $4 \, \text{m s}^{-1}$ round a horizontal circle of radius 2 m. Find:
a) its acceleration;
b) the force which must be exerted upon it to produce this acceleration.

5 Repeat no. **4** if
a) its mass were doubled;
b) its speed were doubled;
c) its radius were doubled.

6 An electron of mass $9 \times 10^{-31} \, \text{kg}$ is moving at $4 \times 10^{6} \, \text{m s}^{-1}$ in a field which produces a force of $2 \times 10^{-15} \, \text{N}$ in the same direction. Find its velocity after an interval of $2 \times 10^{-9} \, \text{s}$.

7 A 4 kg mass has the vector position $\mathbf{r} = 10t\mathbf{i} - 5t^2\mathbf{j}$.
Find its position, velocity, and acceleration, and also the force acting on it when
a) $t = 0$ **b)** $t = 2$ **c)** $t = 10$

8 As a car of mass 1000 kg travels along a level road its motion is opposed by a constant resistance of 100 N. The values of the pull of the engine for given values of the time t s are as follows:

time t(s)	0	1	2	3	4	5	6
pull (N)		1100	1200	1350	1550	1850	2300	3100

Find the acceleration of the car at each of the given values of t. When $t = 0$ the speed of the car is $50 \, \text{km h}^{-1}$. Find graphically the speed of the car in km h^{-1} when $t = 6$.
   State how you would find graphically the distance travelled by the car in the period from $t = 0$ to $t = 6$.

Resultants

As forces are vector quantities, it follows immediately that they are combined by the vector (or parallelogram) law of addition, and their vector sum is usually called their *resultant*.

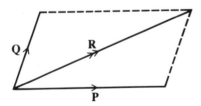

With a number of concurrent forces, this addition is most convenient by means of a *polygon of forces*:

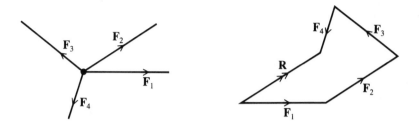

Equilibrium

If a particle is at rest or moving with constant velocity (i.e. it has no acceleration), it is said to be *in equilibrium*. This is clearly so when the resultant force acting upon it is zero and the polygon of forces is *closed*.

In the particular case of two forces in equilibrium, they must clearly be equal and opposite:

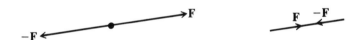

Furthermore, three forces acting on a particle are in equilibrium provided that they can be represented in magnitude and direction by a closed triangle, called a *triangle of forces*:

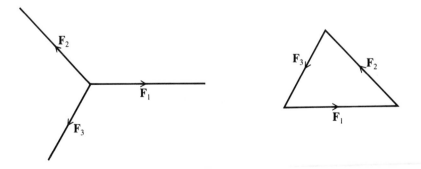

Example 1

A mass of 10 kg is acted upon by two forces, 20 N and 10 N, which are inclined to each other at 60°. Find:

a) the magnitude of the resultant force;
b) the resulting acceleration.

As the two forces are vectors, they must be combined by the vector law of addition.

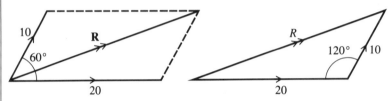

a) The resultant force **R** can be found by any of three methods:
i) scale drawing
ii) use of the cosine rule

$$R^2 = 20^2 + 10^2 - 2 \times 20 \times 10 \times \cos 120°$$
$$= 400 + 100 - 2 \times 20 \times 10 \times -\tfrac{1}{2}$$
$$= 500 + 200 = 700$$
$$\Rightarrow \quad R = 26.5$$

iii) use of components
The two forces can be represented as 20**i**
and $10 \cos 60 \mathbf{i} + 10 \sin 60 \mathbf{j} = 5\mathbf{i} + 5\sqrt{3}\mathbf{j}$

Hence $\mathbf{R} = 20\mathbf{i} + [5\mathbf{i} + 5\sqrt{3}\mathbf{j}]$
$$= 25\mathbf{i} + 5\sqrt{3}\mathbf{j}$$
$$\Rightarrow \qquad R = \sqrt{(625 + 75)} = \sqrt{700} = 26.5$$

b) Hence the resultant force is 26.5 N and so the resultant acceleration has magnitude a, where

$$26.5 = 10\,a$$
$$\Rightarrow \qquad a = 2.65\,\mathrm{m\,s^{-2}}$$

Example 2

If three concurrent forces **P**, **Q**, **R** are in equilibrium, show that their magnitudes and the angles α, β, γ between them are such that

$$\frac{P}{\sin \alpha} = \frac{Q}{\sin \beta} = \frac{R}{\sin \gamma}$$

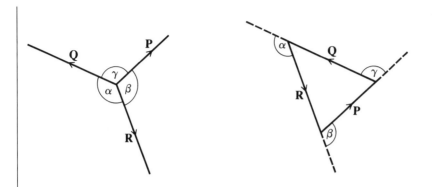

Since the forces are in equilibrium, they can be represented in magnitude and direction by the sides of a triangle as shown. Hence, using the sine rule

$$\frac{P}{\sin(\pi - \alpha)} = \frac{Q}{\sin(\pi - \beta)} = \frac{R}{\sin(\pi - \gamma)}$$

$$\Rightarrow \qquad \frac{P}{\sin \alpha} = \frac{Q}{\sin \beta} = \frac{R}{\sin \gamma}$$

This is usually known as *Lami's theorem*.

Exercise 9.2b

In nos. **1–6** find:
a) the resultant force;
b) the resulting acceleration of the particle;
c) the force necessary to maintain the particle in equilibrium.
(SI units are used throughout.)

1   Forces $8\mathbf{i} - 2\mathbf{j}$, $3\mathbf{i} + 3\mathbf{j}$, $-\mathbf{i} + 4\mathbf{j}$, acting on a particle of mass 5 kg.

2   Forces $3\mathbf{i} + 2\mathbf{j} + \mathbf{k}$, $2\mathbf{i} + 3\mathbf{j} - \mathbf{k}$, $\mathbf{i} + \mathbf{j} + 3\mathbf{k}$, acting on a particle of mass 2 kg.

3   Forces 20 N due East and 30 N due South, acting on a particle of mass 10 kg.

4   Forces 10 N due West and 5 N in direction South–West, acting on a particle of mass 4 kg.

5 Forces 10 N due West and 5 N in direction North–East, acting on a particle of mass 20 kg.

6 Forces 10 N on bearing 080°, 5 N on bearing 165°, and 12 N on bearing 284°, acting on a particle of mass 0.2 kg.

7 A mass of 3 kg is acted on by a constant force of $\begin{pmatrix} 6 \\ 3 \end{pmatrix}$ newtons. Find the acceleration.

At time $t = 0$ it is travelling with velocity $\begin{pmatrix} -3 \\ 2 \end{pmatrix}$ m s^{-1}. Find the displacement from $t = 0$ to $t = 4$ (s), and the velocity at the end of that time.   (SMP)

8 An electron of mass 9×10^{-31} kg is moving at 8×10^6 m s^{-1} when it enters a field, which produces a force of 1.8×10^{-15} N at right angles to its initial direction of motion for a period of 3×10^{-9} s. Find its final velocity.   (MEI)

Acceleration due to gravity: weight

At the Earth's surface, a freely falling body accelerates at roughly 9.8 m s^{-2}. So, using Newton's second law, a mass of 1 kg must be acted upon by a force of approximately 9.8 N vertically downwards. This is called the *weight* of the body, and its magnitude is sometimes called 1 kg wt.

The acceleration due to gravity is usually denoted by **g** (which has magnitude g), and its magnitude varies from place to place. So the weight of a mass of 1 kg will also vary in magnitude:

	acceleration due to gravity (m s^{-2})	weight of 1 kg (N)
at Equator	9.781	9.781
London	9.812	9.812
at North Pole	9.832	9.832
on Moon	1.6	1.6

On the Earth, therefore, 1 N is the weight of a mass of roughly 0.1 kg; or, not inappropriately, of a medium-sized apple.

Example 3

A mass of 10 kg is supported in equilibrium by two strings which are inclined to the vertical at 30° and 20°. Find their tensions.

Taking $g = 9.8\,\text{m s}^{-2}$, the weight of the 10 kg mass is 98 N.

Hence the tensions T_1, T_2 are such that

$$\frac{T_1}{\sin 20°} = \frac{T_2}{\sin 30°} = \frac{98}{\sin 130°}$$

$$\Rightarrow \qquad T_1 = \frac{98 \sin 20°}{\sin 50°} \approx 44\,\text{N}$$

and $\qquad T_2 = \frac{98 \sin 30°}{\sin 50°} \approx 64\,\text{N}$

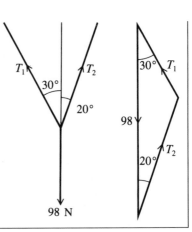

Example 4

A man of mass 100 kg is standing in a lift. What is the force of the floor on the man when the lift has
a) acceleration $2\,\text{m s}^{-2}$ upwards;
b) constant velocity;
c) acceleration $2\,\text{m s}^{-2}$ downwards?
(Take $g = 9.81\,\text{m s}^{-2}$.)

In each case there are two external forces acting on the man:
 i) his weight 981 N vertically down;
 ii) the force of the floor R N vertically up.
Applying Newton's second law in all three cases:

a) Acceleration $2\,\text{m s}^{-2}$ upwards

$$R - 981 = 100 \times 2$$
$$\Rightarrow \qquad R = 1181$$

b) Constant velocity

$$R - 981 = 100 \times 0$$
$$\Rightarrow \qquad R = 981$$

c) Acceleration $2\,\mathrm{m\,s^{-2}}$ downwards

$$R - 981 = 100 \times -2 = -200$$
$$\Rightarrow \qquad R = 781$$

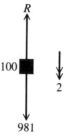

Hence the force of the floor is

     1181 N whilst the lift is accelerating upwards

     981 N whilst it is rising at constant velocity

and   781 N whilst it is decelerating

Example 5

A woman jumps from an aeroplane and falls freely under gravity. Until she opens her parachute the air resistance per unit mass is $-\frac{1}{5}\mathbf{v}$ (measured in newtons), where $\mathbf{v}$ is her velocity (in $\mathrm{m\,s^{-1}}$). What is her terminal velocity (when she ceases to accelerate)?

Suppose that her mass is m (kg) and that at a certain point on her path her acceleration (indicated by a double arrow) is $\mathbf{a}$.

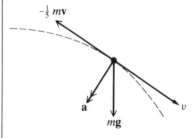

The two forces acting on the woman are   **a)** her weight $m\mathbf{g}$ vertically down and **b)** the resistance $-\frac{1}{5}m\mathbf{v}$.
   By Newton's second law,

$$m\mathbf{g} - \tfrac{1}{5}m\mathbf{v} = m\mathbf{a}$$

$$\Rightarrow \qquad \mathbf{a} = \mathbf{g} - \tfrac{1}{5}\mathbf{v}$$

So as $\mathbf{a} \to 0$ $\mathbf{g} - \tfrac{1}{5}\mathbf{v} \to 0$

$$\Rightarrow \qquad\qquad\qquad \mathbf{v} \to 5\mathbf{g}$$

Hence, taking $g \approx 10$, we see that the terminal velocity is $50\,\mathrm{m\,s^{-1}}$ vertically downwards.

Example 6 (The 'conical pendulum')

A mass m is attached to a fixed point by means of an inextensible string and rotates in a horizontal circle with constant speed v. If the string makes an angle θ with the vertical, show that $v^2 = gr\tan\theta$.

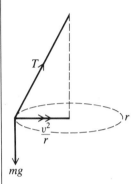

Let the tension in the string have magnitude T and suppose that the mass moves in a circle of radius r. Then the acceleration of m is directed towards the centre and has magnitude v^2/r.

Applying Newton's second law, and resolving horizontally and vertically:

$$T\sin\theta = m\frac{v^2}{r} \tag{1}$$

and $-mg + T\cos\theta = 0$

$\Rightarrow$ $T\cos\theta = mg$ $\tag{2}$

From (1) and (2)

$$\tan\theta = \frac{v^2}{gr} \quad \Rightarrow \quad v^2 = gr\tan\theta$$

Example 7

Find the period of revolution of an artificial satellite which is orbiting the Earth just outside its atmosphere. (Assume the orbit to be approximately circular, the acceleration due to gravity as approximately $10\,\mathrm{m\,s^{-2}}$ and the radius of the Earth as 6400 km.)

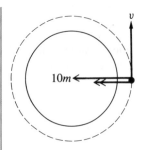

Let the mass of the satellite be m kg, so that its weight will be approximately $10m$ N.

Furthermore, suppose that its speed is v m s^{-1}. Now its distance from the centre of the Earth is approximately 6.4×10^6 m, so its acceleration, which is directed towards the Earth's centre, has magnitude $v^2/(6.4 \times 10^6)$.

Hence, using Newton's second law,

$$10m = m\frac{v^2}{6.4 \times 10^6}$$

$$\Rightarrow \quad v^2 = 64 \times 10^6$$

$$\Rightarrow \quad v = 8 \times 10^3$$

Hence   the satellite's speed ≈ 8 km s^{-1}

But   the circumference of it orbit $\approx 2\pi \times 6400$ km

$$\approx 40\,000 \text{ km}$$

$$\Rightarrow \qquad \text{period of orbit} \approx \frac{40\,000}{8} \approx 5000 \text{ s} \approx 1.4 \text{ h}$$

Exercise 9.2c

(Take $g = 9.8$ m s^{-2} unless otherwise stated.)

1   A miner's cage has mass 1 tonne. What is the tension in its cable if the cage is
   a) accelerating upwards at 2 m s^{-2};
   b) accelerating downwards at 2 m s^{-2};
   c) going up at constant velocity;
   d) going down at constant velocity?

2   If the maximum allowable tension of the cable in no. 1 is 1.2×10^4 N, find the acceleration permitted
   a) when the cage is empty;

b) when it is carrying 5 boys of average mass 40 kg.

3 A ball of mass 2 kg is falling vertically through the air with an acceleration of only 8 m s⁻². Find the air resistance.

When the ball is moving faster, this resistance increases to 10 N. What is then the acceleration of the ball?

4 A mass of 10 kg is supported by two strings inclined to the vertical at 40° and 50°. Find their tensions.

5 A mass of 20 kg is suspended by two wires of lengths 3 m and 4 m attached to two hooks at the same level and a distance 5 m apart. Find the tensions in the wires.

6 **i**, **j**, and **k** are three mutually perpendicular unit vectors; **i** and **j** are horizontal and **k** is vertically upwards. A particle of weight 6 N is in equilibrium, supported by three strings. The tension in one of the strings, in newtons, is represented by $-\mathbf{i} - 2\mathbf{j} + 3\mathbf{k}$, and the other strings are each inclined at an angle θ to **k**, one in the **i**–**k** plane and the other in the **j**–**k** plane. Show that $\theta = 45°$, and find the magnitudes of all three tensions.

(C)

7 Show how the acceleration of a train can be estimated by means of attaching a string with a heavy mass on its end to the ceiling of a coach.
a) What is the acceleration when the string is inclined at a constant angle of 10° to the vertical?
b) At what constant angle will the string be inclined if the acceleration is 1 m s⁻²?

8 A man of mass 100 kg is hanging by a rope attached to a winch in a helicopter which is hovering. The helicopter begins to ascend vertically with an acceleration of 0.75 m s⁻². Determine the tension in the rope in newtons.

If the rope were to be wound in at a constant rate of 1 m s⁻¹, what change would there be in the value of the tension in the rope? Comment briefly.

After a short time the helicopter ceases to rise vertically but subsequently moves horizontally with an acceleration of 0.75 m s⁻². It is observed that with the winch not operating, the rope hangs at a constant angle θ to the vertical where $\tan \theta = \frac{3}{4}$. What is the tension in the rope now, and what is the horizontal force on the man due to the resistance of the air?

(MEI)

9 An aircraft of mass 2×10^4 kg is taking off at 20° to the horizontal with an acceleration of 1.5 m s⁻². Find the forces
a) perpendicular to the plane of its wings (the *lift*);
b) along its direction of motion.

10 A man is swinging a metal ball of mass 0.1 kg in a vertical circle by means of a light wire of length $\frac{1}{2}$ m. Find the tension in the wire if the ball

    a) is at the lowest point and has speed $4\,\text{m s}^{-1}$;

    b) is at the highest point and has speed $3\,\text{m s}^{-1}$.

11 With what period would the Earth need to rotate in order that gravity would appear to vanish at the equator? (Take g as $10\,\text{m s}^{-2}$ and the circumference of the Earth as 4×10^7 m.)

    Under these conditions what would be the angle at latitude $60°$ between a string suspending a weight in relative equilibrium and the 'true' vertical?

    Show that all apparent verticals would be in the same direction in space.

        (MEI)

12 In a 'golf driving practice kit' a ball of mass 0.05 kg is attached by a length of light elastic to the tee, so that when the ball is r metre from the tee the tension in the elastic is $0.2r$ newton. The ball is struck with velocity $\mathbf{u}\,\text{m s}^{-1}$ at time $t = 0$. If $\mathbf{r}$ metre denotes the position vector at a subsequent time t second before it strikes the ground again, and if $\mathbf{j}$ denotes a unit vector vertically upwards and the acceleration of gravity is taken as $10\,\text{m s}^{-2}$, obtain the differential equation

$$\ddot{\mathbf{r}} + 4\mathbf{r} + 10\mathbf{j} = \mathbf{0}$$

Verify that all the conditions of the problem are satisfied by the solution

$$\mathbf{r} = \tfrac{1}{2}\mathbf{u}\sin 2t + \tfrac{5}{2}(\cos 2t - 1)\mathbf{j}\qquad\qquad(\text{SMP})$$

9.3 Reactions: Newton's third law

Some of the objects with which we have so far been concerned, such as the planets, have been free to move anywhere in space. But it frequently happens that we wish to study an object which is under some form of constraint, perhaps merely by having to remain on a particular surface. It is clear that its behaviour would usually be very different if the surface were removed, so the surface must be exerting a force on the object, which we call a *reaction*; and, by Newton's third law, the object must be exerting an equal and opposite force on the surface.

When this force is perpendicular, or normal, to the surface, it is called a *normal reaction* and the surface is said to be *smooth*. Otherwise – to a greater or lesser extent – it is *rough*, and in section 9.4 we shall investigate the consequent frictional forces.

Example 1

A toboggan of mass 50 kg is hauled along horizontal smooth ice by a force of 100 N at 20° to the horizontal. Taking $g \approx 9.8 \, \text{m s}^{-2}$, find the acceleration of the toboggan and the force between ice and toboggan.

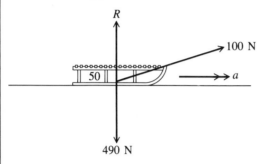

There are three forces acting on the toboggan:
a) its weight 490 N;
b) the normal reaction R N;
c) the tractive force 100 N.
Their resultant has horizontal and vertical components

$$(100 \cos 20°, \qquad R + 100 \sin 20° - 490)$$
$$\equiv \qquad (94, \qquad R - 456)$$

and the acceleration has components $(a, 0)$. So, by Newton's second law,

$$94 = 50a \qquad\qquad R - 456 = 50 \times 0$$
$$\Rightarrow a = \tfrac{94}{50} = 1.88 \qquad\qquad R = 456$$

Hence the acceleration is $1.88 \, \text{m s}^{-2}$ and the normal reaction is 456 N.

Example 2

If a particle is sliding down a smooth plane which is inclined at α to the horizontal, find:
a) its acceleration;
b) the reaction of the plane on the particle.

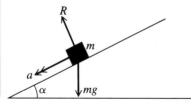

The two forces acting on a particle of mass m are:
a) its weight mg vertically down;
b) a reaction R perpendicular to the plane.

   Letting the acceleration of the particle be a down the plane, we see from Newton's second law

down the plane: $$mg \sin \alpha = ma$$

perpendicular to the plane: $$R - mg \cos \alpha = 0$$

Hence $a = g \sin \alpha$

and $R = mg \cos \alpha$

Exercise 9.3a

In each question state all the forces and accelerations, and draw a figure on which they are clearly indicated.

1 A car of mass 1000 kg is acted upon by a tractive force (i.e. from the road and tangential to the driving wheels) of 500 N. Find its acceleration on a horizontal stretch of road
   a) if road and air resistances are negligible;
   b) if there is a total resistance of 200 N.

2 Repeat no. 1 if the car is going up an incline of 1 in 200 (i.e. at θ to the horizontal, where $\sin \theta = \frac{1}{200}$).

3 Repeat no. 1 if the car is going down an incline of 1 in 200.

4 A mass of 50 kg on a smooth surface is acted upon by a force of 20 N. Find its acceleration and the reaction of the surface when
   a) both the surface and the force are horizontal;
   b) the surface is horizontal, but the force is inclined at 20° to the horizontal;
   c) the force is horizontal, but the surface is inclined at 20° to the horizontal;
   d) both the force and the surface are inclined at 20° to the horizontal.

5 A water-skier can only perform effectively if the thrust of the water acts along the line of her body, assumed straight. If her mass is 50 kg, find the tension in the horizontal tow-rope if she leans back at 10° to the vertical in steady motion.
      What will be her acceleration (assumed horizontal) if the tension is doubled, but she leans back at 15°?   (SMP)

6 A car of mass 800 kg is travelling along a horizontal road, curved in a

circle of radius 200 m and banked at 5° to the horizontal. What is the speed of the car if it has no tendency to slip either inwards or outwards (i.e. with the road as smooth as can be)?

7   A man swings a small bucket of water round in a vertical circle of radius 1 m. If the bucket travels at constant speed without spilling, show that the least speed possible is $\sqrt{g}\,\mathrm{m\,s^{-1}}$.   (MEI)

8   At what angle should a race track corner be banked if its radius is 200 m and vehicles are to have no tendency to side-slip when travelling at $50\,\mathrm{m\,s^{-1}}$?

9   An aircraft travelling at $400\,\mathrm{m\,s^{-1}}$ banks at 10° to turn in a horizontal circle. Assuming that the force of the air on the plane is perpendicular to its wings, find the radius of turn if there is no side-slip.

10   A particle is moving with constant speed v in a horizontal circle on the inside of a smooth hemispherical bowl of radius a. If the line from the particle to the centre of the sphere is inclined to the vertical at an angle θ, show that

$$v^2 = \frac{ag\sin^2\theta}{\cos\theta}$$

11   A charged particle, mass 2×10^{-3} kg, initially at the origin, is attracted to charges at $(4,0)$ and $(0,3)$ by forces of constant magnitude 6×10^{-3} and 3×10^{-3} N respectively, and there are no other forces acting on it.
a) Calculate its initial acceleration, giving the magnitude and direction. Will the particle travel in a straight line?
b) If it were constrained by a smooth tube so that it could only move freely along the line $y = 2x$, what would its initial acceleration then be?
     When the particle is at the point $(2,4)$ show that the attractive force on it is $\begin{pmatrix} 0 \\ -3\sqrt{5} \end{pmatrix} \times 10^{-3}$ N and hence find the force it exerts on the side of the tube.   (SMP)

Connected particles

Newton's third law states that a body cannot exert a force upon another without itself experiencing an equal and opposite force. If a railway engine is pulling a coach forwards, then the coach is inevitably pulling the engine backwards with an equal and opposite force; and just as the Earth is exerting a gravitational attraction on the Moon, so the Moon is exerting an equal and opposite attraction on the Earth. Such pairs are usually called *action and reaction*.

Example 3

An engine of mass 10 tonne is hauling wagons of total mass 50 tonne along a horizontal track, and together they are accelerating at $\frac{1}{10}$ m s^{-2}. Ignoring all resistances, find the tractive force of the engine and the tension in the engine coupling.

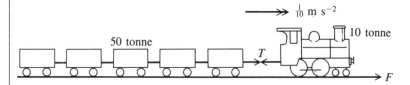

The horizontal forces with which we are concerned are:
a) the tractive force F, which is the horizontal reaction of the rails on the driving wheels (the wheels themselves, thanks to friction, exert a backwards force on the rails);
b) the tension T in the coupling which connects the engine to the rest of the train, so that T is a forwards force on the wagons but a backwards force on the engine.

By Newton's second law,

for the wagons: $\qquad T = 50\,000 \times \frac{1}{10} = 5000$
for the engine: $\quad F - T = 10\,000 \times \frac{1}{10} = 1000$
$\Rightarrow \qquad\qquad\qquad F = 6000$

Hence the tractive force is 6000 N and the tension in the coupling is 5000 N.

If we had merely required the tractive force, this could have been obtained by considering the motion of the train as a whole:

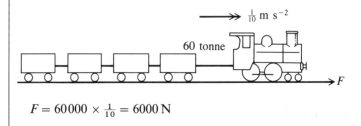

$$F = 60\,000 \times \tfrac{1}{10} = 6000 \text{ N}$$

Example 4

A mass m lies on a plane inclined at α to the horizontal and is connected to a mass M which hangs freely by means of a light inextensible string passing over a pulley at the top of the plane. Supposing that frictional forces are negligible, find the resulting accelerations and the tension in the string.

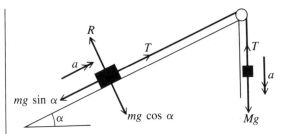

Since the pulley is smooth and the string is light, its tension T is constant along its length; and since the string is inextensible, the magnitudes of the accelerations of the two masses are equal.

Now the forces acting are as follows:

On m **a)** its weight mg, which can be resolved into components $mg \sin \alpha$; and $mg \cos \alpha$, down and perpendicular to the plane;
  b) a normal reaction R;
  c) the tension T of the string.

On M **a)** its weight Mg;
  b) the tension T of the string.

Furthermore, we suppose that the two masses have accelerations of magnitude a. Then, applying Newton's second law,

For m up plane $T - mg \sin \alpha = ma$ (1)
      perpendicular to plane $R - mg \cos \alpha = 0$ (2)
For M $Mg - T = Ma$ (3)

From (1) and (3), $(M - m \sin \alpha)g = (M + m)a$

$$\Rightarrow \qquad a = \frac{M - m \sin \alpha}{M + m} g$$

and $T = ma + mg \sin \alpha$

$$= \frac{m(M - m \sin \alpha)g}{M + m} + mg \sin \alpha$$

$$\Rightarrow \quad T = \frac{Mm(1 + \sin \alpha)g}{M + m}$$

Exercise 9.3b

In each question state all the forces and accelerations which act on every body and draw a figure in which these are clearly indicated.

1   A car of mass 1 tonne is pulling a trailer of mass $\frac{1}{4}$ tonne along a horizontal road. Find their acceleration and the tension in the tow-bar in the following cases:

a) when the tractive force on the car is 2000 N and the road and air resistances are negligible;

b) when the tractive force on the car is 2000 N but it suffers a total resistance of 500 N, and the trailer's resistance is negligible;

c) when the tractive force on the car is 2000 N but it suffers a total resistance of 500 N and the trailer suffers a resistance of 200 N.

In nos. **2** to **7** all surfaces and pulleys are supposed to be smooth and the masses indicated are in kilograms. In each case specify all the forces which are acting on the separate bodies, draw a figure to indicate these and the accelerations and proceed to calculate the tensions and accelerations. (Take $g \approx 10\,\mathrm{m\,s^{-2}}$.)

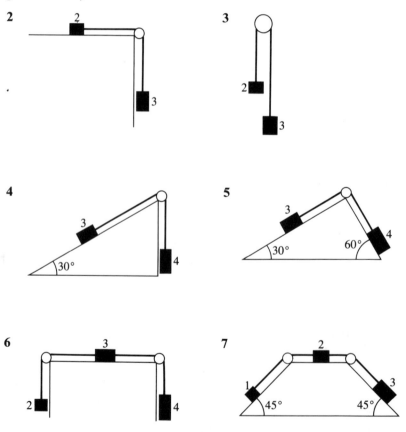

8 Repeat no. **1** when the car is towing the trailer up a slope of 1 in 100 (i.e. at θ to the horizontal, where $\sin \theta = \frac{1}{100}$).

9.4 Friction

We have seen, in the last section, that the reaction of a smooth surface is always perpendicular, or *normal*, to the surface. More usually, however, a surface is rough and capable of exerting a reaction which has a tangential component, called *friction*. In such a case the resultant reaction R has two components:

      a frictional component F
and a normal component N

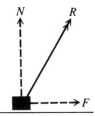

The relations between these quantities depend on the degree of roughness of the surfaces and on whether the particle is moving or at rest. They were first investigated by G. Coulomb who in 1821 summarised his experimental results in the so-called *Laws of statical and dynamical friction*. Though only very approximate, these still form the basis for the study of rough surfaces.

Dynamical friction

When sliding takes place, the component F is in a direction that opposes relative motion and has magnitude μN, where μ is a constant. This constant is called the *coefficient of friction* and depends only on the nature of the surfaces and *not* on the area of contact or the normal reaction.

Example 1

A block of wood of mass 20 kg is being pulled along a rough horizontal surface by means of a rope inclined to the horizontal at 30° and under tension 100 N. If $\mu = 0.3$, find the acceleration of the block. (Take $g \approx 9.8 \, \mathrm{m\,s^{-2}}$.)

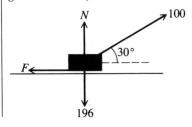

The forces acting on the block are
a) weight 196 N;
b) tension 100 N;
c) normal reaction N;
d) friction F.

As the block is moving,

$$F = 0.3\,N$$

If the acceleration of the block is a, then (by Newton's second law),

Vertically
$$N + 100 \sin 30° = 196$$
$$\Rightarrow \qquad N + 50 = 196$$
$$\Rightarrow \qquad N = 146$$
Hence $F = 0.3 \times 146 = 43.8$

Horizontally
$$100 \cos 30 - F = 20a$$
$$\Rightarrow \qquad 86.6 - 43.8 = 20a$$
$$\Rightarrow \qquad 42.8 = 20a$$
$$\Rightarrow \qquad a = 2.14$$

Hence the block has an acceleration of $2.14\,\mathrm{m\,s^{-2}}$.

Exercise 9.4a

In each question state all forces and accelerations and draw a diagram on which these are clearly indicated. Take $g \approx 10\,\mathrm{m\,s^{-2}}$.

1 A boy of mass 40 kg is sliding across a horizontal sheet of ice ($\mu = 0.1$) at a speed of $2\,\mathrm{m\,s^{-1}}$. Find his deceleration and the distance travelled before he comes to rest.

2 A car travelling at $10\,\mathrm{m\,s^{-1}}$ skids to a halt in a distance of 5 m. Find:
   a) its deceleration (supposed constant);
   b) the coefficient of friction;
   c) its stopping distance from a speed of $20\,\mathrm{m\,s^{-1}}$.

3 A mass of 10 kg is resting on a wooden floor and a horizontal force of 40 N is just sufficient to move it. What is the coefficient of friction? What force is required to give it an acceleration of $2\,\mathrm{m\,s^{-2}}$?

4 An ice-hockey puck is hit with a speed of $20\,\mathrm{m\,s^{-1}}$ and travels 100 m before coming to rest. Calculate μ.

5 A mass of 2 kg is pulled along a horizontal floor by means of a force of 40 N acting at 20° above the horizontal. If $\mu = 0.2$, find the acceleration of the mass. What would be the acceleration if the force acted *downwards* at 20° to the horizontal?

6 A girl at a swimming pool is sliding down a chute which is 10 m long and inclined at 30° to the horizontal. If her coefficient of friction is 0.2, find her acceleration down the chute and her speed on leaving it.

7 A mass m is sliding down a plane inclined at α to the horizontal whose coefficient of friction is μ. Find its acceleration.

8 A mass m rests on a horizontal surface whose coefficient of friction is μ and is connected by a taut string perpendicular to the edge of the table to a mass M which hangs vertically over the edge of the table and which is falling downwards. Find the acceleration of the system and the tension in the string.

9 The triangle ABC, in which the lengths of AB, BC, CA are 4 m, 3 m and 5 m, is the vertical cross-section of a wedge fixed with AB in contact with a horizontal surface. BC is vertical with C above B, and AC is a line of greatest slope. A light inextensible string passes over a smooth pulley at C and connects a mass m on AC with a mass $3m$ hanging freely. The coefficient of friction between the mass m and the wedge is $\frac{1}{2}$. Initially the system is at rest with m at A and $3m$ just below the pulley at C. Show that each mass moves with acceleration $g/2$ until the mass $3m$ hits the horizontal surface. Show that the velocity of the mass m is then $\sqrt{3g}\,\text{m s}^{-1}$ up the slope. Find the further distance which m moves up AC and its velocity just before the string again becomes taut, assuming that the mass $3m$ is then in contact with the horizontal surface.   (MEI)

10 A parcel rests on the horizontal floor at the back of a van which is travelling along a level road at a steady speed of 63 km h^{-1}. The van is brought to rest in a stopping distance of 20 m by a uniform application of the brakes. Show that the parcel will slide forward on the floor if the coefficient of friction between the parcel and the floor is $< \frac{25}{32}$.

If this coefficient is $\frac{3}{4}$, show that the parcel will slide forward $\frac{5}{6}$ m before coming to rest.   (w)

Statical friction

When there is equilibrium, the amount of friction is just sufficient to prevent relative motion and cannot exceed μN (which is therefore called the limiting friction):

$$F \leqslant \mu N$$

Example 2

A block of mass m is at rest on a plane inclined to the horizontal at an angle α. If the coefficient of friction is μ, show that $\mu \geqslant \tan \alpha$.

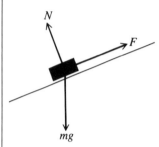

The forces acting on the block are
a) its weight mg;
b) the normal reaction N;
c) the frictional force F.

Since the body is in equilibrium, we obtain by resolving up the plane

$$F = mg \sin \alpha$$

and by resolving perpendicular to the plane

$$N = mg \cos \alpha$$

Hence $\dfrac{F}{N} = \tan \alpha$

But $\dfrac{F}{N} \leqslant \mu, \quad$ so $\quad \mu \geqslant \tan \alpha$

Angle of friction

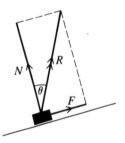

Suppose that a block lies on a rough surface and that its resultant reaction R makes an angle θ with the normal. Then if the normal and frictional components of R are N, F respectively, it follows that

$$\frac{F}{N} = \tan \theta$$

But $\dfrac{F}{N} \leqslant \mu \quad \Rightarrow \quad \tan \theta \leqslant \mu$

If we now introduce an angle λ such that $\mu = \tan \lambda$, it follows that

$$\tan \theta \leqslant \tan \lambda \quad \Rightarrow \quad \theta \leqslant \lambda$$

Hence the reaction must be inclined to the normal at an angle which does not exceed λ. This is therefore called the *angle of friction*, and the possible limiting positions of R are said to lie on the *cone of friction*.

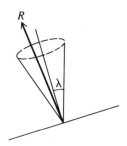

Example 3

A block of weight **W** is placed on a rough plane (coefficient of friction μ) which is inclined to the horizontal at an angle α, where $\tan \alpha > \mu$. Find the magnitude and direction of the *least* force which will prevent the block from slipping down the plane.

Since $\tan \alpha > \mu$, then (as we saw in example 2) the block cannot remain in equilibrium unless another force **P** is applied to prevent it from sliding down the plane.

Hence the forces on the block are
a) its weight **W** vertically down;
b) the reaction of the plane, **R**;
c) the additional force **P**.

   Furthermore, the resultant reaction **R** (if it is just to prevent sliding downwards) must lie on the cone of friction, and so be inclined to the normal at an angle λ, where $\tan \lambda = \mu$.

Hence **R** makes an angle $\alpha - \lambda$ with the vertical and we see from the triangle of forces that **P**, if it is to be as small as possible, must be perpendicular to **R**.

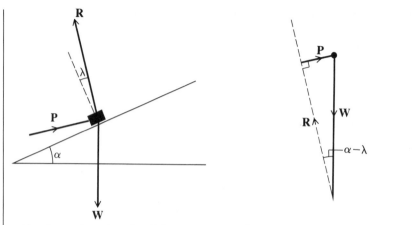

So, from the triangle of forces, $P = W\sin(\alpha - \lambda)$ is the smallest force that will maintain equilibrium and is inclined to the horizontal at an angle $\alpha - \lambda$.

Exercise 9.4b

1   A block of mass 20 kg rests on a horizontal plane whose coefficient of friction is 0.4. Find the least force required to move the block if it acts
   a) horizontally;
   b) at 30° above the horizontal;
   c) at 30° below the horizontal;
   d) in the most favourable direction.

2   A block of mass 50 kg lies on a horizontal plane and can just be moved by a force of 100 N acting horizontally. Find:
   a) the coefficient of friction;
   b) the least force required to move the block if its direction of action can be chosen at will.

3   A block of mass 10 kg rests on a plane which is inclined at 10° to the horizontal and has $\mu = 0.3$. Find the least force required to move the body if the force acts
   a) up the plane;
   b) down the plane.

4   In no. **3** what are the least forces (if there is complete freedom of choice of their direction) required to move the block
   a) up the plane;
   b) down the plane?

5   A block of weight W rests on a horizontal plane with angle of friction λ. Find the minimum force required to move it if the force acts

a) along the plane;
b) at angle θ above the plane;
c) in the most economical direction.

6 Repeat no. **5** when the plane is tilted at an angle α and it is wished to move the block *up* the plane.

7 Repeat no. **5** when the plane is tilted at an angle α and it is wished to move the block *down* the plane.

8 At a fun-fair one of the amusements consists of a circular cylinder of radius 4 m which can be made to rotate with its axis vertical. Inside the cylinder is a horizontal floor which rotates with the cylinder and which can also be lowered and raised a distance of 2 m. When the cylinder is stationary a door in its wall allows access to the floor; the door is closed and forms part of the cylinder before it begins to rotate.

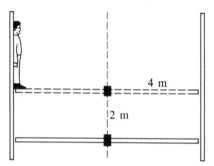

A man enters and stands with his back to the wall of the cylinder. The cylinder is then made to rotate more and more quickly until the friction between the man's back and the wall is sufficient to support his weight. The floor is then lowered 2 m, as in the diagram.

If the coefficient of friction between the man's body and the wall of the cylinder is 0.5, find the least angular velocity of the cylinder from which the man will not slip down. (Take g as $9.8\,\mathrm{m\,s^{-2}}$.)

If the angular velocity of the cylinder is then quickly reduced to four-fifths of its previous value, find how long the man will take to slide down to the floor.   (MEI)

9.5 Impulse and momentum

Motion in a straight line

Suppose that a particle of mass m is pulled in a horizontal line by a constant force of magnitude F, and that in an interval of time t its speed changes from v_0 to v_1 with an acceleration of magnitude a.

Since the acceleration is uniform,

$$v_1 - v_0 = at$$

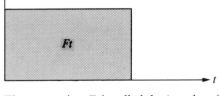

Also, by Newton's second law,

$$F = ma$$

Hence $Ft = mv_1 - mv_0$

The expression Ft is called the *impulse of F during the interval t*; and if, further, the product of mass and velocity is called the *momentum*, we see that $mv_1 - mv_0$ is the increase in momentum.

So ⬛ impulse of force during interval = change of momentum

In SI, the unit of impulse and of momentum is 1 N s.

Example 1

A stone of mass 2 kg is thrown vertically downwards from a high cliff with velocity $4\,\mathrm{m\,s^{-1}}$. Find the impulse of its weight in the first two seconds of its fall and verify that this is equal to its increase in momentum.

Taking $g = 9.8\,\mathrm{m\,s^{-2}}$, the stone's weight is 19.6 N and the impulse of this weight is $19.6 \times 2 = 39.2\,\mathrm{N\,s}$.

Also, its velocity after $2\,\mathrm{s} = 4 + 9.8 \times 2 = 23.6\,\mathrm{m\,s^{-1}}$

So   initial momentum $= 2 \times 4 = 8\,\mathrm{N\,s}$

      final momentum $= 2 \times 23.6 = 47.2\,\mathrm{N\,s}$

and   increase of momentum $= 39.2\,\mathrm{N\,s}$

Impulse of a variable force

It frequently happens when a sudden blow is struck – for instance when a ball is hit by a cricket bat – that a very large force acts for a very small interval of time. In such cases it is often either impossible or unnecessary to investigate either the magnitude of the force or the duration of its action, and attention is confined to the impulse of the force and the consequent change of momentum. Moreover, the force will usually vary with time: the thrust of the bat on ball increases very rapidly from zero (when contact is first made at time t_0) to a maximum value, and then decreases just as rapidly to zero (when the ball finally leaves the bat at time t_1).

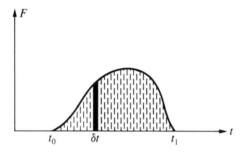

In such cases the total impulse during the interval will be the limit of the sum of small impulses $F\,\delta t$.

So $\text{impulse} = \lim \sum F\,\delta t = \displaystyle\int_{t_0}^{t_1} F\,dt$

Moreover, at any instant $F = m\dfrac{dv}{dt}$

$\Rightarrow \quad \displaystyle\int_{t_0}^{t_1} F\,dt = \int_{t_0}^{t_1} m\dfrac{dv}{dt}\,dt = \left[mv \right]_{v_0}^{v_1} = mv_1 - mv_0$

So, again,

impulse of force during interval = change of momentum

Exercise 9.5a

1   Calculate the momentum (in N s) of:
   a) a car of mass 1 tonne travelling at $30\,\mathrm{m\,s^{-1}}$;
   b) a lorry of mass 10 tonne travelling at $10\,\mathrm{m\,s^{-1}}$;
   c) a pellet of mass 5 g moving at $150\,\mathrm{m\,s^{-1}}$;
   d) an electron of mass $9 \times 10^{-31}\,\mathrm{kg}$ moving at $3 \times 10^7\,\mathrm{m\,s^{-1}}$.

2 Calculate (in SI units) the impulse of a force
   a) 100 N acting for 3 s;
   b) $6t$ N, from $t = 0$ to $t = 2$;
   c) $4t^3$ N, from $t = 1$ to $t = 3$;
   d) $2 \sin t$ N, from $t = 0$ to $t = \frac{1}{2}\pi$.

3 A car of mass 1 tonne moves from rest under the action of a tractive force
   and a resistance whose difference (the *effective force*) has a constant value
   of 400 N. Find:
   a) the impulse of the effective force in the first 5 s;
   b) the resulting momentum;
   c) the resulting velocity.

4 Repeat no. 3 if the effective force diminishes steadily from 400 N to zero.

5 Repeat no. 3 if the effective force after time t is $16(5 - t)^2$ N.

6 A lorry of mass 10 tonne is travelling at $30\,\mathrm{m\,s^{-1}}$ when its brakes are
   applied for a period of 2 s. If the braking force has the constant value
   20 000 N, find:
   a) its impulse;
   b) the final momentum;
   c) the final velocity.

7 Repeat no. 6 if the braking force increases steadily to 20 000 N in the 2 s
   interval.

8 Repeat no. 6 if the braking force increases steadily from zero to 20 000 N
   in one second and then decreases steadily to zero after another second.

9 A ball of mass 1 kg strikes the cushion of a billiard table perpendicularly
   at a speed of $3\,\mathrm{m\,s^{-1}}$ and rebounds at $1\,\mathrm{m\,s^{-1}}$. What is its change of
   momentum? If the total time of impact was 0.02 s and the force of the
   cushion is supposed constant, find the value of this force.

10 Repeat no. 9 on the different supposition that the force of the cushion
   increases steadily from zero and then decreases steadily to zero (all within
   0.02 s). What is the maximum force?

11 A jet of water of radius 1 cm and speed $4\,\mathrm{m\,s^{-1}}$ plays directly on to a steel
   plate which completely destroys its momentum. Find:
   a) the loss of momentum of the water which strikes the plate in 1 s;
   b) the force of the water on the plate.

12 A machine gun fires bullets of mass 30 g with muzzle velocity $500\,\mathrm{m\,s^{-1}}$
   and at a rate of 200 per minute. What average force must the gunner exert
   to prevent recoil?

13 At a waterfall, a stream (which has no vertical velocity at the top of the

waterfall) falls 5 m vertically on to a horizontal rock without rebounding. Given that $0.15\,\mathrm{m^3}$ water pass over the waterfall each second, calculate the force exerted on the rock.   (SMP)

14 A horizontal jet of water delivering $5\,\mathrm{kg\,s^{-1}}$ at a speed of $10\,\mathrm{m\,s^{-1}}$ hits a vertical wall. Assuming that the water does not rebound, calculate the force on the wall. If allowance were made for rebound, would this force be increased or decreased? Give a reason for your answer.   (C)

15 A train of mass 1000 tonne started from rest and the effective horizontal force was recorded at intervals of ten seconds

$t\,(\mathrm{s})$	0	10	20	30	40	50	60
$F\,(10^5\,\mathrm{N})$	1.62	1.45	1.36	1.24	1.07	0.91	0.73

Find the total impulse of this force and hence the final momentum and velocity of the train. (The use of Simpson's rule is recommended.)

* Impulse and momentum as vectors

So far we have confined ourselves to the study of a body moving in a straight line. We can now consider the more general case of a body of mass m moving under the action of a variable force $\mathbf{F}$.

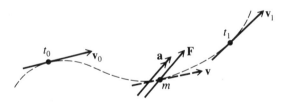

Suppose that after time t the particle has velocity $\mathbf{v}$ and acceleration $\mathbf{a}$ and that when

$$t = t_0 \qquad \mathbf{v} = \mathbf{v}_0$$
$$t = t_1 \qquad \mathbf{v} = \mathbf{v}_1$$

By Newton's second law,

$$\mathbf{F} = m\mathbf{a} = m\frac{\mathrm{d}\mathbf{v}}{\mathrm{d}t}$$

so that we naturally write

$$\int_{t_0}^{t_1} \mathbf{F}\,dt = \int_{t_0}^{t_1} m\frac{d\mathbf{v}}{dt}\,dt = \left[m\mathbf{v} \right]_{t=t_0}^{t=t_1} = m\mathbf{v}_1 - m\mathbf{v}_0$$

This is the first time that we have dared to speak of the integral of a vector quantity, but success with the differentiation of vectors is a strong incentive for such a step; and in any case, the expression

$$\int_{t_0}^{t_1} \mathbf{F}\,dt$$

can be regarded as shorthand for the vector whose components are

$$\int_{t_0}^{t_1} X\,dt \qquad \int_{t_0}^{t_1} Y\,dt \qquad \int_{t_0}^{t_1} Z\,dt$$

(where X, Y, Z are the components of $\mathbf{F}$).

We therefore see that the impulse of a force during an interval of time, like the momentum $m\mathbf{v}$, is essentially a vector quantity; and, again, that

impulse of force during interval = change of momentum

Example 2

The 2 kg mass of Example 1 is now thrown horizontally with a speed of $4\,\mathrm{m\,s^{-1}}$. Investigate the impulse of its weight and the change of its momentum in the first two seconds of its flight.

Taking unit vectors $\mathbf{i}$ horizontally and $\mathbf{j}$ vertically downwards, we see that

$$\text{weight} = 2 \times 9.8\mathbf{j} = 19.6\mathbf{j}$$

$$\Rightarrow \quad \text{impulse} = 19.6\mathbf{j} \times 2 = 39.2\mathbf{j}$$

Also   initial velocity $= 4\mathbf{i}$

and   final velocity $= 4\mathbf{i} + 9.8 \times 2\mathbf{j}$

$$= 4\mathbf{i} + 19.6\mathbf{j}$$

So   initial momentum $= 4\mathbf{i} \times 2 = 8\mathbf{i}$

and   final momentum $= (4\mathbf{i} + 19.6\mathbf{j}) \times 2$

$$= 8\mathbf{i} + 39.2\mathbf{j}$$

Hence change in momentum $= 39.2\mathbf{j}$ and we confirm that this is equal to the impulse of the weight. This can also be summarised diagrammatically:

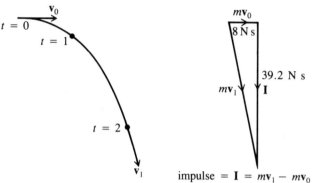

impulse $= \mathbf{I} = m\mathbf{v}_1 - m\mathbf{v}_0$

Example 3

Water flowing in a horizontal pipe of radius 2 cm at a speed of $3\,\text{m s}^{-1}$ first goes round a sharp right-angled bend and then issues as a jet perpendicular to a fixed metal plate which destroys its momentum. Calculate:

a) its thrust on the bend;
b) its thrust on the plate.

In 1 s the stretch of water coming out of the pipe has length 3 m and volume

$$\pi \times (0.02)^2 \times 3 = 12\pi \times 10^{-4}\,\text{m}^3$$

Now $1\,\text{m}^3$ of water has mass 10^3 kg. So the mass of this water is 1.2π kg, and its momentum is

$$1.2\pi \times 3 = 3.6\pi\,\text{N s}$$

Now after going round the bend its momentum changes direction through a right angle, and such a change in momentum must be caused by an impulse **I** such that

$$\mathbf{I} = m\mathbf{v}_1 - m\mathbf{v}_0$$

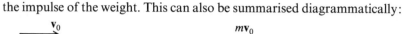

Hence the impulse **I** acts into the bend, at 45° to the two sections of pipe, and has magnitude

$$3.6\pi \times \sqrt{2} \approx 16\,\mathrm{N\,s}$$

But this is the impulse, over an interval of 1 s, of a force of 16 N.

The thrust of the water on the pipe, therefore, is of magnitude 16 N outwards from the bend at 135° with both sections.

When the jet strikes the metal plate, the momentum 3.6π N s is reduced to zero and this therefore necessitates an impulse of 3.6π N s which is caused by a thrust of 3.6π N acting for 1 second. So the thrust of the plate on the jet, and hence of the jet on the plate, has magnitude $3.6\pi \approx 11.4\,\mathrm{N}$.

Exercise 9.5b

1   Find (in SI units) the impulse of
    a) a force $6\mathbf{i} + 2\mathbf{j} + \mathbf{k}$, acting for 2 s;
    b) a force $-\mathbf{i} - 3\mathbf{j} + 2\mathbf{k}$, acting for 3 s;
    c) a force $t\mathbf{i} + (1 - t)\mathbf{j} + 3t^2\mathbf{k}$, from $t = 0$ to 2;
    d) a force $\sin t\,\mathbf{i} + t\mathbf{j}$, from $t = 0$ to $\frac{1}{2}\pi$.

2   Find the final velocity of a mass of 2 kg which has initial velocity $4\mathbf{i} + 3\mathbf{j} + 2\mathbf{k}$ and is acted upon (on separate occasions) by the forces described in no. **1**.

3   A body of mass 4 kg has initial velocity $4\mathbf{i} + 2\mathbf{j} - \mathbf{k}$. What constant force must act upon it so as to produce
    a) velocity $3\mathbf{i} + 2\mathbf{j} + 4\mathbf{k}$ after 2 s;
    b) velocity $\mathbf{i} + \mathbf{k}$ after 5 s?

4   A sudden downpour of r cm of rain, of density $\rho\,\mathrm{kg\,m^{-3}}$, falls in one hour on an area of A km² and p per cent of this water runs straight off the ground into a river b m wide. What is the likely rise in level of the water in the river if this flows at $v\,\mathrm{m\,s^{-1}}$?

    A containing bank turns the river-flow through a right angle. What is the extra force on this bank due to the downpour?   (MEI)

Conservation of momentum

So far we have limited ourselves to the consideration of an impulse upon a single body and its consequent change of momentum. But it is when a number of bodies are connected or interacting that the concept of momentum becomes most valuable.

Suppose that A and B are two particles and that the only forces acting upon them are their mutual reactions $\mathbf{R}$ and $-\mathbf{R}$ which can (as when two billiard balls clash) be changing both in magnitude and direction.

Over any interval of time,

$$\int_{t_0}^{t_1} \mathbf{R}\,\mathrm{d}t = \text{change in momentum of A}$$

$$\int_{t_0}^{t_1} -\mathbf{R}\,\mathrm{d}t = \text{change in momentum of B}$$

But $$\int_{t_0}^{t_1} \mathbf{R}\,\mathrm{d}t + \int_{t_0}^{t_1} -\mathbf{R}\,\mathrm{d}t = \int_{t_0}^{t_1} \mathbf{0}\,\mathrm{d}t = \mathbf{0}$$

So the total change in momentum is zero, and we have verified the *principle of conservation of momentum* for a system which has no external forces acting upon it.

Example 4

A mass of 2 kg moving with velocity $\mathbf{i} - 2\mathbf{j} + 3\mathbf{k}$ coalesces with a mass of 8 kg which is moving with velocity $-\mathbf{i} + 3\mathbf{j} + 6\mathbf{k}$. Find the velocity of the combined mass of 10 kg.

If we let this final velocity be $\mathbf{v}$, we can apply the principle of conservation of momentum and obtain

$$10\mathbf{v} = 2(\mathbf{i} - 2\mathbf{j} + 3\mathbf{k}) + 8(-\mathbf{i} + 3\mathbf{j} + 6\mathbf{k})$$

$$= -6\mathbf{i} + 20\mathbf{j} + 54\mathbf{k}$$

$$\Rightarrow \quad \mathbf{v} = -0.6\mathbf{i} + 2\mathbf{j} + 5.4\mathbf{k}$$

Example 5

A shell of mass m is fired with velocity $\mathbf{v}$ at an angle α to the horizontal. If the gun has mass M and recoils horizontally on smooth rails, find its speed of recoil, V.

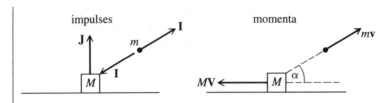

The forces acting for the short interval of the firing are
a) the weights $M\mathbf{g}$ and $m\mathbf{g}$, whose impulses over a very short interval are negligible;
b) the very large explosive force between gun and shell which has equal and opposite impulses $\mathbf{I}$ and $-\mathbf{I}$ on the gun and the shell;
c) the normal reaction between the rails and the gun which exerts a vertical impulse $\mathbf{J}$ on the gun.

So the only external impulse is $\mathbf{J}$, which has no horizontal component. Hence the horizontal component of momentum must be conserved.

$$\Rightarrow \quad mv\cos\alpha - MV = 0$$

$$\Rightarrow \qquad\qquad V = \frac{mv\cos\alpha}{M}$$

Exercise 9.5c

1 Find the velocity of the combined mass formed by the coalescing of
   a) 1 kg moving at $4\mathbf{i} + 3\mathbf{j} - \mathbf{k}$ and   2 kg moving at $\mathbf{i} + 3\mathbf{j} + 2\mathbf{k}$.
   b) 3 kg moving at $\mathbf{i} + \mathbf{j} + \mathbf{k}$,   2 kg moving at $2\mathbf{i} + 3\mathbf{j} - 3\mathbf{k}$,   and   1 kg moving at $3\mathbf{i} + 2\mathbf{j} - 3\mathbf{k}$.

2 A mass of 2 kg moving with velocity $\begin{pmatrix} 3 \\ 5 \end{pmatrix}$ m s^{-1} collides and combines with a mass of 3 kg moving with velocity $\begin{pmatrix} 1 \\ -2 \end{pmatrix}$ m s^{-1}.

   Calculate the velocity of the single mass so formed.
   Find also the impulse which each mass has received.   (SMP)

3 A rocket of mass 10 kg is moving with a velocity whose three components (in m s^{-1}) are $(200, 160, 4)$ and then separates into two parts. If the rear portion has mass 8 kg and velocity $(100, 100, 0)$, find the new velocity of its front portion.

4 A bullet of mass 0.05 kg moving horizontally with velocity 321 m s^{-1} strikes a stationary block of mass 16 kg which is free to slide without rotation on a smooth horizontal plane. The bullet becomes embedded in the block after 0.01 s. Calculate, in newtons, the resistance, assumed uniform, of the block to penetration by the bullet.

5   A rocket of mass 40 tonnes ($= 4 \times 10^4$ kg) is mounted rigidly on a trolley
    of mass 10 tonne which is free to move horizontally. The rocket ejects mass
    1 tonne horizontally in a burst lasting 5 seconds, giving the ejected matter a
    speed which may be taken as 2 km s^{-1} relative to the ground. Calculate the
    velocity of the rocket just after the ejection **a)** neglecting resistances, and **b)**
    assuming resistance of 4×10^4 N.

    If the same operation were carried out with the rocket mounted
    vertically and free to lift off its mounting show that, for a lift-off to occur,
    the duration of the burst (assumed uniform) must not exceed a certain
    critical time, and calculate this time. (The question of resistance does not
    arise in this calculation. Explain why.) (Take $g = 10 \text{ m s}^{-2}$.)   (MEI)

6   Two stones, each of mass $5m$, are moving across a sheet of smooth ice at
    equal speeds of $10v$ in opposite directions on parallel paths, so that no
    collision is involved. A frog of mass m, travelling on one of the stones,
    leaps across to the other one, and in so doing deflects the stone he leaves
    through $30°$ and changes its speed to $8v$.

    Find, by drawing and measurement or by calculation **a)** through what
    angle the other stone is deflected and its subsequent speed **b)** the magnitude
    of the impulse the frog exerts on the stone on which he lands.   (SMP)

9.6 Work, energy, and power

Motion in a straight line

Suppose that a mass m is pulled along a smooth horizontal table by a constant
force of magnitude F and that after moving a distance s its velocity has
increased from v_0 to v_1.

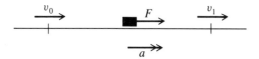

If the magnitude of the constant acceleration of m is a

then $v_1^2 = v_0^2 + 2as$

But, by Newton's second law,

$F = ma$

$\Rightarrow \quad Fs = mas = \tfrac{1}{2}mv_1^2 - \tfrac{1}{2}mv_0^2$

The expression Fs is called the *work done by F in the displacement s*; and if,
further, the expression $\tfrac{1}{2}mv^2$ is called the *kinetic energy* of a mass m moving
with velocity v, we see that $\tfrac{1}{2}mv_1^2 - \tfrac{1}{2}mv_0^2$ is its increase in kinetic energy.

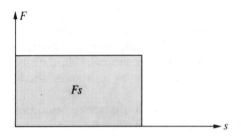

So   **work done by the force = increase of kinetic energy**

In SI units, the unit of work (and therefore of energy) will be 1 N m, which is called 1 joule (1 J).

Example 1

A car of mass 800 kg on a horizontal road accelerates from a speed of $10\,\mathrm{m\,s^{-1}}$ under the action of a tractive force of 500 N and against a resistance of 100 N. Find its speed when it has travelled a distance of 20 m.

Work done by tractive force $= 500 \times 20 = 10\,000\,\mathrm{J}$

Work done by resistance $= -100 \times 20 = -2000\,\mathrm{J}$

Hence total work $= 8000\,\mathrm{J}$

If the final speed $= v$, the gain in kinetic energy

$$= \tfrac{1}{2} \times 800v^2 - \tfrac{1}{2} \times 800 \times 10^2$$

$$= 400(v^2 - 100)$$

So $\qquad 8000 = 400(v^2 - 100)$

$\Rightarrow \quad v^2 - 100 = 20$

$\Rightarrow \qquad\qquad v^2 = 120$

$\Rightarrow \qquad\qquad v = 10.95$

So final speed is $10.95\,\mathrm{m\,s^{-1}}$.

Exercise 9.6a

1   A man pushes a chest of mass 20 kg through a distance of 6 m across a horizontal floor by means of a horizontal force of 50 N. Find:
a) the work done by this force;

b) the final velocity of the chest if the floor is smooth;
c) the final velocity if the coefficient of friction is 0.1.

2 Ignoring all resistances, what mean propulsive force is required to accelerate an aircraft of mass 10 tonne from rest to a take-off speed of $40 \, \text{m s}^{-1}$ on a runway of length 1 km? What force would be required if the take-off speed had to be doubled in the same length of runway?

3 A bullet of mass 20 g is travelling at a speed of $400 \, \text{m s}^{-1}$. If it penetrates a block of wood to a depth of 5 cm, find the average resistance.

4 A stone of mass 2 kg is skidding across a sheet of ice at a speed of $10 \, \text{m s}^{-1}$. If it comes to rest in 80 m, find:
a) the work done by friction;
b) the coefficient of friction.

5 A lorry of mass 8 tonne needs to brake from a speed of $10 \, \text{m s}^{-1}$ in a distance of 80 m. Find:
a) the work done by the braking force;
b) the average braking force;
c) the average braking force required if this distance were to be halved;
d) the average braking force required if the initial speed were to be doubled.

6 Express a vehicle's braking distance d in terms of its speed v and maximum braking force F. Supposing that F is constant, sketch a graph showing d as a function of v.

7 A mass of 2 kg moving at $3 \, \text{m s}^{-1}$ catches up and joins a mass of 3 kg moving at $2 \, \text{m s}^{-1}$ in the same direction. Find:
a) their velocities after impact;
b) their total loss of kinetic energy.

Variable forces

When F is a variable force in the constant direction of motion, the total work done in a displacement can be regarded as the sum of the small amounts $F \, \delta s$, and is therefore represented by the area under the F–s graph:

$$\text{Work} = \lim \sum F \, \delta s = \int_0^s F \, ds$$

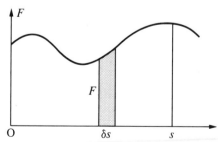

Potential energy

It is clear that if a mass m is at a height h above a given base-level, its weight mg would do an amount of work mgh if it were to return to its original level. This, therefore, is called its *gravitational potential energy* (referred to the given base-level) when at height h.

If there are no other forces (including, for a system, internal forces too) which do work, it follows that the loss of potential energy (p.e.) must be equal to the gain of kinetic energy (k.e.):

loss of p.e. = gain of k.e.

$\Rightarrow$ k.e. + p.e. = constant

which is an example of the *principle of conservation of energy*.

Example 2

Two masses of 1 kg and 4 kg are connected by a light string which passes over a smooth pulley and hangs vertically. If they are released from rest find their speeds when each has moved 2 m.

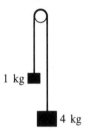

1 kg

4 kg

Let required speed $= v\,\text{m s}^{-1}$

Now   loss in p.e. $= 4g \times 2 - g \times 2$

$= 58.9\,\text{J}$

and   gain in k.e. $= \frac{1}{2} \times 1 \times v^2 + \frac{1}{2} \times 4 \times v^2$

$= \frac{5}{2}v^2\,\text{J}$

But the only forces which do any work are the two weights.

So   gain of k.e. = loss of p.e.

$\frac{5}{2}v^2 = 58.9$

$\Rightarrow$ $v = 4.85\,\text{m s}^{-1}$

Work as a scalar product

So far, we have restricted ourselves to the work done by a force when its displacement is in its own direction, but we are now able to examine the more general case.

If a force **F** undergoes a displacement **s**, the work done is defined as the scalar product **F.s**. We therefore see that work is a scalar of magnitude $Fs \cos \theta$ (where θ is the angle between **F** and **s**) and so can be regarded as

$F \times$ component of **s** along **F**

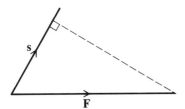

or $s \times$ component of **F** along **s**

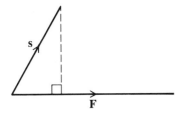

In particular,

a) when $\theta = 0$,   work $= +Fs$

b) when $\theta = \pi$,   work $= -Fs$

c) when $\theta = \frac{1}{2}\pi$,   work $= 0$

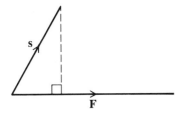

Hence the work done in any displacement perpendicular to the force is always zero. In particular, the work done by a normal reaction at a smooth surface is always zero.

Moreover, we have already seen (in section 8.3) that the operation of scalar multiplication is distributive with respect to addition.

Hence $\mathbf{F}_1.\mathbf{s} + \mathbf{F}_2.\mathbf{s} = (\mathbf{F}_1 + \mathbf{F}_2).\mathbf{s}$ and the sum of the amounts of work done by two forces is equal to the work done by their resultant.

Similarly, $\mathbf{F}.\mathbf{s}_1 + \mathbf{F}.\mathbf{s}_2 = \mathbf{F}.(\mathbf{s}_1 + \mathbf{s}_2)$ and the sum of the amounts of work done by a force in two displacements is equal to the work done by the force in the resultant displacement.

Furthermore, if $\mathbf{F}$ is a variable force, the work done in a small displacement $\delta\mathbf{s}$ will be δW, where

$$\delta W = \mathbf{F}.\delta\mathbf{s}$$

Hence the total work can be written as

$$W = \lim \sum \mathbf{F}.\delta\mathbf{s} = \int \mathbf{F}.d\mathbf{s}$$

Alternatively, if the velocity and acceleration of the particle are $\mathbf{v}$, $\mathbf{a}$ respectively, we can write

$$\frac{\delta W}{\delta t} = \mathbf{F}.\frac{\delta\mathbf{s}}{\delta t} \approx \mathbf{F}.\mathbf{v}$$

$$\Rightarrow \quad \frac{dW}{dt} = \mathbf{F}.\mathbf{v}$$

But, by Newton's second law,

$$\mathbf{F} = m\mathbf{a} = m\frac{d\mathbf{v}}{dt}$$

$$\Rightarrow \quad \frac{dW}{dt} = m\frac{d\mathbf{v}}{dt}.\mathbf{v} = \frac{d}{dt}(\tfrac{1}{2}m\mathbf{v}^2)$$

$$\Rightarrow \quad W = [\tfrac{1}{2}m\mathbf{v}^2]$$

$\Rightarrow$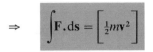

and again we see that the work done by a force is equal to the increase in kinetic energy.

Example 3

A particle of mass 6 kg slides from rest down a rough plane which is inclined at 30° to the horizontal. If the resistance to motion is 12 N throughout, find the work done by the various forces when the mass has moved a distance 8 m and hence find its velocity at this point. (Take $g \approx 10 \, \text{m s}^{-2}$.)

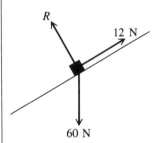

The three forces on the particle are
a) its weight $6 \times 10 = 60 \, \text{N}$;
b) the normal reaction R;
c) the resistance $12 \, \text{N}$;
and the amounts of work done by these in the displacement 8 m are:
a) $60 \times 8 \times \cos 60° = 240 \, \text{J}$;
b) zero (since force is perpendicular to displacement);
c) $-12 \times 8 = -96 \, \text{J}$.

Hence total work $= 144 \, \text{J}$

But final k.e. $= \frac{1}{2} \times 6 \times v^2 = 3v^2$

So $144 = 3v^2$

$\Rightarrow \quad v^2 = 48 \quad \Rightarrow \quad v = 6.9 \, \text{m s}^{-1}$

Exercise 9.6b

1   Find the work done in (SI units) when a force $\mathbf{i} + 3\mathbf{j} + \mathbf{k}$ acts on a particle and displaces it by an amount
   a) $-\mathbf{i} + 2\mathbf{j} + 3\mathbf{k}$ **b)** $3\mathbf{i} - 2\mathbf{j} - \mathbf{k}$ **c)** $\mathbf{i} - \mathbf{k}$

2   A car of mass $800\,\text{kg}$ is travelling at $30\,\text{m s}^{-1}$ with its engine switched off when it comes to a small hill. Find:
   a) its k.e. at the foot of the hill;
   b) the maximum height of hill which the car can surmount (ignoring all resistances).

3   A boy slides down a smooth chute at a swimming pool. If its top is $6\,\text{m}$ above the water and its bottom is $1\,\text{m}$ above the water, find his speed
   a) on leaving the chute;
   b) on entering the water.

4   A box of mass $10\,\text{kg}$ is attached to the end of a rope of length $5\,\text{m}$ and hangs freely. If it is then given a horizontal impulse of $100\,\text{N s}$ find:
   a) its initial speed;
   b) its initial kinetic energy;
   c) the height to which it will rise.

5   A marble of mass $20\,\text{g}$ rests at the top of a smooth circular tube which has diameter $1\,\text{m}$ and is in a vertical plane. If the marble is just displaced from rest, find its speed
   a) when travelling vertically at the end of the horizontal diameter;
   b) when passing through the lowest point of the tube;
   c) when $25\,\text{cm}$ above the lowest point.
   Hence find the reactions of the tube on the marble at these three instants.

6   A boy of mass $40\,\text{kg}$ swings on a rope of length $2\,\text{m}$ which is just taut and attached to a point at the same height. When the rope is vertical, find:
   a) the kinetic energy of the boy;
   b) the speed of the boy;
   c) the tension in the rope.

7   A commando swings on a rope fixed at its upper end A. The rope, initially inclined at $60°$ to the downward vertical, remains taut and may be considered as inextensible and of negligible mass. The commando can be idealised as a particle at the end of the rope. If **a)** his mass is $70\,\text{kg}$,   **b)** the rope is $10\,\text{m}$ long,   **c)** he lets go when the rope is vertical and   **d)** A is $12\,\text{m}$ above the ground,   find his landing speed.
   (Neglect air resistance, and take $g = 10\,\text{m s}^{-2}$.)
   Which of **a)**, **b)**, **c)**, **d)** can be varied, one at a time, without affecting this speed?   (SMP)

8   A particle of mass m is slightly displaced from the top of a smooth

hemispherical dome of radius a. If its speed is v when the radius to the particle makes an angle θ with the vertical, find:
a) an equation connecting v with a, g and θ;
b) the normal reaction N in terms of a, g and θ;
c) the value of θ for which $N = 0$ (i.e. at which the particle leaves the surface).

Power

We know from everyday experience that when a certain task can be accomplished by either of two forces, it frequently happens that one of them can do the necessary work more quickly. Such a force is then said to have 'greater power'.

More precisely, we define the *power P* of a force as its *rate of doing work*

$P = \mathrm{d}W/\mathrm{d}t$

But we have already seen that when a force F is moving its point of application with velocity v in the same direction,

$$\frac{\mathrm{d}W}{\mathrm{d}t} = Fv$$

Hence $\qquad P = \dfrac{\mathrm{d}W}{\mathrm{d}t} = Fv$

So the unit of power in SI is 1 joule per second and is called 1 *watt* (W):

$1\,\mathrm{W} = 1\,\mathrm{J\,s^{-1}}$

Example 4

In Imperial units the standard unit of power was 1 horse power (h.p.), being the rate of working required to lift a mass of 550 pounds through 1 foot in 1 second.

Taking $1\,\mathrm{lb} = 0.454\,\mathrm{kg}$, $1\,\mathrm{ft} = 0.305\,\mathrm{m}$, $g = 9.81\,\mathrm{m\,s^{-1}}$, find 1 h.p. in kilowatts.

In 1 s, a force working at a rate of 1 h.p. will lift 550 lb a height of 1 foot, i.e. $550 \times 0.454\,\mathrm{kg}$ a height of $0.305\,\mathrm{m}$.

So work done in 1 s

$= (550 \times 0.454 \times 9.81) \times 0.305$ J

$= 747$ J

Hence   1 h.p. $= 747$ J s^{-1} $= 747$ W

$\Rightarrow$ 1 h.p. $= 0.747$ kW

Example 5

Find the power of a pump which raises water from a lake at a rate of 5 m^3 s^{-1} and delivers it 6 m higher at a speed of 8 m s^{-1}.

In 1 s, the mass of water moved $= 5 \times 10^3$ kg

So weight of water moved $= 5 \times 10^3 \times 9.81$ N and the work done in raising it through 6 m

$= 5 \times 10^3 \times 9.81 \times 6 = 2.94 \times 10^5$ J

But this water is also given an amount of kinetic energy

$= \frac{1}{2} \times (5 \times 10^3) \times 8^2 = 1.6 \times 10^5$ J

So total work done in 1 s $= 4.54 \times 10^5$ J and the power of the pump

$= 4.54 \times 10^5$ W

$= 454$ kW

Exercise 9.6c

1   A truck has an engine which develops 24 kW at the truck's maximum speed, on a level road, of 25 m s^{-1}. Calculate the total resistance to motion at this speed.   (L)

2   A pump, working at an effective rate of 41 kW, raises 80 kg of water per second from a depth of 20 m. Calculate its speed of delivery.   (L)

3   A pump delivers water at the rate of 900 kg min^{-1} from a reservoir to a nozzle 8 m above the surface of the reservoir. The water emerges from the nozzle with a speed of 12 m s^{-1}. Calculate the power, in kilowatts, of the pump.   (C)

4   A pump raises 1000 kg of water each minute from a depth of 20 m and

delivers it at a speed of $8 \, \mathrm{m \, s^{-1}}$. Assuming that the efficiency of the pump is 40%, calculate its power. (OC)

5   A railway train has mass $1.4 \times 10^6 \, \mathrm{kg}$ and the maximum power of the engine is $1.8 \times 10^6 \, \mathrm{W}$. Calculate the maximum acceleration, in $\mathrm{m \, s^{-2}}$, when the train is travelling on the level at a speed of $12 \, \mathrm{m \, s^{-1}}$, the total resistance being $9.6 \times 10^4 \, \mathrm{N}$. (C)

6   A car of power $50 \, \mathrm{kW}$ has mass $800 \, \mathrm{kg}$, which is equally distributed over the four wheels. The coefficient of friction between the two driving wheels and the road is 0.3. If the road is level, and resistances to motion can be neglected, find the maximum acceleration of the car (in $\mathrm{m \, s^{-2}}$)
a)   at $20 \, \mathrm{km \, h^{-1}}$;
b)   at $10 \, \mathrm{km \, h^{-1}}$. (OC)

7   An escalator takes passengers from a lower level to a higher level through a vertical height of $20 \, \mathrm{m}$. On average two passengers get on the escalator per second and two leave at the top per second. If the average mass of a passenger is taken to be $75 \, \mathrm{kg}$ and the machinery has 60% efficiency, calculate, in watts, the average power output of the motor which drives the escalator. (Take g to be $9.8 \, \mathrm{m \, s^{-2}}$.) (C)

8   A car of mass $1000 \, \mathrm{kg}$ is moving on a level road at a steady speed of $100 \, \mathrm{km \, h^{-1}}$ with its engine working at $60 \, \mathrm{kW}$. Calculate in newtons the resistance to motion, which may be assumed to be constant.
The engine is now disconnected, the brakes applied, and the car comes to rest in $100 \, \mathrm{m}$. Find the additional retarding force arising from the application of the brakes.
If the engine is still disconnected, find the distance the car would run up a hill of inclination $\sin^{-1} \frac{1}{10}$ before coming to rest, starting at $100 \, \mathrm{km \, h^{-1}}$, when the same resistance and braking force are operating. (C)

9   Sand pours over the lip of a chute and falls vertically onto a conveyor belt which is moving horizontally at a steady speed of $0.5 \, \mathrm{m \, s^{-1}}$. If the supply of sand is steady at the rate of $50 \, \mathrm{kg \, s^{-1}}$, what horizontal momentum is given to the sand each second? Find the force that is required to give the sand this momentum.
If $10 \, \mathrm{W}$ is the power required to operate the conveyor belt without a load, find whether a motor of maximum power $25 \, \mathrm{W}$ can deal with sand at the rate of supply of $50 \, \mathrm{kg \, s^{-1}}$.
If the chute were modified so that the sand was delivered with a horizontal component of velocity of $0.5 \, \mathrm{m \, s^{-1}}$ in the direction of motion of the belt, find the maximum speed at which the belt could move the load, assuming that the power required for the belt itself is unaltered and the sand is delivered at the same rate as before. (MEI)

10   A VTOL aircraft hovers by taking in air through a large intake and

ejecting it at high velocity downwards. If the relative velocity of ejection is v m s^{-1} and the mass of the aircraft is M kg, find expressions for
a) the mass of air ejected per second for hovering;
b) the power required to hover.

Later, when the aircraft is flying as a normal aeroplane, the engines work at half this power and the air resistance varies as the square of the speed. The maximum steady speed attainable in level flights is V m s^{-1}. Show that at a speed of $\frac{3}{4}V$ m s^{-1} the engines would supply enough power for the aircraft to climb in normal flight at a rate of $37v/256$ m s^{-1}. (MEI)

11   An engine is fitted to a vehicle of mass m kg. It is known from tests that it will give constant tractive force F N up to a road-speed of U m s^{-1}, and at higher speeds will give constant power FU watt. Show that at speed v m s^{-1} $(v \geqslant U)$ the tractive force is FU/v N. Form the appropriate equations of motion for the vehicle if the resistance to motion at all speeds is given by kv^2 N, k being constant.

Show that there is a maximum attainable speed V m s^{-1} and find an expression for V in terms of k, U and F. Given that $m = 500$, $F = 2500$, $U = 25$ and $V = 50$, determine the value of k. Calculate the acceleration at each of the speeds 5, 15, 25, 35 m s^{-1} and hence estimate the time taken to reach 40 m s^{-1} from rest, giving your result to the nearest second. Explain your method. (MEI)

Miscellaneous problems

1   A particle of mass m is placed on a smooth horizontal surface and is first accelerated in a straight line by a constant horizontal force F_1, then decelerated to rest by a constant horizontal force F_2.

If s is the total distance travelled, prove that the maximum speed is

$$\sqrt{\left(\frac{2F_1 F_2 s}{m(F_1 + F_2)}\right)}$$

2

The diagram shows a particle A of mass 2 kg resting on a horizontal table.

It is attached to particles B of mass 5 kg and C of mass 3 kg by light inextensible strings hanging over light smooth pulleys. Initially the plane ABC is vertical, the strings are taut and pass over the pulleys at right angles to the edges of the table. In the ensuing motion from rest find the common acceleration of the particles and the tension in each string before A reaches an edge

a) when the table is smooth,

b) when the table is rough and the coefficient of friction between the particle A and the table is $\frac{1}{2}$.   (L)

(Take g as $10 \, \text{m/s}^2$.)

3   A particle is attached to one end of a light string, the other end of which is fixed. When the particle moves in a horizontal circle with speed $2 \, \text{m s}^{-1}$, the string makes an angle $\tan^{-1}(5/12)$ with the vertical. Show that the length of the string is approximately 2.5 m.   (L)

4   A dog of mass m dives off the stern of a punt of mass M, which was previously at rest. Immediately after jumping off, the dog has a horizontal speed of V (as seen by an observer on the bank). Calculate the speed with which the punt begins to move.

Prove that the total kinetic energy of the dog and the punt is then

$$\frac{m(M + m) \, V^2}{2 \, M}$$ (SMP)

5   The coordinates of a moving point P are

$$(a \cos \omega t, \, a \sin \omega t)$$

where a and ω are constants. Find the magnitude and direction of the velocity and acceleration of P at time t.

There is a speed limit of $80 \, \text{km h}^{-1}$ round an unbanked curve on a railway track. The curve is in the shape of an arc of a circle and the angle turned through is $30°$ in 500 m.

Find:

a) the radius of the circle;

b) the lateral force on the rails at maximum speed as a fraction of the weight;

c) the least distance before the curve that a warning sign must be placed if the trains cruise at $150 \, \text{km h}^{-1}$ and a comfortable braking retardation is $\frac{1}{2} \, \text{m s}^{-2}$.   (MEI)

6   An object of unit mass is moving in a plane and its position vector from O is $\mathbf{r}$ and its coordinates (r, θ). If

$$\mathbf{r} = \frac{\cos \theta}{2 + \cos \theta} \mathbf{i} + \frac{\sin \theta}{2 + \cos \theta} \mathbf{j}$$

obtain an expression for r in terms of θ.

By differentiating find $\dot{\mathbf{r}}$ in terms of $\dot{\theta}$, $\mathbf{i}$ and $\mathbf{j}$, and if $r^2\dot{\theta} = a$, where a is a constant, prove that

$$\dot{\mathbf{r}} = -2a\sin\theta\mathbf{i} + a(2\cos\theta + 1)\mathbf{j}$$

Hence, by differentiating again, prove that the resultant force acting on the object is of magnitude $2a^2/r^2$ and directed towards O.   (SMP)

7   A particle P of mass m is displaced from rest on the top of a smooth sphere whose centre is O and radius a. Find:
a) the kinetic energy gained by the particle when OP makes an angle θ with the upward vertical;
b) the normal reaction of the sphere on the particle at this instant;
c) the value of θ when the particle leaves the sphere.

8   A planet of mass m moves round the Sun in an approximately circular orbit of radius R and period T. Find:
a) the velocity of the planet;
b) its acceleration towards the Sun;
c) the gravitational attraction of the Sun on the planet.
If, as Kepler discovered, T^2 is proportional to R^3, show that the attraction due to gravity is inversely proportional to the square of distance.

9   Supposing that the Moon moves round the Earth in a circular path of radius R and period T,
a) find the acceleration of the Moon towards the Earth;
b) estimate the acceleration due to gravity at the surface of the Earth, assuming the inverse square law of gravity, $T = 27.3$ days, $R = 3.84 \times 10^8$ m and the radius of the Earth is 6.37×10^6 m.

10 A particle of unit mass is initially at rest at a very great distance and is then attracted to Earth by a gravitational force k/r^2, where r is its distance from the centre of the Earth. Find:
a) the work that has been done by the gravitational force when the particle is at distance r;
b) its velocity at this instant;
c) the value of k, assuming that $g = 9.81\,\mathrm{m\,s^{-2}}$ at the surface of the Earth, when $r = 6.37 \times 10^6$ m;
d) the velocity of the particle, ignoring atmospheric resistance, when it strikes the Earth. (Suppose that $g = 9.81\,\mathrm{m\,s^{-2}}$ at the surface of the Earth.)
Hence find the least velocity (the *velocity of escape*) with which a particle must be projected if it is to move out of the Earth's gravitational field.

11 The Law of Universal Gravitation states that the mutual attraction between masses m_1 amd m_2 at a distance r apart is Gm_1m_2/r^2, where G is the *constant of gravitation*, found by Henry Cavendish in 1788 to be approximately $6.67 \times 10^{-11} \, \text{kg}^{-1} \, \text{m}^3 \, \text{s}^2$.

Taking the radius of the Earth to be 6.37×10^6 m and the value of g to be 9.81 m s^{-2} at the surface of the Earth, calculate the mass of the Earth.

(c)

10 Further calculus

10.1 The logarithmic function, ln x

When we first considered integration, we found that

$$\int t^n \, dt = \frac{t^{n+1}}{n+1} + c, \quad \text{provided that } n \neq -1$$

When $n = -1$, however, this expression contains the meaningless term $t^0/0$. Nevertheless, the problem of finding $\int t^{-1} \, dt$ is certainly not meaningless, and will arise whenever (which is frequently) we need to know the area beneath a section of the curve $f(t) = 1/t$. But readers will probably find that, despite all their efforts, they fail to discover a function whose derivative is $1/t$, and must therefore return to a more careful examination of the area which this integral represents.

We start by considering the area beneath the curve $f(t) = 1/t$ between $t = 1$ and $t = x$, which we call $l(x)$.

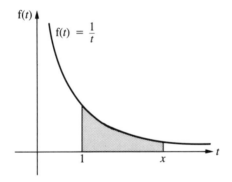

So $\quad l(x) = \displaystyle\int_1^x \frac{1}{t} \, dt, \quad$ which is defined for all $x > 0$

Clearly, if $\qquad x > 1, \quad l(x) > 0$

if $\qquad\qquad 0 < x < 1, \quad l(x) < 0$

and $\qquad\qquad\qquad l(1) = 0$

We can now estimate $l(x)$ for any positive value of x by means of counting squares or by Simpson's rule.

Exercise 10.1a

Count squares or use Simpson's rule (with four intervals in each case) to estimate the values of
a) $l(2)$ **b)** $l(3)$ **c)** $l(6)$ **d)** $l(9)$ **e)** $l(12)$ **f)** $l(\frac{1}{2})$ **g)** $l(\frac{1}{3})$

Properties of $l(x)$

After this last exercise, the reader may have suspected that $l(6) = l(2) + l(3)$, $l(12) = l(2) + l(6)$, and more generally that $l(xy) = l(x) + l(y)$.

This can easily be demonstrated by taking the shaded area between $t = 1$ and $t = y$ and squeezing it, like toothpaste, into the area between $t = x$ and $t = xy$:

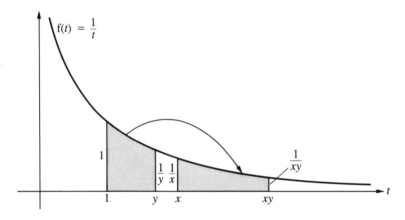

As all horizontal dimensions are stretched by a factor x and all vertical dimensions are compressed by a factor $\dfrac{1}{x}$, the area remains unchanged.

Hence $l(xy) = l(x) + l(y)$

Furthermore, we can now proceed to show that $l(x^n) = nl(x)$.

When $n \in \mathbb{N}$, this is immediately obvious by repeated application of the previous result.

Furthermore, $l(x^n) + l(x^{-n}) = l(x^0) = l(1) = 0$

$$\Rightarrow\quad l(x^{-n}) = -l(x^n) = -nl(x)$$

So the result is also true for negative integers.

Also $l(x^0) = l(1) = 0 = 0l(x)$

Hence the result is true for all $n \in \mathbb{Z}$.

Finally, if $n = p/q$ (where p, q are integers),

$$ql(x^{p/q}) = l(x^{p/q})^q = l(x^p) = pl(x)$$

$$\Rightarrow \quad l(x^{p/q}) = \frac{p}{q}l(x)$$

Hence $\boxed{l(x^n) = nl(x)}$ for all $n \in \mathbb{Q}$

Now these results

$$l(xy) = l(x) + l(y)$$

$$l(x^n) = nl(x)$$

are, of course, also properties of logarithms. So the question immediately arises whether $l(x)$ itself is also a logarithm. But first we must introduce the crucial number e such that $l(e) = 1$, i.e. $t = e$ is the line up to which there is unit area beneath the given curve.

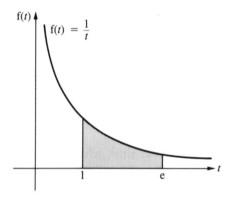

By counting squares, we can show that $e \approx 2.72$, and we shall soon be able to calculate it to a much greater degree of accuracy.

Using the above results, we see that

$$l(e^{l(x)}) = l(x)l(e) = l(x)$$

So the areas under the curve to the value $e^{l(x)}$ and to the value x are equal. These must therefore be the same, and $e^{l(x)} = x$

$$\Rightarrow \quad \boxed{l(x) = \log_e x}$$

ln *x*

$\log_e x$ is usually called the *natural*, or *Napierian*†, *logarithm of x*, and is written ln *x*.

So $\ln xy = \ln x + \ln y$

$\ln x^n = n \ln x$

We shall also usually write $\displaystyle\int \frac{1}{t}\,dt$ and $\displaystyle\int \frac{1}{x}\,dx$ as $\displaystyle\int \frac{dt}{t}$ and $\displaystyle\int \frac{dx}{x}$

So, in summary, if $x > 0$:

$$\int_1^x \frac{dt}{t} = \log_e x = \ln x$$

$$\int \frac{dx}{x} = \ln x + c$$

$$\frac{d}{dx}(\ln x) = \frac{1}{x}$$

and the relationship between $\dfrac{1}{x}$ and ln *x* can be conveniently illustrated by their graphs:

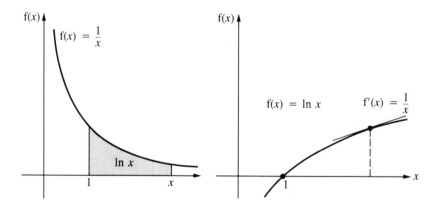

In particular, we notice that

when $x \leqslant 0$,   ln x is not defined

as $x \rightarrow + \infty$,   ln $x \rightarrow + \infty$

as $x \rightarrow 0 +$,   ln $x \rightarrow - \infty$

We are now able to find other areas bounded by such graphs.

Example 1

Investigate the areas represented by

a) $\displaystyle\int_{2}^{5} \frac{dx}{x}$ b) $\displaystyle\int_{-3}^{-2} \frac{dx}{x}$ c) $\displaystyle\int_{-1}^{2} \frac{dx}{x}$

a) $\displaystyle\int_{2}^{5} \frac{dx}{x} = \left[\ln x \right]_{2}^{5}$

$= \ln 5 - \ln 2$

$= \ln \tfrac{5}{2}$

$= \ln 2.5 \approx 0.92$

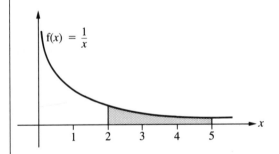

b) $\displaystyle\int_{-3}^{-2} \frac{dx}{x}$ represents an area below the x-axis, and our natural approach would be similar:

$\displaystyle\int_{-3}^{-2} \frac{dx}{x} = \left[\ln x \right]_{-3}^{-2}$

$= \ln(-2) - \ln(-3)$

$= \ln \frac{-2}{-3} = \ln \tfrac{2}{3} \approx -0.41$

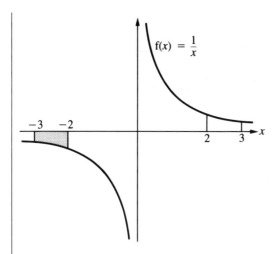

But the reader will rightly object that $\ln(-2) - \ln(-3)$ is completely meaningless, since neither $\ln(-2)$ nor $\ln(-3)$ has been defined. We can, however, easily side-step this obstacle by using the symmetry of the graph about the origin, observing that

$$\int_{-3}^{-2} \frac{dx}{x} = -\int_{2}^{3} \frac{dx}{x} = -\left[\ln x\right]_2^3$$
$$= -(\ln 3 - \ln 2)$$
$$= -\ln 1.5$$
$$\approx -0.41$$

c) $\displaystyle\int_{-1}^{2} \frac{dx}{x}$

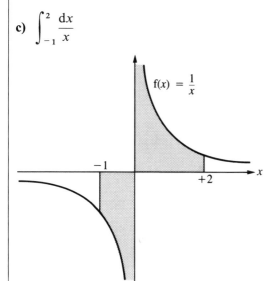

We have already seen that $\ln x \to -\infty$ as $x \to 0+$, so the area beneath the curve between $x = 0$ and $x = 2$ is infinite; and similarly the area to the left of $x = 0$ is infinite.

So $\displaystyle\int_{-1}^{2} \dfrac{dx}{x}$ is composed of two parts, both of which are infinite. It is therefore meaningless and without value.

Example 2

Differentiate:

a) $\ln 5x^3$ **b)** $\ln \dfrac{x^2 + 1}{x}$ **c)** $\lg x$

a) $\ln 5x^3 = \ln 5 + \ln x^3 = \ln 5 + 3\ln x$

$\Rightarrow \quad \dfrac{d}{dx}\ln 5x^3 \quad = \dfrac{d}{dx}(\ln 5 + 3\ln x) = \dfrac{3}{x}$

Or $\quad \dfrac{d}{dx}(\ln 5x^3) = \dfrac{1}{5x^3} \times 15x^2 = \dfrac{3}{x}$

b) $\dfrac{d}{dx}\left(\ln \dfrac{x^2 + 1}{x}\right) = \dfrac{d}{dx}[\ln(x^2 + 1) - \ln x]$

$$= \dfrac{1}{x^2 + 1} \times 2x - \dfrac{1}{x}$$

$$= \dfrac{x^2 - 1}{x(x^2 + 1)}$$

Or $\quad \dfrac{d}{dx}\left(\ln \dfrac{x^2 + 1}{x}\right) = \dfrac{x}{x^2 + 1} \times \dfrac{x \times 2x - (x^2 + 1)}{x^2}$

$$= \dfrac{x^2 - 1}{x(x^2 + 1)}$$

c) Let $\quad y = \lg x$

Then $\quad x = 10^y$

$\Rightarrow \quad \ln x = y \ln 10$

$\Rightarrow \quad y = \dfrac{\ln x}{\ln 10}$

$$\Rightarrow \qquad \frac{dy}{dx} = \frac{1}{(\ln 10)x} \approx \frac{1}{(2.3026)x}$$

So $\quad \dfrac{d}{dx}(\lg x) \approx \dfrac{0.4343}{x}$

Example 3

a) $\displaystyle\int \frac{x^2}{x^3 + 1}\, dx$ **b)** $\displaystyle\int \frac{f'(x)}{f(x)}\, dx$ **c)** $\displaystyle\int_0^{\pi/4} \tan x \, dx$

a) Let $\quad I = \displaystyle\int \frac{x^2 \, dx}{x^3 + 1}$

Now $\quad u = x^3 + 1 \quad \Rightarrow \quad du = 3x^2 \, dx$

So $\quad I = \displaystyle\int \frac{\frac{1}{3}\, du}{u} = \tfrac{1}{3} \ln u + c$

$\Rightarrow \quad \displaystyle\int \frac{x^2 \, dx}{x^3 + 1} = \tfrac{1}{3} \ln (x^3 + 1) + c$

b) Let $\quad I = \displaystyle\int \frac{f'(x)}{f(x)}\, dx$

Here again we let $\quad u = f(x) \quad \Rightarrow \quad du = f'(x)\, dx$

So $\quad I = \displaystyle\int \frac{du}{u} = \ln u + c$

$$= \ln f(x) + c$$

so that we are now able to integrate any quotient whose numerator is the derivative of its denominator.

c) $I = \displaystyle\int_0^{\pi/4} \tan x \, dx$

This can be expressed as

$$I = \int_0^{\pi/4} \frac{\sin x}{\cos x}\, dx = -\int_0^{\pi/4} \frac{-\sin x}{\cos x}\, dx$$

So, using **b)** above,

$$I = \left[-\ln \cos x \right]_0^{\pi/4}$$

$$= -\left(\ln \frac{1}{\sqrt{2}} - \ln 1 \right) = \tfrac{1}{2} \ln 2$$

Hence $\displaystyle\int_0^{\pi/4} \tan x \, dx = \tfrac{1}{2} \ln 2$

Exercise 10.1b

1   Given that $\ln 2 = 0.69315$, $\ln 3 = 1.09861$, $\ln 5 = 1.60944$, calculate to 5 places of decimals:
   a) $\ln 4$ **b)** $\ln 6$ **c)** $\ln 10$ **d)** $\ln 12$ **e)** $\ln \frac{1}{2}$
   f) $\ln \frac{1}{6}$ **g)** $\ln \frac{1}{12}$ **h)** $\ln 10^6$ **i)** $\ln \sqrt{2}$ **j)** $\ln \sqrt[3]{100}$
   Check these answers by use of your calculator.

2   Evaluate: **a)** $\ln e^3$ **b)** $\ln 1$ **c)** $\ln \dfrac{1}{e}$ **d)** $\ln \sqrt{e}$

3   Solve the equations:
   a) $\ln x = 2$ **b)** $\ln x = 100$ **c)** $\ln x = \frac{1}{10}$ **d)** $\ln x = -2$

4   Differentiate with respect to x:
   a) $\ln 2x$ **b)** $\ln x^3$ **c)** $\ln 3x^4$

   d) $\ln \dfrac{x}{3}$ **e)** $\ln \dfrac{1}{x}$ **f)** $\ln \sqrt{x}$

   g) $\ln (x^2 + 1)$ **h)** $\ln \dfrac{x+1}{x}$ **i)** $\ln \sqrt{(x^2 - 1)}$

   j) $\ln x \sqrt{(x+1)}$ **k)** $x \ln x$ **l)** $\dfrac{\ln x}{x}$

5   Differentiate with respect to x:
   a) $\ln \sin x$ **b)** $\ln \cos x$ **c)** $\ln \sec x$
   d) $\ln (\sec x + \tan x)$ **e)** $\ln (3 \sin x)$ **f)** $\ln (\sin 3x)$
   g) $\ln (\sin^2 x)$ **h)** $\ln (4 \sin^2 x)$ **i)** $(\ln x)^2$
   j) $\ln (\ln x)$

6   Find the following integrals:

   a) $\displaystyle\int \dfrac{dx}{3x}$ **b)** $\displaystyle\int \dfrac{dx}{x+2}$ **c)** $\displaystyle\int \dfrac{dx}{2x+3}$

   d) $\displaystyle\int \dfrac{x+1}{x} dx$ **e)** $\displaystyle\int \dfrac{dx}{1-x}$ **f)** $\displaystyle\int \dfrac{2x \, dx}{x^2 + 1}$

g) $\displaystyle\int \frac{x^2\,dx}{1-x^3}$

h) $\displaystyle\int \frac{x-1}{x^2-2x+3}\,dx$

i) $\displaystyle\int \frac{x^2+1}{x^3+3x}\,dx$

j) $\displaystyle\int \cot x\,dx$

k) $\displaystyle\int \tan 3x\,dx$

l) $\displaystyle\int \cot \tfrac{1}{2}x\,dx$

m) $\displaystyle\int \frac{\sin x+\cos x}{\sin x-\cos x}\,dx$

n) $\displaystyle\int \frac{\sec^2 x}{1+\tan x}\,dx$

7 Evaluate (or, if this is impossible, illustrate):

a) $\displaystyle\int_2^5 \frac{dx}{x}$

b) $\displaystyle\int_4^6 \frac{dx}{x+1}$

c) $\displaystyle\int_2^3 \frac{dx}{2x+3}$

d) $\displaystyle\int_1^3 \frac{dx}{x-4}$

e) $\displaystyle\int_1^3 \frac{dx}{x-2}$

f) $\displaystyle\int_2^3 \frac{x\,dx}{x^2-1}$

g) $\displaystyle\int_2^4 \frac{dx}{3-2x}$

h) $\displaystyle\int_0^1 \frac{dx}{1-2x}$

i) $\displaystyle\int_{\frac{1}{6}\pi}^{\frac{1}{3}\pi} \cot \theta\,d\theta$

j) $\displaystyle\int_0^{\pi/4} \frac{\cos\theta-\sin\theta}{\cos\theta+\sin\theta}\,d\theta$

8 Given that $\ln 2 \approx 0.69315$, use differentiation to estimate (to 5 places of decimals):
a) $\ln 2.001$ **b)** $\ln 1.9999$

9 Find the volume of the solids of revolution formed when the area
a) between $y = 1/\sqrt{x}$, $x = 1$, $x = 2$ and $y = 0$ is rotated about the x-axis;
b) between $y = 1/x^2$, $y = 1$ and $y = 4$ is rotated about the y-axis.

10 By using the Newton–Raphson method, or any other suitable iterative method, solve the equation

$$x + \ln x = 3$$

given that the root is close to 2. Obtain the root correct to two decimal places.   (JMB)

11 a) Using the Newton–Raphson method or otherwise, solve the equation

$$\ln x = \tfrac{1}{5}x$$

giving your answers to four significant figures.
b) Find the gradient of the straight line through the origin which touches the graph of $\ln x$, and hence show that

$$\ln x = \alpha x$$

has a solution whenever $\alpha \leqslant 1/e$.

12   Boyle's law states that, at constant temperature, the pressure p and volume V of a gas are such that $pV = c$, where c is a constant. If the work done by the gas when expanding is $\int p\,dV$, show that this is constant ($= c \ln 2$) whenever it doubles its volume.

13   If an animal grows without changing its shape then its length x, surface area S and volume V are such that

$$S = ax^2, \quad V = bx^3 \quad \text{(where } a, b \text{ are constants)}$$

Hence show by differentiation **a)** without use of logarithms **b)** after first taking logarithms, that slight increases δx, δS and δV are related by

$$\frac{\delta S}{S} = 2\frac{\delta x}{x}, \quad \frac{\delta V}{V} = 3\frac{\delta x}{x}$$

i.e. that *fractional* increases of S (and V) are twice (and thrice) the fractional increase of x.

10.2 The exponential function, e^x

We now proceed to investigate the function $f(x) = e^x$. This is clearly a power function, and we start by finding its derivative.

Now $y = e^x$

$\Rightarrow$ $x = \ln y$

$\Rightarrow$ $\dfrac{dx}{dy} = \dfrac{1}{y}$

$\Rightarrow$ $\dfrac{dy}{dx} = y = e^x$

So $\dfrac{d}{dx}(e^x) = e^x$

and we immediately see that e^x is the particular power function whose rate of increase is always equal to its value:

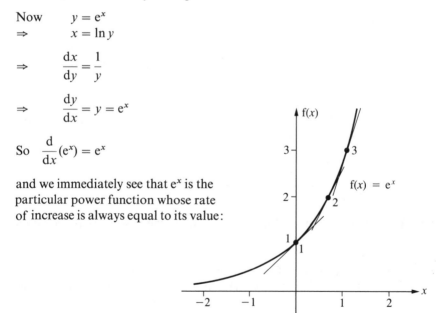

It was mentioned earlier that power functions are sometimes referred to as exponential functions. Because of the supreme importance of e^x it is usually

called *the* exponential function, and is sometimes written exp x.
Similarly we see that

$$\frac{d}{dx}(e^{-x}) = e^{-x} \times -1 = -e^{-x}$$

so that e^{-x} is the corresponding function of exponential decay.

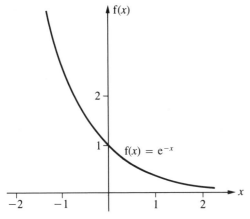

These two functions, and variants of them, occur whenever the rate of growth of a quantity, or its rate of decay, is determined by its present amount. So, for instance, the spread of an infection in the first stage of an epidemic is likely to involve a positive exponential function, whilst the decay of a radioactive element is governed by a negative exponential.

Lastly, just as
$$\frac{d}{dx}(e^x) = e^x$$

so
$$\int e^x dx = e^x + c$$

The function a^x

We have already seen that $\ln a = \log_e a$, i.e. that $\ln a$ is the power to which e has to be raised to produce a:

$$a = e^{\ln a}$$

Hence $a^x = (e^{\ln a})^x$

$\Rightarrow$
$$a^x = e^{x \ln a}$$

so that any function a^x can also be expressed as a power of e.

Hence $\dfrac{d}{dx}(a^x) = \dfrac{d}{dx}(e^{x\ln a}) = e^{x\ln a}\ln a$

$\Rightarrow \qquad \dfrac{d}{dx}(a^x) = a^x\ln a$

$\Rightarrow \qquad \displaystyle\int a^x\,dx = \dfrac{a^x}{\ln a} + c$

Example 1

Differentiate:
a) e^{x^2} **b)** x^x

a) $\dfrac{d}{dx}(e^{x^2}) = e^{x^2} \times 2x = 2xe^{x^2}$

b) If $\qquad y = x^x$

then $\quad \ln y = \ln x^x = x\ln x$

$\Rightarrow \qquad \dfrac{1}{y}\dfrac{dy}{dx} = x.\dfrac{1}{x} + 1.\ln x$

$\qquad\qquad\qquad = 1 + \ln x$

$\Rightarrow \qquad \dfrac{dy}{dx} = x^x(1 + \ln x)$

Example 2

When a charged condenser is discharged through a particular circuit containing an inductance and resistance, the current x at time t obeys the equation

$$\dfrac{d^2x}{dt^2} + 2\dfrac{dx}{dt} + 5x = 0$$

Show that $x = e^{-t}\cos 2t$ satisfies this equation and sketch the graph of the function.

If $\qquad x = e^{-t}\cos 2t$

then $\quad \dfrac{dx}{dt} = e^{-t} \times -2\sin 2t - e^{-t}\cos 2t$

$$= e^{-t}(-\cos 2t - 2\sin 2t)$$

and $\dfrac{d^2x}{dt^2} = e^{-t}(2\sin 2t - 4\cos 2t) - e^{-t}(-\cos 2t - 2\sin 2t)$

$$= e^{-t}(-3\cos 2t + 4\sin 2t)$$

So we verify that

$$\frac{d^2x}{dt^2} + 2\frac{dx}{dt} + 5x$$

$$= e^{-t}\{(-3\cos 2t + 4\sin 2t) + 2(-\cos 2t - 2\sin 2t) + 5\cos 2t\}$$
$$= 0$$

Now $x = e^{-t}\cos 2t$ is the product of an exponential decay and a cosine function

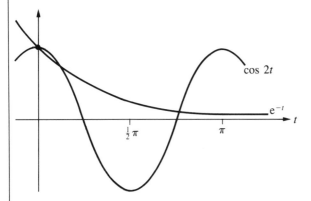

Combining these, we obtain the graph of $x = e^{-t}\cos 2t$ and we readily see that it represents a decaying oscillation.

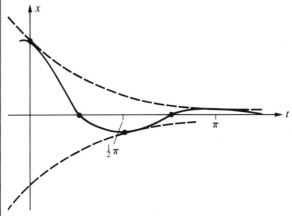

Example 3

A bathful of water is at a temperature of 80 °C in a large room whose temperature is 20 °C. The water cools at a rate which is proportional to its excess temperature and takes 5 min to reach 70 °C.

Find the temperature of the water after t min.

If the *excess* temperature is θ °C, the rate of cooling is $-\dfrac{d\theta}{dt}$.

So $-\dfrac{d\theta}{dt} = k\theta,$ where k is a constant (Newton's law of cooling)

$\Rightarrow \qquad \dfrac{dt}{d\theta} = -\dfrac{1}{k\theta}$

$\Rightarrow \qquad t = -\dfrac{1}{k}\ln\theta + c$

But, when $t = 0$, $\theta = 60$.

So $0 = -\dfrac{1}{k}\ln 60 + c \quad \Rightarrow \quad c = \dfrac{1}{k}\ln 60$

$\Rightarrow \qquad t = -\dfrac{1}{k}\ln\theta + \dfrac{1}{k}\ln 60$

$\qquad\qquad = -\dfrac{1}{k}\ln\dfrac{\theta}{60}$

$\Rightarrow \quad \ln\dfrac{\theta}{60} = -kt$

So far we have only used the general 'law of cooling' and the 'initial condition' that $\theta = 60$ when $t = 0$. But we also know that when $t = 5$, $\theta = 50$.

So $\ln\frac{50}{60} = -k \times 5$

$\Rightarrow \qquad k = -\frac{1}{5}\ln\frac{5}{6} = \frac{1}{5}\ln\frac{6}{5} \approx 0.0365$

So $\ln\dfrac{\theta}{60} = -0.0365t$

$\Rightarrow \qquad \dfrac{\theta}{60} = e^{-0.0365t}$

$\Rightarrow \qquad \theta = 60\,e^{-0.0365t}$

So the temperature of the bath after t minutes will be $20 + 60\,e^{-0.0365t}$

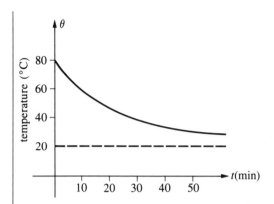

We can now use this function to calculate the temperature of the bath at any subsequent time. For example:

if $t = 10$,   temperature $= 20 + 60\,e^{-0.365} = 61.7\,°C$

if $t = 20$,   temperature $= 20 + 60\,e^{-0.73} = 48.9\,°C$

Exercise 10.2

1 Evaluate e^x when $x = 10$, 100, $\frac{1}{10}$, -10.

2 Find values of x such that
   a) $e^x = 10$, 1000, 10^6 **b)** $e^{-x} = \frac{1}{100}$, 10^{-4}, 10^{-6}

3 Differentiate with respect to x:
   a) e^{3x} **b)** $e^{-\frac{1}{4}x}$ **c)** xe^{-x} **d)** $x^2\,e^x$
   e) $e^{-\frac{1}{2}x^2}$ **f)** e^x/x **g)** $e^{-x}\sin x$ **h)** $e^{2x}\cos 3x$
   i) 3^x **j)** x^{x^2}

4 Find the following integrals:
   a) $\displaystyle\int e^{-2x}\,dx$ **b)** $\displaystyle\int e^{\frac{1}{3}x}\,dx$ **c)** $\displaystyle\int x\,e^{x^2}\,dx$

   d) $\displaystyle\int e^{\sin x}\cos x\,dx$ **e)** $\displaystyle\int x\,e^{-\frac{1}{2}x^2}\,dx$ **f)** $\displaystyle\int 10^x\,dx$

5 Evaluate the following integrals:
   a) $\displaystyle\int_0^2 e^{-\frac{1}{4}x}\,dx$ **b)** $\displaystyle\int_1^3 e^{2x}\,dx$ **c)** $\displaystyle\int_0^1 x\,e^{-\frac{1}{2}x^2}\,dx$ **d)** $\displaystyle\int_0^2 10^x\,dx$

6 Investigate the stationary values and sketch the graphs of the following functions:
   a) $x\,e^{-x}$ **b)** $x^2\,e^{-x}$ **c)** $x^n\,e^{-x}$ **d)** e^{-x^2}

7 Find $\displaystyle\int_0^X e^{-x}dx$

By letting $X \to +\infty$, investigate the area which lies between the curve $y = e^{-x}$ and the two axes (for convenience called $\displaystyle\int_0^\infty e^{-x}dx$).

8 After time t the displacement x of a heavily damped pendulum is given by $x = 2e^{-t}\sin t$.
   a) For what values of t is the displacement zero?
   b) For what values of t is the displacement greatest?
   c) Sketch the graph of x against t.
   d) Prove that

$$\frac{d^2x}{dt^2} + 2\frac{dx}{dt} + 2x = 0$$

9 A biologist knows that a certain organism, of which there is initially 2 g, is growing *continuously* at a rate of $\frac{1}{10}$ g/day for each gram of its mass. If its mass after t days is m g,
   a) show that $\dfrac{dm}{dt} = \frac{1}{10}m$;
   b) deduce that $m = 2e^{t/10}$.
   c) By what percentage does it increase each day?
   d) How long does it take to double its mass?

10 A sample of radium loses mass at a rate which is proportional to the amount present.
   a) If its mass is m after time t, show that

$$\frac{dm}{dt} = -km$$

   b) If its initial mass (when $t = 0$) is m_0, deduce that

$$m = m_0 e^{-kt}$$

   c) If its mass is halved in 1600 years, calculate k.
   d) Calculate the percentage rate of decrease per century.

11 A man with toothache leaves home at noon for the dentist, whose surgery is 1 km away. He starts by running at $10\,\text{km}\,\text{h}^{-1}$, but gradually slows down so that his speed is always proportional to his distance from the surgery.
   a) How far has he gone in the first 5 min?
   b) How long will he take to get within 100 m?
   c) When will he arrive?

10.3 Interlude: infinite series

In Chapter 6 we investigated a number of finite series and also mentioned the 'sum to infinity' of a geometric progression.

It was proved, for instance, that if S_n is the sum of the first n terms of the series

$$1 + \tfrac{1}{2} + \tfrac{1}{4} + \tfrac{1}{8} + \cdots,$$

then $S_n = \dfrac{1 - (\tfrac{1}{2})^n}{1 - \tfrac{1}{2}} = 2 - \dfrac{1}{2^{n-1}}$

Now as $n \to \infty$, $\dfrac{1}{2^{n-1}} \to 0$

So $S_n \to 2$

and we say that the *infinite series*

$$1 + \tfrac{1}{2} + \tfrac{1}{4} + \tfrac{1}{8} + \tfrac{1}{16} + \cdots$$

is *convergent*, and that its *sum* (or *sum to infinity*) is 2.

More generally, we can consider the infinite series

$$u_1 + u_2 + u_3 + \cdots$$

and suppose that the sum of its first n terms is S_n, so that

$$S_n = u_1 + u_2 + u_3 + \cdots + u_n$$

If $S_n \to S$ as $n \to \infty$, we say that the infinite series is *convergent*, and that S is its *sum*.

It is beyond our present scope to investigate the convergence of series, but we can consider some elementary examples.

Example 1

$$1 + \tfrac{1}{2} + \tfrac{1}{3} + \tfrac{1}{4} + \tfrac{1}{5} + \cdots$$

Certainly the terms of this series become very small. If we look at the successive values of S_n, we obtain (approximately)

   1   1.500   1.833   2.083   2.283   …

and before long it seems that the values of S_n are having such a struggle to increase that they must necessarily tend to a limit, which will be the sum of the infinite series. But a little more thought will cause us to write the series as

$$1 + \tfrac{1}{2} + (\tfrac{1}{3} + \tfrac{1}{4}) + (\tfrac{1}{5} + \tfrac{1}{6} + \tfrac{1}{7} + \tfrac{1}{8}) + (\tfrac{1}{9} + \tfrac{1}{10} + \tfrac{1}{11} + \cdots + \tfrac{1}{16}) + \cdots$$
$$> 1 + \tfrac{1}{2} + 2 \times \tfrac{1}{4} + \quad 4 \times \tfrac{1}{8} \quad + \quad\quad 8 \times \tfrac{1}{16} \quad\quad + \cdots$$
$$> 1 + \tfrac{1}{2} + \quad \tfrac{1}{2} \quad + \quad\quad \tfrac{1}{2} \quad\quad + \quad\quad \tfrac{1}{2} \quad\quad + \cdots$$

Now this series, although growing very slowly, is completely unlimited. So, contrary to our expectations, the original series

$$1 + \tfrac{1}{2} + \tfrac{1}{3} + \tfrac{1}{4} + \cdots$$

is also unlimited and does not converge (even though, as we shall later prove, the sum of its first million terms is less than 15).

Example 2

The series

$$\frac{1}{1^2} + \frac{1}{2^2} + \frac{1}{3^2} + \frac{1}{4^2} + \frac{1}{5^2} + \frac{1}{6^2} + \frac{1}{7^2} + \frac{1}{8^2} + \cdots$$

can be written as

$$\frac{1}{1^2} + \left(\frac{1}{2^2} + \frac{1}{3^2}\right) + \left(\frac{1}{4^2} + \frac{1}{5^2} + \frac{1}{6^2} + \frac{1}{7^2}\right) + \left(\frac{1}{8^2} + \cdots + \frac{1}{15^2}\right) + \cdots$$

$$< 1 + 2 \times \frac{1}{2^2} + 4 \times \frac{1}{4^2} + 8 \times \frac{1}{8^2} +$$

$$< 1 + \tfrac{1}{2} + \tfrac{1}{4} + \tfrac{1}{8} + \cdots = 2$$

In this case, therefore, the series

$$1 + \frac{1}{2^2} + \frac{1}{3^2} + \frac{1}{4^2} + \cdots$$

does converge to a sum less than 2 (and actually, though we cannot prove it here, to $\tfrac{1}{6}\pi^2$).

Example 3

As a warning that we must treat infinite series with a certain respect, we now consider the series

$$1 - \tfrac{1}{2} + \tfrac{1}{3} - \tfrac{1}{4} + \tfrac{1}{5} + \cdots$$

It can be shown that this series is convergent and also, perhaps surprisingly, that its sum is $\ln 2 \approx 0.693$. So it would appear to be an innocent operation to rearrange its terms as follows:

$$1 - \tfrac{1}{2} + \tfrac{1}{3} - \tfrac{1}{4} + \tfrac{1}{5} - \tfrac{1}{6} + \tfrac{1}{7} - \tfrac{1}{8} + \cdots$$

$$= 1 - \tfrac{1}{2} - \tfrac{1}{4} + \tfrac{1}{3} - \tfrac{1}{6} - \tfrac{1}{8} + \tfrac{1}{5} - \tfrac{1}{10} - \tfrac{1}{12} + \cdots$$

$$= (1 - \tfrac{1}{2}) - \tfrac{1}{4} + (\tfrac{1}{3} - \tfrac{1}{6}) - \tfrac{1}{8} + (\tfrac{1}{5} - \tfrac{1}{10}) - \cdots$$
$$= \quad \tfrac{1}{2} \quad - \tfrac{1}{4} + \quad \tfrac{1}{6} \quad - \tfrac{1}{8} + \quad \tfrac{1}{10} \quad - \cdots$$
$$= \tfrac{1}{2}(1 - \tfrac{1}{2} + \tfrac{1}{3} - \tfrac{1}{4} + \tfrac{1}{5} - \cdots).$$

But the reader will immediately see that we have 'proved' that

$$\ln 2 = \tfrac{1}{2} \ln 2$$

This paradox, and others like it, show that the theory of infinite series must be developed with very great care: though we cannot attempt this here, it forms a major branch of modern mathematics.

Example 4

We now return to the geometric series

$$1 + x + x^2 + x^3 + \cdots$$

The reader will recall that

$$S_n = \frac{1 - x^n}{1 - x} = \frac{1}{1 - x} - \frac{x^n}{1 - x}$$

and that if $|x| < 1$, $x^n \to 0$ as $n \to \infty$

So if $|x| < 1$ the series is convergent to the sum $\dfrac{1}{1 - x}$

This is a particular case of a *power series*, whose terms are successive powers of x and which converges for a set of values of x called its *domain of convergence*.

The general power series can be written

$$a_0 + a_1 x + a_2 x^2 + a_3 x^3 + \cdots$$

where $a_0, a_1, a_2, \ldots$ are particular constants, and we should expect such a series, if convergent, to have a sum that is a function of x. It will, for instance, be shown in the next section that

$$x - \tfrac{1}{2}x^2 + \tfrac{1}{3}x^3 - \tfrac{1}{4}x^4 + \cdots$$

is convergent to the sum $\ln(1 + x)$, provided that $-1 < x \leqslant 1$; and that

$$1 + x + \frac{x^2}{2!} + \frac{x^3}{3!} + \frac{x^4}{4!} + \cdots$$

is convergent to the sum e^x for all values of x.

10.4 Maclaurin series

We shall begin by reversing the question of the last section and ask whether, if we are given a particular function f(x), we can discover a convergent power series of which it is the sum. For this purpose we shall have to make a number of assumptions.

Firstly we shall restrict ourselves to functions which are differentiable not only once, but any number of times, i.e. whose derived curves are all continuous and smooth.

Secondly, we shall assume the *existence* of such a series for a certain domain of x, and our problem will be simply to *find* the series, granted that it exists. This may seem rather strange but, throughout mathematics, proving such existence is frequently a major problem, and the only statement we shall be able to make here is that, *if* a particular function is expressible as a power series, then this series must be such and such.

Lastly, we assume the fact that if this series exists and is convergent for a certain domain of x, then within this domain it is differentiable 'term by term'. By this we mean simply that the derivative of its sum is equal to the sum of its derivatives, even though there is an infinite series of them. In other words, that just as

$$\frac{d}{dx}(u + v) = \frac{du}{dx} + \frac{dv}{dx}$$

and
$$\frac{d}{dx}(u + v + w) = \frac{du}{dx} + \frac{dv}{dx} + \frac{dw}{dx}$$

and so on for any *finite* series, so also it is true for our *infinite* series that

$$\frac{d}{dx}(u + v + w + \cdots) = \frac{du}{dx} + \frac{dv}{dx} + \frac{dw}{dx} + \cdots$$

Let us suppose that f(x) is a function which, throughout a certain domain including $x = 0$

a) is differentiable any number of times, and

b) is the sum of a convergent power series.

Let this series be

$$f(x) = a_0 + a_1 x + a_2 x^2 + a_3 x^3 + a_4 x^4 + \cdots$$

Differentiating term by term and putting $x = 0$, we obtain successively:

$$
\begin{aligned}
f(x) &= a_0 + a_1 x + a_2 x^2 + a_3 x^3 + a_4 x^4 + \cdots &&\Rightarrow& f(0) &= a_0 \\
f'(x) &= a_1 + 2a_2 x + 3a_3 x^2 + 4a_4 x^3 + \cdots &&\Rightarrow& f'(0) &= a_1 \\
f''(x) &= 2a_2 + 6a_3 x + 12a_4 x^2 + \cdots &&\Rightarrow& f''(0) &= 2!a_2 \\
f'''(x) &= 6a_3 + 24a_4 x + \cdots &&\Rightarrow& f'''(0) &= 3!a_3 \\
f''''(x) &= 24a_4 + \cdots &&\Rightarrow& f''''(0) &= 4!a_4, \text{ etc.}
\end{aligned}
$$

So $a_0 = f(0)$ $a_1 = f'(0)$ $a_2 = \dfrac{f''(0)}{2!}$ $a_3 = \dfrac{f'''(0)}{3!}$

and similarly $a_n = \dfrac{f^{(n)}(0)}{n!}$

Hence
$$f(x) = f(0) + f'(0)x + \frac{f''(0)}{2!}x^2 + \cdots + \frac{f^{(n)}(0)}{n!}x^n + \cdots$$

This is known as Maclaurin's series (after the Scots mathematician Colin Maclaurin, 1698–1746), which we can now illustrate by three very important examples:

Exponential series

If $f(x) = e^x$,

then $f'(x) = f''(x) = \cdots = f^{(n)}(x) = e^x$

So $f(0) = f'(0) = \cdots = f^{(n)}(0) = 1$

Hence
$$e^x = 1 + x + \frac{x^2}{2!} + \frac{x^3}{3!} + \frac{x^4}{4!} + \cdots + \frac{x^n}{n!} + \cdots$$

It can be proved that this series is valid for all values of x; as, indeed, we would expect, since however large x might be (say $x = 1000$), the great size of x^n is eventually overcome by the even greater size of $n!$.

We can now use this series to calculate e to any required degree of accuracy. For, putting $x = 1$,

$$e = 1 + \frac{1}{1!} + \frac{1}{2!} + \frac{1}{3!} + \frac{1}{4!} + \frac{1}{5!} + \cdots$$

and the calculation can be set out to seven places of decimals

1	1.0000000
$\dfrac{1}{1!}$	2) 1.0000000
$\dfrac{1}{2!}$	3) 0.5000000
$\dfrac{1}{3!}$	4) 0.1666667
$\dfrac{1}{4!}$	5) 0.0416667

$\dfrac{1}{5!}$	6) 0.008 333 3
$\dfrac{1}{6!}$	7) 0.001 388 9
$\dfrac{1}{7!}$	8) 0.000 198 4
$\dfrac{1}{8!}$	9) 0.000 024 8
$\dfrac{1}{9!}$	10) 0.000 002 5
$\dfrac{1}{10!}$	0.000 000 2
	2.718 281 5

We see that this series is rapidly convergent and that e $\approx$ 2.718 281 5.

The careful reader will note that this value of e is subject to two sources of error:

a) because individual terms have been calculated only to seven places of decimals.

When, for instance, we state that $1/3! = 0.166 666 7$, we are asserting that it lies between $0.166 666 65$ and $0.166 666 75$ and so is subject to a maximum error of 5 in its eighth decimal place, i.e. 5×10^{-8}. As there are eight such terms which have been corrected in this way, the maximum error from this source is $40 \times 10^{-8} = 4 \times 10^{-7}$.

This is known as the *rounding error*.

b) because the series was cut short at $1/10!$

The remainder after this term is clearly

$$\frac{1}{11!} + \frac{1}{12!} + \frac{1}{13!} + \cdots = \frac{1}{11!}\left(1 + \frac{1}{12} + \frac{1}{12 \times 13} + \cdots\right)$$

$$< \frac{1}{11!}\left(1 + \frac{1}{12} + \left(\frac{1}{12}\right)^2 + \cdots\right)$$

$$< \frac{1}{11!} \times \frac{1}{1 - \frac{1}{12}} = \frac{12}{11 \times 11!}$$

Now $\dfrac{12}{11 \times 11!} < 2.7 \times 10^{-8}$

So the maximum error from this source is less than 2.7×10^{-8}. This is known as the *truncation error*.

We therefore see that the total error is less than the sum of these, and so is less than $4.27 \times 10^{-7} < 5 \times 10^{-7}$.

Hence e lies in the range $2.7182815 \pm 5 \times 10^{-7}$.

$\Rightarrow$ $2.718281 < e < 2.718282$

$\Rightarrow$ $e = 2.71828$,   correct to five places of decimals

(In fact $e = 2.7182818$ to seven places.)

Exercise 10.4a

1 Evaluate the following functions at intervals of 0.5 from $x = -1$ to
   $x = +2$:
   a) 1   **b)** $1 + x$ **c)** $1 + x + \frac{1}{2}x^2$ **d)** $1 + x + \frac{1}{2}x^2 + \frac{1}{6}x^3$
   e) $1 + x + \frac{1}{2}x^2 + \frac{1}{6}x^3 + \frac{1}{24}x^4$ **f)** e^x
   Hence, on a single accurate figure, plot the graphs of these functions from
   $x = -1$ to $x = +2$

2 Find the power series, as far as the x^4 term, of
   a) e^{-x} **b)** e^{2x} **c)** $\frac{1}{2}(e^x + e^{-x})$ **d)** $\frac{1}{2}(e^x - e^{-x})$

3 Use the exponential series to calculate to five decimal places:
   a) $\sqrt{e}$ **b)** $\dfrac{1}{e}$ **c)** $e^{0.1}$

Logarithmic series

If $\ln x = a_0 + a_1 x + a_2 x^2 + a_3 x^3 + \cdots$

it would immediately follow that

$a_0 = \ln 0$

But $\ln 0$ does not exist, so we cannot express $\ln x$ as a power series.
   We can, however, consider the function $\ln(1 + x)$.

Now $f(x) = \ln(1 + x)$ $\Rightarrow$ $f(0) = \ln 1 = 0$

$\Rightarrow$ $f'(x) = \dfrac{1}{1 + x}$ $\Rightarrow$ $f'(0) = 1$

$\Rightarrow$ $f''(x) = -\dfrac{1}{(1 + x)^2}$ $\Rightarrow$ $f''(0) = -1$

$\Rightarrow$ $f'''(x) = +\dfrac{2}{(1 + x)^3}$ $\Rightarrow$ $f'''(0) = +2$

$\Rightarrow$ $f^n(x) = (-1)^{n-1}\dfrac{(n-1)!}{(1 + x)^n}$ $\Rightarrow$ $f^n(0) = (-1)^{n-1}(n-1)!$

So, with the usual assumptions for Maclaurin's series,

$$\ln(1+x) \equiv 0 + 1x + \frac{-1}{2!}x^2 + \frac{2}{3!}x^3 + \cdots + \frac{(-1)^{n-1}(n-1)!}{n!}x^n + \cdots$$

$$\Rightarrow \quad \ln(1+x) = x - \tfrac{1}{2}x^2 + \tfrac{1}{3}x^3 - \tfrac{1}{4}x^4 + \cdots + \frac{(-1)^{n-1}}{n}x^n \cdots \qquad (1)$$

and it can be shown that in this case the domain of convergence is $-1 < x \leqslant 1$.

Putting $-x$ for x, we obtain

$$\ln(1-x) = -x - \tfrac{1}{2}x^2 - \tfrac{1}{3}x^3 - \tfrac{1}{4}x^4 - \cdots \qquad (2)$$

whose domain of convergence is $-1 < -x \leqslant 1$, i.e. $-1 \leqslant x < 1$.

If we now subtract these two results, we obtain

$$\ln(1+x) - \ln(1-x) = 2(x + \tfrac{1}{3}x^3 + \tfrac{1}{5}x^5 + \cdots)$$

$$\Rightarrow \qquad \tfrac{1}{2}\ln\frac{1+x}{1-x} = x + \tfrac{1}{3}x^3 + \tfrac{1}{5}x^5 + \cdots \qquad (3)$$

which is convergent provided both the original series are convergent, i.e. if $-1 < x < 1$.

It is instructive to use these series to calculate $\ln 2$.

If we put $x = 1$ in (1), we obtain

$$\ln 2 = 1 - \tfrac{1}{2} + \tfrac{1}{3} - \tfrac{1}{4} + \cdots$$

But the millionth term is $-1/10^6$ and so is still affecting the sixth decimal place. The series, therefore, converges so slowly that it is not very useful for the calculation of $\ln 2$.

If, on the other hand, we use (3), we see that to calculate $\ln 2$ it will be necessary to let

$$\frac{1+x}{1-x} = 2 \quad \Rightarrow \quad x = \tfrac{1}{3}$$

Putting $x = \tfrac{1}{3}$, we obtain

$$\tfrac{1}{2}\ln 2 = \tfrac{1}{3} + \tfrac{1}{3}(\tfrac{1}{3})^3 + \tfrac{1}{5}(\tfrac{1}{3})^5 + \cdots$$

$$\Rightarrow \quad \ln 2 = 2\{\tfrac{1}{3} + \tfrac{1}{3}(\tfrac{1}{3})^3 + \tfrac{1}{5}(\tfrac{1}{3})^5 + \cdots\}$$

and it is clear that this series converges much more rapidly. How many terms of this series need to be taken in order to determine the sixth decimal place?

Exercise 10.4b

1 Evaluate the following functions at intervals of 0.5 from $x = -1$ to $x = +1$:

a) x **b)** $x - \frac{1}{2}x^2$ **c)** $x - \frac{1}{2}x^2 + \frac{1}{3}x^3$

d) $x - \frac{1}{2}x^2 + \frac{1}{3}x^3 - \frac{1}{4}x^4$ **e)** $\ln(1 + x)$

Hence graph these functions on a single diagram from $x = -1$ to $+1$.

2 Find the power series, as far as the x^4 term of

a) $\ln\left(1 + \dfrac{x}{2}\right)$ **b)** $\ln(1 + 2x)$ **c)** $\ln\dfrac{1}{1 - x^2}$

3 Use the series for $\frac{1}{2}\ln\dfrac{1 + x}{1 - x}$ to evaluate, to three decimal places:

a) $\ln 3$ **b)** $\ln \frac{5}{3}$ **c)** Hence calculate $\ln 5$.

Binomial series

We now turn to the function

$$f(x) = (1 + x)^n \quad (n \in \mathbb{R})$$

Now $f(x) = (1 + x)^n$ $\Rightarrow$ $f(0) = 1$

$\Rightarrow$ $f'(x) = n(1 + x)^{n-1}$ $\Rightarrow$ $f'(0) = n$

$\Rightarrow$ $f''(x) = n(n - 1)(1 + x)^{n-2}$ $\Rightarrow$ $f''(0) = n(n - 1)$

More generally

$$f^{(r)}(x) = n(n - 1)\cdots(n - r + 1)(1 + x)^{n-r}$$

$\Rightarrow$ $f^{(r)}(0) = n(n - 1)\cdots(n - r + 1)$

So, using the Maclaurin series, we obtain the *binomial series*:

$$(1 + x)^n = 1 + nx + \frac{n(n - 1)}{2!}x^2 + \cdots$$

$$\cdots + \frac{n(n - 1)(n - 2)\ldots(n - r + 1)}{r!}x^r + \cdots$$

and it can be shown that the series is convergent, provided $|x| < 1$.

If $n \in \mathbb{Z}^+$, we notice that the series terminates and reduces to the binomial theorem.

For instance, if $n = 4$

$$(1 + x)^4 = 1 + 4x + \frac{4 \times 3}{2!}x^2 + \frac{4 \times 3 \times 2}{3!}x^3 + \frac{4 \times 3 \times 2 \times 1}{4!}x^4$$

$$+ \frac{4 \times 3 \times 2 \times 1 \times 0}{5!}x^5$$

$$+ \frac{4 \times 3 \times 2 \times 1 \times 0 \times -1}{6!}x^6 + \cdots$$

$$= 1 + 4x + 6x^2 + 4x^3 + x^4$$

More generally, if $n \in \mathbb{Z}^+$

$$\frac{n(n-1)(n-2)\ldots(n-r+1)}{r!} = \binom{n}{r} \quad \text{if } r \leqslant n$$

$$\text{and} \quad 0 \quad \text{if } r > n$$

so that the binomial series simply becomes the binomial theorem.

If $n \notin \mathbb{Z}^+$ it is not yet possible to speak of $\binom{n}{r}$, as $\binom{-3}{2}$, $\binom{\frac{1}{2}}{3}$ etc. have no meaning. Nevertheless, we find it convenient to have an abbreviation for the coefficients of the binomial series, so we *define* $\binom{n}{r}$ to be

$$\frac{n(n-1)(n-2)(n-3)\cdots(n-r+1)}{r!}$$

Hence $\binom{\frac{1}{2}}{2} = \frac{\frac{1}{2}(\frac{1}{2} - 1)}{2!} = -\frac{1}{8}$

$$\binom{-2}{3} = \frac{(-2)(-2-1)(-2-2)}{3!} = -4, \quad \text{etc.}$$

With this notation we can write the binomial series, when $|x| < 1$, as

$$(1 + x)^n = 1 + \binom{n}{1}x + \binom{n}{2}x^2 + \cdots + \binom{n}{r}x^r + \cdots$$

Example 1

Find a series for $\left(1 - \dfrac{x}{2}\right)^{-2}$

$$\left(1 - \frac{x}{2}\right)^{-2} = 1 + (-2)\left(-\frac{x}{2}\right) + \frac{(-2)(-3)}{2!}\left(-\frac{x}{2}\right)^2$$

$$+ \frac{(-2)(-3)(-4)}{3!}\left(-\frac{x}{2}\right)^3 + \cdots$$

$$= 1 + x + \tfrac{3}{4}x^2 + \tfrac{1}{2}x^3 + \cdots$$

which is convergent if $\left|-\dfrac{x}{2}\right| < 1 \quad \Leftrightarrow \quad |x| < 2.$

Example 2

Use the binomial series to estimate $\sqrt[3]{999}$ to seven places of decimals.

$$\sqrt[3]{999} = (1\,000 - 1)^{1/3}$$

$$= 10(1 - 0.001)^{1/3}$$

Now $(1 - 0.001)^{1/3} = 1 + \tfrac{1}{3}(-0.001) + \dfrac{\tfrac{1}{3} \times -\tfrac{2}{3}}{2!}(-0.001)^2$

$$+ \frac{\tfrac{1}{3} \times -\tfrac{2}{3} \times -\tfrac{5}{3}}{3!}(-0.001)^3$$

$$+ \frac{\tfrac{1}{3} \times -\tfrac{2}{3} \times -\tfrac{5}{3} \times -\tfrac{8}{3}}{4!}(-0.001)^4 + \cdots$$

The first three terms are

$$+ 1.000\,000\,0000$$
$$- 0.000\,333\,3333$$
$$- 0.000\,000\,1111$$
$$\overline{}$$
$$= + 0.999\,666\,5556$$

It is easy to see that the maximum rounding error is 10^{-10}. Also the remainder is

$$-\frac{5}{81}(0.001)^3 - \frac{10}{243}(0.001)^4 - \cdots$$

so the truncation error is

$$\frac{5}{81}(0.001)^3 + \frac{10}{243}(0.001)^4 + \cdots$$

$$< \frac{5}{81}(0.001)^3 \{1 + (0.001) + (0.001)^2 + \cdots\}$$

$$< \frac{5}{81} \times 10^{-9} \times \frac{1000}{999} < 10^{-10}$$

Hence the total error $< 2 \times 10^{-10}$

and $\sqrt[3]{999} = 9.996665556$

with a maximum possible error of 2×10^{-9}.

So $\sqrt[3]{999}$ lies between 9.996665554 and 9.996665558

Hence $\sqrt[3]{999} = 9.9966656$,   to seven decimal places.

Exercise 10.4c

1 Find the first four terms of the power series of the following functions, in each case stating their domain of convergence:

a) $\sqrt{(1 + x)}$ **b)** $\dfrac{1}{1 + x}$ **c)** $(1 + x)^{-3}$

d) $(1 - 2x)^{-1/2}$ **e)** $(2 + x)^{-1}$ **f)** $\sqrt{(4 - 2x)}$

2 Evaluate, to five places of decimals:

a) $\sqrt{1.01}$ **b)** $\sqrt[4]{0.99}$ **c)** $\sqrt{4.1}$ **d)** $\sqrt[3]{27.27}$ **e)** $\dfrac{1}{(2.01)^2}$

3 Obtain binomial series for $1/(1 - x)^2$ and $1/(1 - x)^3$, and check by means of differentiating the series for $1/(1 - x)$.
   Hence sum the series

$$1 + 2(\tfrac{1}{2}) + 3(\tfrac{1}{2})^2 + 4(\tfrac{1}{2})^3 + \cdots$$

4 **a)** Using $\sqrt{(x^2 + 1)} = x(1 + 1/x^2)^{1/2}$, show that if x is large

$$\sqrt{(x^2 + 1)} \approx x + \frac{1}{2x} - \frac{1}{8x^3} + \cdots \text{(known as an } asymptotic \text{ series)}$$

b) Find the first three terms of similar asymptotic series for $1/(x - 1)^2$ and $\sqrt[3]{(x^3 + 2)}$.

Taylor series

The Maclaurin series enables us to find the value of a function near $x = 0$ in terms of its value and those of its derivatives at $x = 0$. If we now seek to investigate its value in the neighbourhood of another point $x = a$, we can let the point be $x = a + h$ and write

$$f(a + h) \equiv F(h) \quad \Rightarrow \quad f(a) = F(0)$$

so that $\quad f'(a + h) \equiv F'(h) \quad \Rightarrow \quad f'(a) = F'(0)$

$$f''(a + h) \equiv F''(h) \quad \Rightarrow \quad f''(a) = F''(0)$$

But $\quad F(h) = F(0) + hF'(0) + \dfrac{h^2}{2!} F''(0) + \cdots$

$$\Leftrightarrow \quad f(a + h) = f(a) + hf'(a) + \frac{h^2}{2!} f''(a) + \cdots$$

which is known as the Taylor series (after Brook Taylor, 1685–1731), and which enables us to find the value of the function $f(x)$ *near* $x = a$ in terms of its value and those of its derivatives at $x = a$.

In particular, if h is very small

$$f(a + h) \approx f(a) + hf'(a)$$

provides a linear approximation to the function, already familiar in the form

$$f'(a) \approx \frac{f(a + h) - f(a)}{h}$$

Exercise 10.4d

1   Find, as far as the x^4 term, Maclaurin series for
   a) $\sin x$ **b)** $\cos x$ **c)** $\tan x$ **d)** $\sec x$ **e)** $e^x \sin x$ **f)** $\ln \sec x$

2   Obtain the Maclaurin series for $\tan^{-1} x$ and check by integrating the binomial series for $1/(1 + x^2)$. Hence
   a) obtain Gregory's series

$$\frac{\pi}{4} = 1 - \tfrac{1}{3} + \tfrac{1}{5} - \tfrac{1}{7} + \cdots$$

b) use the result (see exercise 5.11a no. **3**) that

$$\frac{\pi}{4} = \tan^{-1}\tfrac{1}{2} + \tan^{-1}\tfrac{1}{3}$$

to show that

$$\frac{\pi}{4} = \{\tfrac{1}{2} - \tfrac{1}{3}(\tfrac{1}{2})^3 + \tfrac{1}{5}(\tfrac{1}{2})^5 - \cdots\} + \{\tfrac{1}{3} - \tfrac{1}{3}(\tfrac{1}{3})^3 + \tfrac{1}{5}(\tfrac{1}{3})^5 - \cdots\}$$

Compare **a)** and **b)** as methods of calculating π.

3 Find Taylor series for
a) $\sqrt{x}$ in the neighbourhood of $x = 9$
b) $\sin x$ in the neighbourhood of $x = \pi/6$
c) $\tan x$ in the neighbourhood of $x = \pi/4$
Hence estimate $\sqrt{10}$, $\sin 31°$, $\tan 46°$.

* 10.5 Integration by parts: reduction formulae

We now return to the problem of integration. As was observed in Chapter 4, a result about differentiation can often be transformed into an equally useful one about integration, and we saw how the chain rule for derivatives led to the method of substitution for finding integrals. We now proceed to look again, in just this way, at the formula for the derivative of a product.

If u and v are two functions of x, we know that

$$\frac{d}{dx}(uv) = u\frac{dv}{dx} + \frac{du}{dx}v$$

So
$$uv = \int u\frac{dv}{dx}\,dx + \int \frac{du}{dx}v\,dx$$

$$\Rightarrow \quad \int u\frac{dv}{dx}\,dx = uv - \int \frac{du}{dx}v\,dx$$

This is the rule for *integrating by parts* and is useful whenever the product $\dfrac{du}{dx}v$ is easier to integrate than $u\dfrac{dv}{dx}$.

If we now replace u and v by U and V, where

$$U = u \quad \text{and} \quad V = \frac{dv}{dx}$$

the result can be written as

$$\int UV \, dx = U \int V \, dx - \int \left(\frac{dU}{dx} \int V \, dx \right) dx \qquad †$$

Example 1

$$\int x \cos x \, dx = x \sin x - \int 1 \sin x \, dx$$

$$= x \sin x + \cos x + c$$

Example 2

$$\int_0^1 x^2 \, e^x \, dx = \left[x^2 \, e^x \right]_0^1 - \int_0^1 2x \, e^x \, dx$$

$$= e - 2 \left\{ \left[x \, e^x \right]_0^1 - \int_0^1 1 \, e^x \, dx \right\}$$

$$= e - 2\{e - (e - 1)\}$$

$$= e - 2$$

Example 3

$$\int \ln x \, dx = \int 1 \times \ln x \, dx$$

$$= x \ln x - \int x \frac{1}{x} \, dx$$

$$= x \ln x - x + c$$

---

† Conveniently remembered as $\int UV = U \int V - \int \left(U' \int V \right)$

Example 4

$$\int e^{-x}\sin x\,dx = (-e^{-x})\sin x - \int(-e^{-x})\cos x\,dx$$

$$= -e^{-x}\sin x + \int e^{-x}\cos x\,dx$$

$$= -e^{-x}\sin x + (-e^{x})\cos x - \int(-e^{-x})(-\sin x)\,dx$$

$$= -e^{-x}(\sin x + \cos x) - \int e^{-x}\sin x\,dx + c$$

$$\Rightarrow \quad 2\int e^{-x}\sin x\,dx = -e^{-x}(\sin x + \cos x) + c$$

$$\Rightarrow \quad \int e^{-x}\sin x\,dx = -\tfrac{1}{2}e^{-x}(\sin x + \cos x) + c'$$

Exercise 10.5a

Find the following integrals:

1 a) $\displaystyle\int x e^{x}\,dx$ **b)** $\displaystyle\int x \sin x\,dx$ **c)** $\displaystyle\int x e^{2x}\,dx$

d) $\displaystyle\int x \ln x\,dx$ **e)** $\displaystyle\int x \cos\frac{x}{2}\,dx$ **f)** $\displaystyle\int \frac{\ln x}{x^2}\,dx$

2 a) $\displaystyle\int x^2 e^{-x}\,dx$ **b)** $\displaystyle\int x^2 \sin x\,dx$

c) $\displaystyle\int \theta \sec^2\theta\,d\theta$ **d)** $\displaystyle\int u\,10^{u}\,du$

3 a) $\displaystyle\int e^{x}\sin x\,dx$ **b)** $\displaystyle\int e^{x}\cos x\,dx$

c) $\displaystyle\int e^{-x}\cos x\,dx$ **d)** $\displaystyle\int e^{x}\sin 2x\,dx$

4 a) $\displaystyle\int (\ln x)^2\,dx$ **b)** $\displaystyle\int \tan^{-1}x\,dx$

c) $\displaystyle\int \sin^{-1}u\,du$ **d)** $\displaystyle\int x \tan^{-1}x\,dx$

5   Evaluate:

a) $\displaystyle\int_0^1 x\,\mathrm{e}^{-x}\,\mathrm{d}x$

b) $\displaystyle\int_0^{\frac{1}{2}\pi} x\sin 2x\,\mathrm{d}x$

c) $\displaystyle\int_0^\pi t^2\cos\tfrac{1}{2}t\,\mathrm{d}t$

d) $\displaystyle\int_1^2 u^2\ln u\,\mathrm{d}u$

6   A mathematical model is constructed to describe the situation in a factory. The number of workers w which an employer will require is a function of the wage rate £r per week for each worker where $w = 100\,\mathrm{e}^{-0.04r}$. If the union wishes to maximize the total pay of its members required by the employer, what rate will it ask for? (Assume that all workers belong to the union.)   (SMP)

Reduction formulae

It is sometimes possible to find a formula which will reduce a given integral to a simpler one of the same type, and then to find the required integral by successive applications of this process.

Example 5

If $I_n = \displaystyle\int x^n\,\mathrm{e}^{-x}\,\mathrm{d}x$, evaluate I_3.

Now $I_n = x^n(-\mathrm{e}^{-x}) - \displaystyle\int (nx^{n-1})(-\mathrm{e}^{-x})\,\mathrm{d}x$

$\qquad\quad = -x^n\mathrm{e}^{-x} + n\displaystyle\int x^{n-1}\mathrm{e}^{-x}\,\mathrm{d}x$

$\Rightarrow\qquad I_n = -x^n\mathrm{e}^{-x} + nI_{n-1}$

Hence $I_3 = -x^3\,\mathrm{e}^{-x} + 3I_2$

$\qquad\quad I_2 = -x^2\,\mathrm{e}^{-x} + 2I_1$

$\qquad\quad I_1 = -x\mathrm{e}^{-x} + I_0$

But $I_0 = -\mathrm{e}^{-x} + c$

So $I_3 = -x^3\,\mathrm{e}^{-x} + 3(-x^2\,\mathrm{e}^{-x} + 2I_1)$

$\qquad\quad = -(x^3 + 3x^2)\mathrm{e}^{-x} + 6(-x\mathrm{e}^{-x} + I_0)$

$\qquad\quad = -(x^3 + 3x^2 + 6x)\,\mathrm{e}^{-x} + 6I_0$

$\qquad\quad = -(x^3 + 3x^2 + 6x + 6)\,\mathrm{e}^{-x} + c$

Example 6

$$I_n = \int_0^{\frac{1}{2}\pi} \sin^n x \, dx$$

$$I_n = \int_0^{\frac{1}{2}\pi} \sin^n x \, dx = \int_0^{\frac{1}{2}\pi} \sin^{n-1} x \times \sin x \, dx$$

$$= \left[\sin^{n-1} x (-\cos x) \right]_0^{\frac{1}{2}\pi}$$

$$- \int_0^{\frac{1}{2}\pi} (n-1) \sin^{n-2} x \cos x (-\cos x) \, dx$$

$$= 0 + (n-1) \int_0^{\frac{1}{2}\pi} \sin^{n-2} x \cos^2 x \, dx$$

$$= (n-1) \int_0^{\frac{1}{2}\pi} \sin^{n-2} x (1 - \sin^2 x) \, dx$$

So $\quad I_n = (n-1) \left\{ \int_0^{\frac{1}{2}\pi} \sin^{n-2} x \, dx - \int_0^{\frac{1}{2}\pi} \sin^n x \, dx \right\}$

$\Rightarrow \quad I_n = (n-1)(I_{n-2} - I_n)$

$\Rightarrow \quad n I_n = (n-1) I_{n-2}$

$\Rightarrow \quad I_n = \dfrac{n-1}{n} I_{n-2}$

and this reduction formula enables us to calculate I_n in terms of I_{n-2}.

But we know that

$$I_1 = \int_0^{\frac{1}{2}\pi} \sin x \, dx = \left[-\cos x \right]_0^{\frac{1}{2}\pi} = 1$$

and $\quad I_0 = \displaystyle\int_0^{\frac{1}{2}\pi} 1 \, dx = \left[x \right]_0^{\frac{1}{2}\pi} = \frac{1}{2}\pi$

So it follows, for instance, that

$$I_5 = \tfrac{4}{5} I_3 = \tfrac{4}{5} \times \tfrac{2}{3} I_1 = \tfrac{4}{5} \times \tfrac{2}{3} \times 1 = \tfrac{8}{15}$$

and $\quad I_6 = \tfrac{5}{6} I_4 = \tfrac{5}{6} \times \tfrac{3}{4} I_2 = \tfrac{5}{6} \times \tfrac{3}{4} \times \tfrac{1}{2} I_0$

$$= \tfrac{5}{6} \times \tfrac{3}{4} \times \tfrac{1}{2} \times \tfrac{1}{2}\pi = \tfrac{5\pi}{32}$$

Similarly, if $J_n = \displaystyle\int_0^{\frac{1}{2}\pi} \cos^n x \, dx$, we can put

$$u = \tfrac{1}{2}\pi - x$$

$$\Rightarrow \quad du = -dx$$

So $\quad J_n = \displaystyle\int_{\frac{1}{2}\pi}^0 \cos^n (\tfrac{1}{2}\pi - u) \times -du$

$$= \int_0^{\frac{1}{2}\pi} \sin^n u \, du = I_n$$

Summarising, if $\quad I_n = \displaystyle\int_0^{\frac{1}{2}\pi} \sin^n x \, dx, \quad J_n = \int_0^{\frac{1}{2}\pi} \cos^n x \, dx$

then $\quad I_n = \dfrac{n-1}{n} I_{n-2}, \quad J_n = \dfrac{n-1}{n} J_{n-2}$

Exercise 10.5b

1 Evaluate:

a) $\displaystyle\int_0^{\frac{1}{2}\pi} \sin^4 \theta \, d\theta$ 　　　**b)** $\displaystyle\int_0^{\frac{1}{2}\pi} \cos^3 \theta \, d\theta$ 　　　**c)** $\displaystyle\int_0^{\frac{1}{2}\pi} \cos^6 \theta \, d\theta$

d) $\displaystyle\int_0^{\frac{1}{2}\pi} \sin^7 \theta \, d\theta$ 　　　**e)** $\displaystyle\int_0^{\frac{1}{2}\pi} \cos^8 \theta \, d\theta$

2 Evaluate:

a) $\displaystyle\int_0^{\pi} \sin^5 \theta \, d\theta$ 　　　**b)** $\displaystyle\int_{-\frac{1}{2}\pi}^{\frac{1}{2}\pi} \cos^3 \theta \, d\theta$ 　　　**c)** $\displaystyle\int_0^{\pi} \cos^3 \theta \, d\theta$

d) $\displaystyle\int_{-\frac{1}{2}\pi}^{\frac{1}{2}\pi} \sin^6 \theta \, d\theta$ 　　　**e)** $\displaystyle\int_{-\frac{1}{2}\pi}^{\frac{1}{2}\pi} \sin^7 \theta \, d\theta$

3 If $\quad I_n = \int (\ln x)^n \, dx$, show that $I_n = x(\ln x)^n - nI_{n-1}$. Hence find I_3.

4 Use $\tan^n \theta = \tan^{n-2} \theta (\sec^2 \theta - 1)$ to find a reduction formula for $I_n = \int \tan^n \theta \, d\theta$, and hence find I_4 and I_6.

5 If $I_n = \int \sec^n \theta \, d\theta$, prove that $(n+1)I_{n+2} = \tan \theta \sec^n \theta + nI_n$.

* 10.6 Partial fractions

At an early age we learnt such simplifications as

$$\frac{2}{3} - \frac{1}{5} = \frac{10 - 3}{15} = \frac{7}{15}$$

and rather later, when we learned that

$$\frac{1}{x - 1} - \frac{1}{x + 1} = \frac{(x + 1) - (x - 1)}{(x - 1)(x + 1)} = \frac{2}{x^2 - 1}$$

we had little doubt that this too represented a simplification, as $2/(x^2 - 1)$ is generally more useful and more convenient than $1/(x - 1) - 1/(x + 1)$.

In the next section we shall frequently be needing to find such integrals as

$$\int \frac{1}{x^2 - 1} \, dx$$

for which there is no immediately obvious answer. But a little meditation might lead us to reverse the above process and say that

$$\frac{1}{x^2 - 1} = \frac{\frac{1}{2}}{x - 1} - \frac{\frac{1}{2}}{x + 1}$$

$$\Rightarrow \quad \int \frac{dx}{x^2 - 1} = \int \left(\frac{\frac{1}{2}}{x - 1} - \frac{\frac{1}{2}}{x + 1} \right) dx$$

$$= \tfrac{1}{2} \ln (x - 1) - \tfrac{1}{2} \ln (x + 1) + c$$

$$= \tfrac{1}{2} \ln \frac{x - 1}{x + 1} + c$$

Obviously we were extremely fortunate to know that

$$\frac{1}{x^2 - 1} = \frac{\frac{1}{2}}{x - 1} - \frac{\frac{1}{2}}{x + 1}$$

and our next task will be to find a process whereby similar rational functions (i.e. the quotients of two polynomials) can be expressed in terms of such partial fractions.

The full theory of partial fractions is beyond our present scope, but we can illustrate the procedure by a number of examples:

Example 1

Express in partial fractions

$$\frac{x + 1}{(x - 1)(x - 2)}$$

Hopefully, we write

$$\frac{x+1}{(x-1)(x-2)} \equiv \frac{A}{x-1} + \frac{B}{x-2}$$

where A, B are constants which are to be determined

$$\Rightarrow \qquad x+1 \equiv A(x-2) + B(x-1)$$

Putting $x = 2$, $3 = A \times 0 + B \times 1$ $\Rightarrow$ $B = 3$

Putting $x = 1$, $2 = A \times -1 + B \times 0$ $\Rightarrow$ $A = -2$

So the required partial fractions would be

$$\frac{-2}{x-1} + \frac{3}{x-2}$$

and it is quickly verified that these can be combined to give the original expression.

Example 2

We now seek to put into partial fractions the more complicated expression

$$\frac{2x^3 - 3x^2 - x + 5}{2x^2 + x - 1}$$

First of all we observe that $2x^3 - 3x^2 - x + 5$ can be divided by $2x^2 + x - 1$ to give a quotient $x - 2$ and remainder $2x + 3$.

So $2x^3 - 3x^2 - x + 5 \equiv (2x^2 + x - 1)(x - 2) + (2x + 3)$

$$\Rightarrow \quad \frac{2x^3 - 3x^2 - x + 5}{(2x^2 + x - 1)} \equiv x - 2 + \frac{2x + 3}{2x^2 + x - 1}$$

Now $\dfrac{2x + 3}{2x^2 + x - 1} \equiv \dfrac{2x + 3}{(2x - 1)(x + 1)}$

so we write

$$\frac{2x + 3}{(2x - 1)(x + 1)} \equiv \frac{A}{2x - 1} + \frac{B}{x + 1}$$

$$\Rightarrow \qquad 2x + 3 \equiv A(x + 1) + B(2x - 1)$$

Putting $x = -1$, $1 = A \times 0 - 3B$ $\Rightarrow$ $B = -\frac{1}{3}$

Putting $x = \frac{1}{2}$, $4 = A \times \frac{3}{2}$ $\Rightarrow$ $A = \frac{8}{3}$

So (as can easily be checked)

$$\frac{2x + 3}{2x^2 + x - 1} \equiv \frac{\frac{8}{3}}{2x - 1} - \frac{\frac{1}{3}}{x + 1}$$

and $$\frac{2x^3 - 3x^2 - x + 5}{2x^2 + x - 1} \equiv x - 2 + \frac{\frac{8}{3}}{2x - 1} - \frac{\frac{1}{3}}{x + 1}$$

In this case the key to our success lay in first dividing the numerator by the denominator, and wherever possible (i.e. unless the degree of the numerator is already less than that of the denominator) this must always be the initial step.

Example 3

We now take a function whose denominator contains a repeated factor:

$$f(x) = \frac{x^2 + 5x + 9}{(x + 3)(x + 2)^2}$$

Again, we first try the simple partial fractions

$$\frac{x^2 + 5x + 9}{(x + 3)(x + 2)^2} \equiv \frac{A}{x + 3} + \frac{B}{(x + 2)^2} \qquad (1)$$

$\Rightarrow \quad x^2 + 5x + 9 \equiv A(x + 2)^2 + B(x + 3)$

Equating coefficients of x^2, x, and the constant terms, we see that it would be necessary to have

$$\left.\begin{array}{l} 1 = A \\ 5 = 4A + B \\ 9 = 4A + 3B \end{array}\right\} \text{ which are clearly inconsistent}$$

So $f(x)$ *cannot* be expressed in terms of partial fractions (1).

For our next attempt, therefore, above the quadratic denominator $(x + 2)^2$ we place a term $Bx + C$, so that

$$\frac{x^2 + 5x + 9}{(x + 3)(x + 2)^2} \equiv \frac{A}{x + 3} + \frac{Bx + C}{(x + 2)^2}$$

$\Rightarrow \quad x^2 + 5x + 9 \equiv A(x + 2)^2 + (Bx + C)(x + 3)$

Putting $x = -3$, $3 = A$ $\Rightarrow \quad A = 3$

Equating coefficients of x^2, $1 = A + B$ $\Rightarrow \quad B = -2$

Equating constants, $9 = 4A + 3C$ $\Rightarrow \quad C = -1$

Hence (as can easily be checked) the required partial fractions are

$$\frac{x^2 + 5x + 9}{(x + 3)(x + 2)^2} \equiv \frac{3}{x + 3} - \frac{2x + 1}{(x + 2)^2}$$

Sometimes it is convenient to split this last fraction still further, by saying that

$$\frac{2x + 1}{(x + 2)^2} \equiv \frac{2(x + 2) - 3}{(x + 2)^2} = \frac{2}{x + 2} - \frac{3}{(x + 2)^2}$$

so that
$$\frac{x^2 + 5x + 9}{(x + 3)(x + 2)^2} \equiv \frac{3}{x + 3} - \frac{2}{x + 2} + \frac{3}{(x + 2)^2}$$

Example 4

Finally, we take a rational function whose denominator contains a quadratic term which cannot be factorised

$$f(x) = \frac{x - 3}{(x - 1)(x^2 + 1)}$$

As it is impossible to express $x^2 + 1$ in real factors, we follow the precedent of the last example and write

$$\frac{x - 3}{(x - 1)(x^2 + 1)} \equiv \frac{A}{x - 1} + \frac{Bx + C}{x^2 + 1}$$

$$\Rightarrow \qquad x - 3 \equiv A(x^2 + 1) + (Bx + C)(x - 1)$$

Putting $x = 1$, $\qquad \Rightarrow \quad -2 = 2A \qquad \Rightarrow \quad A = -1$

Equating x^2 terms, $\quad \Rightarrow \quad 0 = A + B \quad \Rightarrow \quad B = +1$

Equating constants, $\quad \Rightarrow \quad -3 = A - C \quad \Rightarrow \quad C = +2$

So it appears that

$$\frac{x - 3}{(x - 1)(x^2 + 1)} \equiv \frac{-1}{x - 1} + \frac{x + 2}{x^2 + 1}$$

which can quickly be verified.

In summary, therefore, to express a rational function in partial fractions

1 First ensure, by division if necessary, that the numerator is of lower degree than the denominator.

2   Express the given function as the sum of partial fractions, each with arbitrary constants, as follows:

a) to a simple factor $ax + b$ there corresponds a partial fraction $\dfrac{A}{ax + b}$

b) to a repeated factor $(cx + d)^2$ there correspond partial fractions

$$\frac{B}{cx + d} + \frac{C}{(cx + d)^2}$$

c) to an irreducible factor $ex^2 + fx + g$ there corresponds a partial fraction

$$\frac{Dx + E}{ex^2 + fx + g}$$

3   Calculate the arbitrary constants either
   a) by taking particular values of x; or
   b) by equating coefficients

Partial fractions are principally used in integration, but it is also interesting to see how they can help us to find the sums of certain series.

If, for instance, we consider the series

$$\frac{1}{1 \times 3} + \frac{1}{3 \times 5} + \frac{1}{5 \times 7} + \cdots$$

we see that its rth term is $\dfrac{1}{(2r - 1)(2r + 1)}$.

This can easily be expressed in partial fractions as

$$\frac{\frac{1}{2}}{2r - 1} - \frac{\frac{1}{2}}{2r + 1} = \frac{1}{2}\left(\frac{1}{2r - 1} - \frac{1}{2r + 1}\right)$$

Hence the sum of the first n terms is

$$S_n = \frac{1}{1 \times 3} + \frac{1}{3 \times 5} + \frac{1}{5 \times 7} + \cdots + \frac{1}{(2n - 1)(2n + 1)}$$

$$= \frac{1}{2}\left\{\left(\frac{1}{1} - \frac{1}{3}\right) + \left(\frac{1}{3} - \frac{1}{5}\right) + \left(\frac{1}{5} - \frac{1}{7}\right) + \cdots + \left(\frac{1}{2n - 1} - \frac{1}{2n + 1}\right)\right\}$$

$$= \frac{1}{2}\left\{1 - \frac{1}{2n + 1}\right\} = \frac{n}{(2n + 1)}$$

Also, as $n \to \infty$, $S_n \to \frac{1}{2}$

So the sum of the infinite series is $\frac{1}{2}$.

Exercise 10.6a

1 Express in partial fractions:

a) $\dfrac{x+1}{(x-1)(x-2)}$ **b)** $\dfrac{x}{x^2-4}$ **c)** $\dfrac{1}{x(2x+1)}$

d) $\dfrac{x^2}{(x-1)(x-2)(x-3)}$ **e)** $\dfrac{x-1}{x^2+x}$ **f)** $\dfrac{x^2+1}{x^2-2x}$

g) $\dfrac{x^2}{x^2-4}$ **h)** $\dfrac{x^3}{x^2-1}$

2 Express in partial fractions:

a) $\dfrac{1}{x(x-1)^2}$ **b)** $\dfrac{x-1}{x^2(x+1)}$

c) $\dfrac{x}{(x-1)^2(x+1)}$ **d)** $\dfrac{x^3}{(x-1)(x^2-1)}$

3 Express in partial fractions:

a) $\dfrac{1}{x(x^2+1)}$ **b)** $\dfrac{1}{x^4-1}$ **c)** $\dfrac{1}{x^3-1}$ **d)** $\dfrac{x}{x^3+1}$

4 Find the first four terms of the power series for $\dfrac{1}{(1+x)(1+2x)}$ by first expressing it
   a) in partial fractions
   b) as $\{1+(3x+2x^2)\}^{-1}$
   c) as $(1+x)^{-1}(1+2x)^{-1}$

5 Find the rth term, the sum of the first n terms and the sum to infinity of the series

a) $\dfrac{1}{1\times2}+\dfrac{1}{2\times3}+\dfrac{1}{3\times4}+\cdots$

b) $\dfrac{1}{1\times3}+\dfrac{1}{2\times4}+\dfrac{1}{3\times5}+\cdots$

c) $\dfrac{1}{1\times2\times3}+\dfrac{1}{2\times3\times4}+\dfrac{1}{3\times4\times5}+\cdots$

Use of partial fractions in integration

In the next section we shall return to the general problem of integration, but we can already see the use of partial fractions when we need to integrate the quotient of two polynomials (or *rational function*, as it is usually called).

Example 5

$$I = \int \frac{x+1}{x^2 - 4} dx$$

It is quickly seen that

$$\frac{x+1}{x^2-4} \equiv \frac{\frac{3}{4}}{x-2} + \frac{\frac{1}{4}}{x+2}$$

So $I = \int \left(\frac{\frac{3}{4}}{x-2} + \frac{\frac{1}{4}}{x+2} \right) dx$

$$= \tfrac{3}{4} \ln(x-2) + \tfrac{1}{4} \ln(x+2) + c$$

Example 6

$$I = \int \frac{x^2}{(x-1)(x+1)^2} dx$$

We can find the full partial fractions in such a case, with a repeated factor, by letting

$$\frac{x^2}{(x-1)(x+1)^2} \equiv \frac{A}{x-1} + \frac{B}{x+1} + \frac{C}{(x+1)^2}$$

$$\Rightarrow \qquad x^2 \equiv A(x+1)^2 + B(x-1)(x+1) + C(x-1)$$

Putting $x = 1$, $\qquad\qquad\qquad\qquad 1 = 4A \quad \Rightarrow \quad A = \tfrac{1}{4}$

Putting $x = -1$, $\qquad\qquad\qquad 1 = -2C \quad \Rightarrow \quad C = -\tfrac{1}{2}$

Equating coefficient x^2, $\quad 1 = A + B \quad \Rightarrow \quad B = \tfrac{3}{4}$

So $I = \int \left(\frac{\frac{1}{4}}{x-1} + \frac{\frac{3}{4}}{x+1} - \frac{\frac{1}{2}}{(x+1)^2} \right) dx$

$$= \tfrac{1}{4} \ln(x-1) + \tfrac{3}{4} \ln(x+1) + \frac{1}{2(x+1)} + c$$

Example 7

$$I = \int \frac{x^2 \, dx}{(x-1)(x^2+1)}$$

We can express this in partial fractions as

$$\frac{x^2}{(x-1)(x^2+1)} \equiv \frac{A}{x-1} + \frac{Bx+C}{x^2+1}$$

So $\qquad\qquad x^2 \equiv A(x^2+1) + (Bx+C)(x-1)$

Putting $x = 1$, $\qquad\qquad 1 = 2A \qquad \Rightarrow \quad A = \tfrac{1}{2}$

Equating coefficient x^2, $\quad 1 = A + B \quad \Rightarrow \quad B = \tfrac{1}{2}$

Equating constant terms, $\quad 0 = A - C \quad \Rightarrow \quad C = \tfrac{1}{2}$

So $\quad \dfrac{x^2}{(x-1)(x^2+1)} \equiv \dfrac{\tfrac{1}{2}}{x-1} + \dfrac{\tfrac{1}{2}x + \tfrac{1}{2}}{x^2+1}$

Hence $\quad I = \tfrac{1}{2}\displaystyle\int \frac{dx}{x-1} + \tfrac{1}{2}\int \frac{x}{x^2+1}\,dx + \tfrac{1}{2}\int \frac{dx}{x^2+1}$

$\qquad\qquad = \tfrac{1}{2}\ln(x-1) + \tfrac{1}{4}\ln(x^2+1) + \tfrac{1}{2}\tan^{-1}x + c$

In this last example we encountered a quadratic term $x^2 + 1$ which cannot be split into real linear factors (and which is therefore said to be *irreducible*). As such terms, and their corresponding partial fractions, occur quite frequently, our next example will show how they can best be treated.

Example 8

a) $\displaystyle\int \frac{2x-6}{x^2-6x+25}\,dx$ b) $\displaystyle\int \frac{1}{x^2-6x+25}\,dx$ c) $\displaystyle\int \frac{x}{x^2-6x+25}\,dx$

In a) we immediately see that the numerator is the derivative of the denominator.

So $\quad \displaystyle\int \frac{2x-6}{x^2-6x+25}\,dx = \ln(x^2-6x+25) + c$

In b) we first write

$$\int \frac{1}{x^2-6x+25}\,dx = \int \frac{1}{(x-3)^2+16}\,dx$$

Putting $x - 3 = 4\tan\theta \quad \Rightarrow \quad dx = 4\sec^2\theta\,d\theta$, we obtain

$$\int \frac{1}{x^2-6x+25}\,dx = \int \frac{4\sec^2\theta\,d\theta}{16\tan^2\theta+16}$$

$$= \tfrac{1}{4}\int d\theta = \tfrac{1}{4}\theta + c$$

But $x - 3 = 4\tan\theta$

$$\Rightarrow \quad \tan\theta = \frac{x-3}{4} \quad \Rightarrow \quad \theta = \tan^{-1}\frac{x-3}{4}$$

So $\displaystyle\int \frac{1}{x^2 - 6x + 25}\,dx = \tfrac{1}{4}\tan^{-1}\frac{x-3}{4} + c$

In **c)** we express the required integral as a linear combination of the two previous parts:

$$\int \frac{x}{x^2 - 6x + 25}\,dx = \int \frac{\frac{1}{2}(2x-6)}{x^2 - 6x + 25}\,dx + \int \frac{3}{x^2 - 6x + 25}\,dx$$

$$= \tfrac{1}{2}\ln(x^2 - 6x + 25) + \tfrac{3}{4}\tan^{-1}\frac{x-3}{4} + c$$

Exercise 10.6b

Find the following integrals:

1 a) $\displaystyle\int \frac{dx}{x^2 - 1}$
 b) $\displaystyle\int \frac{dx}{x^2 + x}$
 c) $\displaystyle\int \frac{dx}{2x - x^2}$

d) $\displaystyle\int \frac{x^2}{x^2 - 4}\,dx$
 e) $\displaystyle\int \frac{dx}{4x^2 - 1}$
 f) $\displaystyle\int \frac{dx}{2x^2 + x - 1}$

2 a) $\displaystyle\int \frac{dx}{(x+1)^2}$
 b) $\displaystyle\int \frac{dx}{x^2(x+2)}$
 c) $\displaystyle\int \frac{x+1}{x(x-1)^2}\,dx$

d) $\displaystyle\int \frac{x^2\,dx}{(x-1)^2}$

3 a) $\displaystyle\int \frac{2x\,dx}{x^2 + 9}$
 $\displaystyle\int \frac{dx}{x^2 + 9}$
 $\displaystyle\int \frac{x+3}{x^2 + 9}\,dx$

b) $\displaystyle\int \frac{8x\,dx}{4x^2 + 9}$
 $\displaystyle\int \frac{dx}{4x^2 + 9}$
 $\displaystyle\int \frac{2x-3}{4x^2 + 9}\,dx$

c) $\displaystyle\int \frac{2x-6}{x^2 - 6x + 10}\,dx$
 $\displaystyle\int \frac{dx}{x^2 - 6x + 10}$
 $\displaystyle\int \frac{x\,dx}{x^2 - 6x + 10}$

4 a) $\displaystyle\int \frac{x\,dx}{x-1}$
 b) $\displaystyle\int \frac{x^2 + 1}{x^2 - 1}\,dx$
 c) $\displaystyle\int \frac{x^3 + 1}{x^2 - 1}\,dx$

d) $\displaystyle\int \frac{x\,dx}{(x-1)^2}$
 e) $\displaystyle\int \frac{x^2\,dx}{x^3 + 1}$
 f) $\displaystyle\int \frac{x\,dx}{x^3 - 1}$

* 10.7 General integration

In the last section we learned to integrate rational functions by means of partial fractions. We now seek to draw together other methods we have employed for finding integrals.

Faced with a particular function whose integral is required, our primary need is for a very thorough knowledge of differentiation so that we can try to find another function, or 'primitive', whose derivative is the given function. This may not always be possible: for instance, until we invented the function $\ln x$ it was impossible to integrate even such a simple function as $1/x$, and we shall frequently encounter similar examples. Nevertheless, our first step must be to re-state the standard derivatives and the integrals to which they give rise.

$$\frac{d}{dx}(x^n) = nx^{n-1} \qquad \int x^n \, dx = \frac{x^{n+1}}{n+1} + c \quad (n \neq -1)$$

$$\frac{d}{dx}(\ln x) = \frac{1}{x} \qquad \int \frac{dx}{x} = \ln x + c$$

$$\frac{d}{dx}(e^x) = e^x \qquad \int e^x \, dx = e^x + c$$

$$\frac{d}{dx}(\sin x) = \cos x \qquad \int \cos x \, dx = \sin x + c$$

$$\frac{d}{dx}(\cos x) = -\sin x \qquad \int \sin x \, dx = -\cos x + c$$

$$\frac{d}{dx}(\tan x) = \sec^2 x \qquad \int \sec^2 x \, dx = \tan x + c$$

$$\frac{d}{dx}(\sin^{-1} x) = \frac{1}{\sqrt{(1-x^2)}} \qquad \int \frac{dx}{\sqrt{(1-x^2)}} = \sin^{-1} x + c$$

$$\frac{d}{dx}(\tan^{-1} x) = \frac{1}{1+x^2} \qquad \int \frac{dx}{1+x^2} = \tan^{-1} x + c$$

These standard integrals frequently need to be used in conjunction with general methods such as that of substitution or integration by parts and, although these have been investigated in previous sections, considerable practice is necessary in order to sense what is the most appropriate method for each particular example. We have already stressed the crucial importance of integration in finding areas and volumes, and it will soon be seen that it is similarly central in statistics, mechanics, and throughout all branches of applied mathematics.

Example 1

$$I = \int \sqrt{(a^2 - x^2)} \, dx$$

Here a substitution converts the integral into another which is more tractable,

for $x = a \sin \theta \quad \Rightarrow \quad dx = a \cos \theta \, d\theta$

So $I = \int \sqrt{(a^2 - a^2 \sin^2 \theta)} \, a \cos \theta \, d\theta$

$$= a^2 \int \cos^2 \theta \, d\theta$$

$$= \tfrac{1}{2} a^2 \int (1 + \cos 2\theta) \, d\theta$$

$$= \tfrac{1}{2} a^2 (\theta + \tfrac{1}{2} \sin 2\theta) + c$$

$$= \tfrac{1}{2} a^2 (\theta + \sin \theta \cos \theta) + c$$

Finally, it is necessary to revert to the original variable x.

Now $x = a \sin \theta \quad \Rightarrow \quad \theta = \sin^{-1} \dfrac{x}{a}$

$\Rightarrow$ $\qquad\qquad \sin \theta = \dfrac{x}{a}$

$\Rightarrow$ $\qquad\qquad \cos \theta = \sqrt{\left(1 - \dfrac{x^2}{a^2}\right)} = \dfrac{\sqrt{(a^2 - x^2)}}{a}$

So $\int \sqrt{(a^2 - x^2)} \, dx = \tfrac{1}{2} a^2 \sin^{-1} x + \tfrac{1}{2} x \sqrt{(a^2 - x^2)} + c$

Example 2

$$I = \int \tan^{-1} x \, dx$$

Because $\tan^{-1} x$ is a function which we can easily differentiate, there is an immediate attraction in using the method of integration by parts

$$I = \int 1 \tan^{-1} x \, dx$$

$$= x \tan^{-1} x - \int x \frac{1}{1 + x^2} \, dx$$

So $I = x \tan^{-1} x - \frac{1}{2} \int \frac{2x}{1 + x^2} \, dx$

$$= x \tan^{-1} x - \frac{1}{2} \ln(1 + x^2) + c$$

Example 3

$$I = \int \operatorname{cosec} x \, dx$$

It may be recalled that $\sin x$, $\cos x$, and $\tan x$ are easily expressible in terms of $\tan \frac{1}{2} x$:

$$t = \tan \tfrac{1}{2} x$$

$$\Rightarrow \quad \sin x = \frac{2t}{1 + t^2} \qquad \cos x = \frac{1 - t^2}{1 + t^2} \qquad \tan x = \frac{2t}{1 - t^2}$$

So the substitution $t = \tan \frac{1}{2} x$ is frequently convenient for simplifying trigonometric integrals.

Now $t = \tan \dfrac{x}{2} \quad \Rightarrow \quad dt = \frac{1}{2} \sec^2 \frac{1}{2} x \, dx = \frac{1}{2}(1 + t^2) \, dx$

$$\Rightarrow \quad dx = \frac{2 \, dt}{1 + t^2}$$

So $I = \displaystyle\int \frac{dx}{\sin x}$

$$= \int \frac{2 \, dt/(1 + t^2)}{2t/(1 + t^2)} = \int \frac{dt}{t} = \ln t + c = \ln \tan \tfrac{1}{2} x + c$$

Hence $\displaystyle\int \operatorname{cosec} x \, dx = \ln \tan \tfrac{1}{2} x + c$

and it is quickly verified that

$$\frac{d}{dx}(\ln \tan \tfrac{1}{2} x) = \frac{1}{\tan \frac{1}{2} x}(\sec^2 \tfrac{1}{2} x) \times \tfrac{1}{2}$$

$$= \frac{1}{2 \sin \frac{1}{2} x \cos \frac{1}{2} x} = \frac{1}{\sin x} = \operatorname{cosec} x$$

To complete the catalogue of the integrals of elementary trigonometric functions, we recall that

$$\int \tan x \, dx = \int \frac{\sin x}{\cos x} \, dx$$

$$= -\ln \cos x + c$$

$$= \ln \sec x + c$$

and

$$\int \sec x \, dx = \int \frac{\sec x \, (\sec x + \tan x)}{\sec x + \tan x} \, dx$$

$$= \int \frac{\sec x \tan x + \sec^2 x}{\sec x + \tan x} \, dx$$

$$= \ln (\sec x + \tan x) + c$$

Finally, we give two examples which are quickly dealt with by a preliminary rearrangement.

Example 4

$$I = \int \tan^4 x \, dx$$

Now $$I = \int (\tan^2 x)(\tan^2 x) \, dx$$

$$= \int \tan^2 x \, (\sec^2 x - 1) \, dx$$

$$= \int (\tan^2 x \sec^2 x - \sec^2 x + 1) \, dx$$

$$= \tfrac{1}{3} \tan^3 x - \tan x + x + c$$

Example 5

$$I = \int \frac{\sin x}{\sin x + \cos x} \, dx$$

We first notice that we could easily find

$$\int \frac{\sin x + \cos x}{\sin x + \cos x} \, dx = \int dx = x$$

and $\displaystyle\int \frac{\cos x - \sin x}{\sin x + \cos x}\,dx = \ln\left(\sin x + \cos x\right)$

So $\displaystyle\int \frac{\sin x}{\sin x + \cos x}\,dx = \frac{1}{2}\int \frac{\sin x + \cos x}{\sin x + \cos x}\,dx - \frac{1}{2}\int \frac{\cos x - \sin x}{\sin x + \cos x}\,dx$

$\displaystyle\qquad\qquad = \frac{1}{2}x - \frac{1}{2}\ln\left(\sin x + \cos x\right) + c$

Exercise 10.7

Find the following integrals:

1 **a)** $\displaystyle\int 5x^4\,dx$ **b)** $\displaystyle\int t^6\,dt$ **c)** $\displaystyle\int \frac{3\,du}{u^2}$

d) $\displaystyle\int \frac{dx}{2\sqrt{x}}$ **e)** $\displaystyle\int \sqrt{v}\,dv$ **f)** $\displaystyle\int (x+1)^4\,dx$

g) $\displaystyle\int \frac{du}{(2u+1)^3}$ **h)** $\displaystyle\int \frac{dv}{\sqrt{(3v-2)}}$ **i)** $\displaystyle\int x(x^2-1)^5\,dx$

j) $\displaystyle\int \frac{x^2\,dx}{(1-x^3)^2}$ **k)** $\displaystyle\int x\sqrt{(x^2+1)}\,dx$ **l)** $\displaystyle\int \frac{x\,dx}{\sqrt{(1-x^2)}}$

2 **a)** $\displaystyle\int \frac{2}{x}\,dx$ **b)** $\displaystyle\int \frac{3}{x-3}\,dx$ **c)** $\displaystyle\int \frac{du}{1-u}$

d) $\displaystyle\int \frac{v\,dv}{v^2-1}$ **e)** $\displaystyle\int \frac{x^2\,dx}{1-x^3}$ **f)** $\displaystyle\int \frac{2x+1}{x^2+x}\,dx$

g) $\displaystyle\int e^{2x}\,dx$ **h)** $\displaystyle\int 2\,e^{3x}\,dx$ **i)** $\displaystyle\int x\,e^{-x^2}\,dx$

j) $\displaystyle\int \frac{e^u}{1+e^u}\,du$ **k)** $\displaystyle\int 10^t\,dt$ **l)** $\displaystyle\int \frac{\ln x}{x}\,dx$

3 **a)** $\displaystyle\int \cos 2x\,dx$ **b)** $\displaystyle\int \sin \tfrac{1}{2}x\,dx$ **c)** $\displaystyle\int \sin^2\theta\cos\theta\,d\theta$

d) $\displaystyle\int \cos^2\theta\sin\theta\,d\theta$ **e)** $\displaystyle\int \cos^3\theta\,d\theta$ **f)** $\displaystyle\int \sin^3\theta\,d\theta$

g) $\displaystyle\int \sin x\cos x\,dx$ **h)** $\displaystyle\int \sin^2\tfrac{1}{2}x\,dx$ **i)** $\displaystyle\int \cos^2 x\,dx$

j) $\displaystyle\int \frac{\cos x}{\sqrt{(\sin x)}}\,dx$ **k)** $\displaystyle\int \frac{\cos\theta}{\sin\theta}\,d\theta$ **l)** $\displaystyle\int \frac{\sin\theta}{\cos\theta}\,d\theta$

4 a) $\displaystyle\int \sec^2 x \, dx$ **b)** $\displaystyle\int \tan^2 x \, dx$ **c)** $\displaystyle\int \sec\theta \tan\theta \, d\theta$

d) $\displaystyle\int \csc\theta \cot\theta \, d\theta$ **e)** $\displaystyle\int \csc^2\theta \, d\theta$ **f)** $\displaystyle\int \cot^2\theta \, d\theta$

g) $\displaystyle\int \tan\theta \sec^2\theta \, d\theta$ **h)** $\displaystyle\int \sec^3\theta \tan\theta \, d\theta$ **i)** $\displaystyle\int \tan 2x \, dx$

j) $\displaystyle\int \tan^3 x \, dx$ **k)** $\displaystyle\int \cot\tfrac{1}{2}x \, dx$ **l)** $\displaystyle\int \cot^3 x \, dx$

5 a) $\displaystyle\int \frac{dx}{\sqrt{(4 - x^2)}}$ $(x = 2\sin\theta)$ **b)** $\displaystyle\int \frac{dx}{\sqrt{(9 - 4x^2)}}$ $(2x = 3\sin\theta)$

c) $\displaystyle\int \frac{dx}{1 + 4x^2}$ $(2x = \tan\theta)$ **d)** $\displaystyle\int \frac{dx}{4 + 9x^2}$ $(3x = 2\tan\theta)$

e) $\displaystyle\int \frac{du}{\sqrt{(u^2 + 1)}}$ $(u = \tan\theta)$ **f)** $\displaystyle\int \frac{du}{\sqrt{(u^2 - 1)}}$ $(u = \sec\theta)$

g) $\displaystyle\int \sqrt{(1 - x^2)} \, dx$ $(x = \sin\theta)$ **h)** $\displaystyle\int x\sqrt{(1 - x^2)} \, dx$

i) $\displaystyle\int \frac{dx}{4x^2 + 9}$ **j)** $\displaystyle\int \frac{x \, dx}{4x^2 + 9}$

6 a) $\displaystyle\int x\cos x \, dx$ **b)** $\displaystyle\int x\,e^{-x} \, dx$ **c)** $\displaystyle\int t\sin 3t \, dt$

d) $\displaystyle\int x^2\sin x \, dx$ **e)** $\displaystyle\int \sin^{-1}x \, dx$ **f)** $\displaystyle\int \ln x \, dx$

g) $\displaystyle\int \theta\sec^2\theta \, d\theta$ **h)** $\displaystyle\int x^2\,e^{2x} \, dx$ **i)** $\displaystyle\int e^{-t}\sin t \, dt$

j) $\displaystyle\int e^{2t}\cos 3t \, dt$ **k)** $\displaystyle\int x^2\ln x \, dx$ **l)** $\displaystyle\int x\,10^x \, dx$

7 Use the substitution $t = \tan\tfrac{1}{2}x$ to find

a) $\displaystyle\int_0^{\frac{1}{2}\pi} \frac{dx}{1 + \sin x}$ **b)** $\displaystyle\int_0^{\frac{1}{2}\pi} \frac{dx}{1 + \sin x + \cos x}$ **c)** $\displaystyle\int_0^{\frac{1}{2}\pi} \frac{dx}{\sin x + \cos x}$

8 Show that

$$\int_0^a f(x) \, dx = \int_0^a f(a - x) \, dx$$

and hence (using no. 7) evaluate:

a) $\displaystyle\int_0^\pi \frac{x\,dx}{1+\sin x}$ b) $\displaystyle\int_0^{\frac{1}{2}\pi} \frac{x\,dx}{1+\sin x+\cos x}$ c) $\displaystyle\int_0^{\frac{1}{2}\pi} \frac{x\,dx}{\sin x+\cos x}$

9   Find the following integrals:

a) $\displaystyle\int \frac{dx}{x^2+x}$ b) $\displaystyle\int \frac{dx}{x^2-x}$ c) $\displaystyle\int \frac{dx}{x^3-x^2}$

d) $\displaystyle\int \frac{dx}{x^3+x}$ e) $\displaystyle\int \frac{e^x}{1+e^x}\,dx$ f) $\displaystyle\int \frac{dx}{1+e^x}$

g) $\displaystyle\int \frac{x\,dx}{x^4+1}$ h) $\displaystyle\int (\ln x)^2\,dx$ i) $\displaystyle\int x^2\tan^{-1} x\,dx$

j) $\displaystyle\int \frac{dx}{2+\cos x}$ k) $\displaystyle\int \frac{dx}{x\sqrt{(x^2-1)}}$ l) $\displaystyle\int \sqrt{\left(\frac{1-x}{1+x}\right)}\,dx$

10   Evaluate:

a) $\displaystyle\int_0^1 \frac{dx}{(x+1)(x^2+1)}$ b) $\displaystyle\int_0^1 \frac{x\,dx}{(x^2+1)^2}$ c) $\displaystyle\int_0^1 \frac{dx}{(x^2+1)^2}$

d) $\displaystyle\int_0^1 \frac{dx}{(x^2+1)^{3/2}}$ e) $\displaystyle\int_0^{\frac{1}{2}\pi} \sin^5\theta\cos^2\theta\,d\theta$

f) $\displaystyle\int_0^{\frac{1}{2}\pi} \sin^2\theta\cos^6\theta\,d\theta$ g) $\displaystyle\int_1^e x^2\ln x\,dx$

h) $\displaystyle\int_0^1 \sqrt{(4-x^2)}\,dx$ i) $\displaystyle\int_0^{\frac{1}{2}\pi} \frac{d\theta}{2+\cos\theta}$

j) $\displaystyle\int_0^{\frac{1}{4}\pi} \frac{\cos\theta-\sin\theta}{\cos\theta+\sin\theta}\,d\theta$ k) $\displaystyle\int_{-\frac{1}{4}\pi}^{\frac{1}{4}\pi} \frac{d\theta}{3\sin^2\theta+\cos^2\theta}$ (put $t=\tan\theta$)

l) $\displaystyle\int_2^3 \sqrt{\left(\frac{x-2}{4-x}\right)}\,dx$ (put $x=2\sin^2\theta+4\cos^2\theta$)

Miscellaneous problems

1   Calculate $(1+1/n)^n$ for $n=1, 2, 5, 10, 100, 1000, 10^6$. Does it appear to tend to a limit as $n\to\infty$?
a) If $n=1/h$, show that

$$\ln\left(1+\frac{1}{n}\right)^n = \frac{\ln(1+h)}{h} = \frac{\ln(1+h)-\ln 1}{h}$$

Letting $n \to \infty$ (and so $h \to 0$), show that

$$\ln\left(1 + \frac{1}{n}\right)^n \to 1, \text{ and hence that } \left(1 + \frac{1}{n}\right)^n \to e$$

b) Prove similarly that, as $n \to \infty$

$$\left(1 + \frac{x}{n}\right)^n \to e^x$$

2   By the method of two staircases (section 4.2), show that

$$\int_1^{10^6} \frac{dx}{x} \text{ lies between } \sum_{r=2}^{10^6} \frac{1}{r} \text{ and } \sum_{r=1}^{10^6-1} \frac{1}{r}$$

Hence show that the sum of the reciprocals of the first million integers lies between 13.81 and 14.82.

3   By considering the graph of $1/x$ or otherwise, show that

$$\ln n - \ln(n-1) > \frac{1}{n} \quad \text{and} \quad \ln n - \ln(n-1) < \frac{1}{2}\left(\frac{1}{n} + \frac{1}{n-1}\right)$$

where n is an integer greater than 1.
  The function $f(n)$ is defined by

$$f(n) = 1 + \frac{1}{2} + \frac{1}{3} + \cdots + \frac{1}{n} - \ln n$$

Show that $f(n)$ decreases as n increases and that $f(n) - \frac{1}{2n}$ increases as n increases. Deduce that $f(n)$ tends to a finite limit as n tends to infinity. (CS)

4   Show that

$$\int_1^n \ln x \, dx < \sum_{r=1}^n \ln r < \int_1^{n+1} \ln x \, dx$$

and deduce that the geometric mean of the first n positive integers is approximately n/e. (CS)

5   **a)** Investigate the behaviour as $x \to \infty$ of the ratios

$$\frac{\ln x}{x}, \quad \frac{\ln x}{\sqrt{x}}, \quad \frac{\ln x}{\sqrt[1000]{x}}$$

b) Show that if $t > 1$, $\frac{1}{t} < \frac{1}{\sqrt{t}}$

and hence that $\ln x < 2(\sqrt{x} - 1)$ and $\frac{\ln x}{x} < 2\left(\frac{1}{\sqrt{x}} - \frac{1}{x}\right)$

Deduce that as $x \to \infty$, $\dfrac{\ln x}{x} \to 0$

c) Show that $\dfrac{\ln x}{\sqrt[1000]{x}} = 1000\,\dfrac{\ln \sqrt[1000]{x}}{\sqrt[1000]{x}}$

and use b) to deduce its limit as $x \to \infty$.

d) More generally, show that as $x \to \infty$, $\ln x \to \infty$ more slowly than any positive power (however small) of x.

6 a) Investigate the behaviour as $x \to \infty$ of

$$\dfrac{e^x}{x}, \quad \dfrac{e^x}{x^2}, \quad \dfrac{e^x}{x^{1000}}$$

b) Show that

$$\ln \dfrac{e^x}{x} = x\left(1 - \dfrac{\ln x}{x}\right)$$

and use the result of no. 5b) to show that as $x \to \infty$

$$\ln \dfrac{e^x}{x} \to \infty \text{ and hence that } \dfrac{e^x}{x} \to \infty$$

c) Deduce that as $x \to \infty$, $\dfrac{e^x}{x^{1000}} \to \infty$

d) More generally, show that as $x \to +\infty$, $e^x \to +\infty$ faster than *any* positive power of x; and that $e^{-x} \to 0$ faster than *any* negative power of x.

7 If $\qquad I_n(x) = \displaystyle\int_0^x t^n e^{-t}\, dt$

show that $\quad I_n(x) = -x^n e^{-x} + n I_{n-1}(x)$

If $I_n(x) \to I_n$ as $x \to +\infty$, use the result of no. 6 to show that $I_n = n I_{n-1}$

and hence that, if $n \in \mathbb{Z}^+$, $I_n = n!$

8 a) Sketch the graph of the function e^{-x^2} and illustrate the integral

$$I = \int_{-\infty}^{\infty} e^{-x^2}\, dx$$

Can you calculate this integral?
b) Illustrate by a surface the function of two variables $f(x, y)$, where $f(x, y) = e^{-x^2 - y^2}$
c) Show that the volume under this surface can be represented in

Cartesian coordinates as

$$\lim \sum \sum e^{-x^2-y^2} \, \delta x \, \delta y = \int \int e^{-x^2-y^2} \, dx \, dy$$

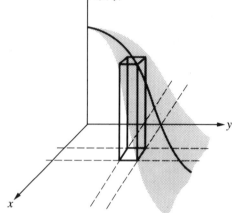

and in polar coordinates as

$$\sum \sum e^{-r^2} (\delta r)(r \, \delta\theta) = \int \int r e^{-r^2} \, dr \, d\theta$$

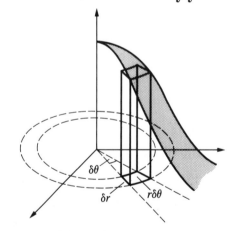

and hence that

$$\left(\int_{-\infty}^{\infty} e^{-x^2} \, dx \right) \left(\int_{-\infty}^{\infty} e^{-y^2} \, dy \right) = \int_{r=0}^{\infty} r e^{-r^2} \, dr \int_{\theta=0}^{2\pi} d\theta$$

d) Finally show that $\int_0^{\infty} r e^{-r^2} \, dr = \frac{1}{2}$ and hence calculate I.

9 Show that, if

$$f(x) = \sum_1^N a_n \sin nx \qquad (1)$$

then $a_n = \dfrac{2}{\pi} \displaystyle\int_0^\pi f(x) \sin nx \, dx \qquad (2)$

Assuming that the function

$$f(x) = x(\pi - x), \quad (0 \leqslant x \leqslant \pi)$$

can be expressed as an infinite series†

$$\sum_1^\infty a_n \sin nx,$$

and that the coefficients are still given by the formula (2), show that in this case

$$a_{2m} = 0 \quad a_{2m+1} = \frac{8}{\pi(2m+1)^3}$$

and hence sum the series

$$1 - \frac{1}{3^3} + \frac{1}{5^3} - \frac{1}{7^3} + \cdots \qquad \text{(CS)}$$

10 Prove that the trapezium bounded by the tangent to the curve $y = f(x)$ at the point $(r, f(r))$, the ordinates $x = r \pm \frac{1}{2}$ and the x-axis has area $f(r)$.
   Sketch the graph of $y = \ln x$ and show that, for $r \geqslant 1\frac{1}{2}$

$$\int_{r-\frac{1}{2}}^{r+\frac{1}{2}} \ln x \, dx < \ln r$$

Deduce that $\displaystyle\int_{1.5}^{500.5} \ln x \, dx < \ln (500!)$

Show also that, for $r \geqslant 1$, $\displaystyle\int_r^{r+1} \ln x \, dx > \frac{1}{2}[\ln r + \ln(r + 1)]$

and deduce that

$$\ln (500!) < \int_1^{500} \ln x \, dx + \tfrac{1}{2} \ln 500$$

Calculate the value, to the nearest integer, of $\ln (500!)$. $\qquad$ (C)

---

† A typical example of a *Fourier series*, so called after their discoverer, J. B. J. Fourier (1768–1830).

Appendix: introduction to coordinate geometry

A.1 Straight lines

In Chapter 1 we introduced the coordinate geometry of the straight line and established that

a) the line through (x_1, y_1) with gradient m has the equation

$$\frac{y - y_1}{x - x_1} = m$$

b) the line joining (x_1, y_1) and (x_2, y_2) has the equation

$$\frac{y - y_1}{x - x_1} = \frac{y_2 - y_1}{x_2 - x_1}$$

c) the equation $ax + by + c = 0$ represents a straight line with gradient $-a/b$.

We now proceed to develop these ideas a little further.

The angle between two lines

Suppose that θ is an angle between the two lines whose gradients are m_1, m_2.

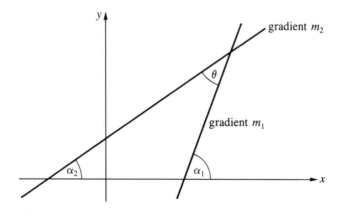

In the diagram

$$\tan \alpha_1 = m_1 \text{ and } \tan \alpha_2 = m_2$$

whilst $\theta = \alpha_1 - \alpha_2$

$\Rightarrow$ $\tan \theta = \tan(\alpha_1 - \alpha_2)$

$$= \frac{\tan \alpha_1 - \tan \alpha_2}{1 + \tan \alpha_1 \tan \alpha_2}$$

$$\tan \theta = \frac{m_1 - m_2}{1 + m_1 m_2}$$

In particular, if the lines are perpendicular, then

$$\theta = \frac{1}{2}\pi \quad \Leftrightarrow \quad m_1 m_2 = -1$$

The perpendicular distance of a point from a line

Given the point $P(x_1, y_1)$ and the line l with equation $ax + by + c = 0$, how can we find the perpendicular distance p from P to l?

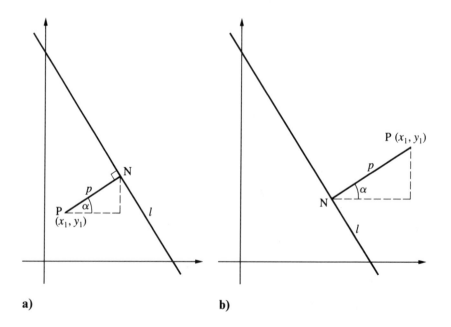

a)   b)

First of all we know that the gradient of l is $-a/b$, so the gradient of the perpendicular PN is $b/a = \tan \alpha$

$$\Rightarrow \quad \cos \alpha = \frac{a}{\sqrt{(a^2 + b^2)}} \quad \text{and} \quad \sin \alpha = \frac{b}{\sqrt{(a^2 + b^2)}}$$

Coordinates of N are (in Figs. **a)**, **b)** respectively)

$$x = x_1 \pm p \cos \alpha = x_1 \pm \frac{ap}{\sqrt{(a^2 + b^2)}}$$

$$\text{and} \quad y = y_1 \pm p \sin \alpha = y_1 \pm \frac{bp}{\sqrt{(a^2 + b^2)}}$$

But N lies on $ax + by + c = 0$

$$\Rightarrow \quad a\left(x_1 \pm \frac{ap}{\sqrt{(a^2 + b^2)}}\right) + b\left(y_1 \pm \frac{bp}{\sqrt{(a^2 + b^2)}}\right) + c = 0$$

$$\Rightarrow \qquad\qquad ax_1 + by_1 + c \pm \sqrt{(a^2 + b^2)}p = 0$$

$$\Rightarrow \quad p = \pm \frac{ax_1 + by_1 + c}{\sqrt{(a^2 + b^2)}}$$

So in both cases

$$p = \left| \frac{ax_1 + by_1 + c}{\sqrt{(a^2 + b^2)}} \right|$$

Example 1

Find the perpendicular distance from the point $(-2, 3)$ to the line $x - 2y + 4 = 0$

$$\text{Here } p = \frac{|-2 - 2(3) + 4|}{\sqrt{(1^2 + (-2)^2)}}$$

$$= \frac{|-4|}{\sqrt{5}} = \frac{4}{\sqrt{5}}$$

Example 2

Find the locus of the point which moves so that its perpendicular distances from the two lines $4x - 3y = 0$ and $5x + 12y = 0$ are equal.

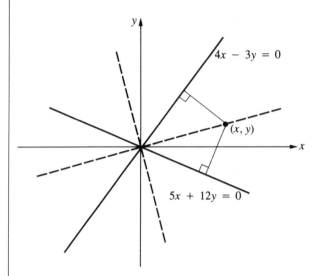

The distances of the point (x, y) from the two lines are

$$\frac{|4x - 3y|}{5} \quad \text{and} \quad \frac{|5x + 12y|}{13}$$

Now these are equal whenever

$$\frac{4x - 3y}{5} = \frac{5x + 12y}{13} \quad \text{or} \quad \frac{4x - 3y}{5} = -\frac{5x + 12y}{13}$$

i.e.   when $3x - 11y = 0$ or $11x + 3y = 0$

Hence, as might have been expected, the required locus is a pair of perpendicular straight lines (gradients $\frac{3}{11}$ and $-\frac{11}{3}$) which are the two bisectors (internal and external) of the angles formed by the original pair.

Exercise A.1

1   Find the tangent of the acute angle between the following pairs of lines:
   a) $y = 2x + 1$ and $y = 3x + 1$
   b) $2x + y + 3 = 0$ and $4x - y - 2 = 0$
   c) $x = 0$ and $y = 4x + 2$
   d) $x + 2y + 5 = 0$ and $2x - 3y - 7 = 0$

2   Find the distances from the given points to the given lines
   a) $(1, 3)$, $2x + y + 1 = 0$
   b) $(2, -1)$, $3x + 2y - 7 = 0$
   c) $(0, 0)$, $3x - 4y + 1 = 0$
   d) $(-1, -4)$, $12x + 5y - 7 = 0$

A.2 Circles

If $P(x, y)$ lies on the circle whose centre is at $C(a, b)$ and whose radius is r, then

$$CP^2 = r^2 \quad \Rightarrow \quad \boxed{(x - a)^2 + (y - b)^2 = r^2}$$

In particular, the equation of the circle with centre at $(0, 0)$ and radius r is

$$\boxed{x^2 + y^2 = r^2}$$

Example 1

Find the equation of the circle with centre at $(2, -3)$ and radius 4.

The equation is

$$(x - 2)^2 + (y + 3)^2 = 4^2$$
$$\Rightarrow \quad x^2 + y^2 - 4x + 6y - 3 = 0$$

Example 2

Show that the equation

$$x^2 + y^2 - 8x + 2y + 8 = 0$$

represents a circle, and find its centre and radius.

We notice that

$$x^2 + y^2 - 8x + 2y + 8 = 0$$

can be rearranged by 'completing the squares' as

$$x^2 - 8x + 16 + y^2 + 2y + 1 = 9$$
$$\Rightarrow \quad (x - 4)^2 + (y + 1)^2 = 3^2$$

Hence the equation represents a circle with centre $(4, -1)$ and radius 3.

More generally, the equation

$$x^2 + y^2 + 2gx + 2fy + c = 0$$

can be rearranged as

$$x^2 + 2gx + g^2 + y^2 + 2fy + f^2 = g^2 + f^2 - c$$

$$\Rightarrow \qquad (x + g)^2 + (y + f)^2 = g^2 + f^2 - c$$

So the equation $x^2 + y^2 + 2gx + 2fy + c = 0$ represents a circle with centre $(-g, -f)$ and radius $\sqrt{(g^2 + f^2 - c)}$, providing this is real.

Example 3

Find the equation of the circle which passes through the points $(4, 5)$, $(2, -1)$ and $(0, 1)$.

Let the required equation be

$$x^2 + y^2 + 2gx + 2fy + c = 0$$

Since the circle passes through $(4, 5)$, $(2, -1)$ and $(0, 1)$, it follows that

$$\left. \begin{array}{l} 16 + 25 + 8g + 10f + c = 0 \\ 4 + 1 + 4g - 2f + c = 0 \\ 0 + 1 + 0 + 2f + c = 0 \end{array} \right\} \Rightarrow \left. \begin{array}{l} 8g + 10f + c + 41 = 0 \\ 4g - 2f + c + 5 = 0 \\ 2f + c + 1 = 0 \end{array} \right\} \qquad \begin{array}{l} (1) \\ (2) \\ (3) \end{array}$$

$$\left. \begin{array}{l} (1) - (3): \quad 8g + 8f + 40 = 0 \\ (2) - (3): \quad 4g - 4f + 4 = 0 \end{array} \right\} \quad \Rightarrow \quad \left. \begin{array}{l} g + f + 5 = 0 \\ g - f + 1 = 0 \end{array} \right\} \qquad \begin{array}{l} (4) \\ (5) \end{array}$$

From (4) and (5), $g = -3$ and $f = -2$

Finally, using (3), $c = 3$

Hence the required equation is

$$x^2 + y^2 - 6x - 4y + 3 = 0$$

Example 4

Find the equation of the tangent at the point $(3, 0)$ to the circle $x^2 + y^2 - 2x + 4y - 3 = 0$.

The centre is at $(1, -2)$, so the radius through $(3, 0)$ must have gradient

$$\frac{0 - (-2)}{3 - 1} = \frac{2}{2} = 1$$

Hence the gradient of the tangent $= -1$, and its equation is

$$\frac{y - 0}{x - 3} = -1 \quad \Rightarrow \quad x + y - 3 = 0$$

Example 5

Find the equation of the circle whose diameter is the line joining $A(1, 2)$ and $B(-3, 4)$.

Clearly we could obtain the required equation by first finding the centre and radius of the circle. More briefly, however, we recall that if $P(x, y)$ lies on the circle, then AP and BP must be perpendicular.

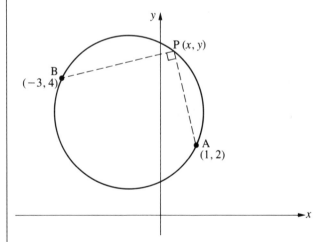

Hence
$$\frac{y - 2}{x - 1} \times \frac{y - 4}{x + 3} = -1$$

$$(x - 1)(x + 3) + (y - 2)(y - 4) = 0$$

$$x^2 + y^2 + 2x - 6y + 5 = 0$$

Example 6

If A is $(-1, 0)$ and B is $(1, 0)$, find the locus of P such that $PA = 2PB$.

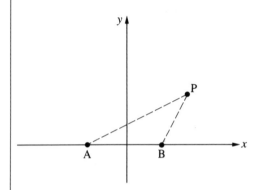

Letting P be (x, y),

$$\sqrt{\{(x + 1)^2 + y^2\}} = 2\sqrt{\{(x - 1)^2 + y^2\}}$$
$$\Rightarrow \qquad (x + 1)^2 + y^2 = 4\{(x - 1)^2 + y^2\}$$
$$x^2 + 2x + 1 + y^2 = 4x^2 - 8x + 4 + 4y^2$$
$$3x^2 + 3y^2 - 10x + 3 = 0$$
$$x^2 + y^2 - \frac{10}{3}x + 1 = 0$$

Hence the locus is a circle with centre $(\frac{5}{3}, 0)$ and radius $= \sqrt{(\frac{25}{9} - 1)} = \frac{4}{3}$. This locus is sometimes called a *circle of Apollonius*.

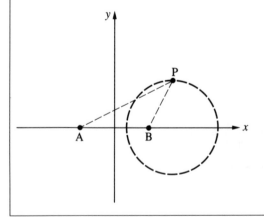

Exercise A.2

1 Find the equations of the circles
   a) with centre $(3, 2)$ and radius 5;
   b) with centre $(-1, -2)$ and radius 3;
   c) with centre $(0, 0)$ and radius 10;
   d) with centre $(0, 4)$ and radius 4.

2 Find the centres and radii of the following circles:
   a) $x^2 + y^2 + 4x + 6y - 3 = 0$
   b) $x^2 + y^2 - 2x - 4y - 4 = 0$
   c) $x^2 + y^2 + 8x + 15 = 0$
   d) $2x^2 + 2y^2 - 4x - 8y + 9 = 0$

3 Find the equations of the circles
   a) through the points $(0, 0)$, $(0, 2)$, $(6, 0)$;
   b) through the points $(4, 5)$, $(4, -3)$, $(0, 5)$;
   c) with a diameter joining $(0, 0)$ and $(5, 8)$;
   d) with a diameter joining $(3, 2)$ and $(-2, -3)$.

* A.3 Parabolas

We already know the locus of a point P which is equidistant from two given points S, T:

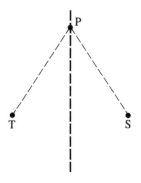

and of a point P which is equidistant from two given lines l, m:

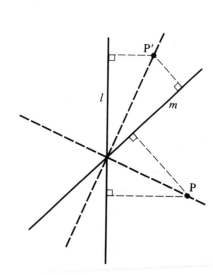

We now ask the hybrid question: what is the locus of a point which remains equidistant from a given point S and a given line l, so that SP = PM?

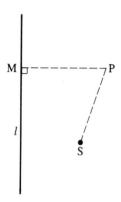

Clearly the point mid-way between S and l is a possible position of P, and if we take this as the origin O and the axis of symmetry as Ox, we can readily find the equation of the locus.

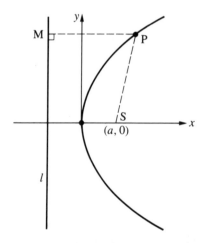

Letting OS $= a$, we see that S is $(a, 0)$ and l is the line $x = -a$.

Then   SP = PM

$\Rightarrow$ SP^2 = PM^2

$\Rightarrow$ $(x - a)^2 + y^2 = (x + a)^2$

$\Rightarrow$ $y^2 = 4ax$

This is taken as the standard form of the parabola, so that its focus is $(a, 0)$ and its directrix $x = -a$.

Furthermore, the point O $(0, 0)$ is called the *vertex* of the parabola, and the line of symmetry $(y = 0)$ its *axis*.

Parametric representation

If $x = at^2$ and $y = 2at$

we notice that $y^2 = 4a^2t^2$

and $4ax = 4a^2t^2$

So, for all values of t, the point $(at^2, 2at)$ lies on the parabola $y^2 = 4ax$.

Hence $x = at^2, \quad y = 2at$

provides a representation of the parabola in terms of a parameter t:

t	-2	-1	0	$+1$	$+2$	$+3$
$x = at^2$	$4a$	$-a$	0	a	$4a$	$9a$
$y = 2at$	$-4a$	$-2a$	0	$2a$	$4a$	$6a$

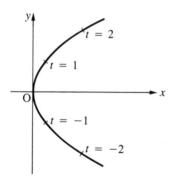

We now proceed to use this parametric form to investigate some of the properties of the parabola.

Chords, tangents, and normals

Suppose that P_1, P_2 are two points on the parabola and that their parameters are t_1, t_2.

Then the line (or *chord*) P_1P_2 has gradient

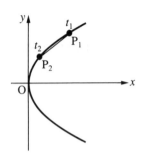

$$\frac{2at_1 - 2at_2}{at_1^2 - at_2^2} = \frac{2a(t_1 - t_2)}{a(t_1^2 - t_2^2)} = \frac{2}{t_1 + t_2}$$

Now the line through P_1 $(at_1^2, 2at_1)$ with this gradient is

$$\frac{y - 2at_1}{x - at_1^2} = \frac{2}{t_1 + t_2}$$

$$\Rightarrow \quad 2x - 2at_1^2 = (t_1 + t_2)y - 2at_1(t_1 + t_2)$$

$$\Rightarrow \quad \boxed{2x - (t_1 + t_2)y + 2at_1t_2 = 0}$$

which is the equation of P_1P_2.

In the particular case when P_1 and P_2 are coincident at a point P and have parameter t, we see that this chord becomes the tangent at P and has the equation

$$2x - 2ty + 2at^2 = 0$$

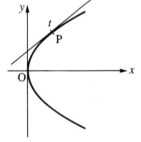

$$\Rightarrow \quad \boxed{x - ty + at^2 = 0}$$

Alternatively, we see that the gradient of the tangent at t is

$$\frac{dy}{dx} = \frac{dy/dt}{dx/dt} = \frac{2a}{2at} = \frac{1}{t}$$

So the equation of the tangent is

$$\frac{y - 2at}{x - at^2} = \frac{1}{t}$$

$$\Rightarrow \quad x - ty + at^2 = 0$$

Similarly the normal at P has gradient $-t$, so that its equation is

$$\frac{y - 2at}{x - at^2} = -t$$

$$\Rightarrow \quad \boxed{tx + y = at^3 + 2at}$$

Example 1

Show that if a perpendicular is drawn from the focus S to any tangent to a parabola, the foot N of this perpendicular lies on the tangent at the vertex.

If P has parameter t,

PN is $x - ty + at^2 = 0$

and   SN has gradient $-t$, so that its equation is

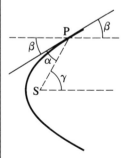

$$\frac{y - 0}{x - a} = -t$$

$\Rightarrow$ $tx + y - at = 0$

Now the point N lies on both PN $x - ty + at^2 = 0$

and on SN $tx + y - at = 0$

Multiplying the second equation by t and adding,

   $(1 + t^2)x + 0y + 0 = 0$ $\Rightarrow$ $x = 0$

Hence N always lies on the tangent at the vertex and, as P varies, the locus of N is the tangent at the vertex.

Example 2

If P is any point on a parabolic mirror, prove that two rays of light through P, one from the focus and the other parallel to the axis, are equally inclined to the tangent at P. (The so-called 'reflector' property of the parabola.)

If P has parameter t, and the angles α, β, γ are as shown in the figure,

$$\tan \beta = \text{gradient of tangent} = \frac{1}{t}$$

and $\tan \gamma = \dfrac{2at - 0}{at^2 - a}$

$= \dfrac{2t}{t^2 - 1}$

But $\tan 2\beta = \dfrac{2 \tan \beta}{1 - \tan^2 \beta} = \dfrac{2\left(\dfrac{1}{t}\right)}{1 - \left(\dfrac{1}{t}\right)^2} = \dfrac{2t}{t^2 - 1} = \tan \gamma$

Hence $2\beta = \gamma$

But $\gamma = \alpha + \beta$

so $\alpha = \beta$

Example 3

Two points on the parabola $y^2 = 4ax$ have parameters t_1, t_2.
a) Find the coordinates of the point of intersection of the tangents at these two points;
b) Prove that perpendicular tangents meet on the directrix.

a) The two tangents are

$$x - t_1 y + at_1^2 = 0$$

and $x - t_2 y + at_2^2 = 0$

Subtracting these, we find that at their point of intersection

$$(t_1 - t_2) y = a(t_1^2 - t_2^2)$$

$\Rightarrow$ $y = a(t_1 + t_2)$

and $x = t_1 a(t_1 + t_2) - at_1^2 = at_1 t_2$

So their point of intersection is

$$x = at_1 t_2 \quad y = a(t_1 + t_2)$$

b) If the two tangents are perpendicular, then

$$\frac{1}{t_1} \frac{1}{t_2} = -1$$

$\Rightarrow$ $t_1 t_2 = -1$

Hence at their point of intersection, $x = -a$, so that this point lies on the directrix.

Alternatively, for part **a)**, we can look more closely at the equation of the tangent to the parabola. Normally we think of this as a linear relation

$$x - ty + at^2 = 0$$

between the two coordinates of a point which lies on a given tangent.

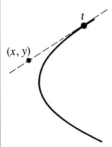

But it can also be written as

$$at^2 - yt + x = 0$$

and we see that this quadratic equation has roots t_1, t_2 which are the parameters of the two points (distinct, coincident or imaginary) whose tangents pass through a particular point (x, y).

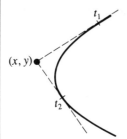

Now we know from the theory of quadratic equations that

$$t_1 + t_2 = \frac{y}{a} \quad \text{and} \quad t_1 t_2 = \frac{x}{a}$$

So, again, $x = at_1 t_2$ and $y = a(t_1 + t_2)$

This second proof, though less obvious than its predecessor, begins to show something of the power of such an alliance between algebra and geometry.

Exercise A.3

1   Find the equation of the parabola whose focus is $(0, 1)$ and directrix $y = 0$.

2   The tangents to a parabola at points P, Q meet at the point T and a line is drawn through T parallel to the axis of the parabola meeting the parabola in another point U. Prove that the tangent at U is parallel to PQ.

3   Two tangents to a parabola are perpendicular. Show that the line joining their points of contact passes through the focus.

4   A chord P_1P_2 of a parabola meets the axis at K, and the tangents at P_1, P_2 meet it at T_1, T_2. Prove that if O is the vertex, then

$$OT_1 \times OT_2 = OK^2$$

5   The normal at a variable point P of a parabola meets its axis at G, and the foot of the perpendicular from P to the axis is N. Prove that NG (the *subnormal*) is of constant length, independent of the position of P.

6   The tangents to a parabola at points P, Q are perpendicular and intersect in a point T. If their corresponding normals meet at a point K, prove that TK is parallel to the axis of the parabola.

7   Show that if the normal at the point t of a parabola passes through a given point (h, k) then t must be a root of the equation

$$at^3 + (2a - h)t - k = 0$$

Hence show that not more than three normals can be drawn through a given point; and that if there are three such normals, the triangle formed by their feet on the parabola has its centroid on the axis.

8   Show that if a circle meets a parabola in four points the sum of their displacements from its axis is zero. Hence (using no. 7) prove that if U, V, W are the three feet of normals to a parabola from a given point, the circle through U, V, W must pass through the vertex of the parabola.

* A.4 Ellipses and hyperbolas

The parabola has been defined as the locus of a point P whose distance from a fixed point S is equal to its distance from a fixed line l (SP = PM), and we now consider the locus of P if these distances, rather than being equal, are proportional.

   $SP = e$PM   (where e is a constant known as the *eccentricity*)

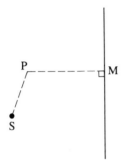

We can clearly plot these loci for particular values of e and shall distinguish between two cases

$e < 1,$ when the locus is an *ellipse*

$e > 1,$ when the locus is a *hyperbola*

Ellipses

$SP = ePM$ $(e < 1)$

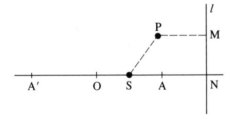

Clearly the locus of P will be symmetrical about the line through S which is perpendicular to l. We let this axis of symmetry meet l in the point N and notice that there will be two possible positions of P on this axis, one on each side of S. We call these A and A′ and take origin O at their mid-point, letting $OA = a$.

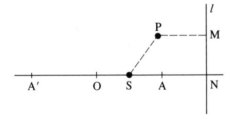

We also let $OS = s$ and $ON = n$.

So $SA = a - s,$ $AN = n - a$

and $SA' = a + s,$ $A'N = n + a$

Now A and A′ are on the locus.

So $a - s = e(n - a)$

and $a + s = e(n + a)$

Hence $s = ae$

and $a = en \implies n = a/e$

We can now express the equation of the ellipse in terms of a and e.

For $SP = ePM$

$\implies$ $SP^2 = e^2 PM^2$

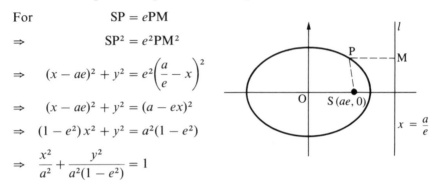

$\implies$ $(x - ae)^2 + y^2 = e^2 \left(\dfrac{a}{e} - x \right)^2$

$\implies$ $(x - ae)^2 + y^2 = (a - ex)^2$

$\implies$ $(1 - e^2)x^2 + y^2 = a^2(1 - e^2)$

$\implies$ $\dfrac{x^2}{a^2} + \dfrac{y^2}{a^2(1 - e^2)} = 1$

If we now write $b^2 = a^2(1 - e^2)$, it follows that

$$\frac{x^2}{a^2} + \frac{y^2}{b^2} = 1$$

which we take as the standard form of the ellipse.

We immediately see that the ellipse is symmetrical both about Ox (its major axis) and Oy (its minor axis), the lengths of these two axes being $2a$, $2b$ respectively; and that it also has point-symmetry about its centre O.

One consequence of these symmetries is that the ellipse must have a focus and directrix not only at S $(ae, 0)$ and $l(x = a/e)$, but also at S′$(- ae, 0)$ and $l'(x = -a/e)$.

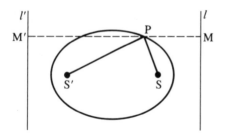

Furthermore, if M and M′ are the feet of the perpendiculars from a point P on to l and l', it follows that

$$SP = e\text{PM}$$

and $\qquad S'P = e\text{PM}'$

$$\Rightarrow \quad SP + S'P = e(\text{PM} + \text{PM}')$$

$$= e\text{MM}' = e\frac{2a}{e} = 2a$$

So as P moves around an ellipse, the sum of its focal distances remains constant. This is the basis of the well-known method of constructing an ellipse by means of a pencil moving within a loop of string tied to two fixed pegs.

Parametric representation

Just as it was convenient to use a parametric representation for the points of a parabola, so the points of the standard ellipse can be expressed in terms of a parameter θ by the equations

$$x = a\cos\theta, \quad y = b\sin\theta$$

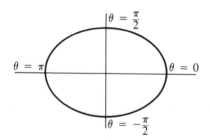

We immediately see that

$$\frac{x^2}{a^2} + \frac{y^2}{b^2} = \frac{a^2\cos^2\theta}{a^2} + \frac{b^2\sin^2\theta}{b^2} = 1$$

so that such a point always lies on the ellipse.

Moreover, we can easily find the equation of the tangent at the point whose parameter is θ.

For $\qquad \dfrac{dy}{dx} = \dfrac{dy/d\theta}{dx/d\theta} = \dfrac{b\cos\theta}{-a\sin\theta} = -\dfrac{b\cos\theta}{a\sin\theta}$

So the required equation is

$$\frac{y - b\sin\theta}{x - a\cos\theta} = -\frac{b\cos\theta}{a\sin\theta}$$

$$\Rightarrow \quad bx\cos\theta + ay\sin\theta = ab(\cos^2\theta + \sin^2\theta) = ab$$

$$\Rightarrow \quad \boxed{\frac{x\cos\theta}{a} + \frac{y\sin\theta}{b} = 1}$$

Exercise A.4a

1   Find the lengths of the axes, the eccentricity, the foci and the directrices of

   a) $\dfrac{x^2}{9} + \dfrac{y^2}{4} = 1$

   b) $4x^2 + 25y^2 = 100$

2   Show that the ellipse

$$\frac{x^2}{4 + \lambda} + \frac{y^2}{1 + \lambda} = 1$$

   always has the same foci, whatever the value of λ (provided $\lambda > -1$). Sketch this family of confocal ellipses.

3   $P_1(x_1, y_1)$ is a point on the ellipse

$$\frac{x^2}{a^2} + \frac{y^2}{b^2} = 1$$

   Find the gradient of the tangent to the ellipse in terms of x_1, y_1, a and b; and hence find the equations of the tangent and normal at P.

4   Find the equation of the normal to the standard ellipse at the point $(a\cos\theta, b\sin\theta)$.

5   The tangent and normal at a point on the standard ellipse meet the major axis at T, N. Prove that $OT \times ON = a^2 - b^2$.

6   Find a linear transformation which, when applied to the points of the circle $x^2 + y^2 = a^2$, transforms them into the points of the standard ellipse. Illustrate the relationship between the ellipse and the circle (known as its *auxiliary* circle).

7   A plane inclined to the horizontal at an angle α contains a hole of radius a, and a vertical beam of light shines through this hole. Show that the patch

of light cast on a horizontal plane is in the shape of an ellipse whose eccentricity is sin α. Hence find the area of the ellipse whose semi-major axes are a, b.

8 a) A man is instructed to go from A to B, first picking up a stone from the straight track *l*. Show that for his path to have minimum length, its two parts must be equally inclined to *l*.

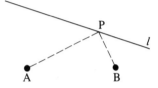

b) At which point on a tangent to an ellipse is the sum of the focal distances a minimum? Why?

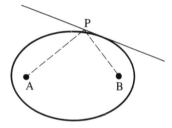

c) Hence show that the lines joining any point of an ellipse to its two foci are equally inclined to the tangent at the given point (the *reflector property* of the ellipse).

Hyperbolas

We have seen that the hyperbola is defined as the locus of a point P such that SP $= e$PM (where $e > 1$)

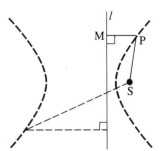

and we have already noticed that this locus has two separate branches.

As with the ellipse, we let A, A′ be the two positions of P which lie on the axis of symmetry (this time both on the same side of S), and take the origin O at their mid-point.

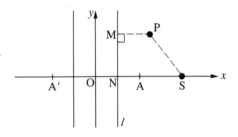

As before, we let N be the point where l meets the line of symmetry, and let

$$OS = s \quad \text{and} \quad ON = n$$

Then $SA = s - a$ $AN = a - n$

$SA' = s + a$ $A'N = a + n$

Now $SA = eAN \Rightarrow s - a = e(a - n)$

and $SA' = eA'N \Rightarrow s + a = e(a + n)$

So $s = ae$

and $a = en \Rightarrow n = \dfrac{a}{e}$

Since $e > 1$, this means that the directrix $x = a/e$ lies between O and the focus $(ae, 0)$.

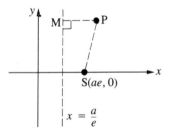

Taking P as the point (x, y),

$$SP = ePM$$

$\Rightarrow$ $SP^2 = e^2PM^2$

$\Rightarrow$ $(x - ae)^2 + y^2 = e^2(x - a/e)^2$

and this, as before, leads to

$$\frac{x^2}{a^2} + \frac{y^2}{a^2(1-e^2)} = 1$$

But as $e > 1$ we write this as

$$\frac{x^2}{a^2} - \frac{y^2}{a^2(e^2-1)} = 1$$

and let $b^2 = a^2(e^2 - 1)$, so that the standard equation for the hyperbola is

$$\frac{x^2}{a^2} - \frac{y^2}{b^2} = 1$$

As before, we see (by symmetry) that there must be another focus at $(-ae, 0)$, and another directrix $x = -a/e$.

Furthermore, if $x = a\sec\theta$ and $y = b\tan\theta$, then

$$\frac{x^2}{a^2} - \frac{y^2}{b^2} = \sec^2\theta - \tan^2\theta = 1$$

So $(a\sec\theta, b\tan\theta)$ is a parametric representation of a point on the hyperbola.

A characteristic feature of the hyperbola, not shared with either the parabola or ellipse, is its pair of asymptotes.

For $$\frac{x^2}{a^2} - \frac{y^2}{b^2} = 1$$

$$\Rightarrow \left(\frac{x}{a} + \frac{y}{b}\right)\left(\frac{x}{a} - \frac{y}{b}\right) = 1$$

So $$\frac{x}{a} - \frac{y}{b} = \frac{1}{x/a + y/b}$$

Now as $x \to +\infty$ and $y \to +\infty$ (or $x \to -\infty$ and $y \to -\infty$) we see that

$$\frac{x}{a} - \frac{y}{b} = \frac{1}{x/a + y/b} \to 0$$

and $$\frac{x}{a} - \frac{y}{b} = 0 \quad \text{is an asymptote}$$

Similarly $$\frac{x}{a} + \frac{y}{b} = \frac{1}{x/b - y/b}$$

So as $x \to +\infty$ and $y \to -\infty$ (or $x \to -\infty$ and $y \to +\infty$), we see that

$$\frac{x}{a} + \frac{y}{b} = \frac{1}{x/a - y/b} \to 0$$

$$\Rightarrow \quad \frac{x}{a} + \frac{y}{b} = 0 \quad \text{is an asymptote}$$

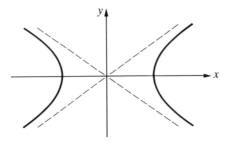

Hence the two lines

$$\frac{x}{a} \pm \frac{y}{b} = 0$$

are asymptotes, to which points on the hyperbola approach more and more closely as they get further and further from its centre.

The rectangular hyperbola

The hyperbola $\dfrac{x^2}{a^2} - \dfrac{y^2}{b^2} = 1$ has a pair of asymptotes $\dfrac{x}{a} = \pm\dfrac{y}{b}$.

If $a = b$, these become the pair of lines $x = \pm y$. Because these are perpendicular, we call the corresponding curve a *rectangular hyperbola*.

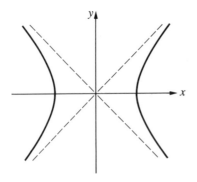

The equation of such a hyperbola is clearly

$$\frac{x^2}{a^2} - \frac{y^2}{a^2} = 1$$

$\Rightarrow \qquad x^2 - y^2 = a^2$

and since $\qquad b^2 = a^2(e^2 - 1)$

we see that $\quad a^2 = a^2(e^2 - 1)$

$\Rightarrow \qquad\qquad 1 = e^2 - 1$

$\Rightarrow \qquad\qquad e = \sqrt{2}$

Hence the rectangular hyperbola has eccentricity $\sqrt{2}$ and when referred to its axes of symmetry has the standard equation

$$x^2 - y^2 = a^2$$

It is, however, frequently convenient to consider the equation of the rectangular hyperbola referred to its asymptotes as axes, and this can be done by means of the linear transformation which rotates the hyperbola through an angle $+\dfrac{\pi}{4}$

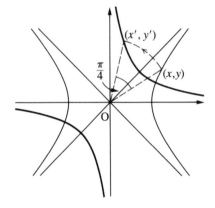

$$\begin{pmatrix} x' \\ y' \end{pmatrix} = \begin{pmatrix} 1/\sqrt{2} & -1/\sqrt{2} \\ 1/\sqrt{2} & 1/\sqrt{2} \end{pmatrix} \begin{pmatrix} x \\ y \end{pmatrix}$$

$\Rightarrow \quad x' = \dfrac{1}{\sqrt{2}}(x - y)$

$\qquad y' = \dfrac{1}{\sqrt{2}}(x + y)$

Hence the equation $\quad x^2 - y^2 = a^2$

is transformed into $\qquad 2x'y' = a^2$

$$\Rightarrow \quad x'y' = \frac{1}{2}a^2$$

We now revert to the use of (x, y) rather than (x', y') and let $c^2 = \frac{1}{2}a^2$, obtaining

$$xy = c^2$$

as the standard form of the rectangular hyperbola when referred to its asymptotes.

If we now consider

$$x = ct \quad y = \frac{c}{t} \quad \text{(where } t \text{ is a parameter)}$$

we see that $\quad xy = ct \times \dfrac{c}{t} = c^2$

Hence a very convenient parametric representation of the rectangular hyperbola $xy = c^2$ is given by

$$x = ct, \quad y = \frac{c}{t}$$

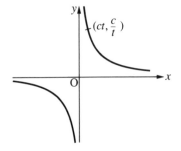

Exercise A.4b

1   Find the asymptotes, eccentricity, foci, and directrices of

   a) $\dfrac{x^2}{4} - y^2 = 1$ b) $4x^2 - 9y^2 = 36$

2   Show that if $-4 < \lambda < -1$, the curve

$$\frac{x^2}{4+\lambda} + \frac{y^2}{1+\lambda} = 1$$

represents a hyperbola, and that the position of its foci is independent of the value of λ. Compare with exercise A.4a, no. **2**, and sketch these two families of curves in one diagram.

3   Find the equation of the tangent to the hyperbola

$$\frac{x^2}{a^2} - \frac{y^2}{b^2} = 1$$

at the point $P\,(a\sec\theta, b\tan\theta)$.

If this tangent meets the two asymptotes at Q, R, show that P is the mid-point of QR.

4   **a)** If P is a point on a hyperbola with foci S, S' and major axis of length $2a$, use the focus-directrix property to prove that

$$|SP - S'P| = 2a$$

b) An explosion takes place at an unknown point and three observers at different points with carefully synchronised watches note the exact time at which its sound reaches them. Use **a)** to show how this information can be used to locate the position of P. [A method once used by artillery in 'sound-ranging', the same principle being the basis of many sea and air navigational systems.]

5   A rectangular hyperbola is represented by

$$x = ct, \quad y = \frac{c}{t}$$

Find the equation of
a) the tangent at the point with parameter t;
b) the normal at the same point;
c) the chord joining points t_1 and t_2.

6   The tangent at a variable point P of the rectangular hyperbola $xy = c^2$ meets the two asymptotes at points Q, R. Prove that $\triangle OQR$ has constant area $2c^2$ and that the locus of its centroid is the hyperbola $xy = \frac{4}{9}c^2$.

7 P_1, P_2, P_3, P_4 are points on the standard rectangular hyperbola which have parameters t_1, t_2, t_3, t_4. If P_1P_2 is perpendicular to P_3P_4, show that $t_1t_2t_3t_4 = -1$. Hence show that if P_1P_2 is perpendicular to P_3P_4, then P_1P_3 is perpendicular to P_2P_4, and P_1P_4 is perpendicular to P_2P_3. Finally show that if three points lie on a rectangular hyperbola, so does their orthocentre.

8 The foot of the perpendicular from O to the tangent at a variable point P
on $xy = c^2$ is Q.
a) Prove that OP × OQ is constant.
b) Find the equation of the locus of Q and sketch it.

* A.5 Polar coordinates

It is frequently convenient to identify a point P not by its Cartesian
coordinates (x, y), but by r, its distance from a fixed point O (the origin, or
pole) and θ, an angle through which a base-line Ox is rotated in order to lie
along OP. (r, θ) are then known as the *polar coordinates* of P, and clearly θ has
a set of values, separated by multiples of 2π.

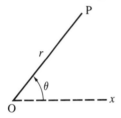

Furthermore, $r = f(\theta)$ usually denotes a curve and is called its *polar equation*.

Example 1

Sketch the curve

$$r = 1 + \cos\theta$$

Taking successive values of θ, we obtain:

θ	0	$\pm\pi/6$	$\pm\pi/3$	$\pm\pi/2$	$\pm 2\pi/3$	$\pm 5\pi/6$	$\pm\pi$
r	2	1.87	1.50	1.00	0.50	0.13	0

from which we obtain the heart-shaped curve called the *cardioid*:

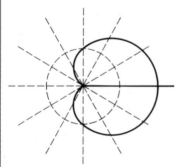

Example 2

$r = \frac{1}{2}\theta$ (where $\theta \geqslant 0$)

In this case,

θ	0	$\pi/4$	$\pi/2$	$3\pi/4$	π	$5\pi/4$	$3\pi/2$	$7\pi/4$	2π	$9\pi/4$
r	0	0.39	0.79	1.18	1.57	1.96	2.36	2.75	3.14	3.53

from which we obtain the so-called *Archimedean spiral*:

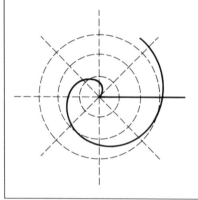

Example 3

Find the polar equation, referred to its focus as pole, of an ellipse with eccentricity e.

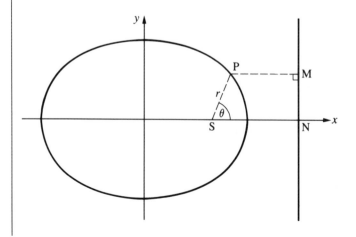

$$SP = e\,PM$$

$$\Rightarrow \quad PM = \frac{SP}{e} = \frac{r}{e}$$

But $r\cos\theta + PM = SN = c$ (a constant)

So $r\cos\theta + \dfrac{r}{e} = c$

$\Rightarrow \qquad r(1 + e\cos\theta) = ec = l,$ say

$\Rightarrow \qquad\qquad r = \dfrac{l}{1 + e\cos\theta}$

Transformations between polar and Cartesian coordinates

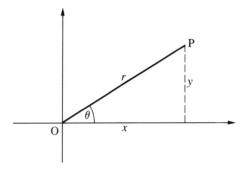

From the diagram we see that

$$x = r\cos\theta \qquad r = \sqrt{(x^2 + y^2)}$$

$$y = r\sin\theta \qquad \theta = \tan^{-1}\frac{y}{x}$$

and these formulae are clearly useful for transforming equations from polar to Cartesian coordinates, and vice versa.

Exercise A.5a

1   Sketch the curves
   a) $r = a\sin\theta$ **b)** $r = a\sin 2\theta$

c) $r^2 = a^2 \cos 2\theta$ (a *lemniscate*)
d) $r = a \cos 3\theta$ (a *trifolium*)

2 Find the polar equations, referred to O and Ox, of:
  a) the circle $x^2 + y^2 = 2y$
  b) the parabola $y = x^2$
  c) the rectangular hyperbola $xy = 1$
  d) the straight line $x + y = 1$

3 Sketch the following curves and find their Cartesian equations:
  a) $r = a \cos \theta$
  b) $r = a \operatorname{cosec} \theta$
  c) $r^2 = a^2 \sin 2\theta$
  d) $r^2 \cos 2\theta = a^2$

Area in polar coordinates

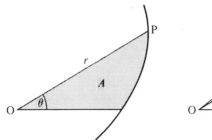 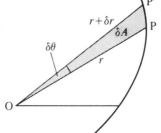

Just as in Cartesian coordinates, so in polar coordinates we can find an expression for the area A, in this case bounded by the given curve and two bounding radii. For, suppose that P is the point (r, θ) of the curve and P′ is a neighbouring point $(r + \delta r, \theta + \delta\theta)$.
  Let δA be the area of the sector OPP′.

Then $\delta A \approx \frac{1}{2} r(r + \delta r)\delta\theta$

$\Rightarrow$ $\dfrac{\delta A}{\delta\theta} \approx \frac{1}{2} r(r + \delta r)$

$\Rightarrow$ $\dfrac{\mathrm{d} A}{\mathrm{d}\theta} = \frac{1}{2} r^2$

$\Rightarrow$ $A = \frac{1}{2} \int r^2 \mathrm{d}\theta$

Example 4

Find the area bounded by the spiral $r = e^\theta$ and the two radii $\theta = 0$ and $\theta = 2\pi$.

$$A = \tfrac{1}{2} \int_0^{2\pi} r^2 d\theta = \tfrac{1}{2} \int_0^{2\pi} e^{2\theta} d\theta = \tfrac{1}{4} \left[e^{2\theta} \right]_0^{2\pi} = \tfrac{1}{4}(e^{4\pi} - 1)$$

Exercise A.5b

Sketch the following curves and find the total areas they enclose:

1    $r = a\cos\theta$,   between $\theta = 0$ and $\theta = \tfrac{1}{2}\pi$.

2    $r = a(1 + \cos\theta)$,   a *cardioid*.

3    $r^2 = a^2 \sin 2\theta$,   a *lemniscate*.

4    $r^2 = a^2 \sin 3\theta$,   a *trefoil*.

5    $r = a\theta$,   an *Archimedean spiral*, between $\theta = 0$ and $\theta = \pi$.

Revision exercises

Chapters 1 and 2

1 Find, without using tables or a calculator, the value of x given that

$$\frac{4^{3+x}}{8^{10x}} = \frac{2^{10-2x}}{64^{3x}}$$ (L)

2 Solve the equation

$$3^{x+1} = 4^{x-1}$$

giving the value of x correct to three significant figures. (C)

3 Given that $\lg y = 3 - \frac{3}{4}\lg x$, express y in terms of x in a form not involving logarithms. (L)

4 a) Solve the equation

$$2^{2/x} = 32$$

b) Solve the equation

$$\log_x 2 \times \log_x 3 = 5$$

giving your answers correct to two decimal points. (L)

5 Functions f and g, each with domain $\mathbb{R}$, are defined as follows

$$f: x \mapsto 3x + 2, \quad g: x \mapsto x^2 + 1$$

For each of f and g, state the range of the function and give a reason to show whether or not the function is one–one.

Give explicit definitions, in the above form, of each of the composite functions $f \circ g$ and $g \circ f$, and find the values of x for which

$$(f \circ g)(x) = (g \circ f)(x)$$

State the domain of the inverse relation $(f \circ g)^{-1}$ and give an explicit definition of this relation. Explain briefly why $(f \circ g)^{-1}$ is not a function. (C)

6 The functions f and g are defined by

$$f: x \mapsto x^2 - 1, \quad (x \in \mathbb{R})$$

$$g: x \mapsto \frac{x+1}{x-1}, \quad (x \in \mathbb{R}, x \neq 1)$$

Sketch the graphs of these functions.

Define, in a similar form,
a) the inverse function g^{-1};
b) the composite function $f \circ g$, stating the range of this function. (c)

7 Given that α and β are the roots of the quadratic equation

$$x^2 + lx + m = 0$$

find the quadratic equation whose roots are $\alpha^3 \beta$ and $\alpha \beta^3$, expressing the coefficients in terms of l and m. (MEI)

8 a) Given that the roots of the equation $x^3 + px^2 + qx + r = 0$ are three consecutive terms of an arithmetic progression, show that

$$2p^3 + 27r = 9pq.$$

b) Given that the roots of the equation $x^3 + px^2 + qx + r = 0$ are three consecutive terms of a geometric progression, find a condition that p, q and r must satisfy. (c)

9 State and prove the remainder theorem.

Given that the quotient when a polynomial $f(x)$ is divided by $(x - a)$ is $g(x)$ and that the quotient when $f(x)$ is divided by $(x - b)$ is $h(x)$, prove that

$$g(b) = h(a)$$ (JMB)

10 The polynomials $P(x)$ and $Q(x)$ are defined by

$$P(x) = x^8 - 1$$

$$Q(x) = x^4 + 4x^3 + ax^2 + bx + 5$$

a) Show that $x - 1$ and $x + 1$ are factors of $P(x)$.
b) It is known that when $Q(x)$ is divided by $x^2 - 1$ a remainder $2x + 3$ is obtained. Find the values of a and b.
c) With these values of a and b find the remainder when the polynomial $[3P(x) + 4Q(x)]$ is divided by $x^2 - 1$. (c)

11 a) Given that two of the roots of the equation $9x^3 + px + 16 = 0$ are α and 2α, show that the third root is -3α.
   Given that α is real, find the value of p.
b) Find, in terms of q and r, the stationary values of

$$x^3 - 3qx + r \quad (q > 0)$$

as x varies.

By considering the product of these values, or otherwise, show that the roots of the equation $x^3 - 3qx + r = 0 \, (q > 0)$ are all real if and only if $4q^3 \geqslant r^2$. (c)

12 **a)** By considering

$$y = \frac{x - 3}{(x - 2)(x + 1)}$$

as a quadratic equation in x, or otherwise, prove that for real values of x the value of y cannot lie between $\frac{1}{9}$ and 1.

Find the values of x for which $y = \frac{1}{9}$ and $y = 1$, and sketch the graph given by the above equation.

b) The equation

$$x^3 - 2x^2 + 3x - 5 = 0$$

has roots α, β, γ. Given that the equation $x^3 + px^2 + qx + r = 0$ has roots $\beta + \gamma, \gamma + \alpha, \alpha + \beta$, obtain numerical values for p, q and r. (c)

13 A student performs an experiment and records the following data for two variables y and x.

x	0.5	2	4	7	8
y	1.4	2.8	4.0	5.3	5.7

She believes that there is a relationship between y and x of the form

$$y = ax^b$$

By plotting $\lg y$ against $\lg x$ show that this relationship holds and find a and b to two significant figures.

After completing the experiment, the student realised that she should have measured y when $x = 6$. Obtain, from your graph, an estimate to two significant figures, of the value of y when $x = 6$. (L)

14 The air resistance, R newtons, to a vehicle is expected to follow a law of the type

$$R = av^b$$

where $v \, \text{m/s}$ is the speed and a and b are constants. Experimental observations give the following results.

v	10	20	30	40
R	73	260	545	920

By drawing a graph on log-log or other graph paper, show that these values support the expectation.

Use your graph to find:

a) the values of a and b;

b) the value of R at $35\,\mathrm{m/s}$;

c) the speed at which the resistance is 400 N   (AEB, 1981)

15   The equation

$$x^3 + px^2 + qx + r = 0$$

has positive roots α, β, γ. The arithmetic mean of α, β, γ is A; the geometric mean is G, where $G = (\alpha\beta\gamma)^{\frac{1}{3}}$; and the harmonic mean is H, where $H^{-1} = \frac{1}{3}(\alpha^{-1} + \beta^{-1} + \gamma^{-1})$. Express p, q, r in terms of A, G, H.

Given that three numbers have arithmetic mean 4, geometric mean 2 and harmonic mean 1, show that the numbers are the roots of the equation

$$x^3 - 12x^2 + 24x - 8 = 0$$

Hence, or otherwise, find the numbers.   (C)

Chapters 3 and 4

1   The graph of

$$y = x^3 + ax^2 + bx + c$$

has stationary points where $x = -1$ and where $x = 3$, and passes through the point $(1, -2)$. Find the values of a, b and c.

Show that the stationary point where $x = 3$ is a minimum point.

Find the coordinates of the point of inflection and sketch the graph.   (C)

2   If $x^2 + xy + y^2 = 3$, find $\dfrac{\mathrm{d}y}{\mathrm{d}x}$, expressing your answer in terms of x and y.

Hence show that there are two stationary points and find their coordinates.   (MEI)

3   Find the coordinates of each turning point on the graph of $y = 3x^4 - 16x^3 + 18x^2$ and determine in each case whether it is a maximum point or a minimum point.

Sketch the graph of $y = 3x^4 - 16x^3 + 18x^2$, and state the set of values of k for which the equation $3x^4 - 16x^3 + 18x^2 = k$ has precisely two real roots for x.   (C)

4   For the curve whose equation is

$$y = \frac{x(x + 2)}{(x - 2)^2}$$

a) find the equations of the asymptotes;
b) find the coordinates and the nature of the stationary point;
c) show that the curve has a point of inflection and find its coordinates.
Sketch the curve, showing on your sketch the tangent to the curve at the point of inflection.   (JMB)

5   Prove that if $(x - a)^2$ is a factor of the polynomial $f(x)$ then $(x - a)$ is a factor of $f'(x)$. Give a counter-example to show that the converse of this result is not true.
a) Solve the equation

$$3x^3 + 29x^2 + 65x - 25 = 0$$

given that two of its roots are equal.
b) Solve the equation

$$54x^4 + 27x^3 - 198x^2 + 164x - 40 = 0$$

given that three of its roots are equal.   (C)

6   Given that $y = \dfrac{x^3}{1 + 3x^4}$, find $\dfrac{dy}{dx}$ in terms of x, and hence find the coordinates of the three stationary points on the graph of y.

By considering the sign of $\dfrac{dy}{dx}$ on either side of the stationary points, or otherwise, determine whether each stationary point is a maximum point, a minimum point or a point of inflection.
Sketch the graph of y.   (C)

7   Obtain the coordinates of the turning points on each of the curves.

$$y = \tfrac{1}{4}x^2(x - 3) \text{ and } y = \frac{4}{x^2(x - 3)}$$

Sketch these curves on separate diagrams and give the equations of the asymptotes of the second curve.
Obtain the set of values of k for which the equation

$$\frac{4}{x^2(x - 3)} = k(x - 3)$$

has   **a)** no real roots;
      b) four real roots;
      c) exactly two real roots.   (C)

8   By comparing 5^3 with 4^3, find whether $\tfrac{5}{4}$ is greater than or less than the cube root of 2.
Taking 1.25 as a first approximation to the real solution of the equation

$$x^3 - 2 = 0$$

use the Newton–Raphson method *once* to find a better approximation.

(SMP)

9 The time T taken by a planet to revolve round the sun and its mean distance r from the sun are related by

$$T = kr^{3/2}$$

where k is constant. Obtain $\dfrac{dT}{dr}$. If the planet's distance from the sun were to be increased by 2%, estimate the percentage increase in the period T.

(L)

10 A curve is given parametrically by the equations

$$x = \frac{1}{2}\left(t + \frac{1}{t}\right) \quad y = \frac{1}{2}\left(t - \frac{1}{t}\right) \quad t \neq 0$$

a) Find the Cartesian equation of the curve.

b) Find $\dfrac{dy}{dx}$ in terms of t and hence find, in its simplest form, the gradient of the tangent to the curve at the point P with parameter p.

c) Show that the equation of the tangent at P is

$$(p^2 + 1)x - (p^2 - 1)y = 2p$$

The tangent to the curve at P meets the line $y = x$ at A and the line $y = -x$ at B and O is the origin. Find the area of the triangle OAB, showing that it is independent of p.

(JMB)

11 The volume of a right circular cone increases at a rate of $120\,\text{cm}^3/\text{s}$ in such a way that the height of the cone is always half the radius of the base.

Find the rate at which the radius is increasing when the volume of the cone is $36\,000\pi\,\text{cm}^3$.

Find also, at the same instant, the rate at which the area of the curved surface of the cone is increasing.

(The area of the curved surface of a cone is πrl, where r is the radius of the base and l is the slant height.)

(AEB)

12 The Highway Code states that for a vehicle under ideal conditions the minimum stopping distance from 20 mph is 40 feet and from 30 mph is 75 feet.

Find this stopping distance s (feet) as a quadratic function of the speed v (mph).

In a model of commuter traffic flow it is assumed that all vehicles are 15 feet long, travelling at constant speed and separated by the minimum stopping distance appropriate to that speed.

Find an expression for the rate of traffic flow R (vehicles per hour) as a function of v. Hence determine the maximum flow rate and the speed at which this occurs.

If conditions are poor and vehicles are spaced at twice the ideal stopping distance appropriate to their speed, find the new maximum flow rate and speed. (MEI)

13 A circular hollow cone of height h and semivertical angle α stands on a horizontal table. A circular cylinder placed in an upright position just fits inside the space between the cone and the table, as shown in the diagram.

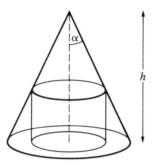

Prove that the maximum possible volume of the cylinder is

$$\frac{4\pi h^3 \tan^2 \alpha}{27}$$
(JMB)

14 A solid right circular cone of greatest volume is to be cut from a solid sphere of radius $3R$. Using calculus, show that the height of this cone is $4R$ and calculate its volume in terms of R. (AEB, 1983)

15 A particle moving along a straight line starts at time $t = 0$ seconds with a velocity $4\,\mathrm{m\,s^{-1}}$. At any subsequent time t seconds the acceleration of the particle is $(6t - 8)\,\mathrm{m\,s^{-2}}$.
Find:
a) the distance the particle moves before first coming to instantaneous rest;
b) the total time T seconds taken by the particle to return to the starting point;
c) the greatest speed of the particle for $0 \leqslant t \leqslant T$. (L)

16 The curves $ay = x^2$ and $y^2 = ax$, where $a > 0$, intersect at the origin O and the point A. Find the coordinates of A.

Obtain, by integration, the area of the region R enclosed by the arcs of the two curves between O and A.

Find the volume of the solid obtained when the region R is rotated through four right angles about the x-axis. (C)

17 Given the curve whose equation is

$$y = x^{-\frac{1}{4}}$$

find:

a) the mean value of $\dfrac{1}{y}$, with respect to x, in the interval $1 \leqslant x \leqslant 4$;

b) the area of the region R bounded by the curve, the x-axis and the lines $x = 1$ and $x = 4$;

c) the volume of the solid generated when the region R is rotated through an angle of 2π radians about the x-axis.   (JMB)

18 Use the trapezium rule with 5 strips of equal width to estimate, to 3 significant figures, the value of $\displaystyle\int_0^1 10^x \, dx$. Show all your working.   (L)

19 Given that $y = \sqrt{(x^3 + 1)}$ show that $\dfrac{dy}{dx}$ and $\dfrac{d^2y}{dx^2}$ are both positive for all $x > 0$.

By using the trapezium rule with 3 ordinates, obtain an approximation for the value of I, where

$$I = \int_0^2 \sqrt{(x^3 + 1)} \, dx$$

giving your answer to three decimal places.

Draw a sketch graph of the curve $y = \sqrt{(x^3 + 1)}$ for $x \geqslant 0$, and hence, or otherwise, determine whether your approximation is greater than or less than the true value of I.   (C)

20 Use Simpson's rule with seven equally spaced ordinates to find an estimate of the mean value of the function $\lg(1 + x^3)$ between the values $x = 1$ and $x = 19$. Show your working in the form of a table and give your final answer to 3 significant figures.   (AEB, 1982)

Chapter 5

1 Prove that, for all values of θ

$$\sin 3\theta - \cos 3\theta = (\sin \theta + \cos \theta)(2 \sin 2\theta - 1)$$

Hence, or otherwise, find the values of θ, such that $0° \leqslant \theta \leqslant 180°$, for which

$$3(\sin 3\theta - \cos 3\theta) = 2(\sin \theta + \cos \theta)$$

[Where necessary give your answers correct to $0.1°$.]   (C)

2 a) By using double angle formulae, or otherwise, show that

$$5\cos^2\theta - 12\sin\theta\cos\theta \equiv A + B\cos(2\theta + \alpha)$$

and determine the constants A, B and $\tan\alpha$.

i) Deduce that

$$9 \geqslant 5\cos^2\theta - 12\sin\theta\cos\theta \geqslant -4$$

ii) Find all the solutions of the equation

$$5\cos^2\theta - 12\sin\theta\cos\theta = 2$$

in the range $0°$ to $360°$ giving your answers to the nearest half degree.
b) Calculate the length of the third side of the triangle with sides of length $4 - \sqrt{3}, 2\sqrt{3}$ which contain an angle $120°$. (L)

3 **a)** Given that

$$7\tan\theta + \cot\theta = 5\sec\theta$$

derive a quadratic equation for $\sin\theta$.

Hence, or otherwise, find all values of θ in the interval $0° \leqslant \theta \leqslant 180°$ which satisfy the given equation, giving your answers to the nearest $0.1°$, where necessary.
b) The acute angles A and B are such that

$$\cos A = \tfrac{1}{2}, \quad \sin B = \tfrac{1}{3}$$

Show, without the use of tables or a calculator, that

$$\tan(A + B) = \frac{9\sqrt{3} + 8\sqrt{2}}{5} \tag{C}$$

4 The roots of the equation $x^3 + px^2 + qx + r = 0$ are $\tan\alpha$, $\tan\beta$, $\tan\gamma$. Find, in terms of p, q, r, the value of $\tan(\alpha + \beta + \gamma)$. (C)

5 Solve the equation

$$\cos 2x = 3 - 5\sin x$$

for values of x satisfying $0 \leqslant x \leqslant \pi$.

Using the same axes, sketch the graphs of

$$y = \cos 2x \quad \text{and} \quad y = 3 - 5\sin x$$

for $0 \leqslant x \leqslant \pi$.

Shade on your sketch the region in which

$$y \geqslant 0, \quad y \leqslant \cos 2x, \quad y \geqslant 3 - 5\sin x \quad \text{and} \quad 0 \leqslant x \leqslant \pi/2.$$

Show that the area of this region is

$$\tfrac{1}{4}(2\pi + 9\sqrt{3} - 14 - 12\sin^{-1}\tfrac{3}{5}) \tag{JMB}$$

6 At time t, the displacement from the origin of a particle of mass m moving on the x-axis is x, and the velocity is v. The particle moves under the action of a force $A \cos 2t$ along Ox, where A is a positive constant. When $t = 0$, the particle is at rest at the origin. Find:

a) v as a function of t;

b) x as a function of t;

c) the maximum distance of the particle from the origin;

d) the time t at which the particle first returns to the origin.   (C)

7 A vertical pole BAO stands with its base O on a horizontal plane, where BA $= c$ and AO $= b$. A point P is situated on the horizontal plane at distance x from O and the angle APB $= \theta$. Prove that

$$\tan \theta = \frac{cx}{x^2 + b^2 + bc}$$

As P takes different positions on the horizontal plane, find the value of x for which θ is greatest.   (AEB, 1982)

8 The parametric equations of a curve are given by

$$x = 2t + \sin 2t, \quad y = \cos 2t \quad \text{for} \quad 0 \leqslant t \leqslant \pi$$

a) Show that

$$\frac{dy}{dx} = -\tan t$$

and deduce that

$$4\frac{d^2y}{dx^2} + \left(1 + \left(\frac{dy}{dx}\right)^2\right)^2 = 0$$

b) Sketch the curve.

c) Find the coordinates of the two points where the curve cuts the x-axis.

d) Find the area of the region bounded by the curve and the part of the x-axis between the two points found in **c)**.   (C)

9 a) Find $\displaystyle\int \frac{2 + \cos x}{\sin^2 x}\,dx$

b) Using the substitution $x = 2 \sin \theta$, or otherwise, evaluate

$$\int_0^1 \frac{x^2}{\sqrt{(4 - x^2)}}\,dx$$

giving two significant figures in your answer.   (C)

10 The illumination I at a point P at the edge of a horizontal circular table (of radius a) from a lamp A vertically above the centre O is given by

$$I = \frac{\lambda \cos x}{(\text{AP})^2}$$

where λ is a constant and x is the angle PAO. Find the value of x within the appropriate domain for which I has a stationary value, and establish whether it is a maximum or a minimum.

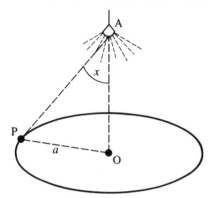

[*Hint.* First express AP in terms of a and x.]   (SMP)

11 The region defined by the inequalities

$$0 \leqslant x \leqslant \pi \quad 0 \leqslant y \leqslant \lg(1 + \sin x)$$

is rotated completely about the x-axis. Using any appropriate rule for approximate integration with five ordinates, find the volume of the solid of revolution formed, giving your answer to 3 significant figures.

(AEB, 1981)

12 Prove, for any positive integers m and n, where $m \neq n$, that

$$\int_0^\pi \sin mx \sin nx \, \mathrm{d}x = 0$$

Prove also that $\int_0^\pi \sin^2 nx \, \mathrm{d}x = \dfrac{\pi}{2}$, if n is a positive integer.

Assume that there are numbers a_n, depending on n but not on x, such that

$$x = \sum_{n=1}^\infty a_n \sin nx \quad \text{for} \quad 0 \leqslant x < \pi$$

By multiplying both sides of this equation by $\sin mx$ and integrating term by term, prove that $a_n = \dfrac{2(-1)^{n+1}}{n}$ for all integers $n \geqslant 1$.

Hence deduce that $\dfrac{\pi}{4} = 1 - \tfrac{1}{3} + \tfrac{1}{5} - \tfrac{1}{7} + \cdots$ (MEI)

Chapter 6

1 An arithmetic progression has a first term of 100 and the common difference is 200. A geometric progression has a first term of 1 and its common ratio is 2. Find:
a) the sum, A_n, of n terms of the arithmetic progression;
b) the sum, G_n, of n terms of the geometric progression.
By trial and error find the least value of n for which G_n is greater than A_n.

(AEB, 1983)

2 **a)** An arithmetic series has first term 1000 and common difference -1.4. Calculate the value of the first negative term of the series, and the sum of all the positive terms.
b) The sum to infinity of a geometric series is equal to fifteen times the sum of the first fifteen terms. Calculate the value of the common ratio, giving your answer correct to 3 decimal places. (C)

3 Explain carefully what is meant by the phrase 'the sum to infinity of the geometric series $a + ar + ar^2 + \cdots$'. Under what circumstances does such a sum exist and what is then its value?
Sum to infinity each of the series
a) $1 + 2x + 4x^2 + 8x^3 + \cdots$
b) $1 + 3x + 5x^2 + 7x^3 + \cdots$
c) $1 + \dfrac{1}{1+x} + \dfrac{1}{(1+x)^2} + \dfrac{1}{(1+x)^3} + \cdots$

stating clearly for each series the conditions which must be imposed on x for it to be convergent. (OC)

4 **a)** Show, by induction or otherwise, that for all positive integers n, $5^n - 2^n$ is divisible by 3.

b) Given that $S_n = \displaystyle\sum_{r=1}^{n} \dfrac{1}{r(r+2)}$, prove, by induction or otherwise, that

$$S_n = \dfrac{3n^2 + 5n}{4(n+1)(n+2)}$$

and find the limit of S_n as $n \to \infty$. (L)

5 Show by induction, or otherwise, that, if n is an integer and $n > 1$, $7^n - 6n - 1$ is divisible by 36, and $5^n - 4n - 1$ is divisible by 16.
Hence, or otherwise, show that $7^n - 5^n - 2n$ is divisible by 4.
Find the highest positive integer which will always divide $2.7^n - 3.5^n + 1$ exactly. (L)

6 Given that $T_n = \displaystyle\sum_{r=1}^{n} r^2$, prove, by induction or otherwise, that

$$T_n = \frac{n(n+1)(2n+1)}{6}$$

and deduce a similar expression for T_{2n}.

Given that $S_n = 1^2 + 3^2 + 5^2 + \cdots + (2n-1)^2$ for $n \geqslant 1$, show by considering $T_{2n} - 4T_n$, that

$$S_n = \frac{n}{3}(4n^2 - 1) \qquad \text{(MEI)}$$

7 By considering the roots of the equation $f'(x) = 0$, or otherwise, prove that the equation $f(x) = 0$, where $f(x) \equiv x^3 + 2x + 4$, has only one real root. Show that this root lies in the interval $-2 < x < -1$.

Use the iterative procedure

$$x_{n+1} = -\frac{1}{6}(x_n^3 - 4x_n + 4), \quad x_1 = -1$$

to find two further approximations to the root of the equation, giving your final answer to 2 decimal places. (L)

8 The equation $x^3 - 12x + 1 = 0$ has two positive roots, α and β, $(\alpha < \beta)$, and one negative root.
a) Prove that $0 < \alpha < 1$ and $3 < \beta < 4$.
b) Use the Newton–Raphson iteration procedure with 3.5 as a starting value to calculate β correct to 2 decimal places.
c) Sketch the curve $y = x^3 - 12x + 1$ and hence, or otherwise, show that the Newton–Raphson iteration procedure applied to the equation $x^3 - 12x + 1 = 0$ will converge to β for any starting value greater than 2, and also for some starting values less than α.
d) Show also with the aid of a separate sketch graph that β can be calculated by the iteration $x_{n+1} = (12x_n - 1)^{\frac{1}{3}}$ $(n = 0, 1, 2, \ldots)$ provided $x_0 > \alpha$. (C)

9 A dinner is to be attended by 6 women and 6 men. The arrangements at the circular dining table specify that the guests must be seated alternately male, female and that Mr Smith, their host, must occupy a particular place. Find the number of different possible seating arrangements.

Two of the women do not wish to sit next to Mr Smith. Calculate the number of the above arrangements which comply with the wishes of both women. (AEB, 1982)

10 Prove that, for any positive integer n

$$(1+x)^n = 1 + \binom{n}{1}x + \binom{n}{2}x^2 + \cdots + \binom{n}{r}x^r + \cdots + x^n$$

and deduce, or prove otherwise, that

a) $1 - \binom{n}{1} + \binom{n}{2} \cdots + (-1)^r \binom{n}{r} + \cdots = 0$

b) $\binom{n+2}{r} = \binom{n}{r} + 2\binom{n}{r-1} + \binom{n}{r-2}$, provided $2 \leqslant r \leqslant n$

Evaluate $\sum_{r=1}^{n} r^2 \binom{n}{r}$ (OC)

Chapter 7

1   Four ball-point pen refills are to be drawn at random without replacement from a bag containing ten refills, of which 5 are red, 3 are green and 2 are blue. Find:
   a) the probability that both blue refills will be drawn;
   b) the probability that at least one refill of each colour will be drawn.
   (JMB)

2   Two grand masters, Chekov and Kassler, play a series of games of chess. In any given game, the probability of a draw is 5/9 and the probability of Chekov winning is three times the probability of Kassler winning.
   Calculate the probability that, after three games,
   a) Chekov has won 2 games and Kassler has won 1;
   b) the series is level. (AEB, 1981)

3   Cards are drawn one at a time, without replacement, from an ordinary pack of 52 cards. Cards drawn have the following values: Aces score 1 point, tens, Jacks, Queens and Kings score 10 points, and cards from two to nine score as many points as the numbers they carry (i.e. twos score 2 points, threes score 3, and so on). Find the probabilities that
   a) the first card scores at least 9 points;
   b) the first two cards drawn each score at least 9 points;
   c) the first two cards drawn score at least 18 points altogether;
   d) the first two cards drawn each score at least 9 points given that they score at least 18 points altogether. (C)

4   An urn contains seven red and five green tickets of which five of the red and two of the green tickets have the number 2 written on them. The rest of the red tickets have the number 1 written on them, and the rest of the green tickets have the number 3 written on them.
   A random sample of three tickets is selected from the urn without replacement. Denoting by R the event that all the tickets selected are red, and by E the event that the sum of the numbers on the tickets selected is equal to 6, find $P(R)$, $P(E)$, $P(R \cap E)$ and $P(R|E)$. (C)

5 **a)** A fair die is thrown three times. Events A, B, C are defined as follows:

A—the total score is an odd number,
B—a six appears at the first throw,
C—the total score is 13.

i) Calculate $P(A)$ and $P(A|B)$ and state whether A and B are independent events.
ii) Calculate $P(C)$ and $P(C|B)$ and state whether B and C are independent events.
b) Find how many different four-figure numbers can be formed from the five digits 1, 2, 3, 4, 5 if repetitions of the digits are allowed.
Find how many of these numbers are less than 2500.   (C)

6 **a)** The events A and B are such that

$$P(A) = 0.4 \quad P(B) = 0.45 \quad P(A \cup B) = 0.68$$

Show that the events A and B are neither mutually exclusive nor independent.
b) A bag contains 12 red balls, 8 blue balls and 4 white balls. Three balls are taken from the bag at random and without replacement. Find the probability that all three balls are of the same colour.
Find also the probability that all three balls are of different colours.
                                                              (L)

7 The probabilities that a man makes a certain dangerous journey by car, motor-cycle or on foot are $\frac{1}{2}, \frac{1}{6}$ and $\frac{1}{3}$ respectively. If the probabilities of an accident when he uses these means of transport are $\frac{1}{5}, \frac{2}{5}$ and $\frac{1}{10}$ respectively, find the probability of an accident occurring in a single journey.
If an accident is known to have happened, calculate the probabilities that the man was travelling
a) by car;
b) by motor-cycle;
c) on foot.   (L)

8 It is estimated that one-quarter of the drivers on the road between 11 p.m. and midnight have been drinking during the evening. If a driver has not been drinking, the probability that he will have an accident at that time of night is 0.004%; if he has been drinking, the probability of an accident goes up to 0.02%. What is the probability that a car selected at random at that time of night will have an accident?
A policeman on the beat at 11.30 p.m. sees a car run into a lamp-post, and jumps to the conclusion that the driver has been drinking. What is the probability that he is right?   (SMP)

9 Three women A, B and C play for a prize by throwing a fair die. The first person to throw a six wins. If they throw the die in the order A, then B, then C, find the probability that

a) A will win on the first throw;

b) B will win on her second throw;

c) A will win eventually;

d) B will win eventually;

e) C will win eventually.

Given that A wins, find the probability that she won on the first throw.

(L)

10 A football cup final between teams A and B ends in a draw and the cup is then to be awarded on the result of a penalty competition. One round of penalties consists of one penalty being taken by each team. If both teams score or if both teams fail to score in one round, then another round is played. The match is won when, in a round, one team scores and the other team fails to score. The probability of team A scoring a goal with a penalty kick is 8/10 and, independently, the probability of team B scoring a goal with a penalty kick is 9/10. Find the probability that the match will be undecided

a) after one round,   **b)** after 3 rounds.

Find the probability that team A will win in

c) the first round,   **d)** the second round.

Find also the probability that team A will win the cup.   (L)

11 A census of married couples showed that 50% of the couples had no car, 40% had one car and the remaining 10% had two cars. Three of the married couples are chosen at random.

a) Find the probability that one couple has no car, one has one car and one has two cars.

b) Find the probability that the three couples have a combined total of three cars.

The census also showed that both the husband and the wife were in full-time employment in 16% of those couples having no car, in 45% of those having one car and in 60% of those having two cars.

c) For a randomly chosen married couple find the probability that both the husband and wife are in full-time employment.

d) Given that a randomly chosen married couple is one where both the husband and wife are in full-time employment, find the conditional probability that the couple has no car.   (JMB)

12 A tyre company tested 100 of their 'Tearaway' radial steel tyres on four-wheel-drive cars in harsh conditions in Australia. They found that 9 of the tyres had a safe life of under 30 000 km, whereas 20 were still safe after 50 000 km.

Assuming that it is reasonable to model the safe life of the tyres by a

Normal probability function, use tables to estimate the standard deviation and to show that the mean safe life is about 42 300 km.

A four-wheel-drive vehicle is fitted with four new 'Tearaway' tyres and is to be driven under similar harsh conditions. Find the probabilities
a) that the first tyre change will not be needed until after 35000 km;
b) that by the time 40000 km have been run, exactly two of the original tyres will have needed replacement. (SMP)

13 The time, in minutes, required to complete a particular task may be assumed to be Normally distributed. Given that there is a probability of 0.01 that the task will be completed in less than 30 minutes and a probability of 0.95 that it will be completed in less than one hour, calculate, correct to two decimal places, the probability that the task will be completed in less than 45 minutes. (JMB)

14 An intelligence test is designed so that, if it is given to a random sample of n 18-year-olds, the *average mark* will be distributed with Normal probability having mean 100 and standard deviation $15/\sqrt{n}$. The test is given to a batch of 25 army recruits aged 18, with the following results:

107	98	106	112	109
89	118	102	91	111
105	123	95	107	135
100	103	116	96	108
125	107	119	79	114

An officer was heard to remark that 'they seem to be an exceptionally intelligent batch of young people'. Use the information about the design of the test given above to analyse the data, and hence evaluate the officer's comment. (SMP)

Chapter 8

1 The lines L_1 and L_2 have equations given respectively by

$$\mathbf{r}_1 = \begin{pmatrix} 2 \\ 2 \\ 2 \end{pmatrix} + t\begin{pmatrix} 2 \\ 1 \\ -1 \end{pmatrix} \quad \text{and} \quad \mathbf{r}_2 = \begin{pmatrix} 0 \\ 3 \\ -1 \end{pmatrix} + s\begin{pmatrix} -2 \\ 1 \\ 1 \end{pmatrix}$$

where t and s are real parameters. Show that L_1 and L_2 do not intersect.

The point P on L_1 has parameter p and the point Q on L_2 has parameter q. Write **PQ** as a column vector in terms of p and q. Given that L_1 and L_2 are both perpendicular to **PQ**, find p and q. Find the length of **PQ**, giving your answer in surd form. (JMB)

2 The line l with equation $\mathbf{r} = \begin{pmatrix} 1 \\ -2 \\ 3 \end{pmatrix} + t\begin{pmatrix} 5 \\ 2 \\ -4 \end{pmatrix}$ meets, at the point A, the

plane π with equation $\mathbf{r} = \begin{pmatrix} -2 \\ 4 \\ 3 \end{pmatrix} + \lambda\begin{pmatrix} -2 \\ 1 \\ 1 \end{pmatrix} + \mu\begin{pmatrix} 0 \\ 1 \\ 2 \end{pmatrix}$. The point C on l has

position vector $\begin{pmatrix} 1 \\ -2 \\ 3 \end{pmatrix}$ and B is the foot of the perpendicular from C to the

plane π.
a) Find the position vectors of the points A and B.
b) Show that $\sin \text{CAB} = \sqrt{(\frac{7}{15})}$.
c) Find the position vector of the reflection of C in the plane π.   (c)

3 Show that

$$\mathbf{r} = \mathbf{a} + s(\mathbf{b} - \mathbf{a}) + t(\mathbf{c} - \mathbf{a})$$

is an equation of the plane which passes through the non-collinear points whose position vectors are $\mathbf{a}$, $\mathbf{b}$, $\mathbf{c}$, where $\mathbf{r}$ is the position vector of a general point on the plane and s and t are scalars. Find a Cartesian equation of the plane containing the points $(1, 1, -1)$, $(2, 0, 1)$ and $(3, 2, 1)$, and show that the points $(2, 1, 2)$ and $(0, -2, -2)$ are equidistant from, and on opposite sides of, this plane.   (L)

4 The position vectors $\mathbf{a}$, $\mathbf{b}$, $\mathbf{c}$ of three points A, B, C respectively are given by

$$\mathbf{a} = \mathbf{i} + \mathbf{j} + \mathbf{k}$$
$$\mathbf{b} = \mathbf{i} + 2\mathbf{j} + 3\mathbf{k}$$
$$\mathbf{c} = \mathbf{i} - 3\mathbf{j} + 2\mathbf{k}$$

Find:
a) a unit vector parallel to $\mathbf{a} + \mathbf{b} + \mathbf{c}$;
b) the cosine of the angle between $\mathbf{a} + \mathbf{b} + \mathbf{c}$ and the vector $\mathbf{a}$;
c) the vector of the form $\mathbf{i} + \lambda\mathbf{j} + \mu\mathbf{k}$ perpendicular to both $\mathbf{a}$ and $\mathbf{b}$;
d) the position vector of the point D which is such that ABCD is a parallelogram having BD as a diagonal.   (c)

5 Referred to an origin O, points A and B have position vectors given respectively by $\mathbf{OA} = \mathbf{i} + 2\mathbf{j} - 2\mathbf{k}$ and $\mathbf{OB} = 2\mathbf{i} - 3\mathbf{j} + 6\mathbf{k}$. The point P on AB is such that $\text{AP:PB} = \lambda : 1 - \lambda$. Show that

$$\mathbf{OP} = (1 + \lambda)\mathbf{i} + (2 - 5\lambda)\mathbf{j} + (-2 + 8\lambda)\mathbf{k}$$

a) Find the value of λ for which OP is perpendicular to AB.
b) Find the value of λ for which angles AOP and POB are equal.   (C)

6   The line l has equation

$$\mathbf{r} = \begin{pmatrix} -1 \\ 5 \\ 3 \end{pmatrix} + \lambda \begin{pmatrix} 2 \\ 4 \\ 1 \end{pmatrix}$$

a) The point A on l is such that OA is perpendicular to l, where O is the origin. Find the position vector of A.
b) Find a unit vector perpendicular to the plane containing O and l.
c) Find the position vectors of the two points P on l such that
$\cos \text{POA} = \dfrac{1}{\sqrt{7}}.$ (C)

7   **a)** Find the length of the perpendicular from the origin O to the plane through the points with position vectors

$$\begin{pmatrix} 3 \\ 0 \\ 0 \end{pmatrix}, \quad \begin{pmatrix} 2 \\ 1 \\ 0 \end{pmatrix}, \quad \begin{pmatrix} 3 \\ 2 \\ -1 \end{pmatrix}.$$

b) The distinct points A, B and C have position vectors $\mathbf{a}$, $\mathbf{b}$ and $\mathbf{c}$ with respect to the origin O, and the plane ABC does not contain O. Given that OA is perpendicular to BC and that OB is perpendicular to AC, prove that
i) OC is perpendicular to AB,
ii) $\text{OA}^2 + \text{BC}^2 = \text{OB}^2 + \text{CA}^2 = \text{OC}^2 + \text{AB}^2.$ (C)

8   At time $t = 0$, a particle is projected from a point O with speed u at an angle of elevation α. At time t, the horizontal and vertical distances of the particle from O are x and y respectively. Express x and y in terms of u, α, t and g. Hence show that

$$y = x \tan \alpha - \frac{gx^2}{2u^2}(1 + \tan^2 \alpha)$$

A golf ball is struck from a point A, leaving A with speed $30 \, \text{m s}^{-1}$ at an angle of elevation θ, and lands, without bouncing, in a bunker at a point B, which is at the same horizontal level as A. Before landing in the bunker, the ball just clears the top of a tree which is at a horizontal distance of 72 m from A, the top of the tree being 9 m above the level of AB. Show that one of the possible values of θ is $\tan^{-1} \frac{3}{4}$ and find the other value. Given that θ was in fact $\tan^{-1} \frac{3}{4}$, find the distance AB.
   [Take g as $10 \, \text{m s}^{-2}$.]   (L)

9   A vertical pole AB of height 2 m stands with its base B on horizontal

ground, and C is a point on the ground 16 m from B. A particle projected from C with speed 14 m s^{-1} and angle of elevation $\tan^{-1} p$ hits the top of the pole. Show that either $p = \frac{3}{4}$ or $p = \frac{7}{4}$.

Find the set of values of p for which the particle can hit the pole.

[Take g to be 9.8 m s^{-2}.]   (c)

10 An equilateral triangle ABC has sides of length 6 units. The three altitudes of the triangle meet at N. Show that $AN = 2\sqrt{3}$ units.

This triangle is the base of a pyramid whose apex V lies on the line through N perpendicular to the plane ABC. Given that $VN = 2$ units, prove that $V\widehat{A}N = 30°$.

The perpendicular from A to the edge VC meets CV produced at R. Prove that $AR = \frac{3}{2}\sqrt{7}$ units, and find the exact value of $\cos A\widehat{R}B$.   (c)

11 A VC10 is flying on a course due North at 800 km/h, and a Concorde is flying at the same height on a bearing of 060° at 2000 km/h. Draw a velocity triangle and hence find the direction of the velocity of the Concorde relative to the VC10.

If, when the VC10 first sights the Concorde, the Concorde is 20 km due West of it, find how close the aircraft will come to each other if neither takes avoiding action.

[Equal credit will be given for solutions by accurate drawing or by calculation.]   (SMP)

12 Two ships, A and B, are travelling with constant velocities of 30 km/h westwards and 40 km/h northwards respectively. Their courses meet at a point O. When B is passing through O the ship A is 15 km from O and moving towards O. Find:

a) the magnitude and direction of the velocity of B relative to A;

b) the least distance between the ships;

c) the distances of the ships from O when they are nearest to one another.

Show that, if A, while maintaining its speed, had changed course when B was passing through O in such a way as to minimise the least distance between the ships, the least distance would have been approximately 9.92 km.   (JMB)

Chapter 9

1 The position vector of a particle P of mass m is given at time t by

$$\mathbf{r} = (t^3 - 2t)\mathbf{i} + t^2\mathbf{j}, \, t \geqslant 0$$

Obtain expressions for the velocity and the acceleration of P at time t, and deduce the two values of t for which the velocity and the acceleration of P are at right angles.

The motion is due to a variable force **F** acting in the x–y plane. Show that, when $t = 0$, the direction of **F** is parallel to the y-axis. Show also that, as t increases, the direction of **F** approaches that of the x-axis. Find the value of t for which the direction of **F** is equally inclined to the axes and give the magnitude of **F** at that instant.   (JMB)

2   A lift, supported by a single cable, has mass M and carries a man of mass m. The lift is moving with an upward acceleration of magnitude a. Obtain an expression for the tension T in the lift cable, and find the vertical force R exerted on the man by the floor of the lift.

Give corresponding expressions for T and R when the lift is moving with a downward acceleration of magnitude a.

Given that $M = 1000\,\text{kg}, g = 10\,\text{m s}^{-2}, a \leqslant 1\,\text{m s}^{-2}$ and $T \leqslant 15000\,\text{N}$, and making any reasonable assumptions, which should be stated, estimate how many people the lift should be licensed to carry. Give full reasons for your answer.   (C)

3   A particle of weight W is placed on a rough plane which is inclined at an angle α to the horizontal. The coefficient of friction between the particle and the plane is μ, where $\mu < \tan \alpha$. Show that the magnitude of the least horizontal force needed to maintain the particle in equilibrium is

$$\frac{\tan \alpha - \mu}{1 + \mu \tan \alpha} W$$

Indicate the direction of this force in a diagram.   (JMB)

4   A particle is projected with initial speed u up a line of greatest slope of a rough plane inclined at an angle α to the horizontal. The coefficient of friction between the particle and the plane is μ, and $\mu = \tan \lambda$. Show that the particle will come to rest after a time

$$\frac{u}{g}\cos \lambda \operatorname{cosec}(\alpha + \lambda)$$

Show also that, if $\lambda < \alpha$, the particle will then immediately start to slide down the plane, and will again be moving with speed u after it has moved a distance of $\dfrac{u^2}{2g}\cos \lambda \operatorname{cosec}(\alpha - \lambda)$ from the highest point it reached.   (C)

5   A body is held at rest on a smooth sloping roof at a distance of $2\,\text{m}$ from the edge of the roof. The roof is inclined at an angle α to the horizontal, where $\sin \alpha = 0.6$. The body is given an initial velocity of $1\,\text{m/s}$ down a line of greatest slope. Show that the body reaches the edge of the roof with a speed of $5\,\text{m/s}$.

After passing the edge of the roof, which is $14\,\text{m}$ above the horizontal ground, the body moves freely under gravity. Calculate:

a) the time taken by the body to reach the ground from the edge of the roof;

b) the direction in which the body is moving when it strikes the ground.

(Take the acceleration due to gravity to be 10 m/s².)   (AEB, 1982)

6   The diagram shows a narrow circular tube OA fixed with its axis vertical.

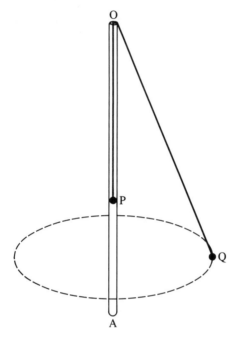

Two particles P and Q, of masses M and m respectively, are attached to the ends of a light inextensible string of length l. The particle P is free to move inside the tube. The particle Q moves in a horizontal circle with constant angular velocity ω and P hangs at rest. The inside of the tube and the lip of the tube, at O, are smooth. Show, in any order, that

a) the angle QOP is independent of ω

b) $M > m$

c) $OQ = \dfrac{Mg}{m\omega^2}$

d) $\omega^2 > \dfrac{Mg}{ml}$ (C)

7   Write down the magnitude and indicate the direction of the acceleration of a particle travelling in a fixed circular path of radius r at a constant speed v.

A satellite is in steady circular orbit at height h above the surface of the earth. At this height the gravitational pull on the satellite has magnitude

$GmM/(R + h)^2$, where m is the mass of the satellite, M is the mass of the earth, R is the radius of the earth, and G is constant. Find the speed of the satellite, and show that it will take a time

$$2\pi(R + h)^{\frac{3}{2}}(GM)^{-\frac{1}{2}}$$

to travel once round the orbit. (MEI)

8 A fixed hollow smooth sphere, of external radius a and centre O, has a small hole at the highest point. A light inextensible string passes through the hole and carries a particle of mass $2m$ which hangs freely, at rest, inside the sphere. A particle P, of mass $3m$, is attached to the other end of the string and moves on the outer surface of the sphere in a horizontal circle of radius $\frac{1}{2}a$ with angular velocity ω. show that

$$\frac{a\omega^2}{g} = \frac{2\sqrt{3}}{9}$$

and find, in terms of m and g, the magnitude of the reaction of the sphere on the particle. (C)

9 A particle of mass $4m$, which is at rest, explodes into two fragments, one of mass m and the other of mass $3m$. The explosion provides the fragments with total kinetic energy E. Find the velocities of the fragments just after the explosion. Hence find, in terms of m and E, the magnitude of the relative velocity of the fragments just after the explosion. (JMB)

10 Liquid of density ρ is released at speed u from a jet of cross-section A which points vertically upwards. A lamina of mass M is supported horizontally by the liquid at a height h above the jet. On striking the lamina, the momentum of the liquid is destroyed. Find an expression for h in terms of u, g, A, ρ, M, and deduce that $M < A\rho u^2/g$. (C)

11 A small sandbag of mass M is suspended from a fixed point by a light inextensible rope of length l. A bullet of mass m is fired horizontally into the sandbag, and the bag with the bullet embedded swings through an angle θ before coming instantaneously to rest. Find the speed of the bullet before impact and the kinetic energy lost during the impact. (L)

12 A car engine works at a rate of 24 kW in driving a car of mass 800 kg along a horizontal road at a steady speed of 40 m/s. Find the resistance to the motion of the car.

Assuming that the resistance remains the same and that the car engine is working at the same rate, calculate:

a) the angle of inclination to the horizontal of the slope up which the car's maximum speed is 15 m/s;

b) the acceleration of the car up a slope of 1 in 16 when moving at a speed of 12 m/s.

If the resistance to motion is now 500 N, find the rate at which the car engine must work in order to produce an acceleration of 1 m/s² when the car is travelling up a slope of 1 in 16 at a speed of 20 m/s.

(Take the acceleration due to gravity to be 10 m/s².)   (AEB, 1982)

13 Prove that, in the usual notation, $\dfrac{dv}{dt} = v\dfrac{dv}{ds}$.

The resistive forces acting on a car of mass 800 kg which is moving along a straight horizontal road are negligible. The car accelerates steadily from speed 15 m/s at the instant when it passes point A to speed 30 m/s at the instant when it passes point B and the power provided by the engine of the car remains constant at 24 kW.

a) Find the time taken by the car to cover the distance AB.

At the moment when the car is s metres from A during its journey from A to B, the speed of the car is v m/s.

b) Show that $v^2\dfrac{dv}{ds} = 30$.

c) Find the distance AB.   (AEB, 1981)

14 A train of mass 2×10^5 kg is running, with the engine shut off, down a hill of inclination α to the horizontal, where $\sin\alpha = \frac{1}{100}$. The acceleration of the train is $0.05\,\mathrm{m\,s^{-2}}$. Taking the value of g to be $10\,\mathrm{m\,s^{-2}}$, find the resistance to motion.

Find the maximum speed of the train when travelling up the same hill against a resistance to motion of 4×10^3 N with the engine working at a power of 3×10^5 W.   (C)

15 A motor truck of mass 15 tonnes travels up a straight slope of gradient $\sin^{-1}(0.05)$ against a resistance of 0.1 N per kilogram. Find the tractive force required to produce an acceleration of $0.1\,\mathrm{m\,s^{-2}}$ and the power in kilowatts which is then developed when the speed is $8\,\mathrm{m\,s^{-1}}$.

Find the steady speed the truck can sustain on the same upward slope if the power exerted by the motor is 135 kW.

[1 tonne = 1000 kg. Take the acceleration due to gravity as $10\,\mathrm{m\,s^{-2}}$.]

          (JMB)

16 At maximum power, a lorry of mass 30000 kg travels at a constant speed of $12\,\mathrm{m\,s^{-1}}$ up a hill of inclination α above the horizontal, where $\sin\alpha = \frac{1}{20}$. The resistance to motion is of magnitude 7000 N. Show that the power at which the engine is working is 264 kW.

Assuming that the resistance remains constant, find the maximum acceleration of the lorry when moving down the same hill at an instant when its speed is $24\,\mathrm{m\,s^{-1}}$.

[Give your answer correct to 3 significant figures.]   (C)

Chapter 10

1 Find the following:

a) $\displaystyle\int_0^2 \frac{x+3}{x^2+4}\,dx$

b) $\displaystyle\int_1^e \frac{(\ln x)^2}{x}\,dx$

c) $\displaystyle\int_0^{\frac{1}{2}\pi} x \sin x\,dx$

d) $\displaystyle\int \frac{x+1}{x^2+5x+6}\,dx$ (MEI)

2 a) Find $\displaystyle\int \frac{x}{(x+1)^2}\,dx$.

b) Use the substitution $t = \tan\theta$, or any other method, to show that

$$\int_{\frac{1}{4}\pi}^{\frac{1}{3}\pi} \frac{1}{\sin 2\theta}\,d\theta = \tfrac{1}{2}\ln 3.$$

c) Use the trapezium rule with ordinates at $x = 0, 1, 2, 3, 4$ to estimate the value of

$$\int_0^4 \frac{1}{1+\sqrt{x}}\,dx$$

giving three significant figures in your answer.   (C)

3 Show graphically, or otherwise, that the equation $\ln x = 4 - x$ has only one real root and prove that this root lies between 2.9 and 3.

By taking 2.9 as a first approximation to this root and applying the Newton–Raphson process once to the equation $\ln x - 4 + x = 0$, or otherwise, find a second approximation, giving your answer to 3 significant figures.   (L)

4 Find the values of the constants a and b for which the expansions, in ascending powers of x, of the two expressions

$$(1+2x)^{\frac{1}{2}} \quad \text{and} \quad \frac{1+ax}{1+bx}$$

up to and including the term in x^2, are the same.

With these values of a and b, use the result

$$(1+2x)^{\frac{1}{2}} \approx \frac{1+ax}{1+bx}$$

with $x = -\dfrac{1}{100}$, to obtain an approximate value of $\sqrt{2}$ in the form p/q, where p and q are positive integers.   (C)

5 Given that $y = \ln(1 + \sin x)$, where $-\tfrac{1}{2}\pi < x < \tfrac{1}{2}\pi$, prove that $y_1 y_2 + y_3 = 0$, where y_r denotes $\dfrac{d^r y}{dx^r}$.

Express y as a series of ascending powers of x up to and including the term in x^5.   (C)

6 Write down the expansion of $\cos x$ in ascending powers of x, up to and including the term in x^6. Write down also the expansion of $\ln(1 + t)$ in ascending powers of t, up to and including the term in t^3. Hence, or otherwise, find the expansion of $\ln(\cos x)$ in ascending powers of x, up to and including the term in x^6. Use these terms to find an approximation to the integral

$$\int_0^{1/2} \frac{\ln(\cos x)}{x} \, dx$$

giving your answer correct to 4 decimal places.   (O)

7 **a)** Find

$$\int \frac{x - 2}{(x + 4)(x + 1)} \, dx$$

b) Evaluate

$$\int_2^4 \frac{\ln x}{x} \, dx$$

i) by using the substitution $u = \ln x$;
ii) by using the trapezium rule with four strips;
giving each answer correct to three significant figures.   (C)

8 Given that, for $n \geqslant 0$,

$$I_n = \int_0^\pi x^n \sin x \, dx$$

prove that, for $n \geqslant 2$,

$$I_n = \pi^n - n(n - 1) I_{n-2}$$

Hence evaluate

$$\int_0^\pi x^5 \sin x \, dx.$$ (L)

9 Given that

$$I_n = \int_0^\alpha \tan^n \theta \sec \theta \, d\theta \quad (0 < \alpha < \tfrac{1}{2}\pi)$$

prove that

$$nI_n = \tan^{n-1}\alpha \sec\alpha - (n-1)I_{n-2} \quad (n \geqslant 2)$$

Evaluate **a)** $\displaystyle\int_0^{\frac{1}{4}\pi} \tan^5\theta\sec\theta\,d\theta$ **b)** $\displaystyle\int_0^{\frac{1}{4}\pi} \tan^4\theta\sec\theta\,d\theta$ (C)

10 Given that

$$I_n = \int_0^1 x^n\sqrt{(1-x^2)}\,dx \quad (n\geqslant 0)$$

show that

$$(n+2)I_n = (n-1)I_{n-2} \quad (n\geqslant 2)$$

Evaluate I_5. (JMB)

11 Given that

$$I_n = \int_0^\pi e^{-x}\sin^n x\,dx \quad (n\geqslant 0)$$

show that

$$(n^2+1)I_n = n(n-1)I_{n-2} \quad (n\geqslant 2)$$

Find the value of I_4. (JMB)

12 Let

$$I_n = \int_0^1 x^{n+1}(1-x^2)^{1/2}\,dx \quad (n > -2)$$

Prove that, for $n > 0$,

$$(n+3)I_n = nI_{n-2}$$

Use this result to evaluate

$$\int_0^{\pi/2} \cos^9\theta\sin^2\theta\,d\theta$$

leaving your answer as a product of fractions. (O)

Appendix

1 Find the radius and the coordinates of the circle S given by the equation

$$x^2 + y^2 + 10x - 6y - 2 = 0$$

A chord AB of S is a tangent to the circle with the same centre but with

radius equal to one half that of S. The tangents to S at A and B intersect at C. Prove that
a) the triangle ABC is equilateral;
b) the point C is on the circle whose equation is

$$x^2 + y^2 + 10x - 6y - 110 = 0 \qquad \text{(JMB)}$$

2   Give the focus–directrix definition of a parabola.
A fixed line l is a tangent to a fixed circle with centre S. A variable circle touches the fixed circle externally and also touches l. Show that the locus of the centre of the variable circle is a parabola with focus S. (C)

3   Show that the tangent to the parabola $y^2 = 4ax$ at the point $P(ap^2, 2ap)$ has the equation

$$py = x + ap^2$$

This tangent meets the line $x + a = 0$ at T. Find, in terms of a and p, the coordinates of the mid-point of PT.
Prove that, for all values of p. this mid-point lies on the curve

$$y^2(2x + a) = a(3x + a)^2 \qquad \text{(C)}$$

4   The tangents to the parabola $y^2 = x$ at the variable points $P(p^2, p)$ and $Q(q^2, q)$ intersect at N. Given that M is the mid-point of PQ, prove that

$$MN = \tfrac{1}{2}(p - q)^2$$

Given also that PN is always perpendicular to QN,
a) show that the locus of N is a straight line, and write down its equation,
b) find the Cartesian equation of the locus of M. (L)

5   Derive the equations of the tangent and the normal to the ellipse $\dfrac{x^2}{a^2} + \dfrac{y^2}{b^2} = 1$ at the point $P(a\cos t, b\sin t)$. The normal at P meets the axes at the points Q and R. Lines are drawn parallel to the axes through the points Q and R; these lines meet at the point V. Find the coordinates of V, and prove that, as P moves round the ellipse, the point V moves round another ellipse, and find its equation. (O)

6   The orbit of Halley's comet is an ellipse which has the sun at one focus. The greatest and least distances of the comet from the sun are 34.989 and 0.587 astronomical units respectively. Calculate the eccentricity of the ellipse, giving your answer to three decimal places. (C)

7   **a)** Find the coordinates of the points A and B where the tangent at the point $P(a\cos\phi, b\sin\phi)$ to the ellipse with equation $\dfrac{x^2}{a^2} + \dfrac{y^2}{b^2} = 1$ cuts the coordinate axes.
b) Q is the position of P in the first quadrant when OA = OB. Find, in

terms of a and b, the coordinates of the point Q.

c) As a and b vary, a family of ellipses is generated. If a and b vary, but such that $a^2 + b^2 = k^2$, where k is a constant, find the locus of Q.   (MEI)

8 The straight line $lx + my + n = 0$ cuts the hyperbola

$$\frac{x^2}{a^2} - \frac{y^2}{b^2} = k^2$$

at the points P and Q. Find the coordinates of the midpoint of PQ, showing that they are independent of k.   (C)

9 a) Prove that the equation of the chord joining the points $P\left(cp, \dfrac{c}{p}\right)$ and $Q\left(cq, \dfrac{c}{q}\right)$ on the rectangular hyperbola $xy = c^2$ is $pqy + x = c(p + q)$.

It is given that PQ subtends a right angle at the point $R\left(cr, \dfrac{c}{r}\right)$ on the curve. Prove that

b) PQ is parallel to the normal at R to the curve;

c) the mid-point of PQ lies on the straight line $y + r^2x = 0$.   (C)

10 The tangent at the point $P\left(ct, \dfrac{c}{t}\right)$, where $t > 0$, on the rectangular hyperbola $xy = c^2$ meets the x-axis at A and the y-axis at B. The normal at P to the rectangular hyberbola meets the line $y = x$ at C and the line $y = -x$ at D.

a) Show that P is the mid-point of both AB and CD.

b) Prove that the points A, B, C and D form the vertices of a square. The normal at P meets the hyperbola again at the point Q and the mid-point of PQ is M.

c) Prove that, as t varies, the point M lies on the curve

$$c^2(x^2 - y^2)^2 + 4x^3y^3 = 0$$ (AEB, 1983)

11 Sketch the curve whose polar equation is

$$r = a + b\sin\theta$$

where $a > b > 0$ and $0 \leqslant \theta < 2\pi$.

a) Find the area of the region enclosed by the curve.

b) AB is a variable chord of the curve which passes through O, the pole of coordinates. Show that the length of AB is independent of its direction and find the polar equation of the locus of the midpoint of AB. Show this locus in a sketch.

c) AB and CD are two perpendicular chords of the curve, each of which passes through O. Show that

$$(AO)(OB) + (CO)(OD) = 2a^2 - b^2$$ (C)

12   A point $P(x, y)$ moves in the Cartesian plane so that the product of its distances from the points $(a, 0)$ and $(-a, 0)$, where $a > 0$, has the constant value a^2.

a) Find a Cartesian equation of the locus of P.

b) Show that the polar equation of this locus is $r^2 = 2a^2 \cos 2\theta$.

c) Sketch the polar graph of this locus.

d) Find the area of one of the loops of the graph.

e) Use the formula for arc length, $\int \sqrt{\left\{ r^2 + \left(\dfrac{dr}{d\theta} \right)^2 \right\}}\, d\theta$, to show that

the arc length of one loop of the graph is given by $2a\sqrt{2} \displaystyle\int_0^{\pi/4} \sqrt{(\sec 2\theta)}\, d\theta$.

Do not attempt to evaluate this integral.   (AEB, 1983)

Answers to exercises

Chapter 1

Exercise 1.1

1. a) $p \Leftarrow q$ b) $p \Rightarrow q$ c) $p \Leftarrow q$ d) $p \Leftrightarrow q$
2. a) $p \not\Rightarrow q$ b) $p \not\Leftarrow q$ c) $p \not\Rightarrow q$ d) $p \not\Rightarrow q, p \not\Leftarrow q$
3. a) $p' \Rightarrow q'$ b) $p' \Leftarrow q'$ c) $p' \Rightarrow q'$ d) $p' \Leftrightarrow q'$
4. a) $p' \Leftarrow q'$ b) $p' \Rightarrow q'$ c) $p' \Leftrightarrow q'$

Exercise 1.2a

1. a) $\{-1, 0, 1, 2, 3\}$ b) $\{1, 2, 3\}$ c) $\{5, 6, 7, \ldots\}$ d) $\{1, 2\}$ e) $\varnothing$
2. a) $\mathscr{E}$ b) $\varnothing$ c) A

Exercise 1.2b

1. a) A b) $\varnothing$ c) $\mathscr{E}$ d) A e) A f) A g) A h) A i) $\mathscr{E}$ j) $\varnothing$
5. a) $A \cap B$ b) $\varnothing$ c) B' d) A

Exercise 1.3a

2. OA: $(2, 2\frac{1}{2})$, $\sqrt{41}$, $\frac{5}{4}$
   OB: $(1\frac{1}{2}, 0)$, 3, 0
   OC: $(0, -1)$, 2, indeterminate
   OD: $(-1, \frac{1}{2})$, $\sqrt{5}$, $-\frac{1}{2}$
   OB, OC

3. AB: $(3\frac{1}{2}, 2\frac{1}{2})$, $\sqrt{26}$, 5
   AC: $(2, 1\frac{1}{2})$, $\sqrt{65}$, $\frac{7}{4}$
   AD: $(1, 3)$, $\sqrt{52}$, $\frac{2}{3}$
   BC: $(1\frac{1}{2}, -1)$, $\sqrt{13}$, $\frac{2}{3}$
   BD: $(\frac{1}{2}, \frac{1}{2})$, $\sqrt{26}$, $-\frac{1}{5}$
   CD: $(-1, -\frac{1}{2})$, $\sqrt{13}$, $-\frac{3}{2}$
   AD, BC are parallel
   AB, BD are perpendicular
   CD is perpendicular to AD, BC

Exercise 1.3b

11 a) $(1, 3)$ **b)** $(-2, 6)$ **c)** $(-1, -1)$ **d)** $(-1, 1), (2, 4)$
    e) $(0, 0), (2, 8), (-2, -8)$

Exercise 1.4a

1 a) $x - 3y = 0$ **b)** $x - y = 0$ **c)** $x - 9y = 0$
2 a) $x + y - 10 = 0$ **b)** $y = 1$ **c)** $2x - y - 5 = 0$
3 $x - 2y - 1 = 0, 4x + y - 25 = 0, 2x + 5y - 23 = 0$
   $G: (5\frac{2}{3}, 2\frac{1}{3})$
4 $x - y - 2 = 0, x = 5, x + 2y - 11 = 0$
   $H: (5, 3)$
5 $x - y - 4 = 0, x = 6, x + 2y - 10 = 0$
   $X: (6, 2)$
6 $x + y = 8$

Exercise 1.4b

3 $(3, 1), (4, 2), (4, 0)$
4 $(2, -2), (-7, -2), (-1, 1)$
5 $(\frac{3}{2}, -\frac{3}{2}), (4, -4), (2, 0)$

Exercise 1.5

2 a) $\frac{3}{10}$ **b)** $\frac{1}{250}$ **c)** $\frac{213}{500}$ **d)** $\frac{243}{100}$

3 a) 0.2 **b)** 0.1 **c)** $0.\dot{3}$ **d)** $0.0\dot{9}$ **e)** $0.\dot{2}8571\dot{4}$
4 a) $\frac{2}{9}$ **b)** $\frac{13}{99}$ **c)** $\frac{52}{111}$

Exercise 1.6

1 a) $2\sqrt{2}$ **b)** $2\sqrt{3}$ **c)** $3\sqrt{2}$ **d)** $5\sqrt{2}$ **e)** $6\sqrt{2}$ **f)** $7\sqrt{2}$
   g) $12\sqrt{2}$ **h)** $10\sqrt{3}$ **i)** $10\sqrt{10}$ **j)** $20\sqrt{6}$
2 a) $\sqrt{2}$ **b)** $3\sqrt{8}$ **c)** $\sqrt{3}$ **d)** $2\sqrt{5}$ **e)** 16 **f)** 100
   g) $\sqrt{2}$ **h)** $2\sqrt{5}$
3 a) $\sqrt{2}$ **b)** $6\sqrt{5}$ **c)** $\sqrt{3}$ **d)** $2\sqrt{3} - 3$
   e) 6 **f)** $6 - 4\sqrt{2}$ **g)** 2 **h)** 11
4 a) $\dfrac{\sqrt{5}}{5}$ **b)** $\sqrt{2}$ **c)** $\dfrac{\sqrt{2}}{2}$ **d)** $\frac{1}{2}(\sqrt{3} + 1)$ **e)** $2(2 - \sqrt{3})$
   f) $\frac{1}{4}(5 - \sqrt{5} - \sqrt{10} + \sqrt{2})$ **g)** $5 - 2\sqrt{6}$ **h)** 2

Exercise 1.7

1 identity   **2** reflection in Oy
3 reflection in O **4** reflection in $y = x$

5 magnification $\times 2$ **6** singular
7 anticlockwise rotation through $90°$
8 clockwise rotation through $90°$ and magnification $\times 2$
9 singular
10 shear parallel to Oy

Exercise 1.8

1   **a) C**   **b) B**   **c) A**

2   **a) I**   **b)** $\begin{pmatrix} -1 & 0 \\ 0 & -1 \end{pmatrix}$

3   **a)** $\begin{pmatrix} 0 & 1 \\ -1 & 0 \end{pmatrix} \begin{pmatrix} 0 & -1 \\ 1 & 0 \end{pmatrix}$ **b) I**

5   No

Miscellaneous problems

2   **a)** $23x - 14y = 0$
    b) $15x + 10y + 44 = 0$
    c) $20x + 10y + 51 = 0$
4 $\frac{1}{2}|ad - bc|$
5   **a)** $(0,0), (a, c), (b, d), (a + b, c + d)$ **b)** $|ad - bc|$

Chapter 2

Exercise 2.1a

1   **a)** $12.56, 0.785, 314, 31\,400$ **b)** 0 **c)** $f(3) = 28.26$ **d)** $f(\frac{1}{3}) = 0.349$
2   **a)** $5\,m, 20\,m, 45\,m$ **b)** $50\,m\,s^{-1}$
3   **a)** $2\,s, 4\,s$ **b)** $\frac{1}{4}\,m, 2.25\,m$
4   **a)** $3.6\,km, 7.2\,km, 36\,km$ **b)** $25\,m, 100\,m$

Exercise 2.1b

1   **a)** $fg: x \mapsto x^3 + 1$ **b)** $gf: x \mapsto (x + 1)^3$ **c)** $f^{-1}: x \mapsto x - 1$
    d) $g^{-1}: x \mapsto \sqrt[3]{x}$ **e)** $(fg)^{-1}: x \mapsto \sqrt[3]{(x + 1)}$ **f)** $g^{-1}f^{-1}: x \mapsto \sqrt[3]{(x - 1)}$
2   **a)** $fg: x \mapsto 3x^2$ **b)** $gf: x \mapsto (3x)^2$ **c)** $f^{-1}: x \mapsto \frac{1}{3}x$ **d)** $g^{-1}: x \mapsto \sqrt{x}$
    e) $(fg)^{-1}: x \mapsto \sqrt{\dfrac{x}{3}}$ **f)** $g^{-1}f^{-1}: x \mapsto \sqrt{\dfrac{x}{3}}$ (if domain of g is $x \geqslant 0$)
3 $(fg)^{-1} \equiv g^{-1}f^{-1}$, provided both exist

Exercise 2.1c

1	**a)** -2	**b)** 7	**c)** 5.5	**d)** -5
2	**a)** 7	**b)** 7	**c)** 7	**d)** 7, even
3	**a)** $-$	**b)** $\frac{1}{9}$	**c)** 0.16	**d)** 1, even
4	**a)** 0	**b)** 27	**c)** 15.625	**d)** -1, odd
5	**a)** 0	**b)** 1.732	**c)** 1.58	**d)** $-$
6	**a)** 0	**b)** $\frac{1}{3}$	**c)** 0.4	**d)** -1, odd
7	**a)** 0	**b)** 0	**c)** 0.5	**d)** 0
8	**a)** 0	**b)** $\frac{1}{9}$	**c)** 0.16	**d)** 1, even
9	**a)** 1	**b)** 1	**c)** 1	**d)** 1, even
10	**a)** 0	**b)** 3	**c)** 2.5	**d)** -1, odd
11	**a)** $-$	**b)** $-\frac{1}{3}$	**c)** $-$	**d)** 1, odd
12	**a)** $-$	**b)** 31	**c)** $-$	**d)** $-$
13	**a)** $-$	**b)** 1	**c)** $-$	**d)** $-$
14	**a)** no	**b)** 0 (if defined)		
15	**a)** even	**b)** odd		
16	**a)** $a = 0$	**b)** $b = 0$		

17 a) $x^3 + 2$ **b)** $x^3 - 3$ **c)** $(x - 1)^3$ **d)** $(x + 4)^3$ **e)** $2x^3$
　　f) $-3x^3$ **g)** $8x^3$ **h)** $-x^3$

Exercise 2.2

1 a) ± 2 **b)** $4, 0$ **c)** $+1, +3$ **d)** 2 **e)** $3.41, 0.59$
2 a) $\frac{1}{2}, -1$ **b)** $0.781, -1.281$ **c)** $\frac{5}{3}, -1$ **d)** $\frac{1}{2}\{p \pm \sqrt{(p^2 + 4q)}\}$
3 a) $f(x) \geqslant -9$: least when $x = 3$
　　b) $f(x) \leqslant \frac{25}{4}$: greatest when $x = -\frac{3}{2}$
　　c) $f(x) \geqslant -1$: least when $x = 3$
4 a) $x < 0$ and $x > 1$ **b)** never **c)** $-1.618 \leqslant x \leqslant 0.618$ **d)** always
5 a) $(\frac{1}{2}, \frac{5}{2}), (\frac{5}{2}, \frac{1}{2})$ **b)** $(1.28, 3.28), (-0.78, 1.22)$
6 a) $m < -7.46, m > -0.54$ **b)** $m = -7.46, -0.54$
　　c) $-7.46 < m < -0.54$
8 2.83 s   **9** 80 m, 20 m **10** 26
11 a) $n > 150$ **b)** 180 **c)** $70, 80$ **d)** 75

Exercise 2.3a

1 a) $2x^3 + 5x^2 - 7x + 2$ **b)** $x^3 + 1$ **c)** $x^4 - 1$
　　d) $4x^4 - 4x^3 - 13x^2 + 23x - 12$
2 a) $2x^2 - x - 2; +9$ **b)** $2x^2 - 2x - 1; +4$
　　c) $x^4 - x^3 + x^2 - x + 1; -1$ **d)** $x^3 + x^2 + x + 1; 0$
　　e) $x^2 + x + 1; 2x + 4$

Exercise 2.3b

1 **a)** 17 **b)** 7 **c)** 7 **d)** -9 **e)** 0 **f)** 5
2 **a)** $(x-1)(x-2)(x-3); 1, 2, 3$ **b)** $x^2(x-1); 0, 1$
   c) $(x-1)(x+1)^2; \pm 1$ **d)** $(x-2)(x+2)(2x-1); \pm 2, \frac{1}{2}$
   e) $-x^2(x-2)(x+2); 0, \pm 2$
   f) $(x+1)(x-1)(x-3)(x-5); -1, 1, 3, 5$
3 **b)** $2, -10, 2x+5$ **4** $-5, -3; (x+1)^2(x-3)$
5 **a)** $(x-a)(x^2+ax+a^2)$ **b)** $(x+a)(x^2-ax+a^2)$

Exercise 2.3c

1 **a)** $-\frac{2}{3}, \frac{5}{3}$ **b)** $-4, -3$ **c)** $-p, q$ **d)** $2, -3$
2 **a)** $x^2-4x+5=0$ **b)** $x^2+5x+4=0$ **c)** $2x^2+x-7=0$
   d) $px^2+p^2x+1=0$
3 **a)** $\frac{34}{9}, -\frac{2}{5}$ **b)** $9x^2-34x+25=0$ **c)** $5x^2+2x-3=0$
   d) $15x^2+34x+15=0$
4 **a)** $6, 11, 6$ **b)** $0, -1, 0$ **c)** $-1, -1, 1$ **d)** $\frac{1}{2}, -4, -2$
5 **a)** $9, 26, 24$ **b)** $x^3-6x^2+11x-6=0$

Exercise 2.4a

1 **a)** $\frac{1}{3}$ **b)** $\frac{1}{16}$ **c)** $\frac{8}{27}$ **d)** $-\frac{1}{5}$ **e)** 64 **f)** $\frac{1}{64}$ **g)** 9
   h) 1   **i)** 1   **j)** 0
2 **a)** 729 **b)** $\frac{1}{4}$ **c)** 64   **d)** $1/100000$ **e)** $1/125$
3 **a)** 6561 **b)** $\frac{1}{64}$ **c)** 1000000 **d)** 16   **e)** 1

Exercise 2.4b

1 **a)** 5 **b)** 2 **c)** 3 **d)** 4 **e)** 27 **f)** 32 **g)** 1 **h** $\frac{1}{2}$ **i)** $\frac{1}{2}$ **j** $\frac{1}{27}$ **k)** $\frac{1}{32}$
   l) $\frac{1}{8}$
2 **a)** $\frac{1}{5}$ **b)** $\sqrt{2}$

Exercise 2.4c

1 **a)** 6.3   **b)** 20.0 **c)** 79.4
2 **a)** 0.70 **b)** 1.43 **c)** 2.08
4 **a)** 1.58 **b)** 1.58 **c)** 1.58

Exercise 2.4d

1 **a)** 10.25% **b)** 15.76% **c)** 21.55%
2 **a)** 121   **b)** 133   **c)** 146   **d)** 195   **3** $23.7°C$
4 **a)** 7.84% **b)** 11.53%

Exercise 2.5a

1 a) $10^5, 5$ **b)** $10^{3/2}, 1.5$ **c)** $10^0, 0$ **d)** $10^{-2}, -2$ **e)** $10^{1/3}, \frac{1}{3}$
f) $10^{-1/2}, -0.5$
2 a) 1.30103 **b)** 2.47712 **c)** $-1.52288\,(\bar{2}.47712)$
d) $-2.69897\,(\bar{3}.30103)$ **e)** 0.77815 **f)** 0.60206
g) $-0.30103\,(\bar{1}.69897)$ **h)** 0.17609
3 a) 6.89 **b)** 2.96 **c)** 921 **d)** 118 **e)** 6.42 **f)** $5.19(5)$ **g)** 0.0125
h) 0.0697 **i)** 37.3 **j)** 0.244

Exercise 2.5b

1 a) 0.602060 **b)** 0.698970 **c)** 0.903090 **d)** 0.954242 **e)** 1
f) 1.079181 **g)** 1.176091 **h)** 1.477121 **i)** 2.255272
j) $-0.602060\,(\bar{1}.397940)$ **k)** 0.176091
l) $-0.176091\,(\bar{1}.823909)$ **m)** $0.690105(5)$ **n)** 0.401373
o) $-238560(5)\,\{\bar{1}.761439(5)\}$
2 a) 4.19 **b)** 19.9 **c)** 8.96 **d)** 0.477
3 a) 1.404 **b)** 1.431 **c)** 5.321 **d)** 0.631
4 a) 1.48 **b)** 6.21
5 $\sqrt{3}$ **6 a)** £163   **b)** £259
7 a) 14.2 years   **b)** 7.26 years   **8** 20.5
9 a) £377   **b)** 4.26 years   **10** 2880 years
11 18.3%   **12** 13%

Exercise 2.5d

1   **b)** $2.72, 1.04$ **c)** 4.3% **d)** 14.5×10^6 **e)** 16.5
2   **b)** $2, 0.5$ **c)** $4.47\,s$ **d)** $0.25\,m$
3 $P = 2000(\frac{4}{5})^n$ **4** $2.0, 0.5$ **5** $y = 2.4x^{1.6}$

Exercise 2.6a

$f(x)$ is continuous
1   everywhere   **2**   everywhere
3 $x \neq 0$ **4**   everywhere
5 $x \geqslant 0$ **6** $x \neq 0$
7 $x \notin \mathbb{Z}$ **8**   everywhere
9   nowhere   **10** $x = 0$

Exercise 2.6b

1   **a)** 0 **b)** 0 **c)** 0, continuous
2   **a)** 0 **b)** 0 **c)** 0, discontinuous

3   **a)** 1 **b)** 0 **c)** does not exist, discontinuous
4   **a)** 0 **b)** 1 **c)** does not exist, discontinuous
5   **a)** 1 **b)** 1 **c)** 1, continuous

Exercise 2.6c

1   $-\infty, -\infty$ 2 $+\infty$, does not exist   3 $+\infty, 0$
4   does not exist $(-\infty$ and $+\infty)$, $+1$
5   neither exists $(-\infty$ and $+\infty)$
6   $+\infty$,   0   7 $+\infty$, does not exist

Exercise 2.7

1   $y \neq 0; x = 1, y = 0$ 2 $y \neq 0; x = -1, y = 0$
3   $y \geqslant 2, \leqslant -2; x = 0, y = x$ 4   all $y; x = 0, y = x$
5   $y \leqslant -4, > 0; x = 1, 2; y = 0$ 6 $0 \leqslant y < 1; y = 1$
7   $y \leqslant 0, > 1; x = \pm 2$
8   $y > 5.83, < 0.17; x = -1, -2; y = 0$
9   $y \leqslant 0, \geqslant 8; x = 1$ 10 $y \geqslant -\frac{1}{4}; x = 1, y = 0$

Miscellaneous problems

3   $-x + 2$
4   $\{P(a) - P(b)\}x + \{aP(b) - bP(a)\}$
5   2.98
6   $-3, \frac{3}{2}, -\frac{1}{3}$
7   $\frac{1}{2}(\sqrt{5} - 1) = 0.618$
8   **a)** 1, 1, 1 **b)** 4, 16, 16 **c)** 27, 19 683, 7 625 597 484 987
9   **a)** 19% **b)** 27.1% **c)** 34.4%, 6.57 min
10   1.06, 323, 384
12   **a)** 2.44 **b)** 2.59 **c)** 2.70 **d)** 2.72
    Approximately 2.72 (or, more precisely 2.718 282 $=$ e)

Chapter 3

Exercise 3.1a

3   80 m, 4 s   4   8 s   6   30, 15, $-15, 0\,\mathrm{m\,s^{-1}}$

Exercise 3.1b

1   **a)** 19.5, 19.95, $20\,\mathrm{m\,s^{-1}}$ **b)** $10 - 5h, 10\,\mathrm{m\,s^{-1}}$ **c)** $-30\,\mathrm{m\,s^{-1}}$
    d) $40 - 10t\,\mathrm{m\,s^{-1}}$

2 a) $20.5, 20.05, 20 \text{ m s}^{-1}$

3 a) -0.09 m **b)** $29.91, -0.09 \text{ m s}^{-1}$ **c)** $10t - \frac{1}{100}t^2 \text{ m s}^{-1}$

Exercise 3.2

1 a) 6 **b)** 0 **c)** -7 **d)** 24 **e)** 0 **f)** $-\frac{3}{4}$ **g)** -2

2 a) $2x; x > 0, x < 0, x = 0$

  b) $6x^2; |x| > 0$, never, $x = 0$

  c) $8x + 1; x > -\frac{1}{8}, x < -\frac{1}{8}, x = -\frac{1}{8}$

  d) $12x^3 + 12; x > -1, x < -1, x = -1$

  e) 0; never, never, always

  f) $-\dfrac{3}{x^2}(x \neq 0)$; never, always, never

  g) $-\dfrac{2}{x^3}(x \neq 0); x < 0, x > 0$, never

Exercise 3.3a

1 $5x^4$ **2** $8x^7$ **3** $\frac{4}{3}x^{1/3}$ **4** $\frac{1}{3}x^{-2/3}$

5 $\frac{3}{2}\sqrt{x}$ **6** $-\dfrac{3}{x^4}$ **7** $-\dfrac{4}{x^5}$ **8** $-\frac{1}{2}x^{-3/2}$

9 $-\frac{3}{2}x^{-5/2}$ **10** $-\frac{1}{3}x^{-4/3}$

Exercise 3.3b

1 $4x^3 + 3x^2$ **2** $5x^4 - 2x$ **3** $6x^5 + 2x$ **4** $4x^3 - 1$

5 $12x^3 + 10x$ **6** $12x^5 - 12x^2$ **7** $6x + 12x^2$ **8** $12x - 3$

9 $2 - \dfrac{1}{x^2}$ **10** $6x - \dfrac{4}{x^3}$ **11** $10x + \dfrac{2}{x^2}$ **12** $-\dfrac{4}{x^3} + \dfrac{3}{x^2}$

13 $\dfrac{1}{\sqrt{x}}$ **14** $-\dfrac{2}{x\sqrt{x}}$ **15** $-\frac{3}{2}x^{-3/2} + x^{-\frac{1}{2}}$

16 $-x^{-3/2} + \frac{3}{2}x^{\frac{1}{2}}$

Exercise 3.4a

1 a) $15x^2, 30x$ **b)** $4x^3 + 1, 12x^2$ **c)** $4x^3 - 2x, 12x^2 - 2$ **d)** $5x^9, 45x^8$

2 a) $-6x^{-7}, 42x^{-8}$ **b)** $-42x^{-4}, 168x^{-5}$ **c)** $-2x^{-3}, 6x^{-4}$

  d) $-15x^{-4}, 60x^{-5}$

3 a) $\frac{1}{3}x^{-2/3}, -\frac{2}{9}x^{-5/3}$ **b)** $\frac{1}{2}x^{-1/2}, -\frac{1}{4}x^{-3/2}$ **c)** $3x^{1/2}, \frac{3}{2}x^{-1/2}$

  d) $-\frac{1}{2}x^{-3/2}, \frac{3}{4}x^{-5/2}$

4 a) $6x^5 + 6x^2, 30x^4 + 12x$ **b)** $2 - \dfrac{3}{x^2}, \dfrac{6}{x^3}$

c) $\frac{1}{4}x^{-1/2} - \frac{1}{4}x^{-3/2}, -\frac{3}{8}x^{-3/2} + \frac{3}{8}x^{-5/2}$ **d)** $1 + x^{-1/2}, -\frac{1}{2}x^{-3/2}$

Exercise 3.4b

1 **a)** $2x - 2, 2$ **b)** $x > 1; x < 1; x = 1$ **c)** always; never; never

2 **a)** $3x^2 - 3; 6x$ **b)** $|x| > 1; |x| < 1; x = \pm 1$ **c)** $x > 0; x < 0; x = 0$

3 **a)** $6x - 6x^2, -6 - 12x$ **b)** $0 < x < 1; x < 0, x > 1; x = 0, 1$
    c) $x < \frac{1}{2}; x > \frac{1}{2}; x = \frac{1}{2}$

4 **a)** $4x^3 - 16x, 12x^2 - 16$
    b) $-2 < x < 0; x > 2; x < -2, 0 < x < 2; x = 0, \pm 2$
    c) $|x| > \dfrac{2}{\sqrt{3}}; |x| < \dfrac{2}{\sqrt{3}}; x = \pm \dfrac{2}{\sqrt{3}}$

5 **a)** $3x^2 - 6x - 9, 6x - 6$ **b)** $x < -1, x > 3; -1 < x < 3; x = -1, 3$
    c) $x > 1; x < 1; x = 1$

6 **a)** $1 - \dfrac{1}{x^2}, \dfrac{2}{x^3}$ **b)** $|x| > 1; |x| < 1; x = \pm 1$ **c)** $x > 0; x < 0$; never

Exercise 3.5

1 **a)** $6x^2$ **b)** $-\dfrac{2}{x^2}$

2 **a)** $3x^2 + 6x, 6x + 6$ **b)** $4, 0$ **c)** $-\dfrac{2}{x^3}, \dfrac{6}{x^4}$ **d)** $-\frac{1}{2}x^{-3/2}, \frac{3}{4}x^{-5/2}$

3 **a)** $4x^3, 12x^2$ **b)** $6x, 6$ **c)** $x^{-1/2}, -\frac{1}{2}x^{-3/2}$ **d)** $-\dfrac{1}{x^2}, \dfrac{2}{x^3}$

4 **a)** $4, 76°, y = 4x - 4, x + 4y = 18$
    b) $1, 45°, y = x - 2, y = -x$
    c) $-1, 135°, x + y = 1, y = x - 1$
    d) $-\frac{1}{2}, 153.4°, x + 2y + 4 = 0, y = 2x + 3$

5 $8x + y + 2 = 0, x - y = \pm 16$

6 $9x - y = 16, (-4, -52)$

7 $250\,\text{m}, 76°$

8 **a)** $9000\,\text{m}$ **b)** $18.4°, 0°, -18.4°$

9 **a)** $(2, 4)$ **b)** $40.6°$

Exercise 3.6

1 $35\,\text{m s}^{-1}, 20\,\text{m s}^{-2}$ **2** $5\,\text{m s}^{-1}, -1\,\text{m s}^{-2}, 32\,\text{m}$
3 **a)** $30\,\text{m s}^{-1}, 42\,\text{m s}^{-2}$ **b)** $6.75\,\text{m}$
4 $4\,\text{s}, 11\,\text{m s}^{-2}$ **5** $\pm 12\,\text{m s}^{-2}$
7 **a)** $180\,\text{m}$ **b)** $15\,\text{s}$ **c)** $24 - \frac{8}{5}t, -\frac{8}{5}$

Exercise 3.7

1 **a)** $\frac{1}{2}x^4 + c$ **b)** $x - \frac{1}{2}x^2 + c$ **c)** $x^3 + 2x + 4$ **d)** $-\dfrac{1}{x} + \dfrac{3}{2}$

2 **a)** $y = 2x^2 - 5x + c$ **b)** $y = 2x^2 - 5x + 4$ **c)** $y = 3x - 1$
  d) $y = -\frac{1}{2}x^2 + \frac{3}{2}$
3 **a)** $5t^2 + c$ **b)** $2t^4 - 2t^3 + c$ **c)** $t^3 + 2t^2 + 2$ **d)** $t - t^2 + 3$
4 **a)** $3t^2 - 8t + 6, t^3 - 4t^2 + 6t + 4$
  b) $-10t + 12, -5t^2 + 12t - 2$
5 $6.75\,\mathrm{m}, 4\,\mathrm{m\,s^{-1}}$ 6 $3\,\mathrm{s}, 9\,\mathrm{m\,s^{-2}}, 162\,\mathrm{m}$
7 **a)** $1.6\,\mathrm{h}$ **b)** $102.4\,\mathrm{km}$ **c)** $96\,\mathrm{km\,h^{-1}}$
8 **a)** $1 - \frac{1}{9}t, \frac{1}{2}t^2 - \frac{1}{54}t^3 + 6$ **b)** $18\,\mathrm{s}, 27\,\mathrm{s}$
9 **a)** $12t - 48$ **b)** $2t^3 - 24t^2 + 42t$
  $1\,\mathrm{s}, 7\,\mathrm{s}, 2.1\,\mathrm{s}$
10 $100\,\mathrm{m\,s^{-1}}, 1333\,\mathrm{m}$ 11 $4.5\,\mathrm{m}, 5\,\mathrm{s}$
12 **a)** $5\,\mathrm{m\,s^{-2}}$ **b)** $15\,\mathrm{s}$ **c)** $56.25\,\mathrm{m\,s^{-1}}$

Exercise 3.8

1 **a)** $f(1) = -9, \text{minimum}$ **b)** $f(1) = 4, \text{maximum}$
  b) $f(-1) = 2, \text{maximum}; f(1) = -2, \text{minimum}$ **c)** none
2 **a)** $(2, 4), \text{minimum}; (-2, -4), \text{maximum}; \text{no p.i.}$
  b) $(1, 3), \text{minimum}; (-2^{\frac{1}{3}}, 0), \text{p.i.}$
  c) $(0, 0), \text{maximum}; (1, -1), \text{minimum}; (\frac{3}{4}, -\frac{81}{128}), \text{p.i.}$
  d) $(0, 0), \text{minimum}; (\pm 1, 2), \text{maximum}; (\pm\sqrt{\frac{3}{5}}, \frac{162}{125}), \text{p.i.}$
3 **a)** $(1, -1), \text{minimum}; (-2, 26), \text{maximum}$
  b) $(2, 1), \text{minimum}; (-1, 28), \text{maximum}$ **c)** $+1, -3$
4 **a)** $y = x^3 - x^2 - x + 1$ **b)** $(-\frac{1}{3}, \frac{32}{27}), \text{maximum}; (1, 0), \text{minimum}$
5 $800\,\mathrm{m}^2$ 6 $128\,\mathrm{cm}^3$
7 $x^2 + \dfrac{16}{x}, 2\,\mathrm{m}$ 8 $4 \times 12 \times 6\,(\mathrm{cm})$
9 $6\frac{2}{3} \times 10 \times 4\,(\mathrm{cm})$ 10 $\frac{2}{27}\mathrm{m}^3$
11 £2400, $20 \times 20 \times 5\,(\mathrm{m})$ 12 $84\,\mathrm{cm}, 66\,\mathrm{cm}$
13 $2a^2, \sqrt{3}a^2$ 14 **a)** £$5\frac{13}{16}\mathrm{min}^{-1}$ **b)** £5
15 $\dfrac{4a}{25}, \dfrac{3a}{25}$ 16 $\dfrac{1}{\sqrt{3}} = 0.577$

Exercise 3.9a

1 **a)** $24x(3x^2 + 1)^3$ **b)** $180x(9x^2 + 4)^9$ **c)** $-12x(1 - x^2)^5$
  d) $5(2x - 1)(x^2 - x)^4$ **e)** $-6x(3x^2 - 1)^{-2}$ **f)** $-\dfrac{2x}{(x^2 - 1)^2}$
  g) $-\dfrac{6x}{(x^2 + 2)^4}$ **h)** $\dfrac{x}{\sqrt{(1 + x^2)}}$ **i)** $\dfrac{2x}{3}(x^2 - 1)^{-2/3}$ **j)** $\frac{1}{2}(1 - x)^{-3/2}$
  k) $\dfrac{5}{\sqrt{x}}(\sqrt{x} + 1)^9$ **l)** $2anx(ax^2 + b)^{n-1}$
2 **a)** $12t^2(t^3 + 1)^3$ **b)** $-30t(1 - 3t^2)^4$ **c)** $-\dfrac{1}{2\sqrt{(1 - t)}}$ **d)** $-\dfrac{t}{(t^2 + 1)^{3/2}}$

4 **a)** $x + 4y - 3 = 0, 8x - 2y - 7 = 0$
b) $3x - 2y + 1 = 0, 2x + 3y - 8 = 0$

Exercise 3.9b

1 **a)** $(1 + 6x)(1 + x)^4$ **b)** $2x(9x + 1)(3x + 1)^3$

c) $(5x + 7)(x + 1)(x + 2)^2$ **d)** $(32x^3 + 1)(2x^3 + 1)^4$ **e)** $\dfrac{2}{(x + 1)^2}$

f) $\dfrac{1 - 2x^2}{(2x^2 + 1)^2}$ **g)** $-\dfrac{4x}{(1 + x^2)^2}$ **h)** $\dfrac{1 - 3x}{(3x + 1)^3}$ **i)** $\dfrac{1}{\sqrt{x}(1 - \sqrt{x})^2}$

j) $-\dfrac{1}{(1 + x)\sqrt{(1 - x^2)}}$

2 **a)** maximum at $(0, 0)$, minimum at $(2, 4)$
b) maximum at $(\sqrt{3}, -\frac{3}{2}\sqrt{3})$, minimum at $(-\sqrt{3}, \frac{3}{2}\sqrt{3})$
3 **a)** $x + y = 4, x - y = 0$
b) $5x - 2y = 2, 2x + 5y = 24$

Exercise 3.10

1 **a)** $\dfrac{x}{y}, 3, y = 3x - 8, x + 3y = 6$

b) $-\dfrac{x + 2y}{2x + y}, -1, x + y = 2, y = x$

c) $-\dfrac{x^2}{y^2}, -\frac{1}{4}, x + 4y = 7, y = 4x + 6$

2 **a)** $\dfrac{1}{3t}, x - 3ty + t^3 = 0, 3tx + y = 6t^4 + t^2$

b) $-\dfrac{1}{4t^2}, x + 4t^2y - 8t = 0, 4t^3x - ty = 16t^4 - 1$

c) $\dfrac{2t}{3t^2 - 1}, 2tx - (3t^2 - 1)y + t^4 + t^2 = 0,$
$(3t^2 - 1)x + 2ty - 3t^5 + 2t^3 - t = 0$
3 $t = 4, x = 40, 76°$

4 **a)** $y^2 = x^3 + 3x^2$ **d)** $\dfrac{3(t^2 - 1)}{2r}, (-2, \pm 2)$

Exercise 3.11a

1 **a)** $6\,\text{m}^2/\text{s}$ **b)** $7.5\,\text{m}^3/\text{s}$
2 $1.25\,\text{cm/s}$
3 **a)** $3.77\,\text{m}^2\,\text{s}^{-1}$ **b)** $0.53\,\text{m}\,\text{s}^{-1}$
4 $0.089\,\text{m}\,\text{s}^{-1}$ **5** $\dfrac{2x}{3}, 2\,\text{m}\,\text{s}^{-1}$

6 $\frac{4}{3}$ m s^{-1} **7** 0.0255 cm s^{-1}
8 0.5 atm s^{-1}

Exercise 3.11b

1 a) 0.048 **b)** -0.0000312 **c)** 0.00025
2 a) 8120 **b)** 3.007   **c)** 3.987
3 $2\pi = 6.28$ m
4 $\delta V = \pi x(2r - x)\,\delta x$, 1.06 cm
5 $2p\%, 3p\%$
6 $\frac{1}{2}$

Exercise 3.12

1 80 m, 50 m s^{-1} **2** 0.2 m s^{-2}, 2.88 km
3 4 m s^{-2}, 5 s   **4** 1 m s^{-2}, 22 m s^{-1}
5 $4\frac{3}{4}$ min   **6** 1.45 m s^{-2}, 417 m
7 $\frac{5}{12}$ m s^{-2}, $40\frac{5}{6}$ m   **8** 7.1 s, 5.2 s

Miscellaneous problems

2 $(0,0), (\pm\sqrt{3}a, -9a^4)$
3 $\dfrac{h(a - x)}{a}$ **4** $h = r, 2r, 2r + a$ **6** 1.47
7 $2 \times 3^{-3/4}$ **9** 1

Chapter 4

Exercise 4.1

1 35   **2** 250   **3** 22   **4** 3.75   **5** 10.5   **6** 4

Exercise 4.2

1 a) $\frac{1}{3}x^3 + c$ **b)** $x^3 - 2x^2 + 2x + c$ **c)** $\frac{1}{3}x^3 - \frac{1}{2}x^4 + c$
  d) $\frac{1}{4}x^4 - \frac{2}{3}x^3 + c$
2 a) $\frac{1}{3}t^3 - t^2 + 3t + c$ **b)** $\frac{1}{3}t^3 + \frac{5}{2}t^2 + c$ **c)** $\frac{1}{3}t^3 + \frac{3}{2}t^2 + 2t + c$
  d) $\frac{4}{3}t^3 + 2t^2 + t + c$
3 a) $\frac{1}{6}u^6 + \frac{1}{2}u^2 + c$ **b)** $\frac{1}{5}u^5 + \frac{2}{3}u^3 + u + c$ **c)** $\frac{1}{3}u^6 - \frac{2}{3}u^3 + c$
  d) $\frac{1}{9}u^9 + \frac{4}{5}u^5 + 4u + c$
4 a) $\frac{2}{7}x^{7/2} + c$ **b)** $\frac{3}{4}x^{4/3} + c$ **c)** $\frac{2}{5}x^{5/2} + \frac{2}{3}x^{3/2} + c$
  d) $\frac{1}{2}x^2 + 4x^{3/2} + 9x + c$

5 a) $-\dfrac{3}{x} + c$ **b)** $-\dfrac{1}{2x^2} - \dfrac{1}{x^3} + c$ **c)** $\frac{2}{3}x^{3/2} + 2x^{1/2} + c$

  d) $-\dfrac{2}{\sqrt{x}} - \dfrac{5}{x} + c$

6 a) $\frac{1}{2}ax^2 + bx + c$ **b)** $\frac{1}{3}ax^3 + \frac{1}{2}bx^2 + cx + d$ **c)** $\frac{2}{3}ax^{3/2} + bx + c$
  d) $\frac{2}{3}ax^{3/2} + 2bx^{1/2} + c$

7 $\dfrac{x^{n+1}}{n+1} + c$ $(n \neq -1)$

8 $\dfrac{n}{n+1}u^{1/n+1} + c$ $(n \neq -1)$

Exercise 4.3a

1 a) $8\frac{2}{3}$ **b)** 44 **c)** $6\frac{1}{5}$ **d)** $2\frac{1}{4}$
2 a) $4\frac{2}{3}$ **b)** $11\frac{1}{4}$ **c)** $\frac{1}{2}$ **d)** $\frac{1}{3}$
3 a) $3\frac{3}{4}$ **b)** $\frac{1}{6}$ **c)** $3\frac{5}{6}$ **d)** $1\frac{1}{3}$
4 a) $10\frac{2}{3}$ **b)** $1\frac{1}{3}$ **c)** $2\frac{2}{3}$ **d)** $\frac{4}{3}$ **e)** $\frac{1}{3}$

Exercise 4.3b

1 a) $-10\frac{2}{3}$ **b)** $5\frac{1}{3}$ **2 a)** 0 **b)** 8
3 a) 45 **b)** -45 **c)** 0 **d)** -80
4 a) $2\frac{14}{15}$ **b)** $3\frac{1}{6}$

Exercise 4.4

1 a) $259\,\text{m}^2$ **b)** $256\,\text{m}^2$
2 a) $116\,\text{m}$ **b)** $117\,\text{m}$
3 a) $2.96\,\text{m s}^{-1}$ **b)** $2.85\,\text{m s}^{-1}$
4 a) $0.7083\,(2.2\%)$ **b)** $0.6945\,(0.2\%)$
5 a) $0.6938\,(0.1\%)$ **b)** $0.6931\,(0.0\%)$
6 a) $3.104\,(1.2\%)$ **b)** $3.127\,(0.5\%)$
7 3.142
8 a) 0.879 **b)** 0.879

Exercise 4.5

1 a) $\dfrac{7\pi}{3}$ **b)** $\dfrac{28\pi}{15}$ **c)** $\dfrac{\pi}{6}$ **d)** 8π **e)** $\dfrac{2\pi}{15}$ **f)** $\dfrac{\pi}{30}$
2 a) 3 **b)** $\frac{14}{3}$ **c)** 12 **d)** $\frac{56}{3}$ **e)** $\frac{8}{3}$
3 a) $\dfrac{28\pi}{3}$ **b)** $\dfrac{15\pi}{2}$ **c)** $\dfrac{96\pi}{5}$ **d)** 6π **e)** $\dfrac{32\pi}{5}$
4 $\frac{1}{3}\pi r^2 h$ **5** $\frac{4}{3}\pi r^3$ **6** $\frac{128}{3}, 128\pi$

7 $(0,0), (2,4), \dfrac{32\pi}{5}$ **8** $\dfrac{5}{3}, \dfrac{20\pi}{7}$ **9** $\dfrac{2}{3}\pi a^3$

10 $\dfrac{16}{3}, \dfrac{96\pi}{5}$ **11** $162\,\text{m}^3$ **12** $\dfrac{256\pi}{3} = 268\,\text{cm}^3$

Exercise 4.6

1 a) 1.5   b) 39   c) 1.22   d) $\frac{1}{3}$
2 a) $2\,\text{m s}^{-1}, 3\,\text{m s}^{-2}$ b) $2\,\text{m s}^{-1}, 0$
3 a) $3142\,\text{kg}$ b) $750\,\text{kg m}^{-3}$
4 7.5 cm

Exercise 4.7

1 a) $\frac{1}{6}(x + 1)^6 + c$ b) $\dfrac{1}{2(1 - 2x)} + c$ c) $-\frac{2}{3}(2 - x)^{3/2} + c$
   d) $\frac{1}{5}(2x + 1)^{5/2} + c$ e) $\frac{3}{10}(2x - 3)(1 + x)^{2/3} + c$
2 a) $\frac{1}{10}(x^2 + 1)^5 + c$ b) $\frac{1}{30}(2x^3 + 3)^5 + c$ c) $-\frac{1}{3}(1 - x^2)^{3/2} + c$
   d) $2\sqrt{(x + x^2)} + c$ e) $\frac{3}{4}(x^2 - 1)^{2/3} + c$
3 a) $\frac{1}{5}$ b) $\frac{1}{35}$ c) $\frac{1}{3}(10\sqrt{10} - 1) = 10.2$ d) $\frac{2}{3}\sqrt{7} = 1.76$
   e) $\frac{4}{3}(3\sqrt{3} - 2\sqrt{2}) = 3.157$

Miscellaneous problems

1 between 13.8 and 14.8   **2** between 666 666 666 and 666 667 667

3 a) 0   b) 4   **4** $\dfrac{8}{15}, \dfrac{\pi}{12}$

6 $(4, \pm 4), \dfrac{32\pi}{5}$ **7** $\frac{1}{6}\pi h^3$

8 $\dfrac{4\pi a b^2}{3}$ **9** approximately 400 000 tonne

Chapter 5

Exercise 5.1a

1 a) 0.57, 0.82, 0.70   b) 0.22, -0.97, -0.23
   c) -0.56, -0.83, 0.67   d) -0.83, 0.56, -1.48
   e) 0.64, 0.77, 0.84   f) -0.53, 0.85, -0.62
3 a) $\sin \theta$, $-\cos \theta$, $-\tan \theta$ b) $-\sin \theta$, $-\cos \theta$, $+\tan \theta$

   c) $-\cos \theta$, $\sin \theta$, $-\dfrac{1}{\tan \theta}$

Exercise 5.1b

1 a) $\dfrac{\sqrt{3}}{2}, -\dfrac{1}{2}, -\sqrt{3}$

b) $-\dfrac{1}{\sqrt{2}}, -\dfrac{1}{\sqrt{2}}, +1$

c) $-\dfrac{1}{2}, \dfrac{\sqrt{3}}{2}, -\dfrac{1}{\sqrt{3}}$

d) $-\dfrac{\sqrt{3}}{2}, \dfrac{1}{2}, -\sqrt{3}$

e) $-\dfrac{1}{2}, -\dfrac{\sqrt{3}}{2}, \dfrac{1}{\sqrt{3}}$

f) $\dfrac{1}{\sqrt{2}}, \dfrac{1}{\sqrt{2}}, 1$

Exercise 5.2

1 a) $21.7°, 158.3°$ **b)** $90°$ **c)** $227.7°, 312.3°$
2 a) $63.3°, 296.7°$ **b)** $127.6°, 232.4°$ **c)** $180°$
3 a) $135°, 315°$ **b)** $0°, 180°, 360°$ **c)** $69.8°, 249.8°$
4 a) $0°, 180°, 270°, 360°$ **b)** $0°, 60°, 300°, 360°$ **c)** $30°, 150°, 210°, 330°$
5 a) $136.8°, 223.2°$ **b)** $243.1°$ **c)** $12.2°, 102.2°, 192.2°, 282.2°$

Exercise 5.3

2 a) $\sin\theta$ **b)** $\tan\theta$ **c)** $\sec\theta$ **d)** $\operatorname{cosec}\theta$
3 a) $\cot^2\theta$ **b)** 2 **c)** $\sec\theta - 1$ **d)** $\sin\theta\cos\theta$ **e)** $\cot^2\theta$ **f)** $\sec^2\theta$
   g) $\sec^2\theta$ **h)** $\sin\theta + \cos\theta$
4 a) $60°, 300°$ **b)** $135°, 315°$ **c)** $0°, 120°, 240°, 360°$
   d) $0°, 45°, 180°, 225°, 360°$ **e)** $41.8°, 138.2°$ **f)** $60°, 300°$
5 a) $\dfrac{5}{3}, \dfrac{4}{3}$

Exercise 5.4

1 a) $b = 2.54, c = 4.77, C = 70°$
   b) $a = 6.71, c = 4.22, C = 37.2°$
2 a) $A = 41.4°, B = 55.8°, C = 82.8°$
   b) $A = 26.7°, B = 36.5°, C = 116.7°$
3 a) $a = 3.09, B = 48°, C = 82°$
   b) $c = 6.58, A = 44°, B = 32.3°$
4 $A = 56.4°, C = 93.6°, c = 5.99$; or $A = 123.6°, C = 26.4°, c = 2.67$
5 $8.6°, 154\,\text{nm}$
6 a) $60°$ **b)** $13\,\text{cm}$
8 a) $33.6°$ **b)** $4.92\,\text{cm}$

Exercise 5.5a

1 a) $\dfrac{\sqrt{3}+1}{2\sqrt{2}} = \dfrac{1}{4}(\sqrt{6} + \sqrt{2}); \dfrac{1 - \sqrt{3}}{2\sqrt{2}} = -\dfrac{1}{4}(\sqrt{6} - \sqrt{2});$

$$\frac{\sqrt{3}+1}{1-\sqrt{3}} = -2 - \sqrt{3}$$

b) $\dfrac{\sqrt{3}-1}{2\sqrt{2}} = \frac{1}{4}(\sqrt{6}-\sqrt{2}); \dfrac{\sqrt{3}+1}{2\sqrt{2}} = \frac{1}{4}(\sqrt{6}+\sqrt{2}); \dfrac{\sqrt{3}-1}{\sqrt{3}+1} = 2 - \sqrt{3}$

2 a) $\cos x, -\sin x, -\cot x$ **b)** $\sin x, -\cos x, -\tan x$

3 a) $\dfrac{1}{\sqrt{2}}(\cos x + \sin x); \dfrac{1}{\sqrt{2}}(\cos x - \sin x); \dfrac{1 + \tan x}{1 - \tan x}$

b) $\frac{1}{2}(\sqrt{3}\cos x - \sin x); \frac{1}{2}(\cos x + \sqrt{3}\sin x); \dfrac{\sqrt{3} - \tan x}{1 + \sqrt{3}\tan x}$

4 a) $\sin 60°$ **b)** $\cos 50°$ **c)** $\sin(2\alpha - \beta)$ **d)** $\cos(2\theta - 3\phi)$
e) $\tan(60° + \theta)$ **f)** $\tan(\theta - 45°)$

5 a) $2\sin A \cos B$ **b)** $-2\sin A \sin B$ **c)** $\cot B$ **d)** $\tan A - \tan B$
e) $\cot B - \tan A$ **f)** -1

7 $\dfrac{\tan A + \tan B + \tan C - \tan A \tan B \tan C}{1 - \tan B \tan C - \tan C \tan A - \tan A \tan B}$
a) $45°$

Exercise 5.5b

1 a) $\sqrt{2}\cos(x - 45°)$ **b)** $\sqrt{13}\cos(x - 33.7°)$ **c)** $5\cos(x - 126.9°)$
d) $\sqrt{10}\cos(x + 71.6°)$

2 a) $0°, 90°, 360°$ **b)** $139.8°, 287.6°$ **c)** $60.5°, 193.3°$ **d)** $57.6°, 159.2°$

Exercise 5.6

1 a) $\sin 30° = \frac{1}{2}$ **b)** $\cos 30° = \dfrac{\sqrt{3}}{2}$ **c)** $\cos 30° = \dfrac{\sqrt{3}}{2}$ **d)** $\cos 30° = \dfrac{\sqrt{3}}{2}$
e) 1 **f)** $\tan 30° = \dfrac{1}{\sqrt{3}}$

2 a) $\frac{24}{25}$ **b)** $\frac{7}{25}$ **c)** $\frac{24}{7}$

3 a) $\sqrt{\left(\dfrac{\sqrt{2}-1}{2\sqrt{2}}\right)}$ **b)** $\sqrt{\left(\dfrac{\sqrt{2}+1}{2\sqrt{2}}\right)}$ **c)** $\sqrt{2}-1$

4 a) $3\sin A - 4\sin^3 A$ **b)** $4\cos^3 A - 3\cos A$

5 a) $\tan\dfrac{A}{2}$ **b)** $\tan x$ **c)** $\cos 2\theta$ **d)** $\tan 2A$ **e)** $\tan\theta$ **f)** 2

6 a) $8\cos^4 A - 8\cos^2 A + 1$ **b)** $\dfrac{4\tan A - 4\tan^3 A}{1 - 6\tan^2 A + \tan^4 A}$

7 $\dfrac{1 + \sqrt{3} - \sqrt{6}}{\sqrt{3} + \sqrt{2} - 1}$

8 a) $30°, 90°, 150°, 270°$ **b)** $90°, 180°, 270°$ **c)** $105°, 165°, 285°, 345°$
d) $0°, 48.6°, 131.4°, 180°, 270°, 360°$

Exercise 5.7

1 **a)** $2\sin 4\theta \cos\theta$ **b)** $2\sin A\cos 3A$ **c)** $2\cos x\cos 60° = \cos x$
  d) $-\sqrt{2}\sin x$

2 **a)** $\cot\dfrac{A-B}{2}$ **b)** $\tan\dfrac{A+B}{2}$ **c)** $-\cot 2A$ **d)** $\cot\dfrac{A}{2}$

5 **a)** $0°, 30°, 60°, 120°, 150°, 180°$ **b)** $45°, 120°, 135°$ **c)** $0°, 90°, 180°$
  d) $0°, 10°, 50°, 90°, 130°, 170°, 180°$

Exercise 5.8a

1 **a)** $\dfrac{\pi}{18}$ **b)** $\dfrac{\pi}{5}$ **c)** $\dfrac{2\pi}{3}$ **d)** π **e)** $\dfrac{3\pi}{2}$ **f)** $\dfrac{3\pi}{4}$ **g)** $\dfrac{\pi}{10800}$ **h)** $\dfrac{\pi}{648000}$

2 **a)** $90°$ **b)** $270°$ **c)** $150°$ **d)** $120°$ **e)** $135°$ **f)** $15°$ **g)** $12°$ **h)** $18°$

3 **a)** $+1$ **b)** $\dfrac{1}{\sqrt{2}}$ **c)** $\sqrt{3}$ **d)** $\tfrac{1}{2}$ **e)** -1 **f)** $\dfrac{1}{\sqrt{2}}$ **g)** $-\tfrac{1}{2}$ **h)** 0 **i)** 0
  j) $-\tfrac{1}{2}$

4 **a)** $\cos x$ **b)** $-\cos x$ **c)** $-\cos x$ **d)** $-\sin x$ **e)** $\cot x$ **f)** $\tan x$

5 **a)** $\dfrac{\pi}{6}, \dfrac{5\pi}{6}$ **b)** $\dfrac{\pi}{4}, \dfrac{3\pi}{4}$ **c)** π **d)** $\dfrac{\pi}{3}, \dfrac{5\pi}{3}$ **e)** $\dfrac{\pi}{4}, \dfrac{5\pi}{4}$ **f)** $\dfrac{2\pi}{3}, \dfrac{5\pi}{3}$

6 **a)** 0.698 **b)** 0.807 **c)** 3.213 **d)** 6.886

7 **a)** $29.8°$ **b)** $99.1°$ **c)** $294.3°$ **d)** $0.6°$

8 **a)** $6\,\text{m}, 9\,\text{m}^2$ **b)** $1.571\,\text{m}, 1.571\,\text{m}^2$ **c)** $1.745\,\text{m}, 3.490\,\text{m}^2$

9 **a)** $20.9\,\text{cm}$ **b)** $43.3\,\text{cm}^2$ **c)** $253\,\text{cm}^2$

10 $3\tfrac{1}{3}\,\text{m}, 33\,\text{m}^2$

11 **a)** $\dfrac{2\pi r}{l}$ **b)** $\pi r l$ **12** 1.86

Exercise 5.8b

1 $\dfrac{n\pi}{2}$

2 $\dfrac{(6n\pm 1)\pi}{3}$

3 $\dfrac{(4n+1)\pi}{12}$

4 $\dfrac{(2n+1)\pi}{2}$

5 $\dfrac{(2n+1)\pi}{4}$

6 $\dfrac{2n\pi}{7}$

7 $\dfrac{(2n+1)\pi}{8}$

8 $\dfrac{(4n+1)\pi}{14}, \dfrac{(4n+1)\pi}{2}$

9 $2n\pi$ or $\dfrac{(2n+1)\pi}{2}$

10 $\dfrac{(4n-1)\pi}{4}$

Exercise 5.9a

2 a) $2\cos 2x$ **b)** $-3\sin 3x$ **c)** $2\pi\cos 2\pi x$ **d)** $-6\sin\dfrac{3x}{2}$

e) $2\sin x\cos x$ **f)** $-3\cos^2 x\sin x$ **g)** $2x\cos 2x + \sin 2x$

h) $-\dfrac{1}{x^2}\cos x - \dfrac{1}{x}\sin x$ **i)** $2\cos 2x$ **j)** $12\sin^2 4x\cos 4x$ **k)** $\dfrac{\cos x}{2\sqrt{\sin x}}$

l) $\dfrac{2x\sin x - x^2\cos x}{\sin^2 x}$

3 b) $\dfrac{x^2}{9} + \dfrac{y^2}{4} = 1$ **c)** $-\dfrac{2}{3}\cot\theta$

4 a) $\dfrac{x^2}{a^2} + \dfrac{y^2}{b^2} = 1$ **b)** $-\dfrac{b}{a}\cot\theta$

5 a) $3\cos t - \sin t\,\mathrm{m\,s}^{-1}; -3\sin t - \cos t\,\mathrm{m\,s}^{-2}$ **b)** $1.25\,\mathrm{s}$
   c) $-3.16\,\mathrm{m\,s}^{-2}, 5.16\,\mathrm{m}$

6 a) $a, 2a$ **b)** $\dfrac{3a}{2}$

7 $1.445, 3.142; 10\,\mathrm{cm}$

8 b) $\sqrt{5} = 2.236\,\mathrm{m}$ **c)** $-2.19\,\mathrm{m\,s}^{-1}$

9 a) $\dfrac{\pi}{3}$, maximum; $-\dfrac{\pi}{3}$, minimum **b)** $\dfrac{\pi}{2}, \dfrac{3\pi}{2}$, maxima; $\dfrac{7\pi}{6}, \dfrac{11\pi}{6}$, minima

10 a) 0.8616 **b)** 0.7108

11 $157\,\mathrm{s}$

12 $\frac{1}{2}(\sqrt{5} - 1) = 0.618$

Exercise 5.9b

1 a) $3\sec^2 3x$ **b)** $4\sec 4x\tan 4x$ **c)** $-\frac{1}{2}\csc^2\dfrac{x}{2}$

d) $-2\csc 2x\cot 2x$ **e)** $x\sec^2 x + \tan x$ **f)** $2\tan x\sec^2 x$

g) $3\sec^3 x\tan x$ **h)** $\dfrac{x\sec x\tan x - \sec x}{x^2}$ **i)** $-6\cot 3x\csc^2 3x$

j) $\dfrac{\sec^2 x}{2\sqrt{\tan x}}$

3 a) $\csc\theta; x^2 - y^2 = 1; \dfrac{x}{y}$ **b)** $\dfrac{b}{a}\sec\theta; \dfrac{x^2}{a^2} - \dfrac{y^2}{b^2} = 1; \dfrac{b^2 x}{a^2 y}$

4 $20\sec\theta - 5\tan\theta + 25, 14.5°, 44.4\,\mathrm{s}$

Exercise 5.10

1 a) $-\frac{1}{2}\cos 2x + c$ **b)** $2\sin\dfrac{x}{2} + c$ **c)** $-\frac{1}{3}$ **d)** 0.402

2 a) $\frac{1}{2}(x - \sin x\cos x) + c$ **b)** $\frac{1}{2}(x + \sin x) + c$ **c)** $\dfrac{\pi}{2}$ **d)** $\frac{1}{2}$

3 a) $-\frac{1}{6}\cos 3x - \frac{1}{2}\cos x + c$ **b)** $\frac{1}{6}\sin 3x + \frac{1}{2}\sin x + c$
c) $\frac{1}{2}\sin x - \frac{1}{6}\sin 3x + c$
4 a) $\frac{1}{2}\sin x^2 + c$ **b)** $-\frac{1}{3}\cos x^3 + c$ **c)** $\frac{1}{3}\sin^3 x + c$ **d)** $-\frac{1}{4}\cos^4 x + c$
5 a) $\tan x + c$ **b)** $\sec x + c$ **c)** $\tan x - x + c$ **d)** $-\cot x - x + c$
6 a) 4π **b)** $9\pi^2$ **7** $4 + 6\pi$

8 $\frac{1}{2}$ **9** $\frac{1}{4} + \frac{\pi}{8}$

10 a) π **b)** $\frac{1}{2}\pi - 1$ **11** $\frac{1}{6}(\pi + 3)$

Exercise 5.11a

1 a) $\frac{\pi}{6}, \frac{\pi}{3}, 0.464$ **b)** $-\frac{\pi}{2}, \pi, \frac{3\pi}{4}$ **c)** $0.775, 0.796, 0.611$

2 $\sin^{-1} x + \cos^{-1} x = \frac{\pi}{2}$

3 a) $\tan^{-1} 1 = \frac{\pi}{4}$ **b)** $\tan^{-1}\frac{11}{3}$ **c)** $\tan^{-1}\frac{x + y}{1 - xy}$

Exercise 5.11b

1 a) $\frac{\pi}{6}$ **b)** $\frac{\pi}{12}$ **c)** $\frac{\pi}{4}$
2 a) $\tan^{-1}(x + 1) + c$ **b)** $\sin^{-1}(x - 1) + c$
c) $\sin^{-1}\sqrt{x} - \sqrt{(x - x^2)} + c$
3 a) $\frac{1}{2}\sin^{-1}\frac{2x}{3} + c$ **b)** $\frac{1}{2}\tan^{-1} 2x + c$

c) $\frac{1}{2}\left\{a^2 \sin^{-1}\frac{x}{a} + x\sqrt{(a^2 - x^2)}\right\} + c$

4 a) $\sec^{-1} x + c$ **b)** $\frac{1}{2}\left(\tan^{-1} x + \frac{x}{1 + x^2}\right) + c$

Miscellaneous problems

1 $1\,\mathrm{m}^2$ **2 b)** $18°, 90°, 162°, 234°, 306°$
6 $273\,\mathrm{km}$

8 maxima at $\frac{\pi}{4}, \pi, \frac{7\pi}{4}$; minima at $0, \frac{3\pi}{4}, \frac{5\pi}{4}, 2\pi$; six inflections

9 a) $a(\theta - \sin\theta), a(1 - \cos\theta)$ **b)** $a(1 - \cos\theta), a\sin\theta, \cot\frac{\theta}{2}$

10 a) $-\tan t$ **b)** $x\sin t + y\cos t = \sin t \cos t$ **c)** $(\cos t, 0), (0, \sin t)$ **d)** 1

Chapter 6

Exercise 6.1a

1 **a)** $8, 9, 10$ **b)** $29, 37, 46$ **c)** $34, 55, 89$ **d)** $128, 256, 512$ **e)** $5, -3, 6$
   f) $40320, 362880, 3628800$
2 **a)** $1, 8, 27, 64$ **b)** $\frac{1}{2}, \frac{2}{3}, \frac{3}{4}, \frac{4}{5}$ **c)** $0, 2, 0, 2$ **d)** $-1, 2, -3, 4$
3 **a)** 7 **b)** 132 **c)** $n + 1$ **d)** $n(n - 1)$

Exercise 6.1b

1 **a)** 30 **b)** $1\frac{17}{60}$ **c)** 255 **d)** 0 **e)** $1\frac{63}{64}$ **f)** 20 **g)** 98 **h)** -10
2 **a)** $\displaystyle\sum_{r=1}^{25} r^2$ **b)** $\displaystyle\sum_{r=2}^{100} \frac{1}{r}$ **c)** $\displaystyle\sum_{r=1}^{10} r^3$ **d)** $\displaystyle\sum_{r=0}^{50} 3^r$ **e)** $\displaystyle\sum_{r=1}^{100} (-1)^{r-1} \times \frac{1}{r}$
   f) $\displaystyle\sum_{r=1}^{n} r(r + 1)$

Exercise 6.2

1 **a)** $4n - 2, 2n^2$ **b)** $\frac{1}{2}(3n - 1), \frac{1}{4}n(3n + 1)$ **c)** $5 - 2n, 4n - n^2$
2 **a)** $50, 2500$ **b)** $41, 2091$ **c)** $31, -155$
3 **a)** 1275 **b)** 1717 4 $220\,\mathrm{m}$
5 $1683, 3367$ 6 34
7 $\frac{3}{4}$ 8 $8, 4, 540$
9 **a)** $4n + 8$ **b)** 4

Exercise 6.3

1 **a)** $1458, 2186$ **b)** $\frac{1}{81}, 4\frac{40}{81}$ **c)** $-256, -170$ **d)** $x^9, \dfrac{x^{10} - 1}{x - 1}$
   e) $(-1)^{n-1} x^n, x\,\dfrac{1 - (-x)^n}{1 + x}$

2 **a)** 3 **b)** $5\frac{2}{5}$ **c)** 10 **d)** $\dfrac{1}{1 + x}$

3 **a)** $\dfrac{4}{33}$ **b)** $\dfrac{41}{333}$

4 £183 000 000 000 000 000 (approximately!)
5 £1260   6   £96.1 (5), £445
7 $7\mathrm{s}, 17\frac{6}{7}\,\mathrm{m}$ 8 $\frac{4}{5}, r = -\frac{1}{2}$, no

Exercise 6.4a

1 $\frac{1}{4}n(n + 1)(n + 2)(n + 3), 210$
2 $\dfrac{n}{n + 1}, \dfrac{4}{5}, 1$

Exercise 6.4b

1 $\frac{1}{3}n(4n^2 - 1), 166\,650$

2 **a)** $r(r + 2), \frac{1}{6}n(n + 1)(2n + 7)$
   b) $2r(2r - 1), \frac{1}{3}n(n + 1)(4n - 1)$
   c) $3r(3r - 2), \frac{3}{2}n(n + 1)(2n - 1)$
   d) $r^3, \frac{1}{4}n^2(n + 1)^2$

Exercise 6.6

1 **a)** $1, 3, 5, 7, 9, 11, 13$
   b) $2, 3, 5, 8, 12, 17, 23$
   c) $4, 13, 40, 121, 364, 1093, 3280$
   d) $-1, 4, -16, 64, -256, 1024, -4096$
   e) $2, 5, 2, 5, 2, 5, 2$
   f) $1, 10, -8, 28, -44, 100, -188$

2 **a)** $u_{n+1} = u_n + 4, \quad u_1 = 3$
   b) $u_{n+1} = u_n/10, \quad u_1 = 1000$
   c) $u_{n+1} = 2(u_n - 1), \quad u_1 = 3$
   d) $u_{n+1} = 0.9u_n, \quad u_1 = 1$
   e) $u_{n+1} = u_n^2, \quad u_1 = 2$
   f) $u_{n+1} = \dfrac{n + 1}{n + 2}u_n, \quad u_1 = \frac{1}{2}$

3 $1, 2, 6, 24, 120, 720, 5040; n!$

4 $5, 13, 35, 97, 275$

5 $2, -1, \frac{1}{2}, 2, -1, \frac{1}{2}, 2$

6 **a)** $3, 1, 0.6, 0.52, 0.504, 0.5008, 0.50016, 0.500032; \text{limit} = 0.5$
   b) $20, 4.4, 1.28, 0.656, 0.5312, 0.50624, 0.501248, 0.5002496$

7 **b)** $4, 0.5, -0.375, -0.59375, -0.6484375, -0.6621093, -0.6655273,$
   -0.6663818

8 3.162278

9 $\sqrt[3]{10}, 2.15$

Exercise 6.7

1 2.618 2 0.382
4 $3.236, -1.236$ 5 $1.83, 0.66, -2.49$
6 0.824 7 0.1078
8 $-1.17, -1.18$ 9 2.54
10 1.926 11 $X = \cos X, 0.73909$
12 $x_{n+1} = (32x_n + 60)^{\frac{1}{4}}; 3.534, 3.627$

Exercise 6.8

1 2.618, 0.382

3 1.879

4 1.166

5 0.64

6 2.055

7 0.66

9 $X = 0.41, N = 4$

Exercise 6.9

1 21, 34, 55, 89

2 $\frac{1}{2}(1 + \sqrt{5})$

Exercise 6.10

1 39916800

2 40320

3 720

4 a) 720 **b)** 1000

5 a) 60 **b)** 20 **c)** 10

6 a) 1024 **b)** 243 **c)** 32

7 4368

8 a) 220 **b)** 243

9 a) $\frac{1}{2}n(n-1)$ **b)** $\frac{1}{6}n(n-1)(n-2)$

    c) $\frac{1}{24}n(n-1)(n-2)(n-3)$

10 5775

11 a) $\binom{n}{r}$ **b)** 2^n

12 a) $\binom{n}{r}$ **b)** $\left(\dfrac{n+1}{r}\right)$

13 90, 6

14 a) 15120 **b)** 59049

15 a) 120 **b)** 60

Exercise 6.11

1 a) $a^4 + 4a^3b + 6a^2b^2 + 4ab^3 + b^4$

    b) $8x^3 + 12x^2y + 6xy^2 + y^3$

    c) $p^5 - 10p^4q + 40p^3q^2 - 80p^2q^3 + 80pq^4 - 32q^5$

    d) $x^4 - 4x^2 + 6 - \dfrac{4}{x^2} + \dfrac{1}{x^4}$

    e) $64x^6 - 192x^5y + 240x^4y^2 - 160x^3y^3 + 60x^2y^4 - 12xy^5 + y^6$

    f) $81x^8 - 216x^6y^2 + 216x^4y^4 - 96x^2y^6 + 16y^8$

2 a) $1 + 8x + 28x^2$ **b)** $1 + 12x + 60x^2$ **c)** $1 - \frac{7}{2}x + \frac{21}{4}x^2$

    d) $32 - 240x + 720x^2$

3 a) $28x^2$ **b)** $-96x^3$ **c)** $40x^4$ **d)** 12 **e)** 15

4 a) 1.105 **b)** 0.94148 **c)** 257.026 **d)** 0.99004

5 a) $1 - 5x + 5x^2$ **b)** $1 + 12x + 78x^2$ **c)** $16 - 32x - 8x^2$

Miscellaneous problems

2 $u_5 = 16$; but $u_n \neq 2^{n-1}$ $(u_6 = 31)$

3 $\frac{1}{2}n(n^2 + 1)$ **4** $20n^2$ **5** 24

6 **a)** $1.6\,\text{cm}^2$, infinite   **b)** $0.4\,\text{cm}^2$, infinite

7 $\dfrac{nx^{n+2} - (n+1)x^{n+1} + x}{(x-1)^2}$

$\dfrac{n^2x^{n+3} - (2n^2 + 2n - 1)x^{n+2} - (n+1)^2 x^{n+1} - x^2 - x}{(x-1)^3}$

8 $\dfrac{\sin\frac{1}{2}nx \sin\frac{1}{2}(n+1)x}{\sin\frac{1}{2}x}$

9 **a)** 2^n **b)** 0 **c)** $n2^{n-1}$ **d)** $\displaystyle\binom{2n}{n} = \dfrac{(2n)!}{(n!)^2}$

10 **a)** $u_1 = 0, u_2 = 1, u_3 = 2, u_4 = 9, u_5 = 44$
b) $u_{n+1} = n(u_n + u_{n-1})$
c) $u_n = nu_{n-1} + (-1)^n$

Chapter 7

Exercise 7.1

1 $\frac{1}{13}, \frac{1}{2}, \frac{4}{13}$ **2** $\frac{1}{6}, \frac{1}{2}, \frac{2}{3}$ **3** $\frac{1}{4}, \frac{1}{4}, \frac{1}{2}$
4 $\frac{1}{8}, \frac{1}{2}, 0$ **5** $\frac{1}{36}, \frac{5}{18}, \frac{25}{36}$ **6** $\frac{1}{9}, \frac{8}{9}, \frac{2}{27}$
7 $\frac{3}{8}, \frac{5}{16}, \frac{1}{4}$ **8** $\frac{1}{10}, \frac{2}{5}$ **9** $\frac{1}{49}, \frac{1}{7}, \frac{12}{49}$

Exercises 7.2a

1 **a)** $\frac{1}{6}, \frac{1}{2}, \frac{1}{3}$ **b)** $\frac{1}{6}, 0, \frac{1}{6}$ **c)** $\frac{2}{3}, \frac{1}{2}, \frac{1}{2}$
3 **a)** C, A **b)** none
4 **a)** $\frac{5}{6}, \frac{1}{2}, \frac{2}{3}$ **b)** $\frac{1}{3}, \frac{1}{2}, \frac{1}{2}$ **c)** $\frac{5}{6}, 1, \frac{5}{6}$
6 **a)** none **b)** $C', A'; A', B'$
7 **a)** $\frac{1}{8}$ **b)** $\frac{1}{10}$
8 $\frac{4}{7}$

Exercise 7.2b

1 $\frac{1}{6}, \frac{1}{6}, \frac{1}{36}$, independent   **2** $\frac{1}{4}, \frac{1}{4}, \frac{1}{16}$, independent
3 $\frac{1}{4}, \frac{1}{4}, \frac{1}{17}$, not independent   **4** $\frac{1}{4}, \frac{1}{5}, \frac{1}{20}$, independent
5 $\frac{1}{4}, \frac{1}{10}, \frac{1}{20}$, not independent   **6** $\frac{15}{52}$
7 13 **8** $6, \frac{4}{27}$

Exercise 7.3a

1 **a)** $\frac{1}{13}, \frac{12}{13}, \frac{1}{2}, \frac{1}{2}$ **b)** $\frac{1}{26}, \frac{1}{26}, \frac{6}{13}, \frac{6}{13}$ **c)** $\frac{1}{13}, \frac{1}{13}, \frac{12}{13}, \frac{12}{13}$ **d)** $\frac{1}{2}, \frac{1}{2}, \frac{1}{2}, \frac{1}{2}$
all independent

2 a) $\frac{1}{6}, \frac{5}{6}, \frac{1}{2}, \frac{1}{2}$ **b)** $\frac{1}{6}, 0, \frac{1}{3}, \frac{1}{2}$ **c)** $\frac{1}{3}, 0, \frac{2}{3}, 1$ **d)** $1, \frac{2}{5}, 0, \frac{3}{5}$
none independent

4 $\frac{1}{4}$

5 a) $0, 0, 0$ **b)** $3p^2$ **c)** $\frac{1}{2}, \frac{1}{3}, \frac{1}{6}$

6 a) $\frac{1}{2}$ **b)** $\frac{5}{8}$

Exercise 7.3b

2 $\frac{1}{8}, \frac{3}{8}, \frac{3}{8}, \frac{1}{8}$ **3** $\frac{29}{40}$

4 $\frac{3}{10}, \frac{3}{10}, \frac{1}{15}, \frac{7}{15}, \frac{8}{15}$ **5** $\frac{117}{125}$

6 $\frac{1}{15}, \frac{31}{90}, \frac{3}{10}$ **7** $\frac{1}{64}, \frac{9}{64}, \frac{11}{32}$

8 b) $\frac{1}{3}, \frac{1}{3}, \frac{1}{12}, \frac{1}{4}$ **c)** $\frac{5}{12}, \frac{7}{12},$ **d)** $\frac{4}{5}, \frac{1}{5}, \frac{4}{7}, \frac{3}{7}$

9 $\frac{2}{3}$ **10 a)** 62% **b)** $\frac{1}{31}$ **c)** $\frac{9}{19}$

11 a) $\frac{3}{40}$ **b)** $\frac{1}{3}$ **c)** $\frac{6}{37}$ **12** $\frac{1}{3}, \frac{1}{4}, \frac{5}{12}, \frac{1}{9}$

Exercise 7.4a

1 $\frac{1}{36}, \frac{2}{36}, \frac{3}{36}, \frac{4}{36}, \frac{5}{36}, \frac{6}{36}, \frac{5}{36}, \frac{4}{36}, \frac{3}{36}, \frac{2}{36}, \frac{1}{36}$

2 $\frac{3}{18}, \frac{5}{18}, \frac{4}{18}, \frac{3}{18}, \frac{2}{18}, \frac{1}{18}$

Exercise 7.4b

1 a) $0.5, 0.5$

b) $0.25, 0.50, 0.25$

c) $0.125, 0.375, 0.375, 0.125$

d) $0.062, 0.250, 0.375, 0.250, 0.062$

e) $0.031, 0.156, 0.312, 0.312, 0.156, 0.031$

f) $0.016, 0.094, 0.234, 0.312, 0.234, 0.094, 0.016$

2 $0.0003, 0.0064, 0.0512, 0.2050, 0.4096, 0.3277$

3 $0.26(5)$ **4** 0.34 **5 a)** 0.26 **b)** 0.66

6 a) 0.107 **b)** 0.376 **c)** 0.302 **d)** 0.323

7 a) 0.38 **b)** $0.14,$ less

8 a) 0.367 **b)** 0.389

9 a) 0.297 **b)** 0.466 **c)** 0.026

10 a) 1.07% **b)** 2.15%

Exercise 7.4c

1 $\frac{1}{9}$ **2** $0.82, 0.15$ **3** $0.77, 0.125$ **4** 7

Exercise 7.5

1 a) 8.8 **b)** 10 **c)** 10 **d)** 5 **e)** 1.94

2 a) 16.8 **b)** 17.5 **c)** 19 **d)** 9 **e)** 2.96

3   **a)** 10 **b)** 9 **c)** 8 **d)** 10 **e)** 3.39
4   **a)** 34 **b)** 35 **c)** 37 **d)** 46 **e)** 11.33
5   **a)** 340 **b)** 350 **c)** 370 **d)** 460 **e)** 113.3
6   **a)** 84 **b)** 85 **c)** 87 **d)** 46 **e)** 11.33
7   Oxford 80.56, 12.65; Cambridge 78.22, 9.06
8   **a)** 53.17, 11.87 **b)** 3 **c)** 7
9   **b)** 2, 4.0 **c)** $9.812 \, \mathrm{m \, s^{-2}}, 0.004 \, \mathrm{m \, s^{-2}}$
10   695 500, 303
11   **a)** $3m, 3s$ **b)** $m + 3, s$
12   **b)** i) 70 ii) 40 iii) 84
13   2.2 minutes
14   151 cm, 12.53 cm
15   25.9, 2.0
16   **a)** i) $3\mu, 9\sigma^2$ ii) $1 - 2\mu, 4\sigma^2$ **b)** i) 2, 23 ii) 11.9, 1.506

Exercise 7.6

1   7.1, 2.44
2   44, 1.46
3   60.75, 21.32
4   30.18, 8.63
5   23.25, 10.90
6   £5090, £1120
7   37.05, 2.77
8   £68.80, £90.20
9   **a)** 55, 21 **b)** 42 **c)** 47
10   **a)** 42 **b)** 69 **c)** 56
11   101, 100.5, 10
12   **b)** 32.6, 15.3, 52.4 **c)** 35.0, 22.9
13   **b)** 31.4, 14.7, 50.5 **c)** 33.6, 21.9
14   **b)** 33.8, 16.0, 54.3 **c)** 36.3, 23.6

Exercise 7.7

1   **a)** 16 **b)** 92 **c)** 33 **d)** 23
2   **a)** 6.7% **b)** 10.6% **c)** 0.6%
3   **a)** 0.319 **b)** 0.145 **c)** 0.030 **d)** 0.537
4   **a)** 6 **b)** 210 **c)** 34
5   66.77, 2.28 **a)** 7.9% **b)** 66.2 cm
6   3.63, 1.71; 2 (since x takes only integer values)
7   **a)** 9.1% **b)** 37% **c)** 530.8
8   63 g
9   **a)** 39.4 min **b)** 44.7 and 49.3 min **c)** 49.3 and 54.6 min
10   **a)** 5.1% **b)** 103.26 mm **c)** less than 1% exceed 107 mm
11   1.004 cm, 0.020 cm
12   14.6, 5.2 months
13   74.35, 3.23
14   7.52 a.m.
15   **a)** 8.48 a.m. **b)** 5 min **c)** 118, 4 entrances
16   7.9%, 15.5%, 24.45 mm, 6.1%

Miscellaneous problems

1 0.145, 0.000 25   **2**   0.067   **3**   0.047   **4** $\frac{29}{84}$

5 0.84   **6**   0.41, 0.57, 23   **7** $\frac{496}{729}, \frac{73}{729}, \frac{160}{729}$

8 0.167, 0.333, 0.042, 0.375, 0.008, 0.367

9 **a)** $\frac{1}{2}, \frac{1}{2}, \frac{1}{2}$ **b)** $\frac{1}{4}, \frac{1}{4}, \frac{1}{4}$ **c)** $\frac{1}{4}$

10 **a)** $\frac{1}{10}, \frac{9}{10}, \left(\frac{9}{10}\right)^{10} = 0.348$

   b) $\frac{1}{100}, \frac{99}{100}, \left(\frac{99}{100}\right)^{100} = 0.363$

   c) approximately 0.368 $\left(\text{more precisely}, 0.3679\ldots = \dfrac{1}{e}\right)$

Chapter 8

Exercise 8.1a

1 **a)** AE **b)** AE **c)** AE **d)** AE
2 **a)** BA + AC, BD + DC, BE + EC
   b) BA + AD + DC, BD + DA + AC,
      BA + AE + EC, BE + EA + AC,
      BD + DE + EC, BE + ED + DC

Exercise 8.1b

1 **a)** 13 cm, 026° **b)** 5.9 cm, 016° **c)** 7.2 cm, 326° **d) e) f)** 11.8 cm, 008°
2 **a)** 6 cm, 180° **b)** 8 cm, 225° **c)** 4 cm, 090°
   d) 5.7 cm, 273° **e)** 7.2 cm, 033° **f)** 11.2 cm, 060° **g)** 5.7 cm, 093°
   h) 7.2 cm, 214° **i)** 11.2 cm, 240° **j)** 1.69 cm, 102°

Exercise 8.1c

1 **a)** 8 cm **b)** 24 cm **c)** 24 cm,   all due N
2 **a)** 9 cm **b)** 12 cm **c)** 21 cm,   all due E
3 **a)** 2 cm due N **b)** 1.5 cm due E **c)** 2.5 cm, 037° **d)** 5 cm, 037°
   e) 2.5 cm, 037°

Exercise 8.2a

2 **a)** $3\mathbf{i} + 2\mathbf{j}, \mathbf{i} + 3\mathbf{j}, -2\mathbf{j}, \frac{1}{2}\mathbf{i} + \frac{1}{2}\mathbf{j}$
   b) $(2, 1), (\frac{1}{2}, 1), (-3, 0), (-1, 1)$
3 **a) b)** $\mathbf{a} - \mathbf{b}, \mathbf{a} - \mathbf{b}$ **b)** $\mathbf{a} + \mathbf{b}$ **c)** $\frac{1}{2}(\mathbf{a} + \mathbf{b})$
4 **a)** $\frac{1}{2}(\mathbf{a} + \mathbf{b}), \frac{1}{2}(\mathbf{b} + \mathbf{c}), \frac{1}{2}(\mathbf{c} + \mathbf{d}), \frac{1}{2}(\mathbf{d} + \mathbf{a})$ **b)** $\frac{1}{2}(\mathbf{c} - \mathbf{a})$
   c) $\frac{1}{4}(\mathbf{a} + \mathbf{b} + \mathbf{c} + \mathbf{d})$
5 **a)** $\mathbf{a} + \mathbf{c}, \mathbf{b} + \mathbf{c}, \mathbf{a} + \mathbf{b}, \mathbf{a} + \mathbf{b} + \mathbf{c}$ **b)** $\mathbf{c} - \mathbf{a}, \mathbf{c} - \mathbf{a}$

c) $\frac{1}{2}(a + c), \frac{1}{2}a + b + \frac{1}{2}c$ d) $\frac{1}{2}(a + b + c)$

6 a) $\frac{1}{2}a, \frac{1}{2}b, \frac{1}{2}c$ b) $\frac{1}{2}(b + c), \frac{1}{2}(c + a), \frac{1}{2}(a + b)$ c) $\frac{1}{4}(a + b + c)$

Exercise 8.2b

1 a) $\sqrt{5,027°}; \sqrt{10,342°}; \sqrt{29,158°}$
   b) $2.12i + 2.12j, 3.54i - 3.54j, -9.56i + 2.92j$
2 a) $\sqrt{14}$ b) $\sqrt{29}$ c) $\sqrt{83}$ d) $\sqrt{3}$
3 $-3i - j, -2i + 3j$
5 $(1 + \frac{2}{3}t)i + \frac{2}{3}tj + (2 + \frac{1}{3}t)k$
   $\frac{2}{3}ti + (\frac{2}{3}t - 1)j + (\frac{1}{3}t + 1)k$
6 $r = (-\frac{6}{5} + \frac{1}{5}t)i + (\frac{16}{5} - \frac{1}{5}t)j + k, t = 16$

Exercise 8.3b

1 a) $3, 18.4°$ b) $1, 70.5°$ c) $-4, 160.5°$
2 a) $2j + 2k, 3j + 2k, 66.9°$
   b) $2j + 2k, -4i + 3j, 64.9°$
   c) $-2i + 3j - k, -2t - 3j + k, 115.4°$
   d) $3j - k, j - 2k, 45°$

Exercise 8.4a

1 a) $r = 2\lambda i + j + 2(1 - \lambda)k; x = \dfrac{y - 1}{0} = 2 - z$

   b) $r = \lambda i + (1 + \lambda)j + (2 + \lambda)k; x = y - 1 = z - 2$
2 a) $2x + y + 3z = 13$ b) $x + 2y + 3z = 8$
   c) $\dfrac{x - 1}{2} = \dfrac{y - 1}{1} = \dfrac{z - 2}{1}$
3 a) $61.9°$ b) $38.2°$
4 lines meet at $(2, 1, 1), 70.5°$
5 $(\frac{3}{2}, \frac{3}{2})$

6 $\begin{pmatrix} 1 - 2u_1 - 2t_1 \\ u_1 - t_1 \\ -4 + u_1 + t_1 \end{pmatrix}; t_1 = u_1 = \frac{3}{5}; \dfrac{7}{\sqrt{5}}$

Exercise 8.4b

1 a) $i + 2j + 3k, i - j - k, 128.1°$ b) $i - j, j + k, 120°$
2 a) $45.6°$ b) $38.6°$
3 a) $3, 1$ b) $7x + 56y - 5 = 0$ and $16x - 2y + 35 = 0$
4 a) $6, 3$ b) $3x + 3y + 4z + 6 = 0$ and $x - y - 8 = 0$
5 a) $35.3°$ b) $54.7°$ c) $\frac{1}{47}$

6 $\sqrt{5}$ 7 3
8 $2x + 2y - z = 9, (4, 1, 1)$
9 $\frac{1}{2}\sqrt{3}, 71.6°, 0.29$

Exercise 8.4c

1 $40.8°$ 2 $25.9°$ 3 $86.4°$
4 **a)** $2\,cm$ **b)** $30°$
5 **a)** $77.13\,cm$ **b)** $45.96\,cm$ **c)** $33.0°$ **d)** $59.1\,cm$ **e)** $134.8°$
6 **a)** $30\,cm, 16\,cm$ **b)** 0.66

Exercise 8.5

1 $2t\mathbf{i} - 5\mathbf{j} + 2t\mathbf{k}, 2\mathbf{i} + 2\mathbf{k}; \sqrt{57}, 2\sqrt{2}$
2 n
3 $2n\pi + \frac{1}{2}\pi$
4 $3\pi \approx 9.42\,s$
$$-4\mathbf{i}\cos\frac{2t}{3} - 4\mathbf{j}\sin\frac{2t}{3} + 8\mathbf{k}\cos\frac{4t}{3}$$
$4\sqrt{5} \approx 8.94\,m\,s^{-2}$
6 $7\,m\,s^{-1}, N\,38.2°\,E, 10\,m\,s^{-1}$
7 **a)** $2t\mathbf{i} + 6t\mathbf{j}, (4t + \frac{1}{2}t^2)\mathbf{i} + (-8t + \frac{1}{2}t^2)\mathbf{j}$
   b) $5\,s$ **c)** $6\,s$

Exercise 8.6

1 **a)** $25t\mathbf{i} + (15t - 5t^2)\mathbf{j}$ **b)** $25\mathbf{i} + (15 - 10t)\mathbf{j}$ **c)** $3\,s$ **d)** $75\,m$
   e) $11.25\,m$
2 **a)** $10t\mathbf{i} + (10\sqrt{3}t - 5t^2)\mathbf{j}, 10\mathbf{i} + (10\sqrt{3} - 10t)\mathbf{j}$
   b) $10.35\,m\,s^{-1}, 15°\,downwards$
   c) after $1.73\,s$
3 **a)** $45\,m$ **b)** $90\,m$
4 $40\,m\,s^{-1}$
6 $19.6\,m$
7 $\frac{1}{2}, 3; 392, 294\,m$
8 $17.9\,m\,s^{-1}$
9 $\tan^{-1}\frac{7}{4}, 86.4\,m$

Exercise 8.7

1 **a)** $4.5\,m\,s^{-2}$ **b)** $0.4\,m\,s^{-2}$ **c)** $16\,m\,s^{-2}$ **d)** $4\,m\,s^{-2}$
   e) $\dfrac{64\pi^2}{9} = 70.2\,m\,s^{-2}$ **f)** $\dfrac{2\pi^2}{9} = 2.19\,m\,s^{-2}$

2 $5 \, \mathrm{m \, s^{-2}}, 500 \, \mathrm{m}$

3 $50 \, \mathrm{m \, s^{-1}}$

Exercise 8.8

1 $337.4°, 30 \, \mathrm{min}$

2 $48.6°, 45 \, \mathrm{s}$

3 $160°$ **4** $99\frac{1}{2}°$ **5** NW

6 $36.9°, 54 \, \mathrm{min}$

7 $v \, \mathrm{km \, h^{-1}}, \mathrm{NE}, 112\frac{1}{2}°$

8 $27 \, \mathrm{m \, s^{-1}}, 266°$

9 $\mathbf{r} = 2t\mathbf{i} + (1 - t)\mathbf{j} + t^2\mathbf{k}, t = \frac{1}{2}$

10 **a)** $-60(6\mathbf{i} + 5\mathbf{j} + \mathbf{k})$

    b) $(30 - 6t)\mathbf{i} + (30 - 5t)\mathbf{j} + (2 - t)\mathbf{k}$

    c) $320 \, \mathrm{s}$

11 **a)** $3.93 \, \mathrm{m \, s^{-1}}, 127°$ **b)** $5.09 \, \mathrm{m \, s^{-1}}, 199°$

12 $T_1, t_0, \dfrac{2Vt_0}{\sqrt{5}}$

13 **a)** $8\mathbf{i} + 15\mathbf{j}, 9\mathbf{i} + 12\mathbf{j}$ **b)** $\mathbf{i} - 3\mathbf{j}$ **c)** $t\mathbf{i} + (10 - 3t)\mathbf{j}$

14 $r = a(2 - \sin \omega t)\mathbf{i} + 2a \cos \omega t\mathbf{j} + a(\sin \omega t - 3)\mathbf{k}$

    $v = -a\omega \cos \omega t\mathbf{i} - 2a\omega \sin \omega t\mathbf{j} + a\omega \cos \omega t\mathbf{k}$

    $t = \dfrac{(2n + 1)\pi}{2\omega}; \sqrt{5}a, 5a$

Miscellaneous problems

3 $\frac{1}{2}\mathbf{j}, 65.9°, 65.9°, 48.2°$

5 **a)** $954 \, \mathrm{m}$ **b)** $0.053 \, W$ **c)** $1.24 \, \mathrm{km}$

7 $s = a + t\mathbf{u} + \frac{1}{2}t^2 \mathbf{g}$

    $v = \sqrt{\left(\dfrac{b^2g^2}{u^2} + \dfrac{u^2}{4}\right)}; u = \sqrt{2bg}$

8 $\dfrac{aU}{V}; \dfrac{2a}{U} \sin^{-1} \dfrac{U}{V}$

9 $\dfrac{2a}{V^2 - v^2}\left(V + \sqrt{(V^2 - v^2)}\right)$

10 $\{\sqrt{(2V^2 - v^2)} + v\}/\{\sqrt{(2V^2 - v^2)} - v\}$

Chapter 9

Exercise 9.2a

1 **a)** 6 N **b)** 120 N **c)** 0.15 N
2 **a)** $3 \, \text{m s}^{-2}$ **b)** $0.6 \, \text{m s}^{-2}$ **c)** $0.0025 \, \text{m s}^{-2}$
3 **a)** 20 kg **b)** $2.5 \times 10^3 \, \text{kg}$ **c)** 0.015 kg
4 **a)** $8 \, \text{m s}^{-2}$ **b)** 24 N
5 **a)** $8 \, \text{m s}^{-2}, 48 \, \text{N}$ **b)** $32 \, \text{m s}^{-2}, 96 \, \text{N}$ **c)** $4 \, \text{m s}^{-2}, 12 \, \text{N}$
6 $8.44 \times 10^6 \, \text{m s}^{-1}$
7 **a)** $\mathbf{0}, 10\mathbf{i}, -10\mathbf{j}, -40\mathbf{j}$
   b) $20\mathbf{i} - 20\mathbf{j}, 10\mathbf{i} - 20\mathbf{j}, -10\mathbf{j}, -40\mathbf{j}$
   c) $100\mathbf{i} - 500\mathbf{j}, 10\mathbf{i} - 100\mathbf{j}, -10\mathbf{j}, -40\mathbf{j}$
8 $1.00, 1.10, 1.25, 1.45, 1.75, 2.20, 3.00 \, \text{m s}^{-2}, 85 \, \text{km h}^{-1}$

Exercise 9.2b

1 $10\mathbf{i} + 5\mathbf{j}, 2\mathbf{i} + \mathbf{j}, -10\mathbf{i} - 5\mathbf{j}$
2 $6\mathbf{i} + 6\mathbf{j} + 3\mathbf{k}, 3\mathbf{i} + 3\mathbf{j} + 1.5\mathbf{k}, -6\mathbf{i} - 6\mathbf{j} - 3\mathbf{k}$
3 $36.1 \, \text{N}, 146.3°; 3.61 \, \text{m s}^{-2}, 146.3°; 36.1 \, \text{N}, 326.3°$
4 $14.0 \, \text{N}, 255.4°; 3.5 \, \text{m s}^{-2}, 255.4°; 14.0 \, \text{N}, 75.4°$
5 $7.37 \, \text{N}, 298.7°; 0.37 \, \text{m s}^{-2}, 298.7°; 7.37 \, \text{N}, 118.7°$
6 $0.54 \, \text{N}, 249.2°; 2.7 \, \text{m s}^{-2}, 249.2°; 0.54 \, \text{N}, 69.2°$
7 $2\mathbf{i} + \mathbf{j}, 4\mathbf{i} + 16\mathbf{j}, 5\mathbf{i} + 6\mathbf{j}$
8 $10^7 \, \text{m s}^{-1}$ at 37° to its original direction

Exercise 9.2c

1 **a)** 11 800 N **b)** 7800 N **c)** 9800 N **d)** 9800 N
2 **a)** $2.2 \, \text{m s}^{-2}$ **b)** $0.2 \, \text{m s}^{-2}$
3 $3.6 \, \text{N}, 4.8 \, \text{m s}^{-2}$
4 75.1 N, 63.0 N
5 157 N, 118 N
6 $3.74 \, \text{N} \, (\sqrt{14}), 1.41 \, \text{N} \, (\sqrt{2}), 2.83 \, \text{N} \, (\sqrt{8})$
7 **a)** $1.73 \, \text{m s}^{-2}$ **b)** 5.8°
8 1055 N, 1225 N, 660 N
9 **a)** $1.84 \times 10^5 \, \text{N}$ **b)** $9.71 \times 10^4 \, \text{N}$
10 **a)** 4.18 N **b)** 0.82 N
11 1.39 h, 30°

Exercise 9.3a

1 **a)** $0.5 \, \text{m s}^{-2}$ **b)** $0.3 \, \text{m s}^{-2}$
2 **a)** $0.45 \, \text{m s}^{-2}$ **b)** $0.25 \, \text{m s}^{-2}$

3 **a)** $0.55\,\mathrm{m\,s^{-2}}$ **b)** $0.35\,\mathrm{m\,s^{-2}}$
4 **a)** $0.4\,\mathrm{m\,s^{-2}}, 490\,\mathrm{N}$ **b)** $0.38\,\mathrm{m\,s^{-2}}, 497\,\mathrm{N}$ or $483\,\mathrm{N}$
   c) $2.98\,\mathrm{m\,s^{-2}}\,467\,\mathrm{N}$; or $3.72\,\mathrm{m\,s^{-2}}, 454\,\mathrm{N}$
   d) $2.98\,\mathrm{m\,s^{-2}}$, or $3.73\,\mathrm{m\,s^{-2}}, 460\,\mathrm{N}$
5 $86.4\,\mathrm{N}, 0.83\,\mathrm{m\,s^{-2}}$
6 $13.1\,\mathrm{m\,s^{-1}}$
8 $52°$
9 $92.6\,\mathrm{km}$
11 **a)** $3.35\,\mathrm{m\,s^{-2}}$ at $26.6°$ with Ox, no
   b) $2.68\,\mathrm{m\,s^{-2}}, 3/\sqrt{5}\,\mathrm{N}$

Exercise 9.3b

1 **a)** $1.6\,\mathrm{m\,s^{-2}}, 400\,\mathrm{N}$ **b)** $1.2\,\mathrm{m\,s^{-2}}, 300\,\mathrm{N}$ **c)** $1.04\,\mathrm{m\,s^{-2}}, 460\,\mathrm{N}$
2 $6\,\mathrm{m\,s^{-2}}, 12\,\mathrm{N}$
3 $2\,\mathrm{m\,s^{-2}}, 24\,\mathrm{N}$
4 $\frac{25}{7} \approx 3.6\,\mathrm{m\,s^{-2}}, \frac{180}{7} \approx 26\,\mathrm{N}$
5 $\frac{1}{7}(20\sqrt{3} - 15) = 2.8\,\mathrm{m\,s^{-2}}; 23\,\mathrm{N}$
6 $\frac{20}{9} = 2.2\,\mathrm{m\,s^{-2}}, 24\frac{4}{9} \approx 24\,\mathrm{N}, 31\frac{1}{9} \approx 31\,\mathrm{N}$
7 $2.4\,\mathrm{m\,s^{-2}}, 14.1, 9.4\,\mathrm{N}$
8 **a)** $1.5\,\mathrm{m\,s^{-2}}, 400\,\mathrm{N}$ **b)** $1.1\,\mathrm{m\,s^{-2}}, 300\,\mathrm{N}$ **c)** $0.94\,\mathrm{m\,s^{-2}}, 460\,\mathrm{N}$

Exercise 9.4a

1 $1\,\mathrm{m\,s^{-2}}, 2\,\mathrm{m}$
2 **a)** $10\,\mathrm{m\,s^{-2}}$ **b)** 1 **c)** $20\,\mathrm{m}$
3 $60\,\mathrm{N}, \mu = 0.4$ **4** 0.1
5 $18.2\,\mathrm{m\,s^{-2}}, 15.4\,\mathrm{m\,s^{-2}}$
6 $5 - \sqrt{3} \approx 3.3\,\mathrm{m\,s^{-2}}, 8.1\,\mathrm{m\,s^{-1}}$
7 $g(\sin\alpha - \mu\cos\alpha)$
8 $\dfrac{(M - \mu m)g}{M + m}, \dfrac{(\mu + 1)\,Mmg}{M + m}$
9 $1.5\,\mathrm{m}, 2.45\,\mathrm{m\,s^{-1}}$

Exercise 9.4b

1 **a)** $78.4\,\mathrm{N}$ **b)** $73.4\,\mathrm{N}$ **c)** $118\,\mathrm{N}$ **d)** $72.8\,\mathrm{N}$
2 **a)** 0.2 **b)** $96\,\mathrm{N}$
3 **a)** $46.0\,\mathrm{N}$ **b)** $11.9\,\mathrm{N}$
4 **a)** $44.0\,\mathrm{N}$ **b)** $11.4\,\mathrm{N}$
5 **a)** $W\tan\lambda$ **b)** $\dfrac{W\sin\lambda}{\cos(\lambda - \theta)}$ **c)** $W\sin\lambda$
6 **a)** $\dfrac{W\sin(\lambda + \alpha)}{\cos\lambda}$ **b)** $\dfrac{W\sin(\lambda + \alpha)}{\cos(\lambda - \theta)}$ **c)** $W\sin(\lambda + \alpha)$

7 **a)** $\dfrac{W\sin(\lambda-\alpha)}{\cos\lambda}$ **b)** $\dfrac{W\sin(\lambda-\alpha)}{\cos(\lambda-\theta)}$ **c)** $W\sin(\lambda-\alpha)$

8 $2.21\,\text{s}^{-1}, 1.06\,\text{s}$

Exercise 9.5a

1 **a)** $3\times10^4\,\text{N}\,\text{s}$ **b)** $10^5\,\text{N}\,\text{s}$ **c)** $0.75\,\text{N}\,\text{s}$ **d)** $2.7\times10^{-23}\,\text{N}\,\text{s}$
2 **a)** $300\,\text{N}\,\text{s}$ **b)** $12\,\text{N}\,\text{s}$ **c)** $80\,\text{N}\,\text{s}$ **d)** $2\,\text{N}\,\text{s}$
3 **a)** $2000\,\text{N}\,\text{s}$ **b)** $2000\,\text{N}\,\text{s}$ **c)** $2\,\text{m}\,\text{s}^{-1}$
4 **a)** $1000\,\text{N}\,\text{s}$ **b)** $1000\,\text{N}\,\text{s}$ **c)** $1\,\text{m}\,\text{s}^{-1}$
5 **a)** $667\,\text{N}\,\text{s}$ **b)** $667\,\text{N}\,\text{s}$ **c)** $0.667\,\text{m}\,\text{s}^{-1}$
6 **a)** $-40000\,\text{N}\,\text{s}$ **b)** $260000\,\text{N}\,\text{s}$ **c)** $26\,\text{m}\,\text{s}^{-1}$
7 **a)** $-20000\,\text{N}\,\text{s}$ **b)** $280000\,\text{N}\,\text{s}$ **c)** $28\,\text{m}\,\text{s}^{-1}$
8 **a)** $-20000\,\text{N}\,\text{s}$ **b)** $280000\,\text{N}\,\text{s}$ **c)** $28\,\text{m}\,\text{s}^{-1}$
9 $4\,\text{N}\,\text{s}, 200\,\text{N}$
10 $400\,\text{N}$
11 **a)** $1.6\pi\approx5\,\text{N}\,\text{s}$ **b)** $5\,\text{N}$
12 $50\,\text{N}$
13 $1500\,\text{N}$
14 $50\,\text{N}$, increase
15 $7.2\times10^6\,\text{N}\,\text{s}, 7.2\times10^6\,\text{N}\,\text{s}, 7.2\,\text{m}\,\text{s}^{-1}$

Exercise 9.5b

1 **a)** $12\mathbf{i}+4\mathbf{j}+2\mathbf{k}$ **b)** $-3\mathbf{i}-9\mathbf{j}+6\mathbf{k}$ **c)** $2\mathbf{i}+8\mathbf{k}$ **d)** $\mathbf{i}+1.23\mathbf{j}$
2 **a)** $10\mathbf{i}+5\mathbf{j}+3\mathbf{k}$ **b)** $\frac{5}{2}\mathbf{i}-\frac{3}{2}\mathbf{j}+5\mathbf{k}$ **c)** $5\mathbf{i}+3\mathbf{j}+4\mathbf{k}$
   d) $4.5\mathbf{i}+3.6\mathbf{j}+2\mathbf{k}$
3 **a)** $-2\mathbf{i}+10\mathbf{k}$ **b)** $-2.4\mathbf{i}-1.6\mathbf{j}+1.6\mathbf{k}$
4 $\dfrac{pAr}{36\,bv}\text{m}; \dfrac{pArv\rho\sqrt{2}}{36}\,\text{N}$

Exercise 9.5c

1 **a)** $2\mathbf{i}+3\mathbf{j}+\mathbf{k}$ **b)** $\frac{5}{3}\mathbf{i}+\frac{11}{6}\mathbf{j}-\mathbf{k}$
2 $\begin{pmatrix}1.8\\0.8\end{pmatrix}; \pm\begin{pmatrix}2.4\\8.4\end{pmatrix}$
3 $(600, 400, 20)$
4 $1600\,\text{N}$
5 $\frac{2}{49}\,\text{km}\,\text{s}^{-1}, \frac{9}{245}\,\text{km}\,\text{s}^{-1}, 5\,\text{s}$
6 **a)** $39°, 5.3v$ **b)** $34\,mv$

Exercise 9.6a

1 **a)** 300 J **b)** 5.48 m s^{-1}; 4.27 m s^{-1}
2 8000 N, 32 000 N
3 3.2 × 10^4 N
4 100 J, 0.064
5 **a)** 4 × 10^5 J **b)** 5000 N **c)** 10000 N **d)** 20000 N
6 $\dfrac{mv^2}{2F}$
7 2.4 m s^{-1}, 0.6 J

Exercise 9.6b

1 **a)** 8 J **b)** −4 J **c)** 0
2 **a)** 3.6 × 10^5 J **b)** 45.9 m
3 **a)** 9.9 m s^{-1} **b)** 10.8 m s^{-1}
4 **a)** 10 m s^{-1} **b)** 500 J **c)** 5.1 m
5 **a)** 3.13 m s^{-1}, 0.39 N **b)** 4.43 m s^{-1}, 0.98 N **c)** 3.83 m s^{-1}, 0.73 N
6 **a)** 784 J **b)** 6.26 m s^{-1} **c)** 1180 N
7 11.8 m s^{-1} **a)** **c)**
8 **a)** $v^2 = 2ag(1 - \cos\theta)$ **b)** $mg(3\cos\theta - 2)$ **c)** $\cos^{-1}\frac{2}{3} \approx 48°$

Exercise 9.6c

1 960 N
2 25 m s^{-1}
3 2.26 kW
4 9.5 kW
5 0.039 m s^{-2}
6 **a)** 1.47 m s^{-2} **b)** 1.47 m s^{-2}
7 4.9 × 10^4 W
8 2 160 N, 1 700 N, 80 m
9 25 N s, 25 N, 0.85 m s^{-1}
10 $\dfrac{Mg}{v}$ kg s^{-1}, $\frac{1}{2}Mgv$ W
11 $m\dot{v} + kv^2 = F\,(v \leqslant U)$; $m\dot{v} = \dfrac{Fu}{v} - kv^2\,(v > U)$;

$\sqrt{\dfrac{F}{k}}\,(F < kU^2)$; $\sqrt[3]{\dfrac{FU}{k}}\ \ (F \leqslant kU^2)$; $k = 0.5$;

4.98, 4.78, 4.38, 2.35 m s^{-2}; 10.6 s

Miscellaneous problems

2 **a)** $2\,\mathrm{m\,s^{-2}}, 40\,\mathrm{N}, 36\,\mathrm{N}$ **b)** $1\,\mathrm{m\,s^{-2}}.\,45\,\mathrm{N}, 33\,\mathrm{N}$

4 mV/M

5 **a)** $954\,\mathrm{m}$ **b)** $0.053\,\mathrm{W}$ **c)** $1.24\,\mathrm{km}$

6 $r = \dfrac{1}{2 + \cos\theta}$

7 **a)** $mga(1 - \cos\theta)$ **b)** $mg(3\cos\theta - 2)$ **c)** $\cos^{-1}\tfrac{2}{3} \approx 48°$

8 **a)** $\dfrac{2\pi R}{T}$ **b)** $\dfrac{4\pi^2 R}{T^2}$ **c)** $\dfrac{4m\pi^2 R}{T^2}$

9 **a)** $\dfrac{4\pi^2 R}{T^2}$ **b)** $9.9\,\mathrm{m\,s^{-2}}$

10 **a)** $-\displaystyle\int_{\infty}^{r} \dfrac{k}{r^2}\,\mathrm{d}r = \dfrac{k}{r}$ **b)** $\sqrt{\left(\dfrac{2k}{r}\right)}$ **c)** 3.98×10^{14} **d)** $11.2\,\mathrm{km/s}$

11 $5.97 \times 10^{24}\,\mathrm{kg}$

Chapter 10

Exercise 10.1a

1 **a)** 0.69 **b)** 1.10 **c)** 1.79 **d)** 2.20 **e)** 2.48 **f)** -0.69 **g)** -1.10

Exercise 10.1b†

1 **a)** 1.38630 **b)** 1.79176 **c)** 2.30259 **d)** 2.48491 **e)** -0.69315
  f) -1.79176 **g)** -2.48491 **h)** 13.81554 **i)** 0.34657(5) **j)** 1.53506

2 **a)** 3 **b)** 0 **c)** -1 **d)** 0.5

3 **a)** e^2 **b)** e^{100} **c)** $\sqrt[10]{e}$ **d)** $\dfrac{1}{e^2}$

4 **a)** $\dfrac{1}{x}$ **b)** $\dfrac{3}{x}$ **c)** $\dfrac{4}{x}$ **d)** $\dfrac{1}{x}$ **e)** $-\dfrac{1}{x}$ **f)** $\dfrac{1}{2x}$ **g)** $\dfrac{2x}{x^2 + 1}$ **h)** $-\dfrac{1}{x(x + 1)}$

   i) $\dfrac{x}{x^2 - 1}$ **j)** $\dfrac{3x + 2}{2x(x + 1)}$ **k)** $1 + \ln x$ **l)** $\dfrac{1 - \ln x}{x^2}$

5 **a)** $\cot x$ **b)** $-\tan x$ **e)** $\tan x$ **d)** $\sec x$ **e)** $\cot x$ **f)** $3\cot 3x$

   g) $2\cot x$ **h)** $2\cot x$ **i)** $\dfrac{2\ln x}{x}$ **j)** $\dfrac{1}{x\ln x}$

6 **a)** $\tfrac{1}{3}\ln x$ **b)** $\ln(x + 2)$ **c)** $\tfrac{1}{2}\ln(2x + 3)$ **d)** $x + \ln x$ **e)** $-\ln(1 - x)$
   f) $\ln(x^2 + 1)$ **g)** $-\tfrac{1}{3}\ln(1 - x^3)$

† In this and subsequent exercises, the arbitrary constant has been omitted from all indefinite
integrals.

h) $\frac{1}{2}\ln(x^2 - 2x + 3)$ **i)** $\frac{1}{3}\ln(x^3 + 3x)$ **j)** $\ln\sin x$ **k)** $\frac{1}{3}\ln\sec 3x$
l) $2\ln\sin\dfrac{x}{2}$ **m)** $\ln(\sin x - \cos x)$ **n)** $\ln(1 + \tan x)$

7 **a)** 0.9163 **b)** 0.3365 **c)** 0.1256 **d)** -1.0986 **e)** impossible
   f) 0.49 **g)** -0.8047 **h)** impossible **i)** 0.6931 **j)** 0.3465
8 **a)** 0.69365 **b)** 0.69310
9 **a)** $\pi\ln 2 \approx 2.18$ **b)** $\pi\ln 4 \approx 4.35$
10 2.21
11 **a)** 1.295, 12.71 **b)** $\dfrac{1}{e}$

Exercise 10.2

1 $2.203 \times 10^4, 2.688 \times 10^{43}, 1.105, 4.54 \times 10^{-5}$
2 **a)** $2.303, 6.908, 12.82$ **b)** $4.605, 9.210$ **c)** 13.82
3 **a)** $3e^{3x}$ **b)** $-\frac{1}{2}e^{-\frac{1}{2}x}$ **c)** $(1 - x)e^{-x}$ **d)** $(x^2 + 2x)e^x$ **e)** $-xe^{-\frac{1}{2}x^2}$
   f) $\dfrac{xe^x - e^x}{x^2}$ **g)** $e^{-x}(\cos x - \sin x)$ **h)** $e^{2x}(2\cos 3x - 3\sin 3x)$
   i $3^x \ln 3$ **j)** $(1 + 2\ln x)x^{x^2+1}$
4 **a)** $-\frac{1}{2}e^{-2x}$ **b)** $3e^{\frac{1}{3}x}$ **c)** $\frac{1}{2}e^{x^2}$ **d)** $e^{\sin x}$ **e)** $-e^{-\frac{1}{2}x^2}$ **f)** $\dfrac{10^x}{\ln 10}$
5 **a)** $2\left(1 - \dfrac{1}{e}\right)$ **b)** $\frac{1}{2}e^2(e^4 - 1)$ **c)** $1 - \dfrac{1}{\sqrt{e}}$ **d)** $\dfrac{99}{\ln 10}$
6 **a)** $\dfrac{1}{e}$ **b)** $0, \dfrac{4}{e^2}$ **c)** $\dfrac{n^n}{e^n}$ **d)** 1
7 $1 - e^{-x}, 1$
8 **a)** $n\pi$ **b)** $n\pi + \frac{1}{4}\pi$
9 **c)** 10.52% **d)** 6.93 days
10 **c)** $\frac{1}{1600}\ln 2 \approx 4.33 \times 10^{-4}$ **d)** 4.3%
11 **a)** 565 m **b)** 13.8 min **c)** never

Exercise 10.4a

2 **a)** $1 - x + \frac{1}{2}x^2 - \frac{1}{6}x^3 + \frac{1}{24}x^4$ **b)** $1 + 2x + 2x^2 + \frac{4}{3}x^3 + \frac{2}{3}x^4$
   c) $1 + \frac{1}{2}x^2 + \frac{1}{24}x^4$ **d)** $x + \frac{1}{6}x^3$
3 **a)** 1.64872 **b)** 0.36788 **c)** 1.10517

Exercise 10.4b

2 **a)** $\frac{1}{2}x - \frac{1}{8}x^2 + \frac{1}{24}x^3 - \frac{1}{64}x^4$ **b)** $2x - 2x^2 + \frac{8}{3}x^3 + 4x^4$ **c)** $x^2 + \frac{1}{2}x^4$
3 **a)** 1.099 **b)** 0.511 **c)** 1.609

Exercise 10.4c

1 **a)** $1 + \frac{1}{2}x - \frac{1}{8}x^2 + \frac{1}{16}x^3$ $|x| < 1$
   b) $1 - x + x^2 - x^3$ $|x| < 1$
   c) $1 - 3x + 6x^2 - 10x^3$ $|x| < 1$
   d) $1 + x + \frac{3}{2}x^2 + \frac{5}{2}x^3$ $|x| < \frac{1}{2}$
   e) $\frac{1}{2} - \frac{1}{4}x + \frac{1}{8}x^2 - \frac{1}{16}x^3$ $|x| < 2$
   f) $2 - \frac{1}{2}x - \frac{1}{16}x^2 - \frac{1}{64}x^3$ $|x| < 2$
2 **a)** 1.00499 **b)** 0.99749 **c)** 2.02485 **d)** 3.00997 **e)** 0.24752
3 4

4 **b)** $\dfrac{1}{x^2} + \dfrac{2}{x^3} + \dfrac{3}{x^4}; x + \dfrac{2}{3x^2} - \dfrac{4}{9x^5}$

Exercise 10.4d

1 **a)** $x - \frac{1}{6}x^3$ **b)** $1 - \frac{1}{2}x^2 + \frac{1}{24}x^4$ **c)** $x + \frac{1}{3}x^3$ **d)** $1 + \frac{1}{2}x^2 + \frac{5}{24}x^4$
   e) $x + x^2 + \frac{1}{3}x^3$ **f)** $\frac{1}{2}x^2 + \frac{1}{12}x^4$
3 **a)** $\sqrt{(9 + h)} = 3 + \frac{1}{6}h - \frac{1}{216}h^2 \ldots;$ 3.162

   b) $\sin\left(\dfrac{\pi}{6} + h\right) = \dfrac{1}{2} + \dfrac{\sqrt{3}}{2}h - \dfrac{1}{4}h^2 \ldots;$ 0.5150

   c) $\tan\left(\dfrac{\pi}{4} + h\right) = 1 + 2h + 2h^2 \ldots;$ 1.036

Exercise 10.5a

1 **a)** $xe^x - e^x$ **b)** $\sin x - x\cos x$ **c)** $\frac{1}{4}(2x - 1)e^{2x}$ **d)** $\frac{1}{2}x^2 \ln x - \frac{1}{4}x^2$
   e) $2x\sin\dfrac{x}{2} + 4\cos\dfrac{x}{2}$ **f)** $-\dfrac{1}{x}(\ln x + 1)$
2 **a)** $-(x^2 + 2x + 2)e^{-x}$ **b)** $2x\sin x + (2 - x^2)\cos x$

   c) $\theta\tan\theta - \ln\sec\theta$ **d)** $\dfrac{u10^u}{\ln 10} - \dfrac{10^u}{(\ln 10)^2}$

3 **a)** $\frac{1}{2}e^x(\sin x - \cos x)$ **b)** $\frac{1}{2}e^x(\sin x + \cos x)$ **c)** $\frac{1}{2}e^{-x}(\sin x - \cos x)$
   d) $\frac{1}{5}e^x(\sin 2x - 2\cos 2x)$
4 **a)** $x\{(\ln x)^2 - 2\ln x + 2\}$ **b)** $x\tan^{-1}x - \frac{1}{2}\ln(1 + x^2)$
   c) $u\sin^{-1}u + \sqrt{(1 - u^2)}$ **d)** $\frac{1}{2}(x^2 + 1)\tan^{-1}x - \frac{1}{2}x$
5 **a)** $1 - \dfrac{2}{e}$ **b)** $\frac{1}{4}\pi$ **c)** $2\pi^2 - 16$ **d)** $\frac{8}{3}\ln 2 - \frac{7}{9}$

6 £25

Exercise 10.5b

1 **a)** $\dfrac{3\pi}{16}$ **b)** $\dfrac{2}{3}$ **c)** $\dfrac{5\pi}{32}$ **d)** $\dfrac{16}{35}$ **e)** $\dfrac{35\pi}{256}$

2 **a)** $\dfrac{16}{15}$ **b)** $\dfrac{4}{3}$ **c)** 0 **d)** $\dfrac{5\pi}{16}$ **e)** 0

3 $x(\ln x)^3 - 3x(\ln x)^2 + 6x\ln x - 6x$

4 $I_n = \dfrac{1}{n-1}\tan^{n-1}\theta - I_{n-2}; \tfrac{1}{3}\tan^3\theta - \tan\theta + \theta;$

$\tfrac{1}{5}\tan^5\theta - \tfrac{1}{3}\tan^3\theta + \tan\theta - \theta$

Exercise 10.6a

1 **a)** $\dfrac{-2}{x-1} + \dfrac{3}{x-2}$ **b)** $\dfrac{\frac{1}{2}}{x-2} + \dfrac{\frac{1}{2}}{x+2}$ **c)** $\dfrac{1}{x} - \dfrac{2}{2x+1}$

d) $\dfrac{\frac{1}{2}}{x-1} - \dfrac{4}{x-2} + \dfrac{\frac{9}{2}}{x-3}$ **e)** $-\dfrac{1}{x} + \dfrac{2}{x+1}$ **f)** $1 - \dfrac{\frac{1}{2}}{x} + \dfrac{\frac{5}{2}}{x-2}$

g) $1 + \dfrac{1}{x-2} - \dfrac{1}{x+2}$ **h)** $x + \dfrac{\frac{1}{2}}{x-1} + \dfrac{\frac{1}{2}}{x+1}$

2 **a)** $\dfrac{1}{x} - \dfrac{1}{x-1} + \dfrac{1}{(x-1)^2}$ **b)** $\dfrac{2}{x} - \dfrac{1}{x^2} - \dfrac{2}{x+1}$

c) $\dfrac{\frac{1}{4}}{x-1} + \dfrac{\frac{1}{2}}{(x-1)^2} - \dfrac{\frac{1}{4}}{x+1}$ **d)** $1 - \dfrac{1}{4(x+1)} + \dfrac{5x-3}{4(x-1)^2}$

3 **a)** $\dfrac{1}{x} - \dfrac{x}{x^2+1}$ **b)** $\dfrac{\frac{1}{4}}{x-1} - \dfrac{\frac{1}{4}}{x+1} - \dfrac{\frac{1}{2}}{x^2+1}$ **c)** $\dfrac{\frac{1}{3}}{x-1} - \dfrac{\frac{1}{3}(x+2)}{x^2+x+1}$

d) $\dfrac{\frac{1}{3}(x+1)}{x^2-x+1} - \dfrac{\frac{1}{3}}{x+1}$

4 $1 - 3x + 7x^2 - 15x^3$

5 **a)** $\dfrac{1}{r(r+1)}, \dfrac{n}{n+1}, 1$

b) $\dfrac{1}{r(r+2)}, \dfrac{n(3n+5)}{4(n+1)(n+2)}, \dfrac{3}{4}$

c) $\dfrac{1}{r(r+1)(r+2)}, \dfrac{1}{4} - \dfrac{1}{2(n+1)(n+2)}, \dfrac{1}{4}$

Exercise 10.6b

1 **a)** $\tfrac{1}{2}\ln\dfrac{x-1}{x+1}$ **b)** $\ln\dfrac{x}{x+1}$ **c)** $\tfrac{1}{2}\ln\dfrac{x}{2-x}$ **d)** $x + \ln\dfrac{x-2}{x+2}$

e) $\tfrac{1}{4}\ln\dfrac{2x-1}{2x+1}$ **f)** $\tfrac{1}{3}\ln\dfrac{2x-1}{x+1}$

2 **a)** $-\dfrac{1}{x+1}$ **b)** $-\dfrac{1}{2x} + \dfrac{1}{4}\ln\dfrac{x+2}{x}$ **c)** $-\dfrac{2}{x-1} + \ln\dfrac{x}{x-1}$

d) $x - \dfrac{1}{x-1} + 2\ln(x-1)$

3 a) $\ln(x^2 + 9), \frac{1}{3}\tan^{-1}\frac{x}{3}, \frac{1}{2}\ln(x^2 + 9) + \tan^{-1}\frac{x}{3}$

   b) $\ln(4x^2 + 9), \frac{1}{6}\tan^{-1}\frac{2x}{3}, \frac{1}{4}\ln(4x^2 + 9) - \frac{1}{2}\tan^{-1}\frac{2x}{3}$

   c) $\ln(x^2 - 6x + 10), \tan^{-1}(x - 3),$
   $\frac{1}{2}\ln(x^2 - 6x + 10) + 3\tan^{-1}(x - 3)$

4 a) $x + \ln(x - 1)$ **b)** $x + \ln\dfrac{x - 1}{x + 1}$ **c)** $\frac{1}{2}x^2 + \ln(x - 1)$

   d) $-\dfrac{1}{x - 1} + \ln(x - 1)$ **e)** $\frac{1}{3}\ln(x^3 + 1)$

   f) $\frac{1}{6}\ln\dfrac{(x - 1)^2}{x^2 + x + 1} + \dfrac{1}{\sqrt{3}}\tan^{-1}\dfrac{2x + 1}{\sqrt{3}}$

Exercise 10.7

1 a) x^5 **b)** $\frac{1}{7}t^7$ **c)** $-\dfrac{3}{u}$ **d)** $\sqrt{x}$ **e)** $\frac{2}{3}v^{3/2}$ **f)** $\frac{1}{5}(x + 1)^5$

   g) $-\dfrac{1}{4(2u + 1)^2}$ **h)** $\frac{2}{3}\sqrt{(3v - 2)}$ **i)** $\frac{1}{12}(x^2 - 1)^6$ **j)** $\dfrac{1}{3(1 - x^3)}$

   k) $\frac{1}{3}(x^2 + 1)^{3/2}$ **l)** $-\sqrt{(1 - x^2)}$

2 a) $2\ln x$ **b)** $3\ln(x - 3)$ **c)** $-\ln(1 - u)$ **d)** $\frac{1}{2}\ln(v^2 - 1)$
   e) $-\frac{1}{3}\ln(1 - x^3)$ **f)** $\ln(x^2 + x)$ **g)** $\frac{1}{2}e^{2x}$ **h)** $\frac{1}{6}e^{3x}$ **i)** $-\frac{1}{2}e^{-x^2}$
   j) $\ln(1 + e^u)$ **k)** $\dfrac{10^t}{\ln 10}$ **l)** $\ln(\ln x)$

3 a) $\frac{1}{2}\sin 2x$ **b)** $-2\cos\dfrac{x}{2}$ **c)** $\frac{1}{3}\sin^3\theta$ **d)** $-\frac{1}{3}\cos^3\theta$

   e) $\frac{1}{12}\sin 3\theta + \frac{3}{4}\sin\theta$ **f)** $-\frac{3}{4}\cos\theta + \frac{1}{12}\cos 3\theta$ **g)** $\frac{1}{2}\sin^2 x$
   h) $\frac{1}{2}(x - \sin x)$ **i)** $\frac{1}{4}\sin 2x + \frac{1}{2}x$ **j)** $2\sqrt{(\sin x)}$ **k)** $\ln\sin\theta$
   l) $-\ln\cos\theta$

4 a) $\tan x$ **b)** $\tan x - x$ **c)** $\sec\theta$ **d)** $-\csc\theta$ **e)** $-\cot\theta$
   f) $-\cot\theta - \theta$ **g)** $\frac{1}{2}\tan^2\theta$ **h)** $\frac{1}{3}\sec^3\theta$ **i)** $-\frac{1}{2}\ln\cos 2x$
   j) $\frac{1}{2}\tan^2 x + \ln\cos x$ **k)** $2\ln\sin\frac{1}{2}x$ **l)** $-\frac{1}{2}\cot^2 x - \ln\sin x$

5 a) $\sin^{-1}\dfrac{x}{2}$ **b)** $\frac{1}{2}\sin^{-1}\dfrac{2x}{3}$ **c)** $\frac{1}{2}\tan^{-1}2x$ **d)** $\frac{1}{6}\tan^{-1}\dfrac{3x}{2}$

   e) $\ln[u + \sqrt{(1 + u^2)}]$ **f)** $\ln[u + \sqrt{(u^2 - 1)}]$

   g $\frac{1}{2}[\sin^{-1}x + x\sqrt{(1 - x^2)}]$ **h)** $-\frac{1}{3}(1 - x^2)^{3/2}$ **i)** $\frac{1}{6}\tan^{-1}\dfrac{2x}{3}$

   j) $\frac{1}{8}\ln(4x^2 + 9)$

6 a) $x\sin x + \cos x$ **b)** $-xe^{-x} - e^{-x}$ **c)** $-\frac{1}{3}t\cos 3t + \frac{1}{9}\sin 3t$
   d) $2x\sin x - (x^2 - 2)\cos x$ **e)** $x\sin^{-1}x + \sqrt{(1 - x^2)}$ **f)** $x\ln x - x$
   g) $\theta\tan\theta + \ln\cos\theta$ **h)** $(\frac{1}{2}x^2 - \frac{1}{2}x + \frac{1}{4})e^{2x}$ **i)** $-\frac{1}{2}e^{-t}(\sin t + \cos t)$

j) $\frac{1}{13}e^{2t}(2\cos 3t + 3\sin 3t)$ **k)** $\frac{1}{9}x^3(3\ln x - 1)$ **l)** $\dfrac{10^x(x\ln 10 - 1)}{(\ln 10)^2}$

7   **a)** 1   **b)** $\ln 2$ **c)** $\dfrac{1}{\sqrt{2}}\ln(3 + 2\sqrt{2})$

8   **a)** $\frac{1}{2}\pi$ **b)** $\frac{1}{4}\pi\ln 2$ **c)** $\dfrac{\pi}{4\sqrt{2}}\ln(3 + 2\sqrt{2})$

9   **a)** $\ln\dfrac{x}{x+1}$ **b)** $\ln\dfrac{x-1}{x}$ **c)** $\ln\dfrac{x-1}{x} + \dfrac{1}{x}$ **d)** $\ln\dfrac{x}{\sqrt{(x^2+1)}}$

   e) $\ln(1 + e^x)$ **f)** $x - \ln(1 + e^x)$ **g)** $\frac{1}{2}\tan^{-1}x^2$

   h) $x(\ln x)^2 - 2x\ln x + 2x$ **i)** $\frac{1}{3}x^3\tan^{-1}x - \frac{1}{6}x^2 + \frac{1}{6}\ln(1 + x^2)$

   j) $\dfrac{2}{\sqrt{3}}\tan^{-1}\left(\dfrac{1}{\sqrt{3}}\tan\dfrac{x}{2}\right)$ **k)** $\sec^{-1}x$ **l)** $\sin^{-1}x + \sqrt{(1 - x^2)}$

10   **a)** $\frac{1}{4}\ln 2 + \frac{1}{8}\pi$ **b)** $\frac{1}{4}$ **c)** $\frac{1}{8}(\pi + 2)$ **d)** $\dfrac{1}{\sqrt{2}}$ **e)** $\dfrac{8}{105}$ **f)** $\dfrac{5\pi}{256}$

   g) $\frac{1}{9}(2e^3 + 1)$ **h)** $\frac{1}{3}\pi + \frac{1}{2}\sqrt{3}$ **i)** $\dfrac{\pi}{3\sqrt{3}}$ **j)** $\frac{1}{2}\ln 2$ **k)** $\dfrac{2\pi}{3\sqrt{3}}$ **l)** $\frac{1}{2}\pi - 1$

Miscellaneous problems

8   d) $\sqrt{\pi}$ **9** $\pi^3/32$ **10** 2611

Appendix

Exercise A.1

1   **a)** $\frac{1}{7}$ **b)** $\frac{6}{7}$ **c)** $\frac{1}{4}$ **d)** $\frac{7}{4}$

2   **a)** $\dfrac{6}{\sqrt{5}}$ **b)** $\dfrac{3}{\sqrt{13}}$ **c)** $\frac{1}{5}$ **d)** 3

Exercise A.2

1   **a)** $x^2 + y^2 - 6x - 4y - 12 = 0$
   b) $x^2 + y^2 + 2x + 4y - 4 = 0$
   c) $x^2 + y^2 = 100$
   d) $x^2 + y^2 - 8y = 0$

2   **a)** $(-2, -3), 4$ **b)** $(1, 2), 3$ **c)** $(-4, 0), 1$ **d)** $(1, 2), 1/\sqrt{2}$

3   **a)** $x^2 + y^2 - 6x - 2y = 0$
   b) $x^2 + y^2 - 4x - 2y - 15 = 0$
   c) $x^2 + y^2 - 5x - 8y = 0$
   d) $x^2 + y^2 - x + y - 12 = 0$

Exercise A.3

1 $y = \frac{1}{2}(x^2 + 1)$

Exercise A.4a

1 **a)** $6, 4; \dfrac{\sqrt{5}}{3}; (\pm\sqrt{5}, 0); x = \pm\dfrac{9}{\sqrt{5}}$

   b) $10, 4; \dfrac{\sqrt{21}}{5}; (\pm\sqrt{21}, 0); x = \pm\dfrac{25}{\sqrt{21}}$

2 $(\pm\sqrt{3}, 0)$

3 $-\dfrac{b^2 x_1}{a^2 y_1}; \dfrac{x x_1}{a^2} + \dfrac{y y_1}{b^2} = 1; \dfrac{a^2 x}{x_1} - \dfrac{b^2 y}{y_1} = a^2 - b^2$

4 $\dfrac{ax}{\cos\theta} - \dfrac{by}{\sin\theta} = a^2 - b^2$

6 $x' = x; y' = \dfrac{by}{a}$

7 πab

Exercise A.4b

1 **a)** $x = \pm 2y; \dfrac{\sqrt{5}}{2}; (\pm\sqrt{5}, 0); x = \pm\dfrac{4}{\sqrt{5}}$

   b) $2x = \pm 3y; \dfrac{\sqrt{13}}{3}; (\pm\sqrt{13}, 0); x = \pm\dfrac{9}{\sqrt{13}}$

2 $(\pm\sqrt{3}, 0)$

3 $\dfrac{x \sec\theta}{a} - \dfrac{y \tan\theta}{b} = 1$

5 **a)** $x + t^2 y = 2ct$

   b) $t^2 x - y = c\left(t^3 - \dfrac{1}{t}\right)$

   c) $x + t_1 t_2 y = c(t_1 + t_2)$

6 $(x^2 + y^2)^2 = 4c^2 xy$ or $r^2 = 2c^2 \sin 2\theta$ (a *lemniscate*)

Exercise A.5a

2 **a)** $r = 2\sin\theta$ **b)** $r = \sec\theta \tan\theta$ **c)** $r^2 = 2\operatorname{cosec} 2\theta$

   d) $r = \dfrac{1}{\cos\theta + \sin\theta}$

3 **a)** $x^2 + y^2 = ax$ **b)** $y = a$ **c)** $(x^2 + y^2)^2 = 2a^2 xy$ **d)** $x^2 - y^2 = a^2$

Exercise A.5b

1 $\frac{1}{8}\pi a^2$ **2** $\frac{3}{2}\pi a^2$ **3** a^2
4 a^2 **5** $\frac{1}{6}\pi^3 a^2$

Revision exercises

Chapters 1 and 2

1 $-\frac{1}{2}$
2 8.64
3 $1000x^{-3/4}$
4 **a)** 0.4 **b)** 1.48
5 $x \geqslant 1; x \mapsto 3x^2 + 5, x \mapsto 9x^2 + 12x + 5; 0, -2;$
  $x \geqslant 5, x \mapsto \pm\sqrt{\{(x-5)/3\}}$
6 **a)** $x \mapsto \dfrac{x+1}{x-1}$ **b)** $x \mapsto \dfrac{4x}{(x-1)^2}$ $(x \in \mathbb{R}, x \neq 1)$
7 $x^2 - m(l^2 - 2m)x + m^4 = 0$
8 **b)** $r = (q/p)^3$
10 **b)** $a = -3, b = -2$ **c)** $8x + 12$
11 **a)** -28 **b)** $r \pm 2q\sqrt{q}$
12 **a)** 5, 1 **b)** $-4, 7, -1$
13 $2, \frac{1}{2}; 4.9$
14 **a)** 1, 1.85 **b)** 700 N **c)** $25\,\mathrm{m\,s}^{-1}$
15 $-3A, 3G^3/H, -G^3; 0.415, 2, 9.585$

Chapters 3 and 4

1 $-3, -9, 9; (1, -2)$
2 $\dfrac{-(2x+y)}{x+2y}; (1, -2), (-1, 2)$
3 $(0, 0)$, minimum; $(1, 5)$, maximum; $(3, -27)$, minimum;
  $\{k : k > 5\} \cup \{k : -27 < k < 0\}$
4 **a)** $x = 2, y = 1$ **b)** $(-\frac{2}{3}, -\frac{1}{8})$, minimum **c)** $(-2, 0)$
5 **a)** $-5, -5, \frac{1}{3}$ **b)** $\frac{2}{3}, \frac{2}{3}, \frac{2}{3}, -\frac{5}{2}$
6 $(-1, -\frac{1}{4})$, minimum; $(0, 0)$, p.i.; $(1, \frac{1}{4})$, maximum
7 $(2, -1)$, minimum; $(0, 0)$, maximum; $(2, -1)$, maximum
  a) $k \leqslant 0$ **b)** $k > 1$ **c)** $0 < k < 1$
8 $\frac{5}{4} < \sqrt[3]{2}, 1.26$
9 3%
10 **a)** $x^2 - y^2 = 1$ **b)** $\dfrac{t^2 + 1}{t^2 - 1}; \dfrac{p^2 + 1}{p^2 - 1}$ **c)** area $= 1$
11 $0.02\,\mathrm{cm\,s}^{-1}, 8.94\,\mathrm{cm^2\,s}^{-1}$

12 $s = \dfrac{v^2}{20} + v; R = \dfrac{105600v}{v^2 + 20v + 300}; R = 1933, v = 17.3;$

$R = \dfrac{52800v}{v^2 + 20v + 150}; R = 1187, v = 12.2$

14 $\dfrac{32\pi R^3}{3}$

15 **a)** 1.19 m **b)** 2 s **c)** 4 m s^{-1} at $t = 0$

16 $(a, a); \frac{1}{3}a^2; \frac{3}{10}\pi a^3$

17 **a)** $1\frac{5}{9}$ **b)** 2 **c)** $2\pi \ln 2$

18 3.98

19 $3.414\,(>I)$

20 2.76

Chapter 5

1 $28.2°, 61.8°, 135°$

2 **a)** $\frac{5}{2}, \frac{13}{2}, \frac{12}{5}; 13.5°, 99.1°, 193.5°, 279.1°$
    b) 5

3 **a)** $6\sin^2\theta - 5\sin\theta + 1 = 0; 19.5°, 30°, 150°, 160.5°$

4 $\dfrac{r - p}{1 - q}$

5 $\dfrac{\pi}{6}, \dfrac{5\pi}{6}$

6 **a)** $v = \dfrac{A}{2m}\sin 2t$ **b)** $x = \dfrac{A}{4m}(1 - \cos 2t)$ **c)** $\dfrac{A}{2m}$ **d)** π

7 $\sqrt{\{b(b + c)\}}$

8 **c)** $(2.57, 0), (3.71, 0)$ **d)** $\left|\dfrac{\pi}{2} - 2\right|$

9 **a)** $-2\cot x - \operatorname{cosec} x + k$ **b)** $\dfrac{\pi}{3} - \dfrac{\sqrt{3}}{2}$

10 0.96, maximum

11 0.490

Chapter 6

1 **a)** $100n^2$ **b)** $2^n - 1; 15$

2 **a)** $-1, 357643$ **b)** 0.995

3 **a)** $\dfrac{1}{1 - 2x}, |x| < \frac{1}{2}$ **b)** $\dfrac{1 + x}{(1 - x)^2}, |x| < 1$ **c)** $\dfrac{x + 1}{x}, x < -2 \text{ and } x > 0$

4 **b)** $\frac{3}{4}$

5 24

6 $T_{2n} = \dfrac{n}{3}(2n + 1)(4n + 1)$

7 -1.18

8 b) 3.42

9 $86400, 34560$

10 $n(n + 1)2^{n-2}$

Chapter 7

1 a) $\frac{2}{15}$ **b)** $\frac{1}{2}$

2 a) $\frac{1}{27}$ **b)** $\frac{215}{729}$

3 a) $\frac{5}{13}$ **b)** $\frac{95}{663}$ **c)** $\frac{127}{663}$ **d)** $\frac{95}{127}$

4 $\frac{7}{44}, \frac{7}{20}, \frac{1}{22}, \frac{10}{77}$

5 a) i) $\frac{1}{2}, \frac{1}{2}$, yes **ii)** $\frac{7}{72}, \frac{1}{6}$, no **b)** $625, 225$

6 b) $\frac{35}{253}, \frac{48}{253}$

7 $\frac{1}{5}$ **a)** $\frac{1}{2}$ **b)** $\frac{1}{3}$ **c)** $\frac{1}{6}$

8 $0.00008, \frac{5}{8}$

9 a) $\frac{1}{6}$ **b)** $\frac{625}{7776}$ **c)** $\frac{36}{91}$ **d)** $\frac{30}{91}$ **e)** $\frac{25}{91}; \frac{91}{216}$

10 a) $\frac{37}{50}$ **b)** 0.405 **c)** $\frac{2}{25}$ **d)** $\frac{37}{625}; \frac{4}{13}$

11 a) $\frac{3}{25}$ **b)** $\frac{23}{125}$ **c)** $\frac{8}{25}$ **d)** $\frac{1}{4}$

12 9174 **a)** 0.384 **b)** 0.346

13 0.37

14 comment justified at 1% level

Chapter 8

1 $\begin{pmatrix} -2(p+q+1) \\ -p+q+1 \\ p+q-3 \end{pmatrix}; \frac{2}{5}, -\frac{3}{5} \cdot \frac{8}{5}\sqrt{5}$

2 a) $\begin{pmatrix} 6 \\ 0 \\ -1 \end{pmatrix}, \begin{pmatrix} 2 \\ 2 \\ 1 \end{pmatrix}$ **c)** $\begin{pmatrix} 3 \\ 6 \\ -1 \end{pmatrix}$

3 $4x - 2y - 3z = 5$

4 a) $(\mathbf{i} + 2\mathbf{k})/\sqrt{5}$ **b)** 0.775 **c)** $\mathbf{i} - 2\mathbf{j} + \mathbf{k}$ **d)** $\mathbf{i} - 4\mathbf{j}$

5 a) $\frac{5}{18}$ **b)** $\frac{3}{10}$

6 a) $\begin{pmatrix} -3 \\ 1 \\ 2 \end{pmatrix}$ **b)** $\frac{1}{\sqrt{6}}\begin{pmatrix} 1 \\ -1 \\ 2 \end{pmatrix}$ **c)** $\begin{pmatrix} 1 \\ 9 \\ 4 \end{pmatrix}, \begin{pmatrix} -7 \\ -7 \\ 0 \end{pmatrix}$

7 $\frac{1}{2}\sqrt{6}$

8 $\tan^{-1}\frac{7}{4}, 86.4\,\text{m}$

9 $\frac{1}{2} < p < \frac{3}{4}, \frac{7}{4} < p < 2$

10 $-\frac{1}{7}$

11 $1744\,\text{km/hr}, 083.4°; 2.30\,\text{km}$
12   **a)** $50\,\text{km/hr}, 036.9°$ **b)** $12\,\text{km}$ **c)** $9.6\,\text{km}, 7.2\,\text{km}$

Chapter 9

1 $\mathbf{v} = (3t^2 - 2)\mathbf{i} + 2t\mathbf{j}; \mathbf{a} = 6t\mathbf{i} + 2\mathbf{j}; t = 0, \frac{2}{3}; t = \frac{1}{3}, |\mathbf{F}| = 2\sqrt{2}m$
2 $T = (M + m)(g + a), R = m(g + a);$
$T = (M + m)(g - a), R = m(g - a); 5$
5   **a)** $1.4\,\text{s}$ **b)** $76.8°$ to the horizontal
7 v^2/r towards centre; $v = \sqrt{\left(\dfrac{GM}{R + h}\right)}$
8 $\dfrac{4\sqrt{3}}{3}mg$
9 $\sqrt{(E/6m)}, 3\sqrt{(E/6m)}; 4\sqrt{(E/6m)}$
10 $Mg = A\rho u\sqrt{(u^2 - 2gh)}$
11 $\dfrac{M + m}{m}\sqrt{\{2gl(1 - \cos\theta)\}}; \dfrac{M(M + m)gl(1 - \cos\theta)}{m}$
12 $600\,\text{N}$ **a)** $\sin^{-1}\frac{1}{8}$ **b)** $1.13\,\text{m s}^{-2}; 36\,\text{kW}$
13   **a)** $11.25\,\text{s}$ **c)** $262.5\,\text{m}$
14 $10^4\,\text{N}, 12.5\,\text{m s}^{-1}$
15 $10500\,\text{N}, 84\,\text{kW}, 15\,\text{m s}^{-1}$
16 $0.633\,\text{m s}^{-2}$

Chapter 10

1   **a)** $\frac{1}{2}\ln 2 + \frac{3}{8}\pi$ **b)** $\frac{1}{3}$ **c)** 1 **d)** $\ln\left|\dfrac{(x + 3)^2}{x + 2}\right| + c$
2   **a)** $\ln|x + 1| + \dfrac{1}{x + 1} + c$ **c)** 1.95
3 2.93
4 $\frac{3}{2}, \frac{1}{2}; 1970, 1393$
5 $y = x - \frac{1}{2}x^2 + \frac{1}{6}x^3 - \frac{1}{12}x^4 + \frac{1}{24}x^5$
6 $1 - \frac{1}{2}x^2 + \frac{1}{24}x^4 - \frac{1}{720}x^6; t - \frac{1}{2}t^2 + \frac{1}{3}t^3; -\frac{1}{2}x^2 - \frac{1}{12}x^4 - \frac{1}{45}x^6; -0.0639$
7   **a)** $\ln\left|\dfrac{(x + 4)^2}{x + 1}\right| + c$ **b)** i) 0.721 ii) 0.719
8 $\pi^5 - 20\pi^3 + 120\pi$
9   **a)** $(7\sqrt{2} - 8)/15$ **b)** $\frac{3}{8}\ln(\sqrt{2} + 1) - \frac{1}{8}\sqrt{2}$
10 $\frac{8}{105}$
11 $24(1 - e^{-\pi})/85$
12 $\frac{8}{11} \times \frac{6}{9} \times \frac{4}{7} \times \frac{2}{5} \times \frac{1}{3}$

Appendix

1 $6, (-5, 3)$

3 $\left(\dfrac{a}{2}(p^2 - 1), \dfrac{a}{2}\left(3p - \dfrac{1}{p}\right)\right)$

4 $y^2 = \frac{1}{2}x - \frac{1}{8}$

5 $ay \sin t + bx \cos t = ab; \; by \cos t - ax \sin t = (b^2 - a^2) \sin t \cos t;$
$\left(\dfrac{a^2 - b^2}{a} \cos t, \dfrac{-(a^2 - b^2)}{b} \sin t\right); \dfrac{x^2}{b^2} + \dfrac{y^2}{a^2} = \dfrac{(a^2 - b^2)^2}{a^2 b^2}$

6 0.967

7 **a)** $(a \sec \phi, 0), (0, b \operatorname{cosec} \phi)$ **b)** $\left(\dfrac{a^2}{(a^2 + b^2)^{\frac{1}{2}}}, \dfrac{b^2}{(a^2 + b^2)^{\frac{1}{2}}}\right)$

  c) $x \pm y = \mp k$

8 $\left(\dfrac{-a^2 \ln}{a^2 l^2 - b^2 m^2}, \dfrac{b^2 mn}{a^2 l^2 - b^2 m^2}\right)$

11 **a)** $\dfrac{\pi}{2}(2a^2 + b^2)$ **b)** $AB = 2a, r = b \sin \theta$

12 **a)** $(x^2 + y^2)^2 = 2a^2(x^2 - y^2)$ **d)** a^2

Contents of Book 2

Preface

Notation

11 Further vectors
11.1 Points of subdivision
11.2 Centres of mass
* 11.3 Vector products
* 11.4 Plane kinematics

12 Complex numbers
12.1 Introduction
12.2 Geometric representation
12.3 Complex conjugates: solution of equations
12.4 De Moivre's theorem
12.5 Loci in the Argand diagram
* 12.6 $e^{i\theta}$
* 12.7 Interlude: hyperbolic functions
* 12.8 Functions of a complex variable

13 Differential equations
13.1 Introduction
13.2 Step-by-step solutions
13.3 First order equations with separable variables
* 13.4 Linear equations of first order
* 13.5 Linear equations with constant coefficients

14 Further mechanics
14.1 Elasticity: Hooke's law and elastic energy
14.2 Oscillations: simple harmonic motion
14.3 Elasticity: Newton's law of impact
14.4 Further statics: moments, couples and equilibrium
* 14.5 Further dynamics: moments of inertia and rotation
14.6 Dimensions

15 Probability distributions and further statistics

15.1 Probability distributions and generators

15.2 Continuous probability distributions

15.3 The Normal distribution

15.4 The Poisson distribution

15.5 Samples

* 15.6 Significance and confidence

* 15.7 Correlation and regression

16 Matrices and transformations, determinants and linear equations

16.1 Matrices

16.2 Linear transformations in two dimensions

16.3 Linear transformations in three dimensions

* 16.4 Inverse matrices

* 16.5 Determinants

* 16.6 Systems of linear equations

* 16.7 Systematic reduction: inverse matrices

Epilogue

Revision exercises

Answers to exercises

Index

Index

Abel, N. H., 48
acceleration, 98, 349
    due to gravity, 351, 357, 376
    relative, 361
altitudes of triangle, 13
angle
    between line and plane, 342
    between planes, 341
Apollonius, circle of, 484
Apollonius' theorem, 183
approximations, 124
arc, length of, 199
area
    beneath curve, 136
    negative, 150
    of sector, 199
    of triangle, 193
arithmetic mean, 285
arithmetic progression or series, 221
associative laws, 5, 26, 321, 323
astroid, 217

binomial distribution, 280
binomial series, 445
binomial theorem, 260
Boolean algebra, 2
Brahé, T., 370

cardioid, 504
centroid of triangle, 13
chain rule, 114
circle, equation of, 481
circular motion, 356
circumcentre of triangle, 13
combinations, 255
commutative laws, 4, 320, 332
contingency tables, 271
continuity, 68
coordinates
    Cartesian, 5

coordinates (contd.)
    polar, 504
Copernicus, N., 370
cosine rule, 181
cycloid, 217

de Morgan, A., 5
derivatives, 83
    higher, 90
Descartes, R., 6
deviation
    mean absolute, 286
    quartile, 285
    standard, 286
differentiation
    of products, 116
    of quotients, 117
    of sums and differences, 89
    of trigonometric functions, 204
    of vectors, 348
distance
    between two points, 6
    from point to line, 478
    from point to plane, 343
distribution
    binomial, 280
    frequency, 293
    geometric, 283
    Normal, 303
    rectangular, 279
distributive laws, 5, 323, 333
domain of function, 28

e, 422
ellipses, 492
    polar equations of, 505
energy
    conservation of, 408
    kinetic, 400
    potential, 408

equations
   of motion, 371
   polynomial, 48
   quadratic, 40
   solution by iteration, 239
   symmetric functions of roots of, 49
   trigonometric, 174, 201
equilibrium, 373
errors, 124
   rounding, 442
   truncation, 442
events, 264
   compound, 268
   exclusive, 269
   exhaustive, 270
   independent, 270
exponential function, 430
exponential series, 441

factor formulae, 195
factorials, 219
Fibonacci numbers, 251
forces, 371
   parallelogram of, 372
   polygon of, 373
   resultant, 372
   triangle of, 373
frequency, 293
   cumulative, 296
   relative, 298
frequency density, 298
frequency distributions, 293
friction, 389
   angle of, 392
   coefficient of, 389
   cone of, 393
function, 28
   codomain of, 30
   composite, 31
   domain of, 28
   error, 306
   even, 35
   exponential, 430
   implicit, 119
   inverse, 31
   linear, 35
   logarithmic, 59, 420
   odd, 35

function (contd.)
   of a function, 114
   polynomial, 43
   power, 54
   quadratic, 36
   range of, 28
   rational, 74
   trigonometric, 171

Galileo, 370
Gauss, K. F., 306
geometric distribution, 283
geometric mean, 285
geometric progression or series, 224
golden section, 78
gravitation, 370
   constant of, 419
   law of universal, 419
gravity, acceleration due to, 351, 357,
   376

harmonic mean, 285
Hero's formula, 193
histogram, 297
hyperbolas, 497
   rectangular, 500

i, 43
impulse of force, 396
induction, mathematical, 231
inflection, points of, 91, 105
integration, 136
   by parts, 450
   by substitution, 165
   definite, 137
   general, 465
   indefinite, 144
   numerical, 152
   of trigonometric functions, 209
irrational numbers, 17
iteration, 234

Kepler's laws of planetary motion, 136,
   370
kinetics, 98, 129

Lami's theorem, 375
Leibniz, G. W., 79

limits, 69
linear transformations, 19
Lissajou's figures, 218
logarithmic function, 59, 420
logarithmic series, 443
logarithms, 59
  Napierian, 423

Maclaurin series, 440
mass, 371
matrices, 22
  identity, 23
  multiplication of, 26
maxima and minima, 104
mean, 162, 284
  arithmetic, 285
  geometric, 285
  harmonic, 285
median of distribution, 284
medians of triangle, 13
mode of distribution, 284
momentum, 396
  conservation of, 402
motion
  circular, 356
  relative, 360
  under gravity, 351
  with constant acceleration, 130

Newton, I., 79, 368
  first law of motion, 368
  second law of motion, 371
  third law of motion, 382
Newton–Raphson method, 247
Normal distribution, 303
normal to a curve, 96
numbers
  irrational, 17
  prime, 253
  rational, 16
  real, 17

orthocentre of triangle, 13

parabolas, 485
  of safety, 366
parameters, 121
partial fractions, 456

Pascal's triangle, 259
permutations, 254
plane
  angle between, 341
  distance from point, 343
  equation of, 339
polynomials, 43
power of a force, 413
prime numbers, 253
probability, 264
  conditional, 274
probability distributions, 279
  binomial, 280
  geometric, 283
  Normal, 303
  rectangular, 279
probability trees, 276
projectiles, 351

quartile, 285
quartile deviation, 286

radian, 198
range
  of distribution, 285
  of function, 28
  interquartile, 285
  semi-interquartile, 286
rational numbers, 16
real numbers, 17
rectangular distribution, 279
recurrence relations, 234
reduction formulae, 453
reflection, transformation of, 20
relative acceleration, 361
relative motion, 360
relative velocity, 360
remainder theorem, 46
rotation, transformation of, 21
rounding errors, 442

scalar product, 331
scalars, 318
sector, area of, 199
sequences, 219
series, 219
  arithmetic, 221
  binomial, 445

series (*contd.*)
  exponential, 441
  finite, 227
  geometric, 224
  infinite, 437
  logarithmic, 443
  Maclaurin, 440
  Taylor, 449
sets, 2
  complementary, 3
  elements of, 2
  intersection of, 4
  null, 3
  sub-sets of, 3
  union of, 4
  universal, 3
shear, transformation of, 21
Simpson's rule, 153
sine rule, 181
standard deviation, 286
stationary values, 84
statistics
  of location, 284
  of spread, 285
Stirling's formula, 220
straight lines
  angle between, 477
  distance of points from, 478
  equation of, 11, 337, 477
surds, 18

tangent to a curve, 95
Taylor series, 449
transformations
  linear, 19
  reflection, 20
  rotation, 21
  shear, 21
trapezium rule, 152
triangle
  area of, 193

triangles (*contd.*)
  centres of, 13
trigonometric functions, 171
  differentiation of, 204
  integration of, 209
  inverse, 213
  of compound angles, 183
  of multiple angles, 191
  of small angles, 203
truncation errors, 442

union of sets, 4

variables
  dependent, 28
  independent, 28
variance, 286
vectors, 318
  addition of, 320
  components of, 327
  differentiation of, 348
  displacement, 319
  free, 320
  localised, 319
  modulus of, 320
  position, 324
  scalar product of, 331
  subtraction of, 321
  unit, 328
velocity
  angular, 356
  average, 79
  definition of, 348
  instantaneous, 80
  relative, 360
Venn diagrams, 4
volumes of revolution, 157

waves, combination of, 187
weight, 371
work done by force, 405